Personalized and Precision Medicine 2022

Personalized and Precision Medicine 2022

Editor

Anne-Marie Caminade

MDPI • Basel • Beijing • Wuhan • Barcelona • Belgrade • Manchester • Tokyo • Cluj • Tianjin

Editor
Anne-Marie Caminade
Laboratoire de Chimie de
Coordination (LCC)
CNRS
Toulouse
France

Editorial Office
MDPI
St. Alban-Anlage 66
4052 Basel, Switzerland

This is a reprint of articles from the Special Issue published online in the open access journal *Journal of Personalized Medicine* (ISSN 2075-4426) (available at: www.mdpi.com/journal/jpm/special_issues/pemed_2021).

For citation purposes, cite each article independently as indicated on the article page online and as indicated below:

LastName, A.A.; LastName, B.B.; LastName, C.C. Article Title. *Journal Name* **Year**, *Volume Number*, Page Range.

ISBN 978-3-0365-7045-7 (Hbk)
ISBN 978-3-0365-7044-0 (PDF)

Contents

About the Editor

Anne-Marie Caminade

Anne-Marie Caminade is Director of Research Exceptional Class at the CNRS (Centre National de la Recherche Scientifique), Deputy Director of the Laboratoire de Chimie de Coordination (LCC) in Toulouse, and head of the "dendrimers and heterochemistry" team. She is co-author of about 510 publications and 18 patents, and previously Editor of two books (2011 and 2018). Her h-index is 77. She recently received Prizes from the French (SCF) and German (GDCh) Chemical Societies. She is a member of the Academia Europaea and of the European Academy of Sciences. Her current research interest is on the synthesis and study of properties of hyperbranched macromolecules named dendrimers, with an emphasis on their properties in nanomedicine.

Preface to "Personalized and Precision Medicine 2022"

The origin of this Special Issue about "Personalized and Precision Medicine 2022" of the Journal of Personalized Medicine is the International Conference on Personalized and Precision Medicine PEMED 2021, which occurred online on 7–9 April 2021, after three editions organized in Paris (25–27 June 2018), Barcelona (15–17 May 2019), and Munich (19–21 February 2020). The diversity of studies carried out in connection with personalized and precision medicine was clearly emphasized during these International Conferences, and is also observed in the 20 papers of this Special Issue. The most represented topics concern brain diseases, cancers, lungs obstruction, and muscles and skeletal diseases. Other topics concern in particular the intestinal microbiome or COVID-19, emerging methods for imaging and using radiotherapy, and a promising new class of therapeutic molecules, the dendrimers.

Contributions by V. Bessi et al. (Alzheimer's disease), K. Hubčíková et al. (Huntington's disease), V. Nocity et al. (multiple sclerosis), R.E. Akins et al. (muscle dysfunction), K. Kim et al. (low back pain), D. Rak et al. (inpatient rehabilitation after knee surgery), J.-W. Kim et al. (3D titanium printing in orthodontics), K.Y. Arga, R. Sinha et al. (cancer hallmark proteins), K. Kodama et al. (interorganizational collaboration in cancer), A. Sherif et al. (2 papers about bladder cancers), S. Boussios et al. (COVID-19 in cancer patients), P. Steiropoulos et al. and D.-K. Kim et al. (two approaches of obstructive sleep apnea), M.-C. Lu et al. (blood-stream infection), W.-T. Hung et al. (auto-immune diseases), A.R. Moschen et al. (microbiome), D. Berritto et al. (ultra-high-frequency ultrasound), V. Nardone et al. (non-oncological radiotherapy), and A.-M. Caminade (dendrimers) illustrate the present and future impact of Personalized and Precision Medicine on all the therapeutic domains.

I wish to thank all the authors who contributed their work, and in so doing made this Special Issue a success. I also wish to thank the staff of JPM for their excellent support throughout the editorial process.

Anne-Marie Caminade
Editor

Journal of *Personalized Medicine*

Editorial

Personalized and Precision Medicine 2022

Anne-Marie Caminade [1,2]

1 Laboratoire de Chimie de Coordination (LCC), CNRS UPR8241, 205 Route de Narbonne, CEDEX 4, 31077 Toulouse, France; anne-marie.caminade@lcc-toulouse.fr
2 LCC-CNRS, Université de Toulouse, CNRS, Toulouse, France

Citation: Caminade, A.-M. Personalized and Precision Medicine 2022. *J. Pers. Med.* **2023**, *13*, 459. https://doi.org/10.3390/jpm13030459

Received: 27 February 2023
Accepted: 28 February 2023
Published: 1 March 2023

This Special Issue, "Personalized and Precision Medicine 2022" (https://www.mdpi.com/journal/jpm/special_issues/pemed_2021 (accessed on 24 February 2023)), in the *Journal of Personalized Medicine*, was first proposed at the International Conference on Personalized and Precision Medicine PEMED 2021, which occurred online on 7–9 April 2021 (https://premc.org/conferences/pemed-personalized-precision-medicine/ (accessed on 24 February 2023)), after three editions organized in Paris (25–27 June 2018), Barcelona (15–17 May 2019), and Munich (19–21 February 2020). The diversity of studies carried out in connection with personalized and precision medicine was clearly emphasized during these international conferences, as also observed in the 20 papers of this Special Issue. The most represented topics concern brain diseases, cancers, lungs obstruction, muscles, and skeletal diseases. Other specific topics include the intestinal microbiome or COVID-19, emerging methods for imaging and using radiotherapy, and a promising new class of therapeutic molecules known as dendrimers.

Alzheimer's disease is the most prevalent brain disease and can take several clinical presentations. The article by V. Bessi et al. [1] compares the empathy deficit and its neural basis for Alzheimer's patients with either logopenic primary progressive aphasia or amnesia. Metabolic disfunctions were observed in different brain regions, depending on the type of the Alzheimer's disease, but they induce the same damage of cognitive empathy and personal distress over time. Contrary to Alzheimer's disease, Huntington's disease has a low prevalence, but it causes a long-lasting burden in affected families. The article by K. Hubčíková et al. [2] discusses the psychosocial impact of this disease on the affected families in the Slowak Republic. Comprehensive genetic counselling, including the possibility of preimplantation genetic diagnosis, can particularly mitigate the psychosocial effects and burden induced by Huntington's disease. Multiple sclerosis is a chronic inflammatory and neurodegenerative disease of the central nervous system, frequently associated with multisystem comorbidities. The review by V. Nocity et al. [3] summarizes the available data on the incidence and prevalence of autoimmune diseases in multiple sclerosis, their effect on the clinical course of the disease, and their impact on the treatment choice.

Spastic-type cerebral palsy is a brain disease that induces a complex neuromuscular disorder, involving altered skeletal muscle microanatomy and growth. The paper by R.E. Akins et al. [4] focuses on the poorly studied mechanisms that contribute to muscle pathophysiology and dysfunction. They examined whether a diagnosis of spastic cerebral palsy is associated with intrinsic DNA methylation differences in myoblasts and myotubes derived from muscle-resident stem cell populations. A comparison of the skeletal muscle biopsies obtained during orthopedic surgeries between patients with and without spastic cerebral palsy demonstrated fundamental differences in DNA methylation, which may reveal new targets for studies of mechanisms that contribute to muscle dysregulation. A very different paper concerning muscles is proposed by K. Kim et al. [5], which investigates the effect of different lumbar belts in patients with nonspecific low back pain. Measurements taken by a three-dimensional motion analysis device, force plate, and surface electromyography demonstrated the positive effect of extensible lumbar belts.

Orthopedic surgeries are frequently followed by inpatient rehabilitation. The paper by D. Rak et al. [6] posed the question of comparing the efficiency between "classical" and expensive rehabilitation or fast-track rehabilitation for inpatients after total knee replacement. It is shown that one year after surgery, impatient rehabilitation does not provide long-term benefits over fast-track rehabilitation. In some cases, orthodontic malocclusion necessitates bone-anchored maxillary protraction, which induces infection or device failures that occur with conventional plates. The paper by J.-W. Kim et al. [7] displays a pilot prospective study using preoperative simulation and the 3D titanium printing of customized plates. Better results were obtained with these customized plates after two years, showing that this method is effective in treating skeletal malocclusion.

Four papers in this Special Issue are based on cancer, which is one of the leading causes of death. The paper by K.Y. Arga and R. Sinha et al. [8] explores cancer hallmark proteins in different cancer types, studied with the aim of discovering measurable indicators. A pan-cancer analysis to map differentially interacting hallmarks of cancer proteins associated with 12 common cancers was carried out. The study presents candidate systems' biomarkers that may be valuable for improving personalized treatment strategies for various cancers. The review by K. Kodama et al. [9] proposes a large overview of the discovery and development of 107 anticancer drugs, in connection with interorganizational collaboration, from 1998 to 2018. It is shown that immune checkpoint blockade agents are a significantly active area in interorganizational transactions, suggesting that such types of agents are a paradigm for cancer treatment, resulting in huge product sales and continuous indication expansion.

Two papers from the same group focus on the treatment of bladder cancers. The first of the two, by A. Sherif et al. [10], uses a clinical multicenter database to demonstrate that a high aspartate transaminase–alanine aminotransferase ratio (De Ritis ratio) before treatment of muscle invasive bladder cancer is associated with increased mortality. However, this ratio cannot be used for downstaging prediction. The second paper by A. Sherif et al. [11] investigates thromboembolic events in patients with muscle-invasive bladder cancer undergoing neoadjuvant chemotherapy. It was suggested that low-molecular-weight heparin is a possible prophylaxis, but a high incidence of decreased renal function was observed in these patients. Amongst the neoadjuvant-chemotherapy-administered patients with thromboembolic events, 41% of patients had decreased renal function, thus, reducing the likelihood of them benefitting from low-molecular-weight heparin prophylaxis.

The review by S. Boussios et al. [12] focuses on three massive healthcare threats (cancer, mucormycosis, and COVID-19), and the danger of one of these diseases becoming predisposed to another. The conclusion was that COVID-19 and mucormycosis pose a larger risk in cancer patients.

Besides the COVID-19 virus, various other obstructive lung diseases are known. The paper by P. Steiropoulos et al. [13] evaluates vitamin D serum levels in patients with obstructive sleep apnea, eventually associated with chronic obstructive pulmonary disease (such an association is defined as the overlap syndrome). Lower vitamin D levels have been observed in patients with overlap syndrome compared with obstructive sleep apnea patients and non-apneic controls, which might be associated with disease severity. Another paper by D.-K. Kim et al. [14] on patients with obstructive sleep apnea proposes an ultra-short-term analysis to detect the balance of autonomic nervous system activity. The overall heart rate variability feature alterations obtained by this method indicated sympathetic overactivity dependent on obstructive sleep apnea severity. The risk factors of mortality in patients with carbapenem-resistant *Klebsiella pneumoniae* bloodstream infection are analyzed in the paper by M.-C. Lu et al. [15]. No effective antimicrobial regimen could be identified, but diverse risk factors were identified, such as lower platelet count and a higher Pitt bacteremia score.

For instance, miscellaneous topics concern the prediction of hepatotoxicity in azathioprine-treated patients with auto-immune diseases. The paper by W.T. Hung et al. [16] studies in Asian patients with thiopurine methyltransferase, which is a rate-limiting enzyme in azathioprine metabolization. Genetic variants in the thiopurine methyltransferase were analyzed, showing that the non-normal metabolizers were associated with hepatotoxicity. Another example discusses the intestinal microbiome. The paper by A.R. Moschen et al. [17] focuses on the short- and long-term effects of capsules of a purified extract from the European black elderberries on the microbiome composition. The supplementation was well tolerated, and changes in species abundance were observed over time. In particular, the relative abundance of *Akkermansia* spp., which may have beneficial effects on inflammation and metabolism, continued to increase in a subset of participants, even beyond the supplementation period.

Two different physical techniques are presented. The clinical applications of ultra-high-frequency ultrasound are proposed by D. Berritto et al. [18] for the study of many superficial targets, within the first 3 cm of skin surface. The high spatial resolution of this technic is especially suitable for innovations to diagnostic imaging of hands, wrists, and feet. A review of the modern approaches of non-oncological radiotherapy is proposed by V. Nardone et al. [19]. Different disorders such as heart tachycardia, soft tissue disorders, muscle–skeletal disorders, osteoarthritis and osteoarthrosis, neurological disorders, and Graves' ophthalmopathy are considered.

Finally, an emerging family of macromolecules with potential therapeutic properties, known as dendrimers, are reviewed by A.-M. Caminade [20], with emphasis on their clinical trials. Many of these clinical trials have recently been reported (2020–2022), with the aim of treating essentially bacterial vaginosis, cancers, and COVID-19.

Overall, the twenty papers in this Special Issue constitute an impressionist overview of the state-of-the-art research in the field of personalized and precision medicine, with some perspectives on the future of this topic.

Acknowledgments: I wish to thank all the authors who contributed their work and made the Special Issue a success in doing so. I also wish to thank the staff of JPM for their excellent support throughout the editorial process.

Conflicts of Interest: The author declares no conflict of interest.

References

1. Giacomucci, G.; Polito, C.; Berti, V.; Padiglioni, S.; Galdo, G.; Mazzeo, S.; Bergamin, E.; Moschini, V.; Morinelli, C.; Nuti, C.; et al. Differences and Similarities in Empathy Deficit and Its Neural Basis between Logopenic and Amnesic Alzheimer's Disease. *J. Pers. Med.* **2023**, *13*, 208. [CrossRef] [PubMed]
2. Hubčíková, K.; Rakús, T.; Mühlbäck, A.; Benetin, J.; Bruncvik, L.; Petrášová, Z.; Bušková, J.; Brunovský, M. Psychosocial Impact of Huntington's Disease and Incentives to Improve Care for Affected Families in the Underserved Region of the Slovak Republic. *J. Pers. Med.* **2022**, *12*, 1941. [CrossRef] [PubMed]
3. Nociti, V.; Romozzi, M. Multiple Sclerosis and Autoimmune Comorbidities. *J. Pers. Med.* **2022**, *12*, 1828. [CrossRef] [PubMed]
4. Robinson, K.G.; Marsh, A.G.; Lee, S.K.; Hicks, J.; Romero, B.; Batish, M.; Crowgey, E.L.; Shrader, M.W.; Akins, R.E. DNA Methylation Analysis Reveals Distinct Patterns in Satellite Cell–Derived Myogenic Progenitor Cells of Subjects with Spastic Cerebral Palsy. *J. Pers. Med.* **2022**, *12*, 1978. [CrossRef] [PubMed]
5. Im, S.-C.; Seo, S.-W.; Kang, N.-Y.; Jo, H.; Kim, K. The Effect of Lumbar Belts with Different Extensibilities on Kinematic, Kinetic, and Muscle Activity of Sit-to-Stand Motions in Patients with Nonspecific Low Back Pain. *J. Pers. Med.* **2022**, *12*, 1678. [CrossRef] [PubMed]
6. Rak, D.; Nedopil, A.J.; Sayre, E.C.; Masri, B.A.; Rudert, M. Postoperative Inpatient Rehabilitation Does Not Increase Knee Function after Primary Total Knee Arthroplasty. *J. Pers. Med.* **2022**, *12*, 1934. [CrossRef] [PubMed]
7. Kim, M.; Li, J.; Kim, S.; Kim, W.; Kim, S.-H.; Lee, S.-M.; Park, Y.L.; Yang, S.; Kim, J.-W. Individualized 3D-Printed Bone-Anchored Maxillary Protraction Device for Growth Modification in Skeletal Class III Malocclusion. *J. Pers. Med.* **2021**, *11*, 1087. [CrossRef] [PubMed]
8. Kori, M.; Ozdemir, G.E.; Arga, K.Y.; Sinha, R. A Pan-Cancer Atlas of Differentially Interacting Hallmarks of Cancer Proteins. *J. Pers. Med.* **2022**, *12*, 1919. [CrossRef] [PubMed]
9. Djurian, A.; Makino, T.; Lim, Y.; Sengoku, S.; Kodama, K. Dynamic Collaborations for the Development of Immune Checkpoint Blockade Agents. *J. Pers. Med.* **2021**, *11*, 460. [CrossRef] [PubMed]

10. Eriksson, V.; Holmkvist, O.; Huge, Y.; Johansson, M.; Alamdari, F.; Svensson, J.; Aljabery, F.; Sherif, A. A Retrospective Analysis of the De Ritis Ratio in Muscle Invasive Bladder Cancer, with Focus on Tumor Response and Long-Term Survival in Patients Receiving Neoadjuvant Chemotherapy and in Chemo Naïve Cystectomy Patients—A Study of a Clinical Multicentre Database. *J. Pers. Med.* **2022**, *12*, 1769. [CrossRef] [PubMed]
11. Rydell, H.; Ericson, A.; Eriksson, V.; Johansson, M.; Svensson, J.; Banday, V.; Sherif, A. Thromboembolic Events in Patients Undergoing Neoadjuvant Chemotherapy and Radical Cystectomy for Muscle-Invasive Bladder Cancer: A Study of Renal Impairment in Relation to Potential Thromboprophylaxis. *J. Pers. Med.* **2022**, *12*, 1961. [CrossRef] [PubMed]
12. Mahajan, I.; Ghose, A.; Gupta, D.; Manasvi, M.; Bhandari, S.; Das, A.; Sanchez, E.; Boussios, S. COVID-19, Mucormycosis and Cancer: The Triple Threat—Hypothesis or Reality? *J. Pers. Med.* **2022**, *12*, 1119. [CrossRef] [PubMed]
13. Archontogeorgis, K.; Voulgaris, A.; Nena, E.; Zissimopoulos, A.; Bouloukaki, I.; Schiza, S.E.; Steiropoulos, P. Vitamin D Levels in Patients with Overlap Syndrome, Is It Associated with Disease Severity? *J. Pers. Med.* **2022**, *12*, 1693. [CrossRef] [PubMed]
14. Ha, S.-S.; Kim, D.-K. Diagnostic Efficacy of Ultra-Short Term HRV Analysis in Obstructive Sleep Apnea. *J. Pers. Med.* **2022**, *12*, 1494. [CrossRef] [PubMed]
15. Liu, K.-S.; Tong, Y.-S.; Lee, M.-T.; Lin, H.-Y.; Lu, M.-C. Risk Factors of 30-Day All-Cause Mortality in Patients with Carbapenem-Resistant *Klebsiella pneumoniae* Bloodstream Infection. *J. Pers. Med.* **2021**, *11*, 616. [CrossRef] [PubMed]
16. Sheu, H.-S.; Chen, Y.-M.; Liao, Y.-J.; Wei, C.-Y.; Chen, J.-P.; Lin, H.-J.; Hung, W.-T.; Huang, W.-N.; Chen, Y.-H. Thiopurine S-Methyltransferase Polymorphisms Predict Hepatotoxicity in Azathioprine-Treated Patients with Autoimmune Diseases. *J. Pers. Med.* **2022**, *12*, 1399. [CrossRef] [PubMed]
17. Reider, S.; Watschinger, C.; Längle, J.; Pachmann, U.; Przysiecki, N.; Pfister, A.; Zollner, A.; Tilg, H.; Plattner, S.; Moschen, A.R. Short- and Long-Term Effects of a Prebiotic Intervention with Polyphenols Extracted from European Black Elderberry—Sustained Expansion of *Akkermansia* spp. *J. Pers. Med.* **2022**, *12*, 1479. [CrossRef] [PubMed]
18. Russo, A.; Reginelli, A.; Lacasella, G.V.; Grassi, E.; Karaboue, M.A.A.; Quarto, T.; Busetto, G.M.; Aliprandi, A.; Grassi, R.; Berritto, D. Clinical Application of Ultra-High-Frequency Ultrasound. *J. Pers. Med.* **2022**, *12*, 1733. [CrossRef] [PubMed]
19. Nardone, V.; D'Ippolito, E.; Grassi, R.; Sangiovanni, A.; Gagliardi, F.; De Marco, G.; Menditti, V.S.; D'Ambrosio, L.; Cioce, F.; Boldrini, L.; et al. Non-Oncological Radiotherapy: A Review of Modern Approaches. *J. Pers. Med.* **2022**, *12*, 1677. [CrossRef] [PubMed]
20. Caminade, A.-M. Dendrimers, an Emerging Opportunity in Personalized Medicine? *J. Pers. Med.* **2022**, *12*, 1334. [CrossRef] [PubMed]

Journal of *Personalized Medicine*

MDPI

Article

Differences and Similarities in Empathy Deficit and Its Neural Basis between Logopenic and Amnesic Alzheimer's Disease

Giulia Giacomucci [1], Cristina Polito [2], Valentina Berti [3,4], Sonia Padiglioni [5,6], Giulia Galdo [1], Salvatore Mazzeo [1,2], Enrico Bergamin [7], Valentina Moschini [8], Carmen Morinelli [8], Claudia Nuti [7], Maria Teresa De Cristofaro [4], Assunta Ingannato [1], Silvia Bagnoli [1], Benedetta Nacmias [1,2], Sandro Sorbi [1,2] and Valentina Bessi [1,*]

1 Department of Neuroscience, Psychology, Drug Research and Child Health, University of Florence, 50134 Florence, Italy
2 IRCCS Fondazione Don Carlo Gnocchi, 50143 Florence, Italy
3 Department of Biomedical, Experimental and Clinical Sciences "Mario Serio", University of Florence, 50134 Florence, Italy
4 Nuclear Medicine Unit, Azienda Ospedaliero-Universitaria Careggi, 50134 Florence, Italy
5 Regional Referral Centre for Relational Criticalities—Tuscany Region, 50134 Florence, Italy
6 Research and Innovation Centre for Dementia-CRIDEM, AOU Careggi, 50134 Florence, Italy
7 University of Florence, 50134 Florence, Italy
8 SOD Neurologia I, Dipartmento Neuromuscolo-Scheletrico e degli Organi di Senso, AOU Careggi, 50134 Florence, Italy
* Correspondence: valentina.bessi@unifi.it; Tel.: +39-05-7948660; Fax: +39-05-7947484

Abstract: The aims of the study were to assess empathy deficit and neuronal correlates in logopenic primary progressive aphasia (lv-PPA) and compare these data with those deriving from amnesic Alzheimer's disease (AD). Eighteen lv-PPA and thirty-eight amnesic AD patients were included. Empathy in both cognitive and affective domains was assessed by Informer-rated Interpersonal Reactivity Index (perspective taking, PT, and fantasy, FT, for cognitive empathy; empathic concern, EC, and personal distress, PD, for affective empathy) before (T0) and after (T1) cognitive symptoms' onset. Emotion recognition was explored through the Ekman 60 Faces Test. Cerebral FDG-PET was used to explore neural correlates underlying empathy deficits. From T0 to T1, PT scores decreased, and PD scores increased in both lv-PPA (PT $z = -3.43$, $p = 0.001$; PD $z = -3.62$, $p < 0.001$) and in amnesic AD (PT $z = -4.57$, $p < 0.001$; PD $z = -5.20$, $p < 0.001$). Delta PT (T0–T1) negatively correlated with metabolic disfunction of the right superior temporal gyrus, fusiform gyrus, and middle frontal gyrus (MFG) in amnesic AD and of the left inferior parietal lobule (IPL), insula, MFG, and bilateral superior frontal gyrus (SFG) in lv-PPA ($p < 0.005$). Delta PD (T0-T1) positively correlated with metabolic disfunction of the right inferior frontal gyrus in amnesic AD ($p < 0.001$) and of the left IPL, insula, and bilateral SFG in lv-PPA ($p < 0.005$). Lv-PPA and amnesic AD share the same empathic changes, with a damage of cognitive empathy and a heightening of personal distress over time. The differences in metabolic disfunctions correlated with empathy deficits might be due to a different vulnerability of specific brain regions in the two AD clinical presentations.

Keywords: Alzheimer's disease; logopenic variant primary progressive aphasia; empathy

Citation: Giacomucci, G.; Polito, C.; Berti, V.; Padiglioni, S.; Galdo, G.; Mazzeo, S.; Bergamin, E.; Moschini, V.; Morinelli, C.; Nuti, C.; et al. Differences and Similarities in Empathy Deficit and Its Neural Basis between Logopenic and Amnesic Alzheimer's Disease. *J. Pers. Med.* **2023**, *13*, 208. https://doi.org/10.3390/jpm13020208

Academic Editor: Anne-Marie Caminade

Received: 23 November 2022
Revised: 17 January 2023
Accepted: 18 January 2023
Published: 25 January 2023

1. Introduction

Empathy is widely recognized as the capacity to "put oneself in another's shoes", and it can be defined as the crucial ability to both feel and comprehend what others feel [1]. Empathy is conceptualized as a multidimensional construct regulated by unconscious affective and conscious cognitive processes [2]. Indeed, Decety and Jackson proposed the current model of empathy and suggested that empathy may be divided into two major components: affective empathy is the capacity to experience affective reactions to the observed experiences of others, while cognitive empathy is the capacity to recognize and understand another's emotional state in order to enable the observer to adopt the other's point of view [3].

Empathy seems to be impaired in several neurodegenerative diseases. Severe loss of empathy has been widely described as a diagnostic criterion of the behavioral variant of frontotemporal dementia (bv-FTD) [4–6]. On the other hand, despite studies that explored empathy deficit in Alzheimer's disease (AD) not showing conclusive results [7], at the state of the art, researchers agree that empathy is impaired in AD, with a predominant loss of cognitive empathy, while the affective domain seems to be spared. Recently, it has been shown that there is a peculiar involvement of brain regions related to empathy in the AD continuum: the impairment of cognitive empathy seems to start at the mild cognitive impairment (MCI) stage and may be related to a progressive involvement starting from right middle frontal gyrus in the prodromal stage, extending to the insula and superior temporal gyrus in the dementia stage. On the other hand, an increase of emotional contagion (part of affective empathy) was already found in preclinical phases, and this alteration might be related to the derangement of mirror neurons systems in parietal regions in prodromal stages and to impairment of the temporal emotion inhibition system in advanced phases [8].

Despite the growing interest in empathy and social cognition in bv-FTD and A, little is known about the potential impairment of empathy in primary progressive aphasia (PPA). The term PPA defines a group of neurodegenerative syndromes characterized by progressive, selective decline in speech and language functions [9]. According to the current classification, three main clinical variants have been identified on the basis of specific linguistic features: non-fluent/agrammatic (nfv-PPA) and semantic (sv-PPA) variants, which are considered as part of the fronto-temporal lobar degeneration (FTLD) spectrum, and logopenic variant primary progressive aphasia (lv-PPA), which is considered as an atypical presentation of AD [9]. Interestingly, compromised brain regions across variants extend beyond classical language areas, and none of them is exclusively devoted to linguistic functions [10–12]. Deficits in both cognitive and affective empathy have been described in sv-PPA and may be independent from language dysfunction [13,14]. On the other hand, while affective empathy seems to be spared in nfv-PPA, findings about cognitive empathy are far from conclusive [15]. Finally, only few reports explore empathy and emotion recognition in lv-PPA, showing a decrease in cognitive empathy outcomes when compared to the pre-morbid results alongside a marginal reduction of affective empathy [15,16]. Evidence regarding empathy in lv-PPA is still sparse and inconclusive, and the neural correlates of empathy impairment have never been explored before.

In this scenario, the aims of our study were: (1) to investigate possible empathy deficits in lv-PPA; (2) to explore the metabolic pattern associated with empathy deficit in lv-PPA by means of [18F]fluorodeoxyglucose positron emission tomography ([18F]FDG-PET); and (3) to compare empathy and its neural correlates in lv-PPA and in prototypical amnesic AD.

2. Materials and Methods

2.1. Participants

Eighteen patients with a clinical diagnosis of lv-PPA [9] and thirty-eight patients with a diagnosis of amnesic AD [17] were longitudinally included in this study. All participants underwent a comprehensive family and clinical history, general and neurological examination, extensive neuropsychological assessment, and evaluation of empathy through the Interpersonal Reactivity Index (IRI) [18,19]. Facial emotion recognition ability was assessed through the Ekman 60 Faces (EK-60 F) Test in 47 patients (34 AD and 13 lv-PPA) [20,21]. IRI and EK-60 F data of AD and lv-PPA patients were compared to those obtained in 31 subjects with subjective cognitive decline (SCD) [22]. Age at empathy assessment was defined as the age of the subject when the IRI and/or EK-60 F tests were administered. Age at onset was defined as age at the onset of complaints of cognitive symptoms. Seventy-one subjects underwent APOE genotyping.

2.2. Neuropsychological Evaluation, Empathy, and Facial Emotion Recognition Assessment

All subjects were evaluated by means of an extensive neuropsychological battery standardized and described in further detail elsewhere [23]. Mini-Mental State Examination was used as global measurement. Short-term verbal and spatial memory were explored through Digit Span and Corsi Tapping Test and verbal long-term memory through Rey Auditory Verbal Learning Test, Immediate Recall RVLT-I and Delayed Recall RVLT-D, and Babcock Short Story Immediate and Delayed Recall. Token Test and Category Fluency Task were used to evaluate language [23–25]. Rey–Osterrieth Complex Figure copy and recall of Rey–Osterrieth Complex Figure test were used to explore visual-spatial abilities and visuo-spatial long-term memory, respectively [26]. Dual Task [27], Phonemic Fluency Test [25], Trail Making Test (TMT) [28], and Visual Search [29] were used to explore attention/executive function. Everyday memory was assessed by means of Rivermead Behavioral Memory Test (RBMT) [30]. All raw test scores were adjusted for age, education, and gender according to the correction factor reported in validation studies for the Italian population [23,24,26–30]. Language was further assessed by SAND (Screening for Aphasia in NeuroDegeneration) [31] or ENPA (Neuropsychological Evaluation for Aphasia) [32]. The presence and severity of depressive symptoms were evaluated by means of the 22-item Hamilton Depression Rating Scale (HRSD) [33].

Empathy was evaluated by Interpersonal Reactivity Index (IRI) [18,19], which consists of a 28-item questionnaire divided in four different 7-item subscales: perspective taking (PT), fantasy (FT), empathic concern (EC), and personal distress (PD). Perspective taking and fantasy subscales explore cognitive empathy, while empathic concern and personal distress subscales are used to assess affective empathy. In. more details, PD subscale is a measure of emotional contagion [34], which is a primitive structure of affective empathy and defined as the automatic, total identification with another's behavior in order to encourage altruistic behavior [3]. Each item of IRI consists of an affirmation, and the individual has to express the degree of agreement on a 5-point Likert scale from 1 (does not describe me/the patient at all) to 5 (describes me/the patient very well). IRI was administered to the informants [35]. Caregivers had to rate patients' empathy before (T0) and after (T1) cognitive symptoms' onset. The differences from T0 to T1 of the scores of each scale were quantified as Delta (Δ): ΔFT, ΔPT, ΔEC, and ΔPD.

Facial emotion recognition was assessed by Ekman 60 Faces (EK-60 F) Test, which consists of 60 black and white pictures of the Ekman and Friesen series of Pictures of Facial Affect [20], representing the faces of ten actors (six women and four men), each of which shows one of six basic emotions (anger, sadness, happiness, fear, disgust, or surprise). A global score (EK-60 F global score) of 60 indicates the best possible performance. Each basic emotion has a sub-score of a maximum of 10 points. Images were shown each for 5 s according to the Ekman and Friesen procedure [20] via Power Point presentation on a computer. Patients were asked to indicate which of the basic emotions better represents the facial emotion shown on the display [21].

2.3. Apolipoprotein E (APOE) Genotyping

Patients' DNA was extracted from peripheral blood samples by use of a standard automated method (QIAcube, QIAGEN Hilden, Germany). APOE genotypes were investigated by HRMA [36]. Two sets of PCR primers were designed to amplify the regions encompassing rs7412 (NC_000019.9:g.45412079C > T) and rs429358 (NC_000019.9:g.45411941T > C). The APOE genotype was coded as APOE ε4− (no APOE ε4 alleles) and APOE ε4+ (presence of one or two APOE ε4 alleles).

2.4. Amyloidosis and Neurodegeneration Biomarkers Analysis

Amyloidosis biomarkers analysis was performed in 69 patients. Sixty-four patients (27 amnesic AD, 17 lv-PPA, 20 SCD) underwent cerebrospinal fluid (CSF) biomarkers analysis.

The CSF samples were collected at 8 a.m. by lumbar puncture, immediately centrifuged, and stored at −80 °C until performing the analysis. Aβ1-42, Aβ1-42/1-40 ratio, t-tau, and p-tau were measured using a chemiluminescent enzyme immunoassay (CLEIA) analyzer LUMIPULSE G600 (Fujirebio). Fujirebio guidelines determined cut-off values (Diagnostic sensitivity and specificity using clinical diagnosis and follow-up golden standard. 19 November 2018), which were Aβ1–42 > 670 pg/mL, Aβ42/40 ratio > 0.062, t-tau < 400 pg/mL, and p-tau < 60 pg/mL [37].

Twenty-three patients (7 amnesic AD, 6 lv-PPA, 10 SCD) underwent cerebral amyloid PET. Amyloid PET imaging was performed according to national and international guidelines [38], with any of the available fluorine18-labeled tracers (18Florbetaben (FBB)—Bayer-Pyramal, 18Flutemetamol (FMM)—General Electric). Images were rated as either positive or negative according to criteria defined by the manufacturers.

According to ATN classification [39], patients were classified as A+ if at least one of the amyloid biomarkers (CSF or amyloid PET) revealed the presence of Aβ pathology and as A- if none of the biomarkers revealed the presence of Aβ pathology. All amnesic AD and lv-PPA patients were A+.

2.5. FDG-PET Brain Imaging

All amnesic AD and lv-PPA patients underwent brain [18F]FDG-PET. Scans were performed using advanced hybrid PET-CT scanner in 3D list mode. Patients were instructed to fast for 6 h before the study, and blood sugar level was tested to be lower than 120 mg/dL. Patients were injected with 185 MBq of [18F]-FDG via a venous cannula. After the injection, patients were left in a dimly lit, quiet room and told to keep their eyes closed. All [18F]FDG-PET scans were acquired following the EANM guidelines [40]. All PET images were corrected for photon attenuation, scatter, and radioactive decay and reconstructed using 3D iterative algorithm. [18F]FDG-PET scans pre-processing and statistical analysis are described in Section 2.6.

2.6. Statistical Analysis

IBM SPSS Statistics Software Version 25 (SPSS Inc., Chicago, IL, USA) and computing environment R 4.0.3 (R Foundation for Statistical Computing, Vienna, 2013) were used to perform all statistical analysis. All p-values were two-tailed, and the significance level for all analyses was set at $\alpha = 5\%$, corresponding to a threshold p of 0.05. All variables are described as mean and standard deviation. Shapiro–Wilk test was used to assess the distribution of all variables. Chi-square test was used to compare categorical data. One-way ANOVA followed by Bonferroni post hoc test was used to evaluate differences among groups in continuous variables. Variation of IRI scores over time, from before to after the onset of cognitive symptoms (T0-T1), was explored through Wilcoxon signed-rank test. The influence of demographic variables and neuropsychological scores on current empathy and on facial emotion recognition ability was investigated through Spearman's correlation. Bonferroni correction for multiple comparisons was applied for correlations between each IRI subscales and demographic features ($p = 0.0019$), neuropsychological measures ($p = 0.0019$), and SAND scores ($p = 0.0016$); similarly, it was applied for correlations between facial emotion recognition ability and demographic features ($p = 0.002$), neuropsychological measures ($p = 0.002$), and SAND scores ($p = 0.0017$).

2.7. SPM Analysis

In order to assess the metabolic pattern related to empathy changes in lv-PPA and how it differs between prototypical amnesic AD and lv-PPA, a total of 42 patients were considered (26 amnesic AD and 16 lv-PPA patients). Each patient had positive amyloid biomarkers. [18F]FDG-PET images were normalized to the MNI space using a validated procedure. Images were smoothed with an isotropic 3D Gaussian kernel with a FWHM of 8 mm in each direction and then were used for a single-subject SPM-based routine for diagnostic purposes [41]. Age was included as a covariate in the two-sample t-test analysis. The correlation between IRI subscales, resulting from behavioral data analysis, and brain hypometabolism in the amnesic AD and lv-PPA groups was explored using the

SPM multiple regression design, including age and MMSE as nuisance variables in the linear model. The threshold was set at p-value < 0.001, uncorrected, to test for correlations also in the small subsamples of amnesic AD and lv-PPA. Only clusters containing more than 50 voxels were considered significant.

2.8. Data Availability

The data that support the findings of this study are available from the corresponding author upon reasonable request.

3. Results

3.1. Demographic Features and Biomarkers Analysis

Table 1 shows the demographic variables of the cohort. Considering the whole sample, 57 patients were females and 30 males. Age at onset was significantly different among the three groups (F [2, 82] = 17.527, $p < 0.001$): indeed, SCD (54.40 ± 10.08, $p < 0.001$) were younger than amnesic AD (66.53 ± 6.708, $p < 0.001$) and lv-PPA (64.50 ± 7.60, $p < 0.001$) patients. Age at empathy assessment was significantly different among groups (F [2, 83] = 3.739, $p = 0.028$): in detail, at empathy evaluation, SCD were younger (65.61 ± 9.48) than amnesic AD (71.04 ± 7.33, $p < 0.024$) but not younger than lv-PPA (68.65 ± 7.02, $p = 1.000$) patients. Mini Mental State Examination (MMSE) was different among the groups (F [2, 81] = 48.897 $p < 0.001$), with lower scores in amnesic AD (17.03 ± 5.28) as compared to SCD (27.73 ± 2.04, $p < 0.001$) but not to lv-PPA (16.69 ± 6.37, $p = 1.000$) patients. Considering the subsample who underwent APOE genotype analysis, 36.62% resulted to be APOE ε4 carriers.

Table 1. Demographic features in Alzheimer's disease (AD) and logopenic variant primary progressive aphasia (lv-PPA) groups.

	SCD n **= 31**	**AD** n **= 38**	**lv-PPA** n **= 18**
Gender (M/F)	4/27 $^{*\alpha}$	17/21 *	9/9 $^{\alpha}$
Age at onset (years)	54.94 ± 10.08 $^{\beta\gamma}$	66.53 ± 6.71 $^{\beta}$	64.50 ± 7.60 $^{\gamma}$
Age at empathy (years)	65.61 ± 9.48 $^{\delta}$	71.04 ± 7.33 $^{\delta}$	68.65 ± 7.02
Disease duration (years)	9.62 ± 7.51 $^{\varepsilon\eta}$	4.44 ± 3.59 $^{\varepsilon}$	2.87 ± 1.54 $^{\eta}$
Family history of dementia	22/6 $^{\theta}$	16/17 $^{\theta}$	6/11
Years of education	12.58 ± 3.40 $^{\iota}$	9.80 ± 4.65 $^{\iota\kappa}$	13.22 ± 4.45 $^{\kappa}$
MMSE	27.43 ± 2.04 $^{\lambda\mu}$	17.03 ± 5.28 $^{\lambda}$	16.69 ± 6.37 $^{\mu}$
***APOE* ε4+**	30.77%	50.00%	50.00%
A+/A−	8/12 (40%) $^{\nu\circ}$	30/0 (100%) $^{\nu}$	18/0 (100%) $^{\circ}$

Values are reported as mean and standard deviation for continuous variables and as frequencies or percentages for categorical variables. Statistical differences among groups are underlined. M, males; F, females; MMSE, Mini Mental State Examination. * $\chi^2 = 8.172$, $p = 0.008$; α $\chi^2 = 6.126$, $p = 0.019$; β $p < 0.001$; γ $p = 0.001$; δ $p = 0.024$; ε $p < 0.001$; η $p < 0.001$; θ $\chi^2 = 5.838$, $p = 0.019$; ι $p = 0.025$; κ $p = 0.018$; λ $p < 0.001$; μ $p < 0.001$; ν $\chi^2 = 24.92$, $p < 0.001$; ° $\chi^2 = 16.71$, $p < 0.001$.

Sixty-four patients (27 amnesic AD, 17 lv-PPA, 20 SCD) underwent CSF biomarkers analysis. Twenty-three patients (7 amnesic AD, 6 lv-PPA, 10 SCD) were subjected to cerebral amyloid PET, which was positive in 17 patients (6 amnesic AD, 6 lv-PPA, 5 SCD). Based on the positivity for at least one cerebral amyloidosis biomarker, 30 amnesic AD and 18 lv-PPA patients and 8 subjects with SCD were classified as A+ (56 out of 69, 82.35%).

3.2. IRI Empathy Results

Significant differences were detected neither in premorbid empathy, in IRI total score T0 (F [2, 79] = 0.397, $p = 0.674$), nor in any subscale FT-T0 (F [2, 79] = 1.599, $p = 0.209$), PT-T0

(F [2, 79] = 1.945, $p = 0.150$), EC-T0 (F [2, 79] = 0.530, $p = 0.590$), and PD-T0 (F [2, 79] = 1.014, $p = 0.368$) among the three groups.

As regards current empathy, one-way ANOVA showed significant differences in FT-T1 (F [2, 79] = 5.046, $p = 0.008$) and PD-T1 (F [2, 79] = 10.004, $p < 0.001$) among groups. At Bonferroni post hoc test, amnesic AD patients' FT-T1 scores were significantly lower than those of SCD (14.91 ± 6.50 vs. 19.54 ± 4.45, $p = 0.009$). Both lv-PPA (28.62 ± 6.03, $p < 0.001$) and amnesic AD (26.09 ± 6.00, $p = 0.003$) presented higher scores than SCD (21.04 ± 5.51) on the PD-T1 subscale, while no differences were found between lv-PPA and amnesic AD patients ($p = 0.481$). No significant differences were found in PT-T1 and EC-T1 scores among groups (Figure 1).

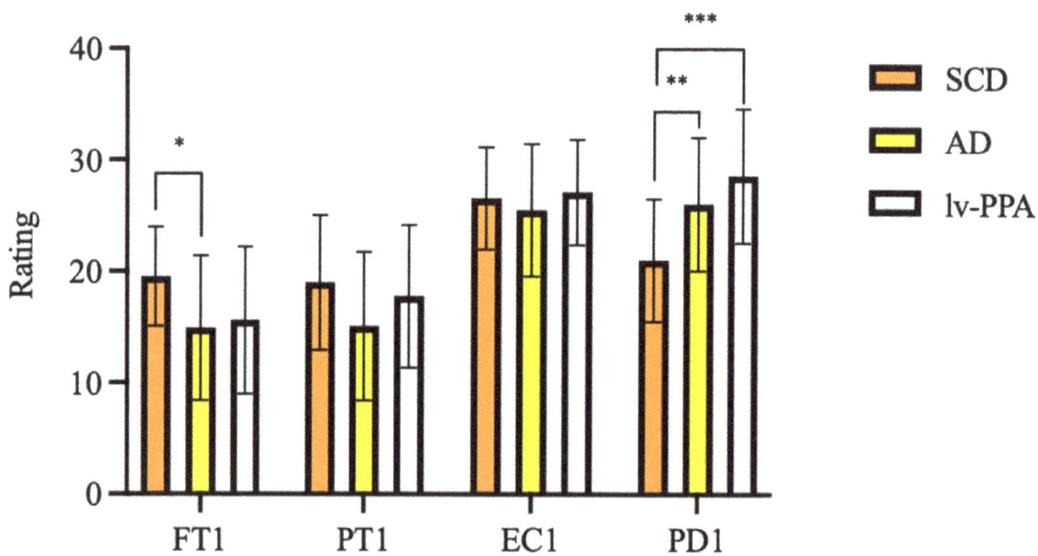

Figure 1. Empathy assessed by Interpersonal Reactivity Index (IRI) in subjective cognitive decline (SCD), Alzheimer's disease (AD) and logopenic variant primary progressive aphasia (lv-PPA). * $p < 0.05$, ** $p < 0.005$, *** $p < 0.001$.

To estimate changes of empathy from before to after the onset of cognitive symptoms in amnesic AD and lv-PPA, Wilcoxon signed-rank test was used (Table 2). A significant decrease of FT (17.11 ± 5.09 vs. 15.05 ± 6.21, z = −2.464, $p = 0.014$) was detected in amnesic AD subgroup. A decrease in PT scores was found both in amnesic AD (21.08 ± 6.15 vs. 15.42 ± 6.66, z = −4.753, $p < 0.001$) and in lv-PPA patients (23.94 ± 6.40 vs. 15.82 ± 6.61, z = −3.435, $p < 0.001$). Finally, a significant increase of PD was found both in amnesic AD (17.08 ± 4.99 vs. 26.03 ± 5.72, z = −5.204, $p < 0.001$) and in lv-PPA patients (15.76 ± 5.47 vs. 28.35 ± 6.07, z = −3.623, $p < 0.001$).

Table 2. Change of empathy capacity from before to after the onset of cognitive symptoms in Alzheimer's disease (AD) and logopenic variant primary progressive aphasia (lv-PPA) groups.

	AD			lv-PPA		
	Mean ± SD	*z*	*p*	**Mean ± SD**	*z*	*p*
IRI 0 IRI 1	81.26 ± 13.31 81.95 ± 13.35	−0.828	0.408	83.41 ± 11.12 83.76 ± 14.90	−0.166	0.868
FT 0 FT 1	17.11 ± 5.09 15.05 ± 6.21	−2.464	**0.014**	17.18 ± 5.89 15.24 ± 6.21	−1.574	0.115
PT 0 PT 1	21.08 ± 6.15 15.42 ± 6.66	−4.752	**<0.001**	23.94 ± 6.40 15.82 ± 6.61	−3.435	**0.001**
EC 0 EC 1	26.39 ± 4.45 25.76 ± 5.70	−1.185	0.236	25.94 ± 5.12 24.35 ± 6.47	−1.250	0.211
PD 0 PD 1	17.08 ± 4.99 26.03 ± 5.72	−5.204	**<0.001**	15.76 ± 5.47 28.35 ± 6.07	−3.623	**<0.001**

Values are reported as mean and standard deviation. Statistical differences between groups are in **bold**.

3.3. EK-60 F Emotion Recognition Results

Considering emotion recognition ability assessed by EK-60F test, all the variables were significantly different among the three groups (Table 3). At EK-60F global score, lv-PPA and amnesic AD patients performed significantly poorer than SCD (32.25 ± 8.09 vs. 33.29 ± 7.56 vs. 47.58 ± 5.14, respectively, $p < 0.001$). Concerning the single emotions' recognition, lv-PPA and amnesic AD had worse performances in the recognition of anger, disgust, sadness, surprise, and happiness detection as compared to SCD (Table 3).

Table 3. Facial emotion recognition ability in subjective cognitive decline (SCD), Alzheimer's disease (AD), and logopenic variant primary progressive aphasia (lv-PPA) as assessed by the Ekman 60 Faces Test (EK-60 F).

	SCD	AD	lv-PPA	F	*p*	*p* between SCD and AD	*p* between SCD and lv-PPA	*p* between AD and lv-PPA
EK-60 F total score	47.58 ± 5.14	33.29 ± 7.56	32.25 ± 8.09	42.863	***p* < 0.001**	***p* < 0.001**	***p* < 0.001**	*p* = 1.000
Execution time (seconds)	290 ± 56.19	445.74 ± 124.03	425.08 ± 103.60	21.427	***p* < 0.001**	***p* < 0.001**	***p* < 0.001**	*p* = 1.000
Anger	7.35 ± 1.80	5.29 ± 2.33	5.08 ± 2.11	9.480	***p* < 0.001**	***p* = 0.001**	***p* = 0.006**	*p* = 1.000
Disgust	8.10 ± 1.45	5.26 ± 2.49	5.25 ± 2.60	16.134	***p* < 0.001**	***p* < 0.001**	***p* = 0.001**	*p* = 1.000
Fear	5.10 ± 2.68	2.94 ± 2.00	3.67 ± 2.43	6.862	***p* = 0.002**	***p* = 0.001**	*p* = 0.235	*p* = 1.000
Happiness	9.84 ± 0.45	8.94 ± 1.043	7.33 ± 2.15	21.962	***p* < 0.001**	***p* = 0.006**	***p* < 0.001**	***p* < 0.001**
Sadness	7.97 ± 2.01	4.71 ± 2.51	4.75 ± 2.80	17.476	***p* < 0.001**	***p* < 0.001**	***p* < 0.001**	*p* = 1.000
Surprise	9.48 ± 1.03	6.24 ± 2.75	6.17 ± 2.25	21.666	***p* < 0.001**	***p* < 0.001**	***p* < 0.001**	*p* = 1.000

Values are reported as mean and standard deviation. Statistical differences between groups are in **bold**.

SCD scores for fear recognition were higher than those obtained by amnesic AD patients (5.10 ± 2.68 vs. 2.94 ± 2.00, $p = 0.001$) but not higher than those obtained by lv-PPA (3.67 ± 2.43, $p = 0.235$). On the other hand, no differences were detected between lv-PPA and amnesic AD patients except for happiness, with lower scores found in the lv-PPA subgroup as compared to the amnesic AD one (7.33 ± 2.15 vs. 8.94 ± 1.04 vs., $p = 0.01$) (Figure 2).

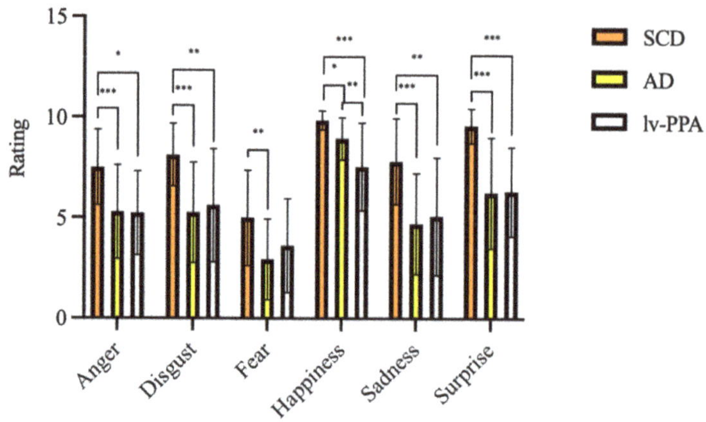

Figure 2. Emotional recognition ability as assessed by Ekman 60 Faces Test in subjective cognitive decline (SCD), Alzheimer's disease (AD), and logopenic variant primary progressive aphasia (lv-PPA). * $p < 0.05$, ** $p < 0.005$, *** $p < 0.001$.

3.4. Correlations between Demographic Data, Neuropsychological Variables, Empathy, and Emotion Recognition

Correlations were found neither between each IRI subscales, age at empathy evaluation, years of education, and HDRS scores; nor in the whole cohort; nor in lv-PPA; nor in amnesic AD subgroups. On the other hand, significant differences between women and men in the whole group were found: women had higher scores than men in IRI-T0 (84.02 ± 11.69 vs. 78.46 ± 10.71, p = 0.034), IRI-T1 (86.67 ± 11.88 vs. 79.79 ± 13.94, p = 0.044), FT-T0 (18.45 ± 4.74 vs. 16.54 ± 4.85, p = 0.031), FT-T1 (17.86 ± 5.56 vs. 14.29 ± 6.79, p = 0.004), and PD-T0 (21.16 ± 6.29 vs. 20.92 ± 5.36, p = 0.024). These gender differences were not found when we analyzed lv-PPA and amnesic AD except for IRI-T1 scores, which were higher in amnesic AD women as compared to men (84.29 ± 14.04 vs. 77.53 ± 11.69, p = 0.021). We also analyzed correlations between neuropsychological tests, SAND evaluation, and each IRI subscale: correlations were found neither in the whole cohort, nor in lv-PPA, nor in amnesic AD subgroups.

Correlations were found between each EK-60 F scores, age at empathy evaluation, and years of education neither in the whole cohort, nor in lv-PPA, nor in amnesic AD subgroups. Similar to IRI subscales, we found significant difference between women and men in the whole group: in more detail, women obtained higher scores as compared to men in EK-60 F total (40.16 ± 9.92 vs. 34.75 ± 8.40, p = 0.012), disgust (6.20 ± 2.50 vs. 5.17 ± 2.46, p = 0.004), sadness (6.45 ± 2.65 vs. 4.71 ± 2.88, p = 0.011), and surprise (7.86 ± 2.53 vs. 6.50 ± 2.80, p = 0.017). Moreover, EK-60 F execution time was lower in women as compared to men (361.69 ± 121.13 vs. 438.13 ± 110.53, p = 0.001). These gender differences were not found when we analyzed lv-PPA and amnesic AD subgroups. Moreover, no correlations were found between neuropsychological tests, SAND evaluation, and emotion recognition ability (both EK-60 F total score and single emotion recognition scores) in the whole cohort, in logopenic, or in amnesic Alzheimer's disease subgroups.

3.5. SPM Results

Significant correlations between IRI subscales and brain metabolism in amnesic AD and lv-PPA groups were found through the SPM multiple regression analysis.

In detail, a negative correlation between ΔPT and brain metabolism was found:

- In amnesic AD in the right superior temporal gyrus, fusiform gyrus, and middle frontal gyrus ($p < 0.005$);

- In lv-PPA in the left inferior parietal lobule, left insula, left middle frontal gyrus, right posterior paracentral lobule, and bilateral superior frontal gyrus ($p < 0.005$) (Table 4 and Figure 3).

Table 4. Correlation between changes in empathy over time and cerebral hypometabolism in in logopenic primary progressive aphasia and amnesic Alzheimer's disease patients at 18F-FDG-PET SPM analysis.

Negative Correlation between ΔPT (PT-T0 - PT-T1) and Brain Metabolism					
	Cluster Extent	**Talairach Coordinates (mm)**			**T Score**
		x	y	z	
Amnesic AD					
R Superior Temporal Gyrus	130	32.0	16.0	−24.0	4.90
R Middle Frontal Gyrus	90	24.0	18.0	43.0	3.69
R Fusiform Gyrus	67	57.0	−43.0	−8.0	3.38
lv-PPA					
L Inferior Parietal Lobule	138	−30.0	−50.0	45.0	4.95
L Superior Frontal Gyrus	125	−8.0	−1.0	66.0	4.83
L Middle Frontal Gyrus		−10.0	−11.0	58.0	3.15
R Posterior Paracentral Lobule	156	20.0	−42.0	56.0	4.74
R Superior Frontal Gyrus	68	10.0	24.0	52.0	4.48
L Insula	64	−38.0	14.0	10.0	4.13
L Insula		−38.0	22.0	12.0	4.11
Positive Correlation between ΔPD (PD-T0 - PD-T1) and Brain Metabolism					
	Cluster Extent	**Talairach Coordinates (mm)**			**T Score**
		x	y	z	
Amnesic AD					
R Inferior Frontal Gyrus	121	42.0	7.0	33.0	5.43
lv-PPA					
L Inferior Parietal Lobule	85	−50.0	−31.0	33.0	6.77
R Superior Frontal Gyrus	68	14.0	−12.0	67.0	4.94
L Insula	168	−44.0	4.0	5.0	4.93
R Precuneus	158	14.0	−48.0	56.0	4.51
L Superior Frontal Gyrus	77	−12.0	−14.0	67.0	4.42
L Superior Frontal Gyrus		−6.0	−6.0	67.0	3.36

Abbreviations: AD, Alzheimer's disease; lv-PPA, logopenic primary progressive aphasia; L, left; R, right. Significant differences at $p < 0.005$ in both amnesic AD and in lv-PPA for negative correlation with ΔPT. Significant differences at $p < 0.001$ in amnesic AD and $p < 0.005$ in lv-PPA for positive correlation with ΔPD.

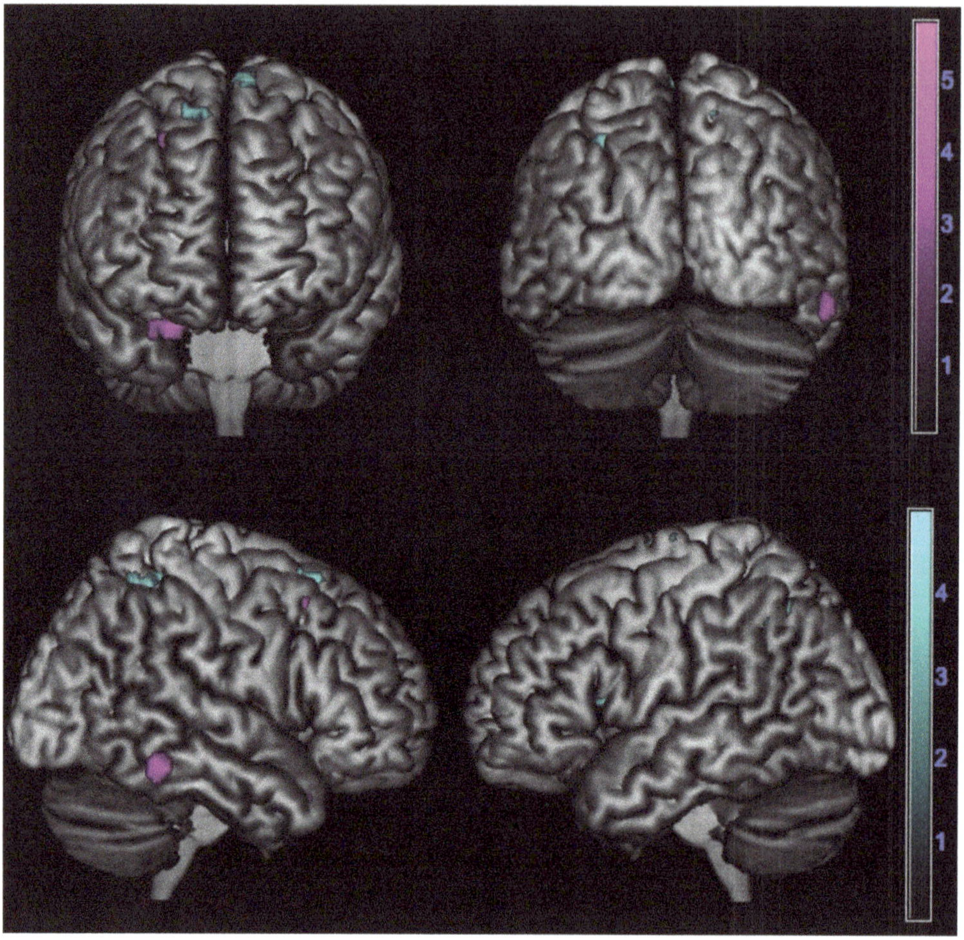

Figure 3. Negative correlation between changes in perspective taking ability over time assessed by ΔPT (PT-T0 - PT-T1) and brain metabolism in logopenic primary progressive aphasia and amnesic Alzheimer's disease patients at 18F-FDG-PET SPM analysis. Significant clusters projected on the standardized Montreal Neurological Institute (MNI) magnetic resonance imaging (MRI) render surface. Color grading: cyan, lv-PPA; violet, amnesic AD.

A positive correlation between ΔPD and brain metabolism was found:

- In amnesic AD in the right inferior frontal gyrus ($p < 0.001$);
- In lv-PPA in the left inferior parietal lobule, insula, and superior frontal gyrus and in the right precuneus and superior frontal gyrus ($p < 0.005$) (Table 4 and Figure 4).

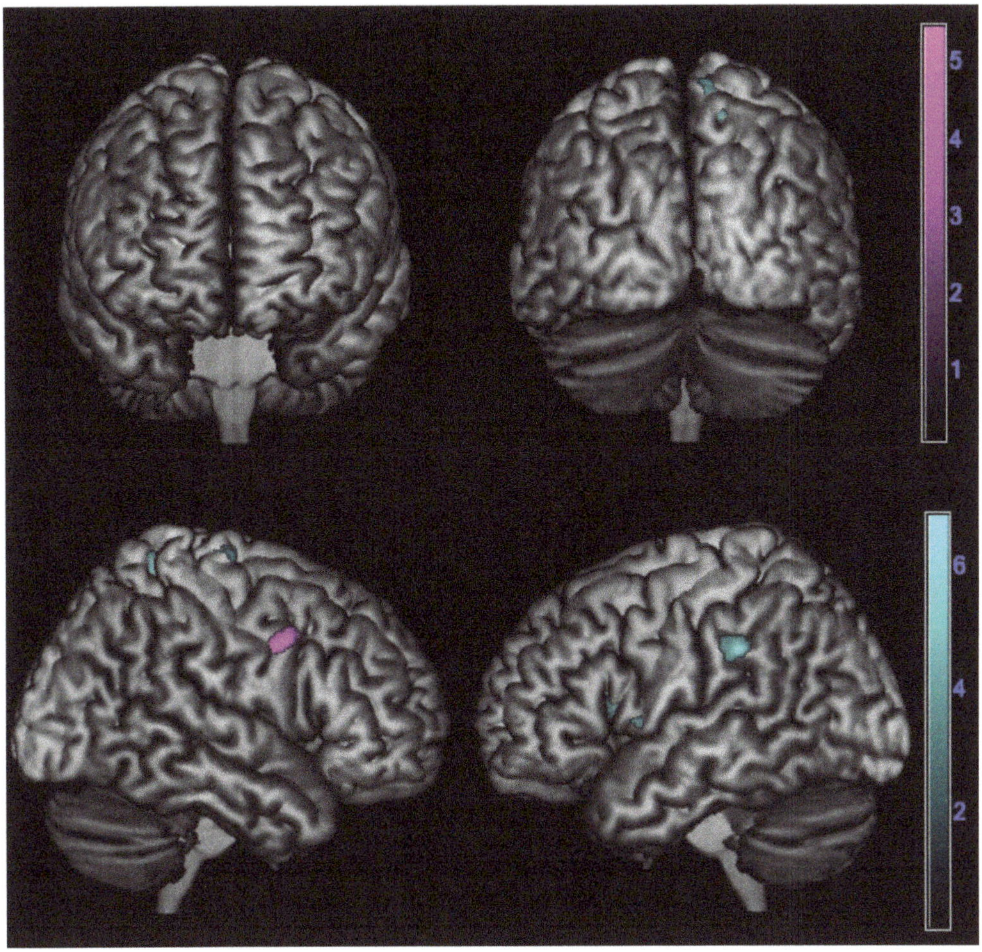

Figure 4. Positive correlation between changes in emotional contagion over time assessed by ΔPD (PD-T0 - PD-T1) and brain metabolism in logopenic primary progressive aphasia and amnesic Alzheimer's disease patients at 18F-FDG-PET SPM analysis. Significant clusters projected on the standardized Montreal Neurological Institute (MNI) magnetic resonance imaging (MRI) render surface. Color grading: cyan, lv-PPA; violet, amnesic AD.

4. Discussion

Our study is the first that deeply explored empathy in lv-PPA while also trying to define the neural correlates of empathy impairment in this variant of PPA, which is considered as an atypical presentation of Alzheimer's disease. Moreover, for the first time, we compared the behavioral and metabolic data of empathy between lv-PPA and prototypical amnesic AD in order to define similarities and differences between these two presentations of Alzheimer's disease.

Our study showed several similarities in empathic impairment between lv-PPA and amnesic AD. In more detail, both lv-PPA and amnesic AD patients presented a decrease in PT over time, suggesting a significant damage to cognitive empathy. On the other hand, a relative sparing of EC (part of affective empathy) was found in both subgroups.

In particular, a significant decline in PT was found in both logopenic and in amnesic AD, suggesting that both clinical phenotypes of AD present a significant change of cognitive empathy from before to after the onset of cognitive symptoms. Although results about empathy deficits in AD are far from conclusive, recent studies suggested a selective impairment of cognitive empathy [35,42–44]. Moreover, a decrease of perspective taking ability from before to after the onset of cognitive disturbance has been recently described in AD, and this change seems to be present also in the prodromal stage of the disease [8]. Considering lv-PPA, Hazelton et al. already reported a disruption of cognitive empathy, showing a decrease of perspective taking ability by comparing premorbid and present scores of the PT subscale [15].

Affective empathy seems to be spared both in lv-PPA and in amnesic AD patients since no significant changes were found in EC scores from before to after the onset of cognitive symptoms. Moreover, no differences were found in EC-T1 among lv-PPA, amnesic AD, and also SCD patients, which may be considered as the preclinical phase of cognitive decline [45]. However, we found changes in PD in both lv-PPA and in amnesic AD. Personal distress has been widely defined as a measure of emotional contagion, which could be considered as a primitive structure of emotional empathy and indicates the tendency to automatically adopt the behavior of another person [2]. Both lv-PPA and amnesic AD patients presented higher PD scores as compared to SCD. When we analyzed the trend of each component of empathy over time, we found a significant increase in PD from before to after the onset of cognitive disturbs both in amnesic AD and in lv-PPA patients, leading to the further hypothesis that emotional contagion heightens with the development of cognitive decline.

Concerning amnesic AD, our results are in line with some studies that previously described a higher emotional contagion as compared to healthy controls [7,34]: Sturm et al. hypothesized that emotional contagion might increase linearly from healthy controls to MCI and AD patients, who presented the highest degree of emotional contagion [34]. Moreover, a previous work of our group has already shown a significant increase of PD scores from before to after the onset of cognitive disturbance [8], suggesting that this change in emotional contagion might be a peculiar feature of AD patients.

Changes in PD scores in PPA have also been previously described by Hazelton et al., who explored empathy deficit in lv-PPA and nfv-PPA, showing an increase in personal distress in both variants following the disease onset [15].

We also explored emotion recognition ability through the EK-60 F test. According to our results, both lv-PPA and amnesic AD patients presented difficulties in facial emotion recognition, showing lower scores in recognition of almost all six basic emotions as compared to SCD subjects. The only exception was the recognition of fear since no differences were detected between lv-PPA and SCD subgroups, while amnesic AD patients performed more poorly than SCD. Our results are in line with the current literature reporting severe difficulties in identification of facial emotions in AD patients [46–49]. On the other hand, emotion recognition has been only recently studied in lv-PPA, and results are far too few to be clearly conclusive. A previous study reported spared performance in this variant through a task requiring identification of emotional expressions in static faces [16]. On the other hand, another work reported better performance in an emotion recognition task involving naturalistic scenes (i.e., TASIT) in lv-PPA than nfv-PPA, in which emotion recognition seems to be impaired [15,16,50]. Nevertheless, this difference was not statistically significant. Moreover, it has been shown that lv-PPA significantly outperforms those with semantic variant [51]. Thus, considering these findings, it has been hypothesized that emotion recognition deficits in lv-PPA (if present) may be milder than in other variants [10]. However, considering our results, we might suggest that emotion recognition ability may be impaired in lv-PPA. However, these data need to be confirmed.

Globally, no clear differences were detected in emotion recognition ability between lv-PPA and amnesic AD. Although several studies reported incongruent results, emotion recognition ability seems to be impaired in AD [8,47,48]. Indeed, it has been suggested that the recognition of specific emotions (i.e., disgust) may be spared in AD probably due to the relative sparing of basal ganglia [52]. However, these findings have not been confirmed in other works [8]. To the best of our knowledge, no previous studies have analyzed this aspect by comparing these two presentations of Alzheimer's disease. Our results lead to the hypothesis that this emotion processing function is similarly impaired in the two presentation of the disease. Notably, the only difference was found in the recognition of happiness, with more difficulties found in lv-PPA as compared to amnesic AD patients. This result is challenging to discuss. In fact, this finding has never been previously described. Piguet et al. analyzed emotion recognition ability through the EK 60-F test by comparing healthy controls, nfv-PPA, and lv-PPA patients, showing significant group differences for the negative (anger, disgust, fear, and sadness) but not the positive (surprise and happiness) emotions due to the low performance of the non-fluent (but not logopenic) group compared to healthy controls. In addition, non-fluent patients also performed worse than logopenic patients for the emotion fear, which is considered as a negative emotion [16]. Such difficulties in happiness recognition in our lv-PPA patients have never been described so far: this subtle difference may be highlighted in our work through the comparison between the two presentation of AD. However, this result needs to be confirmed in further studies with larger cohorts.

Interestingly, we did not find any correlation between demographic data, neuropsychological tests, language evaluation, IRI subscale, and emotion recognition in either lv-PPA or in amnesic AD patients. This is partially in contrast with a previous report according to which the reduction of PT scores in lv-PPA was correlated with visuospatial abilities [15]. However, no further studies have explored this topic yet as well as correlations between emotion recognition ability and neuropsychological evaluation. Moreover, it has been recently highlighted that PD positively correlates with depression in healthy subjects [53]: indeed, no correlations were found between HDRS and PD in our cohort either in lv-PPA or in amnesic AD patients. Although our work is a preliminary study, our results might suggest that the empathic impairment and difficulties in emotion recognition might not be correlated with cognitive decline, with depressive symptoms, and in particular with language impairment in lv-PPA patients. Nevertheless, further studies are needed to confirm this hypothesis.

The further aims of our study were to explore the metabolic pattern associated with empathy changes in lv-PPA and to compare empathy and its neural correlates in logopenic and in prototypical amnesic AD. First of all, to the best of our knowledge, this is the first study that explored the neural bases of empathy deficits in lv-PPA. Surprisingly, despite the similar empathy impairment, the metabolic correlates of empathy deficits were different between lv-PPA and amnesic AD patients.

Indeed, while in amnesic AD patients, the empathic dysfunction was related to hypometabolism of specific brain regions mainly located on the right hemisphere, in lv-PPA, it was correlated with hypometabolism of both hemispheres. This involvement of both hemispheres in lv-PPA might be explained by the fact of the presence of a specific and peculiar left lateralization of neurodegeneration in this disease [9].

Cognitive empathy changes over time were correlated with the involvement of the right superior temporal gyrus, middle frontal gyrus, and fusiform gyrus in amnesic AD and with involvement of the left inferior parietal lobule, insula, middle frontal gyrus, and bilateral superior frontal gyrus in lv-PPA.

Previous works have already described the correlation between cognitive empathy deficits and the involvement of superior temporal and middle frontal gyri in AD patients [8]. This correlation might be explained by the fact that superior temporal gyrus plays a role in mentalizing activity and perspective-taking tasks [54,55] through its connections with the temporal poles and medial prefrontal cortex [56]. Similarly, the middle frontal gyrus, as part of the dorsolateral prefrontal cortex (DLPFC), is involved perspective-taking tasks [55,57]. Moreover, it seems to intentionally inhibit self-perspective in order to consider the other's point of view [35] and to be involved in emotion evaluation [58]. Interestingly, we also found an involvement of the fusiform gyrus: Rankin et al. already described the role of the fusiform gyrus in cognitive empathy, in particular in perspective-taking ability, probably related to its direct involvement in facial perception and recognition [35].

On the other hand, loss of perspective taking over time in lv-PPA has been related to the impairment of the inferior parietal lobule, insula, middle frontal gyrus, and superior frontal gyrus. As regards the inferior parietal lobule, although it has been described as directly involved in the network of emotional contagion, it has been clearly stated that perspective-taking tasks engage several brain regions that also include the inferior parietal lobule [55]. Similarly, the superior frontal gyrus, in particular the supplementary motor area, seems to be involved in cognitive empathy tasks [55]. Concerning the insula, it is well-known that this brain region plays a major role in generating forward models of feeling states for others in order to predict and understand the social and affective behavior of other people [2,55,59].

Such neurophysiological bases might explain the association of the impairment of these brain regions and the perspective-taking deficits over time. Nevertheless, our findings might suggest that loss of perspective-taking ability over time might be related to the selective impairment of different empathy-related brain regions in the two clinical presentation of Alzheimer's disease.

We also found that the amplification of emotional contagion over time was correlated with the involvement of the inferior parietal lobule, insula, superior frontal gyrus, and precuneus in lv-PPA and with the involvement of the right inferior frontal gyrus in prototypical amnesic AD. The involvement of frontal and parietal regions (in particular of the inferior frontal gyrus and inferior parietal lobule) may be explained by the fact that these areas are part of the mirror neurons system (MNS) network [2]. Emotional contagion is rooted in the MNS, which transforms sensory representations of others' behavior into one's own visceromotor representations and allows understanding others' actions according to the perception–action model [59]. In more detail, it has been suggested that the inferior frontal gyrus identifies the goals or intentions of actions by their resemblance to stored representations for these actions [60]. Despite the unquestioned involvement in cognitive empathy, several studies have also described a putative role of the insula in the mirror mechanism of emotions: in fact, it has been demonstrated that there is a clear overlap between insular activation elicited by one's own and others' emotions, such as disgust [59,61,62]. Similarly, the insula is also involved in the empathic process of "shared pain": in fact, empathizing with people in pain is associated with hemodynamic activity in the brain that is similar to the activity that occurs when people feel pain themselves, leading to the activation of the "empathy to pain network", which involves the anterior cingulate cortex and insula [63]. It has been suggested that empathy for pain of others may be considered, at least in part, as an automatic, primitive, bottom-up process of affective empathy since it may aid in the immediate perception and avoidance of a threat to oneself [64]. The superior frontal gyrus and supplementary motor area also seem to be involved not only in cognitive empathy tasks but also in affective empathy [55].

The most interesting data show that similar changes in empathy correspond to the involvement of different brain regions in the two phenotypes of Alzheimer's disease. This may be explained by the fact that empathy is a complex construct: a cognitive function that implies several interconnected brain regions [3]. On the other hand, it is well-known that neurodegeneration in lv-PPA and in prototypical amnesic AD involve different brain regions due to the selectivity of vulnerability of these two presentations [65,66]. Considering these premises, we might hypothesize that the selective neurodegeneration of the two clinical subtypes of Alzheimer's disease might damage different brain regions and networks related to empathy, leading to the same impairment.

Our study has some remarkable strengths. First of all, this is one of the first studies deeply analyzing empathy changes in lv-PPA and comparing empathy impairment with that found in amnesic AD patients in order to detect differences and similarities between these two clinical presentations of the disease. Another strength is the research of the neural correlates of empathy deficits in lv-PPA, which have never been explored before, by means of FDG-PET.

The main limitation of our study is the relatively small sample size and thus the lack of corrections for multiples comparisons in the correlation analysis between empathy deficits and hypometabolism in FDG PET analysis. Indeed, we chose to use a more exploratory threshold in order to explore the metabolic correlates of empathy also in the small subsamples of the lv-PPA and amnesic AD patients individually considered. Another limitation of the study is the use of a caregiver-report questionnaire even though the IRI is the most used, validated instrument for the evaluation of empathy. In fact, even if observer-based measures are more ecologically valid and have yielded valuable data previously [67], they are nevertheless limited by their dependence on informants' varying reliability [68]. On the other hand, the scores attributed by the informants present the advantage of capturing real-life empathic behavior independently from the patients' anosognosia [10]. The last limitation is the absence of healthy controls and the need to use SCD to compare the behavioral data although despite this category of subjects has been already used in previous work exploring empathy in AD [69].

5. Conclusions

In conclusion, to the best of our knowledge, our study analyzed for the first time differences and similarities between lv-PPA and amnesic AD, which are two clinical presentations of this neurodegenerative disease. We described a peculiar involvement of specific brain areas in empathy changes in lv-PPA, comparing them with that found in amnesic AD.

Interestingly, lv-PPA and amnesic AD presented the same changes in empathy, showing a damage of perspective-taking ability (part of cognitive domain) together with a heightening of emotional contagion (the most primitive structure of affective empathy) over time. Nevertheless, these similar changes in empathy correspond to the involvement of different brain regions in the two phenotypes of Alzheimer's disease. This could be explained by the fact that neurodegeneration might damage different brain regions and networks related to empathy in these two presentations of Alzheimer's disease due to the selectivity of vulnerability of the brain areas of these variants.

Moreover, our study described empathy changes and difficulties in emotion recognition in lv-PPA. In fact, while clinical diagnosis of PPA is basically anchored in linguistic disturbances, a proportion of cases cannot be easily classified based solely on language tests. In this scenario, emerging experimental works are revealing impairments in several socio-cognitive processes in these patients in association with brain measures. While some studies have already described deficits in emotion recognition, the theory of mind, and empathy in nfv-PPA and sv-PPA, evidence about lv-PPA are sparse and far from conclusive. Our work showed for the first time that lv-PPA patients presented deficits in empathy and in emotion recognition, too. Further studies are needed to confirm our results and better explore these socio-cognitive aspects of PPA, which might be very helpful in improving early diagnostic accuracy.

Author Contributions: Conceptualization, G.G. (Giulia Giacomucci), S.S. and V.B. (Valentina Bessi); methodology, G.G. (Giulia Giacomucci) and S.M.; software, C.P.; validation, G.G. (Giulia Giacomucci), C.P., V.B. (Valentina Berti), S.P., G.G. (Giulia Galdo), S.M., E.B., V.M., C.M., C.N., M.T.D.C., A.I., S.B., B.N., S.S. and V.B. (Valentina Bessi); formal analysis, G.G. (Giulia Giacomucci) and C.P.; investigation, G.G. (Giulia Giacomucci) and C.P.; resources, V.B. (Valentina Berti), S.P., S.M., E.B., V.M., C.M., C.N., M.T.D.C., A.I., S.B., B.N. and S.S.; data curation, G.G. (Giulia Giacomucci), C.P., V.B. (Valentina Berti), S.P. and G.G. (Giulia Galdo); writing—original draft preparation, G.G. (Giulia Giacomucci) and C.P.; writing—review and editing, G.G. (Giulia Giacomucci) and V.B. (Valentina Bessi); visualization, G.G. (Giulia Giacomucci), C.P., V.B. (Valentina Berti), S.P., G.G. (Giulia Galdo), S.M., E.B., V.M., C.M., C.N., M.T.D.C., A.I., S.B., B.N., S.S. and V.B. (Valentina Bessi); supervision, S.S. and V.B. (Valentina Bessi); project administration, V.B. (Valentina Bessi); funding acquisition, B.N. All authors have read and agreed to the published version of the manuscript.

Funding: This study was funded by RICATEN22 (Ateneo Università di Firenze, fondi Ateneo 2022). The funding source had no such involvement in the study design.

Institutional Review Board Statement: The ethics committee of the Azienda Ospedaliero-Universitaria Careggi approved the protocol of the study. All procedures involving experiments on human subjects have been done in accordance with the ethical standards of the Committee on Human Experimentation of the institution and with the Helsinki Declaration of 1975. Specific national laws have been observed.

Informed Consent Statement: All participants gave informed consent to participate in the study.

Data Availability Statement: The data that support the findings of this study are available from the corresponding author upon reasonable request.

Conflicts of Interest: The authors declare no conflict of interest.

References

1. Bartochowski, Z.; Gatla, S.; Khoury, R.; Al-Dahhak, R.; Grossberg, G.T. Empathy changes in neurocognitive disorders: A review. *Ann. Clin. Psychiatry* **2018**, *30*, 220–232. [PubMed]
2. Bernhardt, B.C.; Singer, T. The neural basis of empathy. *Annu. Rev. Neurosci.* **2012**, *35*, 1–23. [CrossRef] [PubMed]
3. Decety, J.; Jackson, P.L. The functional architecture of human empathy. *Behav. Cogn. Neurosci. Rev.* **2004**, *3*, 71–100. [CrossRef]
4. Rankin, K.P.; Kramer, J.H.; Mychack, P.; Miller, B.L. Double dissociation of social functioning in frontotemporal dementia. *Neurology* **2003**, *60*, 266–271. [CrossRef] [PubMed]
5. Cerami, C.; Dodich, A.; Iannaccone, S.; Marcone, A.; Lettieri, G.; Crespi, C.; Gianolli, L.; Cappa, S.F.; Perani, D. Right Limbic FDG-PET Hypometabolism Correlates with Emotion Recognition and Attribution in Probable Behavioral Variant of Frontotemporal Dementia Patients. *PLoS ONE* **2015**, *10*, e0141672. [CrossRef] [PubMed]
6. Rascovsky, K.; Hodges, J.R.; Knopman, D.; Mendez, M.F.; Kramer, J.H.; Neuhaus, J.; van Swieten, J.C.; Seelaar, H.; Dopper, E.G.P.; Onyike, C.U.; et al. Sensitivity of revised diagnostic criteria for the behavioural variant of frontotemporal dementia. *Brain* **2011**, *134*, 2456–2477. [CrossRef]
7. Fischer, A.; Landeira-Fernandez, J.; Sollero de Campos, F.; Mograbi, D.C. Empathy in Alzheimer's Disease: Review of Findings and Proposed Model. *J. Alzheimers Dis.* **2019**, *69*, 921–933. [CrossRef]
8. Giacomucci, G.; Galdo, G.; Polito, C.; Berti, V.; Padiglioni, S.; Mazzeo, S.; Chiaro, E.; De Cristofaro, M.T.; Bagnoli, S.; Nacmias, B.; et al. Unravelling neural correlates of empathy deficits in Subjective Cognitive Decline, Mild Cognitive Impairment and Alzheimer's Disease. *Behav. Brain Res.* **2022**, *428*, 113893. [CrossRef]
9. Gorno-Tempini, M.L.; Hillis, A.E.; Weintraub, S.; Kertesz, A.; Mendez, M.; Cappa, S.F.; Ogar, J.M.; Rohrer, J.D.; Black, S.; Boeve, B.F.; et al. Classification of primary progressive aphasia and its variants. *Neurology* **2011**, *76*, 1006–1014. [CrossRef]
10. Fittipaldi, S.; Ibanez, A.; Baez, S.; Manes, F.; Sedeno, L.; Garcia, A.M. More than words: Social cognition across variants of primary progressive aphasia. *Neurosci. Biobehav. Rev.* **2019**, *100*, 263–284. [CrossRef]
11. Gallese, V. Before and below "theory of mind": Embodied simulation and the neural correlates of social cognition. *Philos. Trans. R. Soc. Lond. B Biol. Sci.* **2007**, *362*, 659–669. [CrossRef] [PubMed]
12. Adolfi, F.; Couto, B.; Richter, F.; Decety, J.; Lopez, J.; Sigman, M.; Manes, F.; Ibáñez, A. Convergence of interoception, emotion, and social cognition: A twofold fMRI meta-analysis and lesion approach. *Cortex* **2017**, *88*, 124–142. [CrossRef]
13. Binney, R.J.; Henry, M.L.; Babiak, M.; Pressman, P.S.; Santos-Santos, M.A.; Narvid, J.; Mandelli, M.L.; Strain, P.J.; Miller, B.L.; Rankin, K.P.; et al. Reading words and other people: A comparison of exception word, familiar face and affect processing in the left and right temporal variants of primary progressive aphasia. *Cortex* **2016**, *82*, 147–163. [CrossRef]
14. Irish, M.; Kumfor, F.; Hodges, J.R.; Piguet, O. A tale of two hemispheres: Contrasting socioemotional dysfunction in right- versus left-lateralised semantic dementia. *Dement. Neuropsychol.* **2013**, *7*, 88–95. [CrossRef] [PubMed]
15. Hazelton, J.L.; Irish, M.; Hodges, J.R.; Piguet, O.; Kumfor, F. Cognitive and Affective Empathy Disruption in Non-Fluent Primary Progressive Aphasia Syndromes. *Brain Impair.* **2017**, *18*, 117–129. [CrossRef]
16. Piguet, O.; Leyton, C.E.; Gleeson, L.D.; Hoon, C.; Hodges, J.R. Memory and emotion processing performance contributes to the diagnosis of non-semantic primary progressive aphasia syndromes. *J. Alzheimers Dis.* **2015**, *44*, 541–547. [CrossRef]
17. McKhann, G.M.; Knopman, D.S.; Chertkow, H.; Hyman, B.T.; Jack, C.R.; Kawas, C.H.; Klunk, W.E.; Koroshetz, W.J.; Manly, J.J.; Mayeux, R.; et al. The diagnosis of dementia due to Alzheimer's disease: Recommendations from the National Institute on Aging-Alzheimer's Association workgroups on diagnostic guidelines for Alzheimer's disease. *Alzheimers Dement.* **2011**, *7*, 263–269. [CrossRef]
18. Davis, M.H. Measuring individual differences in empathy: Evidence for a multidimensional approach. *J. Personal. Soc. Psychol.* **1983**, *44*, 113–126. [CrossRef]
19. Albiero, P.; Lo Coco, A.; Ingoglia, S. Contributo all'adattamento italiano dell'Interpersonal Reactivity Index. *Test. Psicometria Metodol.* **2007**, *13*, 107–125.
20. Ekman, P.; Friesen, W. *Pictures of Facial Affect*; Consulting Psychologists Press: Palo Alto, CA, USA, 1976.

21. Dodich, A.; Cerami, C.; Canessa, N.; Crespi, C.; Marcone, A.; Arpone, M.; Realmuto, S.; Cappa, S.F. Emotion recognition from facial expressions: A normative study of the Ekman 60-Faces Test in the Italian population. *Neurol. Sci.* **2014**, *35*, 1015–1021. [CrossRef]
22. Jessen, F.; Amariglio, R.E.; van Boxtel, M.; Breteler, M.; Ceccaldi, M.; Chételat, G.; Dubois, B.; Dufouil, C.; Ellis, K.A.; van der Flier, W.M.; et al. Subjective Cognitive Decline Initiative (SCD-I) Working Group A conceptual framework for research on subjective cognitive decline in preclinical Alzheimer's disease. *Alzheimers Dement.* **2014**, *10*, 844–852. [CrossRef] [PubMed]
23. Bracco, L.; Amaducci, L.; Pedone, D.; Bino, G.; Lazzaro, M.P.; Carella, F.; D'Antona, R.; Gallato, R.; Denes, G. Italian Multicentre Study on Dementia (SMID): A neuropsychological test battery for assessing Alzheimer's disease. *J. Psychiatr. Res.* **1990**, *24*, 213–226. [CrossRef]
24. Carlesimo, G.; Buccione, I.; Fadda, L.; Graceffa, A.; Mauri, M.; Lorusso, S.; Bevilacqua, G.; Caltagirone, C. Normative data of two memory tasks: Short-Story recall and Rey's Figure. *Nuova Riv. Neurol.* **2002**, *12*, 1–13.
25. Spinnler, H.; Tognoni, G. *Standardizzazione e Taratura Italiana di Test Neuropsicologici: Gruppo Italiano per lo Studio Neuropsicologico Dell'invecchiamento*; Masson Italia Periodici: Milano, Italy, 1987.
26. Caffarra, P.; Vezzadini, G.; Dieci, F.; Zonato, F.; Venneri, A. Rey-Osterrieth complex figure: Normative values in an Italian population sample. *Neurol. Sci.* **2002**, *22*, 443–447. [CrossRef] [PubMed]
27. Baddeley, A.; Della Sala, S.; Papagno, C.; Spinnler, H. Dual-task performance in dysexecutive and nondysexecutive patients with a frontal lesion. *Neuropsychology* **1997**, *11*, 187–194. [CrossRef]
28. Giovagnoli, A.R.; Del Pesce, M.; Mascheroni, S.; Simoncelli, M.; Laiacona, M.; Capitani, E. Trail making test: Normative values from 287 normal adult controls. *Ital. J. Neurol. Sci.* **1996**, *17*, 305–309. [CrossRef]
29. Basso, A.; Capitani, E.; Laiacona, M. Raven's coloured progressive matrices: Normative values on 305 adult normal controls. *Funct. Neurol.* **1987**, *2*, 189–194.
30. Brazzelli, M.; Della Sala, S.; Laiacona, M. Calibration of the Italian version of the Rivermead Behavioural Memory Test: A test for the ecological evaluation of memory. *Bollet. Psicol. Appl.* **1993**, *206*, 33–42.
31. Catricalà, E.; Gobbi, E.; Battista, P.; Miozzo, A.; Polito, C.; Boschi, V.; Esposito, V.; Cuoco, S.; Barone, P.; Sorbi, S.; et al. SAND: A Screening for Aphasia in NeuroDegeneration. Development and normative data. *Neurol. Sci.* **2017**, *38*, 1469–1483. [CrossRef]
32. Capasso, R. *Esame Neuropsicologico per l'Afasia—E.N.P.A.*; Springer: Berlin/Heidelberg, Germany, 2001.
33. Hamilton, M. A rating scale for depression. *J. Neurol. Neurosurg. Psychiatry* **1960**, *23*, 56–62. [CrossRef]
34. Sturm, V.E.; Yokoyama, J.S.; Seeley, W.W.; Kramer, J.H.; Miller, B.L.; Rankin, K.P. Heightened emotional contagion in mild cognitive impairment and Alzheimer's disease is associated with temporal lobe degeneration. *Proc. Natl. Acad. Sci. USA* **2013**, *110*, 9944–9949. [CrossRef] [PubMed]
35. Rankin, K.P.; Gorno-Tempini, M.L.; Allison, S.C.; Stanley, C.M.; Glenn, S.; Weiner, M.W.; Miller, B.L. Structural anatomy of empathy in neurodegenerative disease. *Brain* **2006**, *129*, 2945–2956. [CrossRef] [PubMed]
36. Sorbi, S.; Nacmias, B.; Forleo, P.; Latorraca, S.; Gobbini, I.; Bracco, L.; Piacentini, S.; Amaducci, L. ApoE allele frequencies in Italian sporadic and familial Alzheimer's disease. *Neurosci. Lett.* **1994**, *177*, 100–102. [CrossRef] [PubMed]
37. Alcolea, D.; Pegueroles, J.; Muñoz, L.; Camacho, V.; López-Mora, D.; Fernández-León, A.; Bastard, N.L.; Huyck, E.; Nadal, A.; Olmedo, V.; et al. Agreement of amyloid PET and CSF biomarkers for Alzheimer's disease on Lumipulse. *Ann. Clin. Transl. Neurol.* **2019**, *6*, 1815–1824. [CrossRef] [PubMed]
38. Minoshima, S.; Drzezga, A.E.; Barthel, H.; Bohnen, N.; Djekidel, M.; Lewis, D.H.; Mathis, C.A.; McConathy, J.; Nordberg, A.; Sabri, O.; et al. SNMMI Procedure Standard/EANM Practice Guideline for Amyloid PET Imaging of the Brain 1.0. *J. Nucl. Med.* **2016**, *57*, 1316–1322. [CrossRef] [PubMed]
39. Jack, C.R.; Bennett, D.A.; Blennow, K.; Carrillo, M.C.; Dunn, B.; Haeberlein, S.B.; Holtzman, D.M.; Jagust, W.; Jessen, F.; Karlawish, J.; et al. NIA-AA Research Framework: Toward a biological definition of Alzheimer's disease. *Alzheimers Dement.* **2018**, *14*, 535–562. [CrossRef]
40. Varrone, A.; Asenbaum, S.; Vander Borght, T.; Booij, J.; Nobili, F.; Någren, K.; Darcourt, J.; Kapucu, Ö.L.; Tatsch, K.; Bartenstein, P.; et al. EANM procedure guidelines for PET brain imaging using [18F]F.D.G.; version 2. *Eur. J. Nuclear Med. Mol. Imaging* **2009**, *36*, 2103–2110. [CrossRef]
41. Perani, D.; Della Rosa, P.A.; Cerami, C.; Gallivanone, F.; Fallanca, F.; Vanoli, E.G.; Panzacchi, A.; Nobili, F.; Pappatà, S.; Marcone, A.; et al. EADC-PET Consortium. Validation of an optimized SPM procedure for FDG-PET in dementia diagnosis in a clinical setting. *Neuroimage Clin.* **2014**, *6*, 445–454. [CrossRef]
42. Dermody, N.; Wong, S.; Ahmed, R.; Piguet, O.; Hodges, J.R.; Irish, M. Uncovering the Neural Bases of Cognitive and Affective Empathy Deficits in Alzheimer's Disease and the Behavioral-Variant of Frontotemporal Dementia. *J. Alzheimers Dis.* **2016**, *53*, 801–816. [CrossRef]
43. Narme, P.; Mouras, H.; Roussel, M.; Devendeville, A.; Godefroy, O. Assessment of socioemotional processes facilitates the distinction between frontotemporal lobar degeneration and Alzheimer's disease. *J. Clin. Exp. Neuropsychol.* **2013**, *35*, 728–744. [CrossRef]
44. Bond, R.L.; Downey, L.E.; Weston, P.S.J.; Slattery, C.F.; Clark, C.N.; Macpherson, K.; Mummery, C.J.; Warren, J.D. Processing of Self versus Non-Self in Alzheimer's Disease. *Front. Hum. Neurosci.* **2016**, *10*, 97. [CrossRef] [PubMed]
45. Jessen, F.; Amariglio, R.E.; Buckley, R.F.; van der Flier, W.M.; Han, Y.; Molinuevo, J.L.; Rabin, L.; Rentz, D.M.; Rodriguez-Gomez, O.; Saykin, A.J.; et al. The characterisation of subjective cognitive decline. *Lancet Neurol.* **2020**, *19*, 271–278. [CrossRef] [PubMed]

46. Pernigo, S.; Gambina, G.; Valbusa, V.; Condoleo, M.T.; Broggio, E.; Beltramello, A.; Moretto, G.; Moro, V. Behavioral and neural correlates of visual emotion discrimination and empathy in mild cognitive impairment. *Behav. Brain Res.* **2015**, *294*, 111–122. [CrossRef] [PubMed]
47. Spoletini, I.; Marra, C.; Di Iulio, F.; Gianni, W.; Sancesario, G.; Giubilei, F.; Trequattrini, A.; Bria, P.; Caltagirone, C.; Spalletta, G. Facial Emotion Recognition Deficit in Amnestic Mild Cognitive Impairment and Alzheimer Disease. *Am. J. Geriatr. Psychiatry* **2008**, *16*, 389–398. [CrossRef]
48. Weiss, E.M.; Kohler, C.G.; Vonbank, J.; Stadelmann, E.; Kemmler, G.; Hinterhuber, H.; Marksteiner, J. Impairment in Emotion Recognition Abilities in Patients With Mild Cognitive Impairment, Early and Moderate Alzheimer Disease Compared With Healthy Comparison Subjects. *Am. J. Geriatr. Psychiatry* **2008**, *16*, 974–980. [CrossRef]
49. Pietschnig, J.; Aigner-Wöber, R.; Reischenböck, N.; Kryspin-Exner, I.; Moser, D.; Klug, S.; Auff, E.; Dal-Bianco, P.; Pusswald, G.; Lehrner, J. Facial emotion recognition in patients with subjective cognitive decline and mild cognitive impairment. *Int. Psychogeriatr.* **2016**, *28*, 477–485. [CrossRef] [PubMed]
50. Johnen, A.; Reul, S.; Wiendl, H.; Meuth, S.G.; Duning, T. Apraxia profiles-A single cognitive marker to discriminate all variants of frontotemporal lobar degeneration and Alzheimer's disease. *Alzheimers Dement.* **2018**, *10*, 363–371. [CrossRef] [PubMed]
51. Multani, N.; Galantucci, S.; Wilson, S.M.; Shany-Ur, T.; Poorzand, P.; Growdon, M.E.; Jang, J.Y.; Kramer, J.H.; Miller, B.L.; Rankin, K.P.; et al. Emotion detection deficits and changes in personality traits linked to loss of white matter integrity in primary progressive aphasia. *Neuroimage Clin.* **2017**, *16*, 447–454. [CrossRef]
52. Henry, J.D.; Ruffman, T.; McDonald, S.; O'Leary, M.-A.P.; Phillips, L.H.; Brodaty, H.; Rendell, P.G. Recognition of disgust is selectively preserved in Alzheimer's disease. *Neuropsychologia* **2008**, *46*, 1363–1370. [CrossRef]
53. Kim, H.; Han, S. Does personal distress enhance empathic interaction or block it? *Personal. Individ. Differ.* **2018**, *124*, 77–83. [CrossRef]
54. Frith, U.; Frith, C.D. Development and neurophysiology of mentalizing. *Philos. Trans. R. Soc. Lond. B* **2003**, *358*, 459–473. [CrossRef] [PubMed]
55. de Waal, F.B.M.; Preston, S.D. Mammalian empathy: Behavioural manifestations and neural basis. *Nat. Rev. Neurosci.* **2017**, *18*, 498–509. [CrossRef] [PubMed]
56. Wong, C.; Gallate, J. The function of the anterior temporal lobe: A review of the empirical evidence. *Brain Res.* **2012**, *1449*, 94–116. [CrossRef] [PubMed]
57. Miller, B.L.; Cummings, J.L. *The Human Frontal Lobes: Third Edition: Functions and Disorders;* Guilford Press: New York, NY, USA.
58. Nejati, V.; Majdi, R.; Salehinejad, M.A.; Nitsche, M.A. The role of dorsolateral and ventromedial prefrontal cortex in the processing of emotional dimensions. *Sci. Rep.* **2021**, *11*, 1971. [CrossRef] [PubMed]
59. Rizzolatti, G.; Sinigaglia, C. The mirror mechanism: A basic principle of brain function. *Nat. Rev. Neurosci.* **2016**, *17*, 757–765. [CrossRef]
60. Rizzolatti, G.; Fabbri-Destro, M.; Cattaneo, L. Mirror neurons and their clinical relevance. *Nat. Clin. Pract. Neurol.* **2009**, *5*, 24–34. [CrossRef]
61. Hennenlotter, A.; Schroeder, U.; Erhard, P.; Castrop, F.; Haslinger, B.; Stoecker, D.; Lange, K.W.; Ceballos-Baumann, A.O. A common neural basis for receptive and expressive communication of pleasant facial affect. *Neuroimage* **2005**, *26*, 581–591. [CrossRef]
62. Wicker, B.; Keysers, C.; Plailly, J.; Royet, J.P.; Gallese, V.; Rizzolatti, G. Both of us disgusted in My insula: The common neural basis of seeing and feeling disgust. *Neuron* **2003**, *40*, 655–664. [CrossRef]
63. Shamay-Tsoory, S.G. The neural bases for empathy. *Neuroscientist* **2011**, *17*, 18–24. [CrossRef]
64. Yamada, M.; Decety, J. Unconscious affective processing and empathy: An investigation of subliminal priming on the detection of painful facial expressions. *Pain* **2009**, *143*, 71–75. [CrossRef]
65. Rabinovici, G.D.; Rosen, H.J.; Alkalay, A.; Kornak, J.; Furst, A.J.; Agarwal, N.; Mormino, E.C.; O'Neil, J.P.; Janabi, M.; Karydas, A.; et al. Amyloid vs FDG-PET in the differential diagnosis of AD and FTLD. *Neurology* **2011**, *77*, 2034–2042. [CrossRef] [PubMed]
66. Rabinovici, G.D.; Jagust, W.J.; Furst, A.J.; Ogar, J.M.; Racine, C.A.; Mormino, E.C.; O'Neil, J.P.; Lal, R.A.; Dronkers, N.F.; Miller, B.L.; et al. Abeta amyloid and glucose metabolism in three variants of primary progressive aphasia. *Ann. Neurol.* **2008**, *64*, 388–401. [CrossRef] [PubMed]
67. Rankin, K.P.; Kramer, J.H.; Miller, B.L. Patterns of cognitive and emotional empathy in frontotemporal lobar degeneration. *Cogn. Behav. Neurol.* **2005**, *18*, 28–36. [CrossRef] [PubMed]
68. Shany-Ur, T.; Poorzand, P.; Grossman, S.N.; Growdon, M.E.; Jang, J.Y.; Ketelle, R.S.; Miller, B.L.; Rankin, K.P. Comprehension of insincere communication in neurodegenerative disease: Lies, sarcasm, and theory of mind. *Cortex* **2012**, *48*, 1329–1341. [CrossRef]
69. Singleton, E.H.; Fieldhouse, J.L.P.; van't Hooft, J.J.; Scarioni, M.; van Engelen, M.-P.E.; Sikkes, S.A.M.; de Boer, C.; Bocancea, D.; van den Berg, E.; Scheltens, P.; et al. Social cognition deficits and biometric signatures in the behavioural variant of Alzheimer's disease. *Brain*, 2022; awac382, *ahead of print*.

Journal of
*Personalized
Medicine*

MDPI

Article

Psychosocial Impact of Huntington's Disease and Incentives to Improve Care for Affected Families in the Underserved Region of the Slovak Republic

Katarína Hubčíková [1,2,*], Tomáš Rakús [1,2,3], Alžbeta Mühlbäck [4,5,6], Ján Benetin [1], Lucia Bruncvik [2,7], Zuzana Petrášová [1,2], Jitka Bušková [2,8] and Martin Brunovský [2,8]

1 Neuropsychiatric Department, Psychiatric Hospital of Philipp Pinel in Pezinok, 90201 Pezinok, Slovakia
2 Third Faculty of Medicine, Charles University in Prague, 10000 Prague, Czech Republic
3 Department of Psychiatry, Slovac Medical University, 83303 Bratislava, Slovakia
4 Department of Neuropsychiatry, kbo-Isar-Amper-Klinikum, 84416 Taufkirchen (Vils), Germany
5 Department of Neurology and Center of Clinical Neuroscience, 1st Faculty of Medicine, Charles University and General University Hospital in Prague, 12821 Prague, Czech Republic
6 Department of Neurology, University Hospital of Ulm, 89081 Ulm, Germany
7 Landesklinikum Hainburg, 2410 Hainburg an der Donau, Austria
8 National Institute of Mental Health, 25067 Klecany, Czech Republic
* Correspondence: k.hubcikova@gmail.com; Tel.: +421917111575; Fax: +421336482121

Abstract: Introduction: Huntington's disease (HD) is often on the margin of standard medical practice due to its low prevalence, the lack of causal treatment, and the typically long premanifest window prior to the onset of the symptoms, which contrasts with the long-lasting burden that the disease causes in affected families. Methods: To capture these socio-psychological aspects of HD and map the experiences of affected individuals, persons at risk of HD, and caregivers, we created a questionnaire using a qualitative research approach. The questionnaire containing 16 questions was conducted online for a period of three months through patient associations in Slovakia and their infrastructures. Results: In total, we received 30 responses. The survey results, in particular, indicate insufficient counselling by physicians with explicitly missing information about the possibility of preimplantation genetic diagnostic. There was also a necessity to improve comprehensive social and health care in the later stages of the disease, raise awareness of the disease in the general health community, and provide more information on ongoing clinical trials. Conclusion: The psychosocial effects, as well as the burden, can be mitigated by comprehensive genetic counselling as well as reproductive and financial guidelines and subsequent therapeutic programs to actively support patients, caregivers, children, and adolescents growing up in affected families, preferably with the help of local HD community association. Limitations: We have used online data collection to reach a wider HD community, but at the same time, we are aware that the quality of the data we would obtain through face-to-face interviews would be considerably better. Therefore, future studies need to be conducted to obtain more detailed information.

Keywords: Huntington's disease; psychosocial effects; underserved region; disease burden; genetic counselling; genetic testing

Citation: Hubčíková, K.; Rakús, T.; Mühlbäck, A.; Benetin, J.; Bruncvik, L.; Petrášová, Z.; Bušková, J.; Brunovský, M. Psychosocial Impact of Huntington's Disease and Incentives to Improve Care for Affected Families in the Underserved Region of the Slovak Republic. *J. Pers. Med.* **2022**, *12*, 1941. https://doi.org/10.3390/jpm12121941

Academic Editor: Anne-Marie Caminade

Received: 17 August 2022
Accepted: 14 November 2022
Published: 22 November 2022

1. Introduction

Huntington's disease (HD) is an autosomal dominant, neurodegenerative disorder leading to the progressive impairment of motor skills, cognitive function, and psychiatric disorders [1]. HD is caused by the expansion mutation of the cytosine, adenine, and guanine (CAG) triplet on chromosome 4, as identified in 1983 [2]. The huntingtin gene (*HTT*, OMIM 613004), formerly IT-15, was then identified in 1993 [3] Huntington's disease Collaborative Research. The *HTT* gene is responsible for coding the Huntingtin (HTT) protein. The expansion mutation of the *HTT* gene leads to the production of mutant HTT (mHTT) protein accumulating in the striatum of the basal ganglia, leading to progressive neuronal dysfunction and death, usually within a few decades after the onset of symptoms [4]. The number of CAG repeats in *HTT* is inversely associated with disease onset; the greater the CAG number, the earlier the onset of HD [5]. The onset of the disease is defined as a manifestation of significant motor or neurological symptoms and occurs on average around the age of 40 [6].

The number of CAG repeats within the HTT gene varies from 6 to 35 in the general population [7]. CAG repeats in the range of 36 to 39 refer to reduced penetrance repeat length [8]. CAG repeats with less than 27 present a stable transmission, and no manifestation is expected. The CAG repeat lengths of 27 to 35 refer to the intermediate allele and are also not associated with HD development. However, there is the possibility of expansion upon transmission, resulting in genetic anticipation [9].

Especially in paternal transmission, the number of repetitions in sperm increases due to the meiotic instability of the abnormal CAG repeat, often leading to Juvenile Huntington's Disease (JHD) with a presence of more than 60 CAG repeats and occurs before the age of 21, which comprises about 5% of all HD cases [10,11]. For this reason, the children of an affected parent (common father) may be at risk of developing the disease earlier than their parent with HD [12].

Current HD management focuses on symptomatic treatment in terms of pharmacological and non-pharmacological interventions, such as psychotherapy, physiotherapy, and speech and occupation therapies by a multidisciplinary team [13]. Several promising methods are designed to modify the natural history of HD and are currently under investigation in research studies and clinical trials aiming to reduce the reproduction of mutant *HTT* gene products [14]. The natural history of HD with the progression of symptoms in individual stages of HD about disability and life milestones, including suicidal tendencies, is illustrated in Figure 1.

Caring for patients with HD represents a gradually increasing burden for their caregivers, involving family members, partners, and other relatives lasting several decades [17]. The situation is further complicated due to the reconciliation with a 50% risk of disease in their offspring, which significantly increases emotional burden and suffering compared to other non-hereditary conditions. The genetic nature of HD means that it is not uncommon for children to care for their parents before they develop symptoms, or for a parent to care for a partner and then for the child, and sometimes for several generations to care for them at the same time. Such a chronic and widespread role of care in families with HD can deepen isolation, feelings of loss of their family members and their future, feelings of guilt, anger, burden, and helplessness [18]. The diagnosis of HD also affects the wider blood relatives.

Physicians and healthcare professionals consulting families with HD should do their utmost to ensure that all family members obtain the necessary information to make an informed decision about genetic testing, reproduction, and future financial and life plans [19,20]. Appropriate genetic counselling plays an important role in HD management. The requirements for genetic counselling differ on the condition of a client or patient, depending on the type of consultation. Two types of procedures are available: differential diagnostic and presymptomatic (predictive) testing. The predictive genetic test allows for measuring CAG repeat length in a blood laboratory sample and predicting whether an at-risk individual will develop HD later in life [19]. Predictive testing should be reserved for adults who have participated in a thorough consultation with a genetic counsellor or other trained professional or HD expert about their genetic risks and the potential risks and benefits of the test itself, as well as post-test support [21,22]. The participant needs to prepare for an unfavourable result by receiving information about the disease and the social consequences of the genetic test [22]. The most important predictive and diagnostic genetic testing procedure is appropriate counselling; informed consent from the patient must be obtained before testing [19,21]. The implications of a positive result for the patient and their family must be clearly outlined. Both the nature of HD itself and its autosomal dominant inheritance must be explained to make it clear that if the test is positive for HD, any of their children would have a 50% chance of inheriting it [20,23,24]. The usual standard process of genetic counselling and testing is summarised in Table 1.

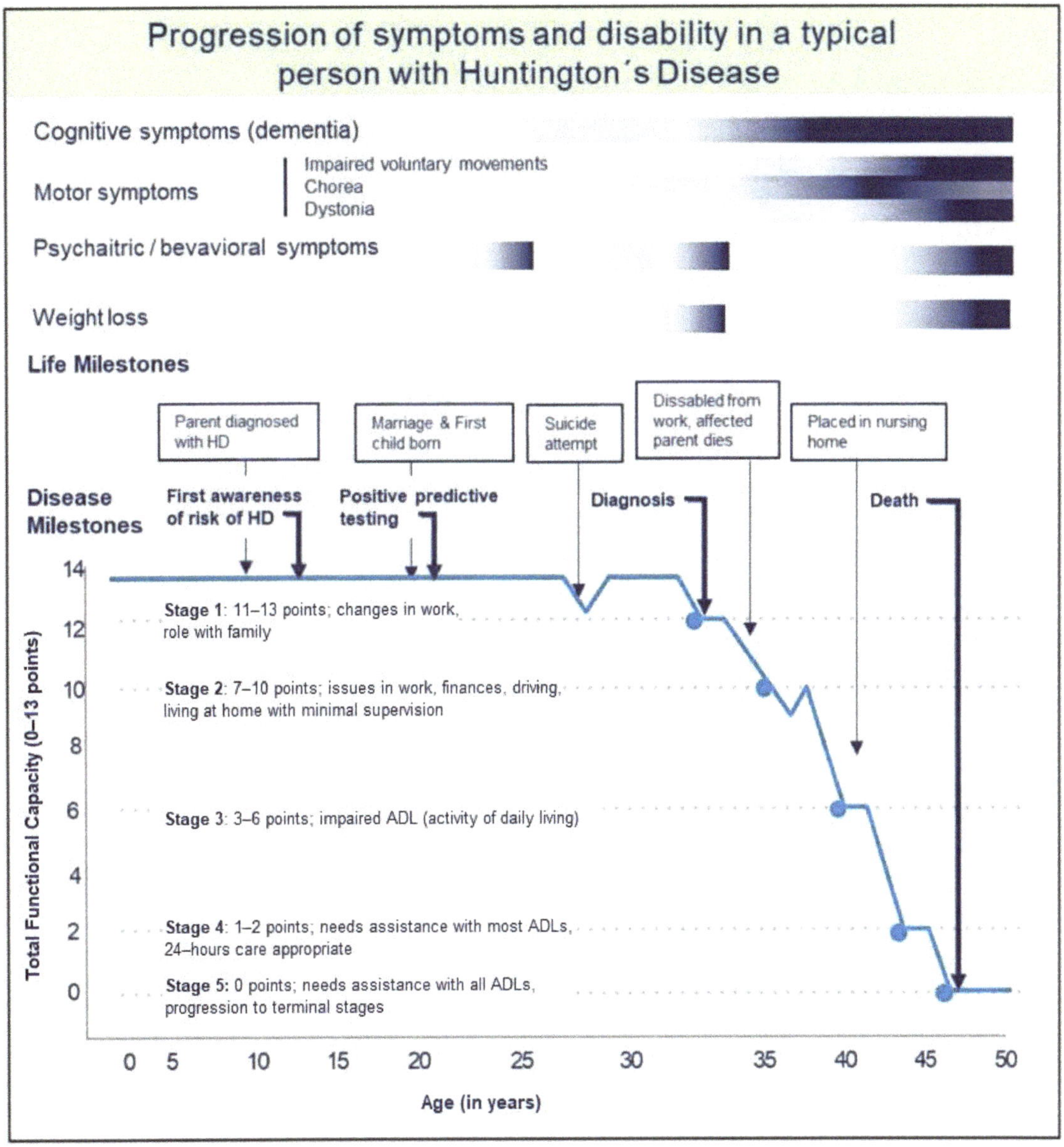

Figure 1. Progression of symptoms and disability in a typical person with HD, adapted from A Physician Guide to Management of HD [15] and onset of cognitive impairment adjusted by Langley et al., 2021 [16].

Table 1. The process of genetic counselling and testing for HD, adapted from [21]. Abbreviations: CAG—cytosine, adenine, guanine.

The Process of Genetic Counselling and Testing for HD	
Initial session	• Providing details of family history to the doctor or genetic counsellor at the session, who then attempts to confirm the diagnosis history • A doctor or genetic counsellor provides information about Huntington's disease, the genetic test process and possible results • Discussion on reasons for requesting a test at this stage and their approach to the possible outcomes. There will also be a chance to discuss, if wanted, the reproductive options available if the person wants to avoid the possibility of passing the disease on to the next generation • Allowing identifying someone who will support them through the process • Encourages them to consider the impact of any result on life, family or friends. This can be particularly important concerning family members who may not wish to be tested, but the test could also reveal their status • Encouraging to consider financial implications and other issues such as life insurance and employment
Reflection period	• The initial session is followed by a summary letter and a reflection period • If the person wants to continue with the process, a second session is arranged
Second/third session	• A review of the information discussed at the first session is undertaken. • The doctor/genetic counsellor, by genetic testing guidelines, may consider a neurological and psychologicaö examination to evaluate the person. However, it is not a requirement for participation in predictive testing.
Genetic test	• Discussion about preparing for genetic tests and results • The date is arranged for a blood sample to be taken • The results will not be provided on that day
The results	• At a follow-up appointment (usually a month after), there will be a face-to-face discussion concerning the test results.
Follow-up sessions	• Relevant follow-up sessions are arranged as required after results are given
Advice and support	• Regardless of the result, the person must have access to support and advice available. • It is appropriated to referral to local associations for HD, thus ensuring contact with others who are going through the same situation; there are mostly support groups and online communities

In addition, it is appropriate to enrich the education with knowledge about the expression of HD within an intermediate allele of 27–35 CAG repeats and the impact of reduced penetration between 36–39 CAG repeats [25]. At the same time, information about technological advances in preimplantation genetic diagnostic testing (PGD) and prenatal testing improved collaborative research networks need to be discussed. A better understanding of the psychosocial aspect of living with HD needs to be provided. The HD expansion mutation carrier, manifest patients and persons at risk of HD need to be informed of their reproductive potential in preference to planned pregnancy according to the regulation of the respective countries [26]. Determining genetic status confronts people with new dilemmas in their reproductive decisions.

It is important to highlight the HD community's local peer support. Anyone affected by HD can benefit from participating in the activities of local HD associations, such as support group seminars on the latest information, and thus be in touch with others experiencing the same situation. In our European area is the European Huntington Association (EHA), an umbrella organisation that provides an extensive network of people coping with the disease—either in person or professionally. It accelerates the information flow and support to people with HD in their country and builds cross-border relationships [27]. The patient community comprises people personally affected by HD (with a local presence in most European countries), and the research community includes experts working with HD. The situation of affected patients and their relatives was unsatisfactory in the Slovak Republic for a long time. Although the Slovak self-help group was officially active from 1994 to 2016, it was founded and maintained by only a few volunteers in Slovakia who, despite their heroic efforts, did not receive enough support from the government and health professionals and burnt out over time. A functioning infrastructure was not developed, and even such simple things as a functioning website and communication channels with members were not established and were only based on personal contacts. This was in great contrast to the situation in the Czech Republic, with which Slovakia was a common state until 1993, so the issue of the HD community had practically the same starting point at that time. Furthermore, due to the cultural and linguistic proximity of the two countries, the affected families in Slovakia had the opportunity to compare the situation and were understandably frustrated. The Czech Huntington's Association quickly became a member of the EHA, and at the same time, several specialised centres and/or several EHDN clinics were established in the country. Comprehensive care with options such as in vitro fertilisation (IVF), including preimplantation diagnostics, is covered by statutory health insurance for patients with Huntington's disease. At the time of conducting this survey, there was no EHDN clinic in Slovakia and IVF, including preimplantation genetic diagnosis, has to be paid for by the patients themselves. At the end of 2019, the Slovak HD Society resumed its activities at the request of patients and their families and shortly afterwards became a member of the EHA. The support group created a dynamic exchange platform to promote and provide better access to the needs of those affected. Work began on building an infrastructure with a website. The main goal is to provide support and information about the nature of the disease, its inheritance, the possibilities of genetic diagnosis, treatment and care for patients and family members, and the establishment of support groups [28].

This survey was the first opportunity to officially express experiences and suggestions for improving the situation of the until then unheard HD community. It was the starting point for defining the first and next steps for the restoration of the Slovak HD association. The presented paper aims to explore and evaluate the psychosocial impacts of HD and incentives to improve the care of affected patients and families in the underserved region of Slovakia.

2. Material and Methods

2.1. Study Assessment

The project was set up to map the situation of patients and families affected by HD in the Slovak Republic (SR). This project is part of a larger national initiative to improve the care of affected patients and families. For this purpose, two types of questionnaires with 16 questions were designed to explore the main psychological aspects and domains. The questionnaire was implemented and accessible via the online platform www.survio.com accessed on 1 November 2022. The first questionnaire was aimed at people who have tested positive for Huntington's disease, regardless of the stage of the disease. The second questionnaire focused on people at risk, not affected by HD and caregivers and partners of affected individuals. Each questionnaire consisted of 13 open questions and three multiple-choice questions with the possibility to answer in the text. The questionnaire covered the main domains of genetic counselling about Huntington's disease knowledge (initial information, main source of information, reproductive counselling, experimental treatment, lack of knowledge), genetic testing (reasons for testing or not), the possibility of psychological counselling, the impact of Huntington's disease (the most burdensome symptom, financial burden, impact on relationships) and social and health services in Slovakia. The questionnaire was available online for three months in the year 2020. The questionnaire was designed by an expert group of representatives of the patient HD community and health care professionals, such as the neurologist, the psychiatrist, and the psychologist and was reviewed to ensure proper understanding and clear language This survey, including a questionnaire, has been approved by a constituted Ethics Committee of the Psychiatric Hospital of Philipp Pinel in Pezinok, Slovakia (Approval Date: 30 September 2020) and conforms to the provisions of the Declaration of Helsinki in 1995 (as revised in Edinburgh in 2000).

2.2. Study Participants

The regular members of the Slovak HD Society—specifically patients, persons at risk and their relatives have received the possibility to participate in this project. Membership in the newly renewed Slovak HD society was obtained through online form through the newly opened website huntington.sk in the course of 2020, while information about Slovak HD society and the possibility of membership was also sent in the form of a leaflet to all genetic clinics and all neurological, psychiatric and neuropsychiatric inpatient facilities in the Slovak Republic at the beginning of 2020. As part of filling out the membership application form, it was possible, among other things, to voluntarily choose whether the applicant for membership is a patient, a person at risk, a relative, or a medical professional or write his or her other motivation to join verbally. The links to the questionnaire were sent to the regular members of the Slovak HD Patient Society, who stated that they are patients, persons at risk or relatives when applying for membership. They previously provided an email address and consent to participate in the research project to the patient's network administrator.

In total, 60 regular members have been contacted and received an email with an explanation text and a questionnaire link. The connection to the patient questionnaire was sent on two occasions to the members to participate in the project. It was possible to fill out two questionnaires per family by repeatedly accessing the link. In the questionnaire for patients, it was possible to indicate in the opening multiple questions that the answers were recorded by another close person since the patient was no longer able to fill them in himself. Three patients used this option. The data have been proceeding anonymously, without any identifiable parameters.

3. Results

In this project, the qualitative methodological approach was used to evaluate the finding of two questionnaires with 13 open-ended questions and as well the quantitative analysis to evaluate three multiple-choice questions. The survey data from the open-ended questions were analysed based on the grounded theory approach [29,30]. Repeated reading and coding of the results of the open-ended questions resulted in several themes related to the questions, and the answers were clustered into the four main domains covering:

(I) Knowledge about HD, including initial information about HD, the main source of information, reproductive options, information on experimental treatment, and additional information.
(II) Genetic testing and counselling, including the process of testing and psychological support and reasons why an individual decided to take a genetic test or not to be tested.
(III) Impact of HD on the relationships, financial burden and burden to HD (symptoms).
(IV) Social and health care services in Slovakia, including experience and proposal for improvement.

In total, 60 links to access the questionnaire were sent to the regular members of the Slovak HD Society. The responsiveness rate was 50% (30 answers). In this study, we obtained responses from the following individuals: nine individuals tested positive for HD, four of whom were asymptomatic and five of whom had the manifest disease, five individuals at risk for HD, and 16 responses from caregivers and partners of asymptomatic and manifest patients, as summarised in the first line of Table 2.

Table 2. Questions and answers of the respondents on the questionnaire administered via the Slovak HD association to patients, persons at risk, their relatives and partners, including their experiences and suggestions for improving the situation for the HD community in SR. The questions have been administered as multiple-choice or as open questions. The table shows an overview and a summary of the available answers. Abbreviations: GT—Genetic Testing; HD—Huntington Disease; N- number of participants; n.n. not known. PGD—Preimplantation genetic diagnosis; SSE—social services establishment; SR—Slovak Republic.

Questions in Individual Categories of the Questionnaire	Answers and Particular Groups of Respondents (Number of Participants)		
	Patients with HD (N = 5)/Asymptomatic HD Carriers (N = 4)	**Individuals at Risk (N = 5)**	**Partners of Asymptomatic Persons (N = 6)/Partners of Manifest HD Patients (N = 10)**
Knowledge about HD			
Initial information How and when did you first learn about the risk or presence of HD in your family or partner?	At the age: 14, 38, 42, 39, 50/15, 35, 50, n.n. From family: 6, From doctor: 3	At the age: 0, 14, 20, 38, n.n.	At the age: 7, 15, 25, 34, 39, n.n/25, 26, 30, 35, 38, 40, 46, 58, n.n, n.n From family: 5/1, From doctor: 1/5, From partner:0/4
The main source of information Where did you mainly get information about HD? How did your doctor teach you about HD? (open question)	From the family: 3; From doctor: 1, Internet: 7	From the family: 2, From doctor: 0, Internet: 3	From family:1, From doctor: 1, internet: 16
Reproductive options Has a doctor provided information on reproductive options, including PGD (your child would be healthy even if you/your partner is affected)?	Yes: 2 No, I am not interested anyway: 2 No, I searched by myself: 2 No, first time to hear: 3	Yes: 2 No, I am not interested anyway: 2 No, I searched by myself: 1 No, first time to hear: 0	Yes: 2 No, I am not interested anyway: 2 No, I searched by myself: 7 No, first time to hear: 3
Experimantal treatment Did you receive information on ongoing clinical trials for HD? How do you get any updated information?	Yes: 1 (doctor), 3 (Internet), family (2) No: 2 I am interested /I am not interested in updates: 8/0	Yes: 4 (Internet) No: 1 I am interested /I am not interested in updates: 5/0	Yes: 1 (doctor), 14 (Internet) No: 1 I am interested /I am not interested in updates: 15/1
Lack of knowledge What information do you think a person at risk for HD should receive? What did you find out over time, and would you appreciate it if you knew it before? (open questions)	PGD: 3 Consulting HD expert: 1 Early knowledge of HD risk (early diagnosis of parent/myself/not hiding HD in the family): 3; I don't know: 2	Reprod. options, PGD: 1 Better awareness among health professionals: 1 Entitlement to a disability pension: 1; I don't know: 2	PGD: 3 Information on research: 4 More support for HD community: 2 Consulting HD expert: 1; I don't know: 6
Genetic testing (GT)			
Reasons for (not) being tested Please write down why you decided/consider being tested, resp. not to be tested? (open question)	For GT: know the truth: 2 Emerging symptoms: 3 Disability pension: 1 Future planning: 3	For GT: because of the children: 1 Against GT: live a fulfilled life: 1 I don't know: 2	For GT: know the truth: 3 Future planning: 7 Disability pension: 1 Emerging symptoms: 3; I don't know: 2
Psychologist Consultation Have you talked to a psychologist during genetic testing? Was it/was it not beneficial for you?	Yes: 2 (beneficial), 1 (not helpful) No: 7	Yes: 1 (beneficial), 1 (not beneficial) No: 3	Yes: 3 (beneficial), 2 (not beneficial) No: 11
Impact of HD			
Most aggravating symptom Which symptoms of HD are/were the most burdensome for you? (open question)	Positive test: 2 Motor skills (chorea): 5 Psychiatric (nervousness, cognitive decline): 2	Motor skills (chorea): 2 Psychiatric (depression, cognitive decline, aggression): 4 I don't know: 1	Positive test: 1 Motor skills (chorea): 3 Psychiatric (cognitive decline, aggression): 7 Dysphagia: 3 Independence loss: 4
Financial burden How has HD affected your family budget? Was the help of the state sufficient for you? (open question)	Higher financial burden: 7 State support limited: 4, we are modest: 2, family help: 2. I don't know: 2	Higher financial burden: 4 State support limited: we are modest 2, family help: 1 Any support yet: 1 I don't know: 1	Higher financial burden: 13 State support limited: we are modest 3, family help: 5, extra job:1 Any support yet: 3 I don't know: 3
Impact on relationships (answers resulted from other open-ended questions)	Blaming parents for hiding HD: 1 Fear of transmitting HD: 1 Hesitation/anxiety to say a positive result to the family: 2 Suspicion of alcoholism due to chorea: 2 Criminal prosecution: 2	Decision with the partner to ignore the fears of the HD and live the life that will come: 1	Anxiety about the gradual loss of a partner: 1 Fear of destroying the life of my healthy partner: 1 Fear of divorce: 2 Feeling my life is ruined due to illness of relatives: 1

Table 2. *Cont.*

Questions in Individual Categories of the Questionnaire	Answers and Particular Groups of Respondents (Number of Participants)		
	Patients with HD (N = 5)/Asymptomatic HD Carriers (N = 4)	**Individuals at Risk (N = 5)**	**Partners of Asymptomatic Persons (N = 6)/Partners of Manifest HD Patients (N = 10)**
Social and health care services in SR			
Experiences How do you evaluate social and nursing services for patients with HD in SR?	Insufficient—SSE staff and paramedics do not know HD: 5 I don't know: 4	Insufficient—SSE staff and paramedics do not know HD: 2 I don't know: 3	Insufficient—SSE staff and paramedics do not know HD: 9 Sufficient: 1; I don't know: 6
Suggestions for improvement What do you think should be improved in social/health services for patients with HD? (open questions)	Better awareness among health professionals: 3 Support home care of HD patients: 1 Experts/Psychologists for HD: 1 I don't know: 4	Awareness among health professionals and the public: 2 I don't know: 3	Aids for later stages: 2 Special SSE for HD: 3 Better awareness: 5 Preferential appointment: 2 Support groups for both carriers and patients: Acceptance of the HD in the context of disability 1

For individual categories of questions, we also present citations of selected answers that authentically illustrate the situation and care of affected individuals with HD and its psychosocial effects. As stated above, the categories were clustered into domains I -IV to ease the overview.

3.1. First Information on HD

Only three out of thirty respondents mentioned receiving a good source of initial information about the HD from their physician. However, all participants lacked information on reproductive options, instruction on preimplantation testing in in vitro fertilisation, and a generally more empathic approach. Seven participants even learned about the possibility and accessibility of the PID in Slovakia for the first time only from our questionnaire. Moreover, all respondents would better appreciate at least some information on the ongoing clinical trials.

To the question *'How did your physician approach you?'* we provide some examples: *"There was no information ... I searched for everything on the internet myself". "The neurologist didn't pay attention and immediately wrote us off as an incurable diagnosis". "The neurologist told me not to make a 'big head' out of it". "I even had to actively explain to the general practitioner what the disease was and what it was causing". "It was told me that I will die".*

We addressed the questions about the information that a person at risk for HD should receive and what to know earlier upon the genetic testing or disease onset. We received a detailed answer stating that *"greater emphasis should be placed on possible symptomatic treatment so that one does not take it for final."* The participants would also appreciate *information on support possibilities for caregivers, "we have to unite to help each other; there is no other support for us."* The participants were missing information on how to approach the affected person: *"I have often heard from a husband that I am healthy and do not understand him".*

3.2. Genetic Counselling and Testing

In the section about genetic testing, we explore the reasons for testing and experience with genetic counselling. The participants stated they mainly sought this information as they needed to know the truth about planning for the future and emerging symptoms.

Of particular interest are the answers of persons at risk who either did not yet know their genetic status, as they either did not take the genetic test or were waiting for the test results at the time of participating in the survey. Some answers pointed out the option not to know about the genetic status, "We decided not to know" *on the other hand, some participants stated in more detail, "when planning the future, we chose the path of PGD with prenatal diagnosis, but without success. Subsequently, we decided to ignore the fear of the disease and live life as it is and as it will come. I don't want to be tested; I know what awaits me in the case of a positive test, and I don't want to wrap my life in the gloom of Huntington's disease".* This questionnaire could also see the burden of the COVID-19 pandemic and how it affects families. *"My husband and I are waiting for test results. Although I went to the tests with him, the doctor did not let me into the office or tell me anything—allegedly for the corona reason. At the same time, she was not wearing a face mask. I was sorry because it is difficult for both my husband and me"*.

It is important to understand why the persons decide to take a test and disclose their genetic status; we also respond to individuals who have witnessed HD in their relatives since childhood/adolescence. *"When I discovered that there was a 50% chance I could inherit this stuttering gene, I was determined to know what I was up to. Everything stopped making sense, and my only goal was to find out if I was sick or not"*. *"So that I wouldn't ruin my partner's life . . . And I would be ready for the institution and that I would be insane"* *"To stop it in our family"*.

3.3. The Impact of HD

We divide this section into three subcategories to better understand the burden for caregivers, the effect on relationships in general, and the impact on families' financial situation with HD.

3.3.1. Burden for Caregivers

The results present the high burden of caregivers, showing higher burdens when coping with the mental symptoms rather than the motoric ones of their affected family members; *"My father and sister were HD patients. I saw aggression, the desire to kill, they ruined my life, nobody will bring me back my youth ages"*. It also affects the partnership *"The hardest thought is that my beloved husband, who is very wise, rational, just, will become someone who cannot take care of himself and lose his identity"*. The partners and caregivers at the same time are often overwhelmed with the care of the affected person in the family, and sometimes the care of the rest family comes short: *"I had a mother-in-law who was affected, and same time I have a husband and a son. I was caring for all; I would welcome any help"*.

3.3.2. Relationships

The relationships are mostly affected within the family and involve different levels. The communication between parents and children can be damaged at different levels, as communication about HD implies a lot of challenges. Some respondents stated: *"My parents didn't tell me anything about HD, although they knew about it. I blame them"*. The affected members experience problems communicating their genetic status in their own family and experience the feeling of guilt, *"Knowing about the HD is good when young people plan to start a family; I went to test myself for the kids, and now I don't know how to tell them, that I am affected, I feel guilty"*.

The own perspective of the positive genetic result plays a role and reflects further in the relationship: *"The positive result destroyed me, my partnership; I don't want to do it to them either"*. *some even stated*, *"everything about this disease puts me in depression so far . . . I try not to think about it and live. Still, every encounter with such a person takes away my zest for life"*.

3.3.3. Impact on the Financial Situation of Families with HD

The impact on families' financial situation with HD is very important, as most respondents describe it as unsatisfactory. It is very difficult as the partner or family needs to compensate for an affected member, *"I had to work full-time at retirement age, to be able to pay for my husband's private caring facility"*. Most caregivers experience a difficult time as they have a full-time job and care for an affected individual and the rest f family. *"My husband was left without a job; the disability pension is so small. I have to go to work, and we don't know what will happen next"*. The affected families need to rely on the help of further family members and wider relatives. *"We wouldn't have been able to manage it without the help of our relatives"*. The situation is even more complicated for the early-manifest respondents who experience the problems due to the symptoms of HD: *"I haven't been able to find a job for a long time, I act like I'm an alcoholic with my involuntary movements. And once I got a job,*

I almost everywhere had an accident at work. Currently, I am at the unemployment office, and I receive benefits, which, of course, are not even enough to cover the basic living costs like household and food; my family helps me".

3.4. Social and Health Services in the Slovak Republic

The emerging fact is the lack of support for social and health services in the Slovak Republic. Most respondents consider them to be insufficient, with the most common blame regarding inadequate knowledge about HD among the health professionals and staff working in the social services and nursing facilities. The respondents summarise their perspectives in the following strong statements:

"The patient is just waiting to die".

"No support in the system".

"Psychiatrists know nothing about it".

"Inadequate, no one knows what to do".

"Doctors and staff have minimal knowledge. Rather they take the disease as a sensation".

The participants were asked what could improve the health care and social systems. The majority of participants proposed rational and possibly easy solutions to enable better maintenance of affected individuals, such as *Better prescription of medical devices, prioritisation during the doctor visits, advice on how to deal with the dentist; insurance companies should also pay for the recreational stays for caregivers as well (not only affected person).* It became clear that all respondents wish for more information and education about HD.

4. Discussion

This is the first survey of its kind to examine the needs and problems of the Huntington's landscape in the underserved region of Slovakia. Our survey results on the psychosocial impact of Huntington´s suggest similar conclusions to other studies in this field [18,31,32]. For overview purposes, we will discuss the different themes, which were clustered into subthemes based on responses.

4.1. Genetic Counselling and Testing

This survey highlighted the importance of proper guidelines for genetic testing and counselling [21], as implemented in many European countries. Physicians conducting the genetic test need to provide information about HD in a timely, private, and sensitive manner to those who wish to do so while respecting the interests of those who do not want the test [33]. The liberty of choice for the patient and the possibility to opt-out of the procedure at any step and re-enter the protocol if the patients wish so at the later stage of life [22,34]. The respondents at risk were determined not to undergo predictive genetic testing. They are afraid to receive irreversible information for the future and are worried about possible stigma [35]. However, higher importance should be given to genetic counselling on reproductive options as the option is regularly available in Slovakia. Consultation on reproductive options should be public for all interested persons, regardless of whether they have previously had to undergo predictive testing, or it should usefully be included in consultations on genetic testing if the person so wishes [20,36].

In addition, it should not be forgotten that individuals who learn that they will develop HD in the future needs to be offered support not only in a short time post-testing but even from the long time perspective [35]. An earlier survey of attitudes to predictive testing suggested that 11–15% of people at risk would consider suicide if given an unfavourable test result [35]. In contrast, Quaid et al. report that only four (2.1%) of the 189 individuals in the US predictive test sample were hospitalised after receiving unfavourable results [37]. The study of Almquist et al. even described only 0.97% of all participants who experienced a catastrophic response to predictive testing [38]. In summary, all reported frequencies are still high and require attention.

However, rather rare negative reactions and impacts after the predictive testing should be compared to thousands of individuals who have been relieved of the anxiety of not knowing whether HD will develop [34,39]. Furthermore, in preventing fatal reactions after predictive testing, it is necessary to emphasise the importance of ongoing counselling regardless of the test result, identifying problematic reconciliation techniques and active support in reconciliation strategies [21]. Despite all recommendations about the appropriate psychological interviews as a part of predictive genetic testing, only a few respondents stated that they had attended an interview with a psychologist during this process.

Moreover, in our survey, we specifically process the responses of individuals who have witnessed HD in their relatives since early childhood and adolescence.

4.2. Family Relationships

HD has a major impact on family systems; impairments shape caregiver roles in the affected family member and change in time upon disease progression [40]. Our results confirmed the earlier findings of the role shifts in many HD families, including parenting (children overtaking the role of a parent in a family partly) [40]. In almost all areas, teenagers growing up in a family with HD actively assist in complex nursing care for a parent with HD. The topics that concern them include responsibilities and early decision-making, personal exposure and burden due to individual risk for HD. [41]. Some of our respondents describe their early experience with HD as negative; *my father and sister were HD patients. I saw aggression and desire to kill; they ruined my life, and nobody will bring me back to my youth.* Our findings are in line with interviews conducted with 32 young people living in a family with HD, where the following main problems were mentioned: the possibility of starting a family as a distressing decision; feelings of loneliness in the family and among peers, the experience of the affected parent being unable to have more reasonable conversations, and the non-affected parent being mostly described as absent and unreachable because they are taking care of a partner and often has several jobs for financial reasons; family life is so difficult and marked by the conflicts [42].

Our results are in line with numerous other studies highlighting the most affected domains in the family system involving family functioning, emotions and reactions, social functioning, and state and social services [41–45].

The respondents experience negative and traumatic experiences (witnesses of domestic violence, aggression, suicide and suicide attempts), causing long-term impacts on their own lives [42].

The hereditary aspect is also important as some respondents were, for a long time, not aware of their own risk as proper communication did not take place in the family [46] as also stated in our interviews: *"my parents didn't tell me anything about HD"," we did not talk about it", "I went to test me for the kids, and now I don't know how to tell them, that I am affected, I feel guilty"* .

Overall, the respondents also reported feelings of being overburdened and feeling alone without any support and, at the same time, caring for multiple jobs and households.

4.3. Future and Family Planning

Future and family planning is one of the most important topics and reasons for undergoing predictive testing in clinical practice [47]. There is significant distress already when considering the predictive testing, and multiple factors, such as parenteral experience with disease and risk perception, play a role in decision-making [48]. Decruyenaere et al. observed that more than the majority of HD expansion mutation carriers, after undergoing predictive genetic testing, had chosen to have children with a prenatal or preimplantation genetic diagnosis. About one in three (35%) decided to have no children anymore. A minority (7%) was undecided or had no children for other reasons [49]. The factors impacting the decision not to have children were the personal experience of growing up in a family with HD and the ethical issues related to prenatal and preimplantation genetic diagnoses [49]. The factors associated with the non-use of prenatal testing and the preimplantation genetic diagnosis were aversion to terminating the pregnancy and optimism associated with treatment options and ongoing clinical trials for HD [50]. As we have already assumed, several factors play a role in the decision-making process, such as gender, knowledge and ethical issues about PD and PGD, the availability of these methods, the strength of the desire to have children, and personal experiences with HD in the family [45,49,50]. The results illustrate the complexity of the decision-making process and the necessity of in-depth genetic counselling. A group of those who knew about the risk and decided to have children had their decision mostly based on the hope of a cure that would come in time to help their children if needed; a certain role played the "wish thinking", in which they believe that their children will not develop HD [49]. Findings similar to our responses were presented in the group of those who had children before they knew about the risk. They reflected the lack of information about HD and inheritance patterns [51]. In the group of those who choose not to have children because of the risk, the main reason was their own experience, how they saw the decline and death of family members for HD in generations [51].

4.4. Social and Health Care System Support

Our findings also highlight the central role of the physician or general practitioner who might consider working with families on strategies to improve emotional engagement, communication, and problem-solving and ensure better care for affected members [52]. The findings of studies point out the importance of support needed for families with HD facing different problems. There is an urgent need to develop strategies to help and support genetic counsellors, healthcare providers, and school workers [53]. Our respondents reported major complaints about the problematic and lacking support for patients with HD in the presymptomatic and especially advanced stage of the disease and reflecting the need for a multidisciplinary approach—the key role in helping build and develop the infrastructure played here patients and lay organisations. There is an urgent need for an interdisciplinary approach to HD management to ensure adequate systematic care during all stages of the disease. It is important to provide appropriate care at all stages of Huntington's disease. Starting in the period before predictive testing with psychological support, involving the general practitioner, the specialist and the family, to balance early between fatalism ("I already have HD, and there is nothing I can do about it") and denial ("These symptoms are not HD" or "I have no symptoms") [21,23]. Available supports should include educational materials, home health care or skilled nursing support, respite care, advocacy organisations, and local or online support groups. Educating patients and caregivers about the availability of these services and providing access to them is imperative for neurologists and other healthcare providers [54].

5. Limitations

Although there are so far no epidemiological studies available in Slovakia, the prevalence of the HD is expected to be as in other European regions 6–15: 100,000 [55,56], with an anticipated increase of another 17% by 2030 [57]. Based on this information by the population of five million inhabitants in Slovakia, we emphasise that there are more affected families and individuals that we could not reach with our survey and would need to be identified in future.

The one limitation, especially from the perspective of the qualitative research approach, was the online form of data collection. We preferred the questionnaire form with the possibility of answers in their own words to an interview via phone call or zoom, mainly to preserve the anonymity of the respondents and obtain answers with a lower degree of self-censorship. However, we are aware that the optimal way to obtain information would be a personal interview with the trained local HD medical specialist, who already knows the respondent and has a therapeutic relationship with them. However, this was not possible due to the lack of HD specialists in Slovakia and non-existent HD centres where patients would naturally concentrate. The number and type of participants could also be significantly influenced by the necessity of sufficient cognitive ability of patients to use the Internet, which we tried to cover with the possibility that the family could fill in the questionnaire on behalf of the patient. Another source of influence on the number and type of respondents could be the lack of access to the Internet, which according to data from Eurostat for the year 2020 [58]—at the time of data collection, on the level of access of households to the Internet represented only 14% of households in Slovakia that did not have access to the Internet.

In addition, the COVID-19-related isolation periods and restrictive measures may influence some responses. It should also be noted that the survey took place between the first two COVID-19 pandemic waves, although the first wave in Slovakia was very mild, and the measures were considered mild compared to other European countries. All of the responses were received within four weeks of publication in early October 2020, so the survey was completed quickly.

6. Conclusions

In summary, Huntington´s disease requires a multidisciplinary approach due to the different needs of those affected at different stages, which is still scarce in Slovakia. In routine clinical practice, health professionals often assume that after informing the patient and his/her blood relatives about the possibility of diagnosis by genetic testing and the real limitations of currently available treatment, no other kind of support is possible than symptomatic control of motor and psychiatric symptoms. The results of our survey on the experiences and suggestions for improvement of the HD community in Slovakia point mainly to the lack of basic education of doctors and the explicit absence of information about the possibility of prenatal and preimplantation genetic testing. Strategies are needed to improve the knowledge of health professionals and to provide better counselling. Furthermore, information on ongoing clinical trials, as well as research progress, is seen as important. The results also show that comprehensive social and health services need to be provided for the later stages of the disease. Experts and multidisciplinary teams should be formed, and public awareness of the disease should be raised to improve the situation in any underserved region.

Author Contributions: Conceptualization, K.H.; Data curation, A.M.; Formal analysis, J.B. (Ján Benetin) and Z.P.; Methodology, K.H., L.B. and Z.P.; Resources, K.H. and A.M.; Supervision, J.B. (Ján Benetin), J.B. (Jitka Bušková) and M.B.; Validation, Z.P.; Visualization, T.R. and L.B.; Writing—original draft, K.H., T.R. and L.B.; Writing—review & editing, A.M. and M.B. All authors have read and agreed to the published version of the manuscript.

Funding: This work was supported by MSMT CR programme Cooperatio 38—Neuroscience, Charles University.

Institutional Review Board Statement: The study was conducted according to the guidelines of the Declaration of Helsinki and approved by the Ethics Committee of the Psychiatric Hospital of Philipp Pinel in Pezinok, Slovakia (Approval Date: 30 September 2020).

Informed Consent Statement: Informed consent was obtained from all subjects involved in the study. Written informed consent has been obtained from the patients to publish this paper.

Data Availability Statement: The data presented in this study are available upon request from the corresponding author. The data are not publicly available because the database contains patient personal data.

Acknowledgments: We would like to thank all patients, partners, caregivers and families for participating in the project. We also thank all persons involved in the restoration process of the Slovak Society for Help in Huntington's Disease and acknowledge Vladimir Vaclavik, Bea de Shepper, and Astri Arnesen for their support to the Slovak HD community.

Conflicts of Interest: The authors declare no conflict of interest.

References

1. Harper, P.S. *The Natural History of Huntington's Disease*; W.B. Saunders Company Ltd.: Philadelphia, PA, USA, 1991; pp. 127–139.
2. Gusella, J.F.; Wexler, N.S.; Conneally, P.M.; Naylor, S.L.; Anderson, M.A.; Tanzi, R.E.; Watkins, P.C.; Ottina, K.; Wallace, M.R.; Sakaguchi, A.Y.; et al. A polymorphic DNA marker genetically linked to Huntington's disease. *Nature* **1983**, *306*, 234–238. [CrossRef] [PubMed]
3. Huntington's Disease Collaborative Research Group. A novel gene containing a trinucleotide repeat that is expanded and unstable on Huntington's disease chromosomes. *Cell* **1993**, *72*, 971–983. [CrossRef]
4. Roos, R.A.C. Huntington's disease: A clinical review. *Orphanet J. Rare Dis.* **2010**, *5*, 40. [CrossRef] [PubMed]
5. Snell, R.G.; Macmillan, J.C.; Cheadle, J.P.; Fenton, I.; Lazarou, L.P.; Davies, P.; Shaw, D.J. Relationship between trinucleotide repeat expansion and phenotypic variation in Huntington's disease. *Nat. Genet.* **1993**, *4*, 393–397. [CrossRef] [PubMed]
6. Gusella, J.; MacDonald, M. Huntington's disease. *Semin. Cell Biol.* **1995**, *6*, 21–28. [CrossRef]
7. Nopoulos, P.C. Huntington disease: A single-gene degenerative disorder of the striatum. *Dialogues Clin. Neurosci.* **2016**, *18*, 91–98. [CrossRef]
8. Rubinsztein, D.; Leggo, J.; Coles, R.; Almqvist, E.; Biancalana, V.; Cassiman, J.; Hayden, M.R. Phenotypic characterisation of individuals with 30–40 CAG repeats in the Huntington disease (HD) gene reveals HD cases with 36 repeats and apparently normal elderly individuals with 36–39 repeats. *Am. J. Hum. Genet.* **1996**, *59*, 16–22. [PubMed]
9. Ha, A.D.; Jankovic, J. Exploring the correlates of intermediate CAG repeats in Huntington disease. *Postgrad. Med.* **2011**, *123*, 116–121. [CrossRef] [PubMed]
10. Duyao, M.; Ambrose, C.; Myers, R.; Novelletto, A.; Persichetti, F.; Frontali, M.; Macdonald, M. Trinucleotide repeat length instability and age of onset in Huntington's disease. *Nat. Genet.* **1993**, *4*, 387–392. [CrossRef]
11. Quarrell, O.W.J.; Nance, M.A.; Nopoulos, P.; Paulsen, J.S.; Smith, J.A.; Squitieri, F. Managing juvenile Huntington's disease. *Neurodegener. Dis. Manag.* **2013**, *3*, 267–276. [CrossRef]
12. Apolinário, T.A.; Paiva, C.L.A.; Agostinho, L.A. Intermediate alleles of Huntington's disease HTT gene in different populations worldwide: A systematic review. *Genet. Mol. Res.* **2017**, *16*, 1–16. [CrossRef] [PubMed]
13. Radder, D.L.M.; Nonnekes, J.; Van Nimwegen, M.; Eggers, C.; Abbruzzese, G.; Alves, G.; Bloem, B.R. Recommendations for the Organization of Multidisciplinary Clinical Care Teams in Parkinson's Disease. *J. Park. Dis.* **2020**, *10*, 1087–1098. [CrossRef]
14. Mühlbäck, A.; Lindenberg, K.S.; Saft, C.; Priller, J.; Landwehrmeyer, G.B. Genselektive Therapieansätze bei der Huntington-Krankheit. *Der Nervenarzt.* **2020**, *91*, 303–311. [CrossRef]
15. Nance, M.A.; Paulsen, J.S.; Rosenblatt, A.; Wheelock, V. *A Physician's Guide to the Management of Huntington Disease*, 3rd ed.; The Huntington Society of Canada (HSC): Waterloo, ON, Canada, 2013.
16. Langley, C.; Gregory, S.; Osborne-Crowley, K.; O´Callaghan, C.; Zeun, P.; Lowe, J.; Johnson, E.B.; Papoutsi, M.; Scahill, R.; Rees, G.; et al. Fronto-striatal circuits for cognitive flexibility in far from onset Huntington's disease: Evidence from the Young Adult Study. *J. Neurol. Neurosurg. Psychiatry* **2021**, *92*, 143–149. [CrossRef]
17. Ready, R.E.; Mathews, M.; Leserman, A.; Paulsen, J.S. Patient and caregiver quality of life in Huntington's disease. *Mov Disord.* **2008**, *23*, 721–726. [CrossRef]
18. Aubeeluck, A.V.; Buchanan, H.; Stupple, E.J. 'All the burden on all the carers': Exploring quality of life with family caregivers of Huntington's disease patients. *Qual. Life Res.* **2012**, *21*, 1425–1435. [CrossRef]
19. Craufurd, D.; Macleod, R.; Frontali, M.; Quarrell, O.; Bijlsma, E.K.; Davis, M.; Roos, R.A. Diagnostic genetic testing for Huntington's disease. *Pract. Neurol.* **2015**, *15*, 80–84. [CrossRef]
20. Nance, M.A. Genetic counseling and testing for Huntington's disease: A historical review. *Am. J. Med. Genet. Part B: Neuropsychiatr. Genet.* **2017**, *174*, 75–92. [CrossRef]
21. Macleod, R.; Tibben, A.; Frontali, M.; Evers-Kiebooms, G.; Jones, A.; Martinez-Descales, A.; Roos, R. Recommendations for the predictive genetic test in Huntington's disease. *Clin. Genet.* **2013**, *83*, 221–231. [CrossRef]
22. Ramond, F.; Quadrio, I.; Le Vavasseur, L.; Chaumet, H.; Boyer, F.; Bost, M.; Ollagnon-Roman, E. Predictive testing for Huntington disease over 24 years: Evolution of the profile of the participants and analysis of symptoms. *Mol. Genet. Genom. Med.* **2019**, *7*, e00881. [CrossRef]
23. Stopford, C.; Ferrer-Duch, M.; Moldovan, R.; MacLeod, R. Improving follow up after predictive testing in Huntington's disease: Evaluating a genetic counselling narrative group session. *J. Community Genet.* **2020**, *11*, 47–58. [CrossRef]
24. MacLeod, R.; Moldovan, R.; Stopford, C.; Ferrer-Duch, M. Genetic Counselling and Narrative Practices: A Model of Support following a "Negative" Predictive Test for Huntington's Disease. *J Huntingtons Dis.* **2018**, *7*, 175–183. [CrossRef]
25. Semaka, A.; Creighton, S.; Warby, S.; Hayden, M.R. Predictive testing for Huntington disease: Interpretation and significance of intermediate alleles. *Clin. Genet.* **2006**, *70*, 283–294. [CrossRef]
26. Van Rij, M.C.; de Koning Gans, P.A.; van Belzen, M.J.; Roos, R.A.; Geraedts, J.P.; De Rademaeker, M.; de Die-Smulders, C.E. Genetic diagnosis for Huntington's disease in the Netherlands (1998–2008). *Clin. Genet.* **2014**, *85*, 87–95. [CrossRef]
27. EHA. European Huntington Association. 2022. Available online: http://eurohuntington.org/an-overview-of-the-huntingtons-disease-community/ (accessed on 16 August 2022).

28. Huntington Slovakia. Spoločnosť pre pomoc pri Huntingtonovej chorobe. 2022. Available online: https://huntington.sk/o-nas/ (accessed on 30 July 2022).
29. McAllister, M. Grounded Theory in Genetic Counseling Research. *J. Genet. Couns.* **2001**, *10*, 233–250. [CrossRef]
30. Slade, M.E.; Priebe, S.E. *Choosing Methods in Mental Health Research: Mental Health Research from Theory to Practice*; Routledge/Taylor & Francis Group: New York, NY, USA, 2006.
31. Avila-Giron, R. (1973). Medical and social aspects of Huntington's chorea in the state of Zulia, Venezuela. *Adv. Neurol.* **1973**, *1*, 261–266.
32. Batista, P.; Pererira, A. Quality of Life in Patients with Neurodegenerative Diseases. *J. Neurol. Neurosci.* **2016**, *7*, 1–7. [CrossRef]
33. Manrique de Lara, A.M.; Soto-Gómez, L.; Núñez-Acosta, E.; Saruwatari-Zavala, G.; Rentería, M.E. Ethical issues in susceptibility genetic testing for late-onset neurodegenerative diseases. *Am. J. Med. Genetics. Part B Neuropsychiatr. Genet.* **2019**, *180*, 609–621. [CrossRef]
34. Hayden, M.R.; Bombard, Y. Psychosocial effects of predictive testing for Huntington's disease. *Adv. Neurol.* **2005**, *96*, 226–239.
35. Kenen, R.H.; Schmidt, R.M. Stigmatisation of carrier status: Social implications of heterozygote genetic screening programs. *Am. J. Public Health* **1978**, *68*, 1116–1120. [CrossRef]
36. Simpson, S.A.; Harper, P.S. Prenatal testing for Huntington's disease: Experience within the UK 1994–1998. *J. Med. Genet.* **2001**, *38*, 333–335. [CrossRef]
37. Quaid, K.A. Presymptomatic testing for Huntington disease in the United States. *Am. J. Hum. Genet.* **1993**, *53*, 785–787.
38. Almqvist, E.W.; Bloch, M.; Brinkman, R.; Craufurd, D.; Hayden, M.R. A Worldwide Assessment of the Frequency of Suicide, Suicide Attempts, or Psychiatric Hospitalization after Predictive Testing for Huntington Disease. *Am. J. Hum. Genet.* **1999**, *64*, 1293–1304. [CrossRef]
39. Tibben, A.; Duivenvoorden, H.J.; Niermeijer, M.F.; er Vlis, M.V.-v.; Roos, R.A.; Verhage, F. Psychological effects of presymptomatic DNA testing for Huntington's disease in the Dutch program. *Psychosom. Med.* **1994**, *56*, 526–532. [CrossRef]
40. Røthing, M.; Malterud, K.; Frich, J.C. Caregiver roles in families affected by Huntington's disease: A qualitative interview study. *Scand. J. Caring Sci.* **2014**, *28*, 700–705. [CrossRef] [PubMed]
41. Sparbel, K.J.; Driessnack, M.; Williams, J.K.; Schutte, D.L.; Tripp-Reimer, T.; McGonigal-Kenney, M.; Paulsen, J.S. Experiences of teens living in the shadow of Huntington Disease. *J. Genet. Couns.* **2008**, *17*, 327–335. [CrossRef]
42. Kjoelaas, S.; Tillerås, K.H.; Feragen, K.B. The Ripple Effect: A Qualitative Overview of Challenges When Growing Up in Families Affected by Huntington's Disease. *J. Huntingt. Dis.* **2020**, *9*, 129–141. [CrossRef]
43. Etchegary, H. Healthcare experiences of families affected by Huntington disease: Need for improved care. *Chronic Illn.* **2011**, *7*, 225–238. [CrossRef]
44. Winkler, E.; Ausserhofer, D.; Mantovan, F. The life as a caregiver of a person affected by Chorea Huntington: Multiple case study. *Pflege Z.* **2012**, *65*, 608–611.
45. Mand, C.M.; Gillam, L.; Duncan, R.E.; Delatycki, M.B. "I'm scared of being like mum": The Experience of Adolescents Living in Families with Huntington Disease. *J. Huntingt. Dis.* **2015**, *4*, 209–217. [CrossRef]
46. Ellison, M. The impact of Huntington disease on young people. *Handb. Clin. Neurol.* **2017**, *144*, 179–182. [CrossRef]
47. Evers-Kiebooms, G.; Nys, K.; Harper, P.; Zoeteweij, M.; Dürr, A.; Jacopini, G.; Simpson, S. Predictive DNA-testing for Huntington's disease and reproductive decision making: A European collaborative study. *Eur. J. Hum. Genet.* **2002**, *10*, 167–176. [CrossRef]
48. Decruyenaere, M.; Evers-Kiebooms, G.; Boogaerts, A.; Cassiman, J.J.; Cloostermans, T.; Demyttenaere, K.; Fryns, J.P. Psychological functioning before predictive testing for Huntington's disease: The role of the parental disease, risk perception, and subjective proximity of the disease. *J. Med. Genet.* **1999**, *36*, 897–905.
49. Decruyenaere, M.; Evers-Kiebooms, G.; Boogaerts, A.; Philippe, K.; Demyttenaere, K.; Dom, R.; Fryns, J.P. The complexity of reproductive decision-making in asymptomatic carriers of the Huntington mutation. *Eur. J. Hum. Genet.* **2007**, *15*, 453–462. [CrossRef] [PubMed]
50. Richards, F.H.; Rea, G. Reproductive decision making before and after predictive testing for Huntington's disease: An Australian perspective. *Clin. Genet.* **2005**, *67*, 404–411. [CrossRef] [PubMed]
51. Quaid, K.A.; Swenson, M.M.; Sims, S.L.; Harrison, J.M.; Moskowitz, C.; Stepanov, N.; Westphal, B.J. What were you thinking?: Individuals at risk for Huntington Disease talk about having children. *J. Genet. Couns.* **2010**, *19*, 606–617. [CrossRef]
52. Jona, C.M.H.; Labuschagne, I.; Mercieca, E.C.; Fisher, F.; Gluyas, C.; Stout, J.C.; Andrews, S.C. Families Affected by Huntington's Disease Report Difficulties in Communication, Emotional Involvement, and Problem Solving. *J. Huntingt. Dis.* **2017**, *6*, 169–177. [CrossRef] [PubMed]
53. Van Walsem, M.R.; Howe, E.I.; Iversen, K.; Frich, J.C.; Andelic, N. Unmet needs for healthcare and social support services in patients with Huntington's disease: A cross-sectional population-based study. *Orphanet J. Rare Dis.* **2015**, *10*, 124. [CrossRef]
54. Tarolli, C.G.; Chesire, A.M.; Biglan, K.M. Palliative Care in Huntington Disease: Personal Reflections and a Review of the Literature. *Tremor Other Hyperkinet Mov.* **2017**, *11*, 454. [CrossRef]
55. Bates, G.P.; Dorsey, R.; Gusella, J.F.; Hayden, M.R.; Kay, C.; Leavitt, B.R.; Tabrizi, S.J. Huntington disease. *Nat. Rev. Dis. Prim.* **2015**, *1*, 15005. [CrossRef]

56. Rawlins, M.D.; Wexler, N.S.; Wexler, A.R.; Tabrizi, S.J.; Douglas, I.; Evans, S.J.W.; Smeeth, L. The Prevalence of Huntington's Disease. *Neuroepidemiology* **2016**, *46*, 144–153. [CrossRef]
57. Squitieri, F.; Griguoli, A.; Capelli, G.; Porcellini, A.; D'Alessio, B. Epidemiology of Huntington disease: First post-HTT gene analysis of prevalence in Italy. *Clin. Genet.* **2016**, *89*, 367–370. [CrossRef]
58. Eurostat. 2022. Available online: http://ec.europa.eu/eurostat/databrowser/view/tin00134/default/table?lang=en (accessed on 13 November 2022).

Journal of
Personalized Medicine

MDPI

Review

Multiple Sclerosis and Autoimmune Comorbidities

Viviana Nociti [1,2,*] and Marina Romozzi [1]

1 Centro Sclerosi Multipla, Fondazione Policlinico Universitario 'Agostino Gemelli' IRCCS, 00168 Rome, Italy
2 Università Cattolica del Sacro Cuore, 00168 Rome, Italy
* Correspondence: viviana.nociti@policlinicogemelli.it

Abstract: Multiple sclerosis (MS) is a chronic inflammatory and neurodegenerative disease of the central nervous system characterized by broad inter- and intraindividual heterogeneity and different prognoses. Multisystem comorbidities are frequent features in people with MS (PwMS) and can affect treatment choices, quality of life, disability and mortality. In this scenario, autoimmune comorbidities play a cardinal role for several reasons, such as the implication on MS pathogenesis, diagnostic delay, disease activity, disability progression, brain atrophy, and treatment choice. However, the impact of an autoimmune comorbid condition on MS is not fully elucidated. This review aims to summarize the currently available data on the incidence and prevalence of autoimmune diseases in PwMS, the possible effect of this association on clinical and neuroradiological MS course and its impact on treatment choice.

Keywords: multiple sclerosis; personalized medicine; autoimmune; comorbidities

Citation: Nociti, V.; Romozzi, M. Multiple Sclerosis and Autoimmune Comorbidities. *J. Pers. Med.* **2022**, *12*, 1828. https://doi.org/10.3390/jpm12111828

Academic Editor: Anne-Marie Caminade

Received: 1 October 2022
Accepted: 29 October 2022
Published: 3 November 2022

1. Introduction

Multiple sclerosis (MS) is a chronic, inflammatory and degenerative demyelinating disease of the central nervous system (CNS), of unknown etiology, affecting individuals in early adulthood [1]. The disease is characterized by broad inter- and intraindividual heterogeneity with various clinical presentations, different subtypes and prognoses and consequently variable responses to treatments [1]. MS treatment guidelines have changed significantly over the past decades due to the growing number of approved disease-modifying therapies (DMTs), which can influence the disease course by preventing relapses and disability progression in people with MS (PwMS) [2].

Although the etiology and pathogenesis of MS remain unclear, it is believed to be an immune-mediated disease. The pathophysiology of MS seems to be characterized by an aberrant immune activation [3]. This immune dysregulation leads to neuroinflammation in which both immune cells from the periphery and resident cells of the CNS (e.g., microglia and astrocytes) are involved [3].

MS may share similar underlying pathogenesis with certain comorbidities, particularly autoimmune pathologies, which may arise from genetic susceptibility to autoimmunity and overlapping pathogenetic mechanisms [4].

Comorbidity is defined as the co-existence of multiple diseases or medical conditions in a patient, which may share similar overlapping pathophysiology and affect the course of the disease itself [5].

Multisystem comorbidities are frequent features in PwMS, particularly neurological disturbance, other comorbidity autoimmune conditions, psychiatric comorbidities, cardiovascular diseases, chronic lung diseases and metabolic disorders. Among neurological disturbances, epilepsy, migraine and restless leg syndrome are more frequent in patients with MS compared with the general population [6].

Psychiatric comorbidities (i.e., depression, anxiety, and bipolar disorder) have been associated with poor quality of life (QoL), reduced adherence to DMTs and fatigue [7].

Cardiovascular comorbidities such as abnormalities in blood pressure, heart rate, heart rhythm, and left ventricular systolic function are common in PwMS, and a diagnosis of MS increases the risk of myocardial infarction, stroke and heart failure [8,9].

Many comorbidities are present before or at MS symptom onset, but it seems that the prevalence increases over time [10]. For example, some comorbidities such as depression and anxiety may represent preclinical symptoms of MS resulting from immunological and inflammatory changes in the CNS [7].

Over the last decades, the average age of the MS population has increased correlating with increasing general life expectancy, advances in DMTs, the adoption and subsequent widespread use of magnetic resonance imaging (MRI) and improved health and social care for these patients. Further, MS has become a lifelong condition. The risk of multi-morbidity increases with age and in patients with a chronic disease such as MS [11]. However, over time, the accrual of comorbidities can be explained by several other factors such as physical invalidity, a background of common risk factors and the use of DMTs [4].

Several studies have shown that the presence of comorbidities can delay diagnosis. Furthermore, comorbidities can affect treatment choices, QoL, disability and mortality [12].

Marrie et al., using the North American Research Committee on Multiple Sclerosis Registry, evaluated the association between comorbidities and both the diagnostic delay and severity of the disability at MS diagnosis. They found that the presence of comorbidities was associated with greater diagnostic delays and increased disability at diagnosis [13]. The presence of comorbid conditions increases the complexity of patient management and bears important clinical and socioeconomic implications for PwMS [12].

This review aims to describe the current evidence regarding the range of comorbidities in PwMS, focusing on autoimmune comorbidities, and widen the current knowledge about the influence of these comorbidities on the clinical features and therapeutic choices in MS.

2. Neurological Comorbidities and MS

Several studies have suggested that comorbid neurological conditions such as epilepsy and migraine are more frequent in PwMS [6].

2.1. Epilepsy

Epileptic seizures are more common in PwMS than in the general population. A systematic review found that the incidence of seizure disorders (18 studies) ranged from 0.6% to 6% after MS onset, and the prevalence (24 studies) ranged from 0.9% to 8%, both higher than reported in the general population. However, heterogeneity among these studies was substantial [14].

Seizures may occur at any point during the course of MS and have also been described as the presenting symptom of MS [15–17]. Seizures have been observed in patients with relapsing-remitting MS (RRMS) as well as in progressive forms [15,18].

Several studies suggested that focal onset seizures are more common than generalized seizures and secondary generalization of focal seizures is also frequent [19,20].

A systematic review demonstrated that patients with MS who had seizures are more likely to have a younger age of onset compared to MS patients without seizures, and there was also a trend demonstrating higher disability scores in patients with epilepsy [18]. A longitudinal study by Calabrese et al. showed an association between the severity of cortical pathology, namely the number and volume of cortical lesions evaluated by double-inversion recovery (DIR) imaging, and the occurrence of epilepsy in PwMS. Moreover, in patients with RRMS and epilepsy, cognitive dysfunction was four-times higher than in the RRMS group without epilepsy [21].

In a study by Martinez-Lapiscina et al., MS patients with seizures had a higher mean number of cortico-juxtacortical lesions on T2-weighted/fluid attenuation inversion recovery magnetic resonance imaging as compared to MS controls [22].

The pathologic substrate of epilepsy in MS is not fully elucidated. Still, the involvement of grey matter, along with brain inflammation, are presumably the reasons behind the increased incidence of epilepsy [23].

2.2. Migraine

Migraine affects up to 50% of PwMS with a significantly higher prevalence than in the general population [24].

PwMS are more than twice as likely to report migraine than controls, with a significant association for migraine without aura [25]. Migraine can precede MS onset, be associated with relapses, or manifest during the disease course [26,27]. However, the prevalence of comorbid migraine seems to be higher in patients with RRMS compared to those with progressive forms [28].

The mechanism explaining the association between MS and migraine remains to be clarified, but several hypotheses have been postulated. The inflammatory demyelinating MS lesions may induce migraine by disrupting the pathways involved in the pathogenesis of migraine, deputed to pain stimuli processing. It was observed that lesions within the midbrain, especially in the periaqueductal grey matter, were frequently associated with comorbid migraine in PwMS [29,30].

Furthermore, a study on rodent models of autoimmune-induced cortical demyelination showed that cortical demyelination was associated with accelerated cortical spreading depression (CSD), which has been implicated in migraine pathophysiology, both with and without aura [31,32].

Only a few pieces of evidence support the hypothesis that CSD may lead to a subtle increase in permeability of the blood–brain-barrier and neuroinflammation, thereby exposing antigens in the CNS to circulating T cells [33]. Graziano et al. found an increased frequency of contrast-enhancing lesions in MS patients with migraine, specifically within the RRMS disease subtype, and postulated that having migraine comorbidity may increase the level of blood–brain-barrier disruption in these patients [27].

Regarding the overall impact on MS severity, some studies have found no significant correlation between the level of disability and the presence of migraine [28], while the presence of comorbid migraine negatively affects some aspects of QoL in these patients [34].

3. Autoimmune Comorbidities and MS

The coexistence of autoimmune/inflammatory comorbidities has recently engendered significant interest in the MS research field. First, MS and specific comorbidities may have underlying pathogenesis. Second, the association between autoimmune comorbidities and MS may reveal common genetic and environmental risk factors [4].

For example, smokers with multiple sclerosis had an increased risk of developing comorbid autoimmune disease after MS onset [35].

Several articles describe the association between autoimmune/inflammatory conditions and MS, but the results are still inconsistent. A population-based case-control study on autoimmune comorbidities and MS showed that before MS diagnosis, uveitis occurred 3-fold more often, and inflammatory bowel disease (IBD) occurred 1.7-fold more often in patients with MS [36]. A nationwide cohort study in Denmark found that PwMS were at an increased risk of developing ulcerative colitis and pemphigoid [37]. Nevertheless, other studies, including a recent multicentric population-based study, did not find any increased risk of autoimmune diseases among MS patients [38].

A systematic review of the incidence and prevalence of autoimmune disease in PwMS found that psoriasis and thyroid disease were the most prevalent autoimmune comorbidities. The findings also supported an increased risk of IBD, uveitis, and pemphigoid [39]. However, the authors stated that they could not draw conclusions because the study was mainly heterogeneous with respect to the populations studied, methods of ascertaining comorbidities, and reporting of findings [39].

A systematic review and meta-analysis on the overall risk of other autoimmune diseases in PwMS and their first-degree relatives showed that the OR of thyroid disease was increased in both PwMS (OR 1.66) and their relatives (OR 2.38). A comparable association was observed between MS and IBD (OR 1.56) and psoriasis (OR 1.31; $p < 0.0001$), though not in relatives [40].

Table 1 lists relevant studies on MS and concomitant autoimmune conditions.

3.1. Type 1 Diabetes Mellitus

The co-occurrence of type 1 diabetes mellitus (DM) and MS is supported by several studies. A population-based cohort study found that patients with type 1 DM were at 3-fold greater risk for developing MS. Furthermore, the risk for type 1 DM in first-degree relatives of patients with MS was increased by about 40% [41]. The authors suggested that similarities in immunological features between type 1 DM and MS and/or unknown environmental factors might contribute significantly to the co-occurrence of these two diseases [41]. Similarly, a Sardinian study observed a 5-fold and 2-fold higher prevalence of type 1 DM in patients with MS and their first-degree relatives compared with the general population. It concluded that common genes might contribute to susceptibility to both diseases [42].

Furthermore, Bechtold et al. conducted a cohort study on a pediatric and adolescent population affected by type 1 DM in Germany and Austria. They demonstrated a considerably higher risk of MS co-occurrence in the diabetic population [43].

3.2. Autoimmune Thyroid Disease

Autoimmune thyroid disease seems to affect PwMS and their first-degree relatives more than the general population, regardless of treatment with IFN-β and alemtuzumab, but the results are variable. A controlled prospective study found that thyroid disorders were at least three-times more common in women with MS than in female controls [44]. Another study on 353 patients not treated with IFN-β found a statistically significant higher prevalence of autoimmune thyroiditis in male MS patients compared with male controls but not in female patients [45]. In a study on a Spanish cohort of 93 untreated PwMS, MS patients had a higher prevalence of antithyroid antibodies compared with the general population [46].

3.3. Inflammatory Bowel Disease

Previous studies have established an association between MS and IBD, including Crohn's disease and ulcerative colitis (UC) [47,48]. A systematic review and meta-analysis showed that IBD and PwMS seem to be associated with a 50% increased risk of MS or IBD comorbidity, respectively, with no apparent differences between patients with Crohn's disease or ulcerative colitis [49].

3.4. Psoriasis

An association between MS and psoriasis has not been clarified, and several studies have tried to determine whether an association exists. Marrie et al. found that the risk of incident psoriasis was 54% higher in PwMS (HR 1.54; 95%CI: 1.07–2.24) [50]. One case-control study investigating whether patients with a diagnosis of MS had higher rates of concomitant psoriasis found a higher-than-expected frequency of psoriasis among PwMS [51]. In contrast, other studies did not corroborate the association between psoriasis and MS [52]. In a study on 658 consecutive patients, the prevalence of psoriasis in MS patients compared to the general population did not differ significantly [48]. Similarly, a large multicenter study on autoimmune disease risk in MS patients found that the frequency of psoriasis in MS patients did not differ from spousal controls (5.8% of the MS population and 5.4% of controls) [38]. The association between the two conditions may reflect shared genetic, environmental factors and immune pathways, and the effectiveness of fumarates in both conditions may yield etiologic insights into MS [53].

Table 1. Relevant studies on the co-occurrence of MS and autoimmune diseases.

Author (Date) [Reference]	Concomitant AID	Study Population	Main Findings
Nielsen (2006) [41]	Type 1 DM	11,862 PwMS, 6078 type 1 DM cases	RR for MS in type 1 DM patients: 3.26 (95% CI: 41.30–7.97)
Marrosu (2002) [42]	Type 1 DM	1090 PwMS, 2180 parents, 3300 siblings	Type 1 DM prevalence in MS is 3-fold greater than in healthy siblings and 5-fold the general population
Bechtold (2014) [43]	Type 1 DM	19 PwMS, 56,653 type 1 DM cases	RR of MS in type 1 DM patients: 3.35 (95% CI: 1.56–7.21) to 4.79 (2.01–11.39)
Edwards (2005) [48]	Type 1 DM	658 PwMS	Type 1 DM prevalence of 0.9%; OR 18.14 (95% CI: 6.40–2.17)
Ramagopalan (2007) [38]	UC	5032 PwMS, 2707 controls	No difference between PwMS and controls
Roshanisefat (2012) [47]	UC	20,276 PwMS, 203,951 controls	Increased risk in PwMS: HR 1.49 (95% CI: 1.22–1.82)
Roshanisefat (2012) [47]	CD	20,276 PwMS, 203,951 controls	Increased risk in PwMS: HR 1.45 (95% CI: 1.17–1.81)
Edwards (2005) [48]	IBD	658 PwMS	IBD prevalence of 1.2%; OR 3.17 (95% CI: 6.40–2.17)
Farez (2014) [52]	CD	211 PwMS, 211 controls	No difference between PwMS and controls
Ramagopalan (2007) [38]	Psoriasis	5032 PwMS, 2707 controls	No difference between PwMS and controls
Roshanisefat (2012) [47]	Psoriasis	20,276 PwMS, 203,951 controls	Increased risk in PwMS: HR 1.73 (95% CI: 1.42–2.10)
Edwards (2005) [48]	Psoriasis	658 PwMS	No higher prevalence compared to controls
Marrie (2017) [50]	Psoriasis	4911 PwMS, 23,274 controls	Risk of psoriasis was 54% higher in PwMS (HR 1.54; 95%CI: 1.07–2.24).
Fellner (2014) [51]	Psoriasis	214 PwMS, 192 controls	Psoriasis prevalence of 4.21%; OR: 8.39 (95% CI: 1.05–66.81)
Farez (2014) [52]	Psoriasis	211 PwMS, 211 controls	No difference between PwMS and controls
Niederwieser (2003) [45]	AITD	353 PwMS, 308 controls	Higher prevalence of AITD in male MS patients (9.4%) than in male controls (1.9%; $p = 0.03$)
Edwards (2005) [48]	AITD	658 PwMS	AITD prevalence of 3.2%; OR 1.80 (95% CI: 3.02–1.07)

AID, autoimmune disease; AITD, autoimmune thyroid disease; CD, Crohn's disease; CI, confidence interval; DM, diabetes mellitus; HR, hazard ratio; IBD, inflammatory bowel disease; OR, odds ratio; PwMS, people with multiple sclerosis; RR, relative risk; UC, ulcerative colitis.

3.5. Drug-Related Autoimmune Disorders

The possibility of comorbid autoimmune disorders developing secondary to DMTs has also been a concern. Autoimmunity following alemtuzumab therapy is a well-recognized adverse effect of alemtuzumab, with thyroid disease being the most common in more than a third of patients. Other antibody-mediated autoimmune disorders are associated with alemtuzumab, including idiopathic thrombocytopenic purpura and anti-glomerular basement membrane disease [54]. IFN-β might be associated with an increased risk of thyroid autoimmunity and dysfunction, particularly within the first year of treatment [55].

4. Effects of Autoimmune Comorbidities on the Clinical and Neuroradiological Course of MS

Several studies investigated the effect of different autoimmune conditions on the clinical course of MS. Patients with MS, and a concurrent autoimmune disorder seem to have lesser disabilities compared to those with isolated MS after an average of 5.91 years. It has been suggested that the concurrent autoimmune disorder may increase body tolerance against autoantigens [56]. A study performed by Fanouriakis et al. on 9 patients with both systemic lupus erythematosus and MS showed that the coexistence of the two diseases does not seem to be associated with a severe phenotype for either entity. In particular, systemic lupus erythematosus remained quiescent in all patients while on standard immunomodulatory MS therapy [57]. A study on 66 patients with concomitant MS and IBD showed that these patients have a milder course of the disease than patients with MS alone, after a median of 12 years of disease duration [58].

Regarding the effects of comorbid autoimmune diseases on the radiological outcome, Zivadinov et al. analyzed magnetic resonance (MR) imaging findings in MS patients with autoimmune comorbidities. They found that comorbidities in patients with MS were associated with more severe MR imaging outcomes of demyelination and neurodegeneration, evidenced by several nonconventional MR imaging measures, including brain atrophy, magnetization transfer imaging, and diffusivity. The findings were significant for psoriasis, type 2 DM, and thyroid disease [59]. However, the study included 40 patients with type 2 DM, which is considered more a metabolic-acquired disease rather than an autoimmune condition, and only 3 patients with type 1 DM.

Lorefice et al. performed a case-control MRI study on 286 PwMS, of which 30 subjects had DM type 1, 53 patients had autoimmune thyroiditis, and 4 had celiac disease. Multiple regression analysis found an association between type 1 DM and lower grey matter and cortical grey matter volumes, independent from MS clinical features and related to type 1 DM duration [60].

The correlation between brain atrophy and autoimmune comorbidities may have significance in the progression of disability, as several reports have shown that brain and spinal atrophy may represent reliable biomarkers of neurodegeneration that correlate with physical and cognitive impairment in PwMS [61,62].

5. Effect of Autoimmune Comorbidities on Treatment Choices

The treatment choice in PwMS and concomitant autoimmune comorbidities that share common immunological interfaces is still challenging. Numerous therapeutic strategies are either approved for different autoimmune disorders or may be repurposed for several diseases, with the possibility of using a drug that can be effective in both conditions. Contrarily, other medications may exacerbate pre-existing autoimmune disorders or trigger their onset. For example, tumor necrosis factor-alpha (TNFα) blockers, established as effective agents in the treatment of several autoimmune conditions such as rheumatoid arthritis, psoriasis and IBD, can induce or worsen demyelinating diseases and are contraindicated in the treatment of MS [63].

Similarly, interferons can unmask silent autoimmune processes or induce de novo autoimmune diseases, and their use in MS with concomitant autoimmune diseases is not recommended [64–66].

Alemtuzumab is not recommended in the case of MS and autoimmune comorbidities for its association with autoimmunity. The cause of autoimmunity is not entirely understood and is likely related to the pattern of T- and B-cell repopulation after their depletion [67].

Natalizumab, a monoclonal anti-α4 antibody, inhibits the trafficking of lymphocytes from the blood into CNS. There have been several cases of patients with MS treated with natalizumab who developed rheumatoid arthritis or experienced an onset or exacerbation of psoriasis [68,69]. Natalizumab modifies the composition of lymphocyte subpopulations and alters the migration of leukocytes across the blood–brain-barrier, shifting the inflammatory response from the CNS towards other tissues [70]. Furthermore, natalizumab has been demonstrated to be an effective drug for the induction and maintenance of remission in patients with Crohn's disease [71,72]. Therefore, natalizumab may be an alternative treatment option for patients with MS and Crohn's disease but should be used cautiously in patients with comorbid psoriasis or rheumatoid arthritis.

Anti-CD20 antibodies used for MS include rituximab, ocrelizumab and ofatumumab. Ocrelizumab has been demonstrated to be an effective and safe treatment option for rheumatoid arthritis [73]. In contrast, ocrelizumab and rituximab were associated with an increased risk of developing psoriasis and IBD. Therefore, it should be avoided in patients with MS plus psoriasis or IBD [74–76].

The sphingosine-1-phosphate receptor modulators (fingolimod, siponimod, ozanimod) are potentially useful in IBD, rheumatoid arthritis and psoriasis, as demonstrated in animal models and few reports [77–79]; however, further applications in clinical trials are still needed to ascertain the effectiveness in these pathologies. Ozanimod has been proven safe and effective in patients with moderate to severe ulcerative colitis and may represent an option for patients with concomitant highly active MS and ulcerative colitis [80].

Among first-line therapies, dimethyl fumarate enhances the nuclear factor erythroid 2 related factor 2 (Nrf2) transcriptional pathway. It is approved for both relapsing-remitting mild/moderate MS and moderate/severe psoriasis and represents the treatment of choice in case of co-occurrence of these pathologies [81,82].

The DMTs approved for multiple sclerosis and other common autoimmune conditions are synthesized in Table 2. A pragmatic treatment approach to MS and autoimmune disorders should foresee first a check for contraindications, second, an evaluation of the disease/severity of the concomitant conditions, to tailor therapies for the individual patient [83].

Table 2. DMTs approved for MS and their use in other autoimmune disorders.

DMTs	MS	IBD	PsO	RA
INF	+	−	−	−
DMF	+	−	*	−
TER	+	−	−	*
S1PR Ozanimod	+	* + (UC)	−	−
CLAD	+	−	−	−
NAT	+	+	−	−
ALE	+	−	−	−
Anti-CD20	+	−	−	+

+, FDA approved; * used off-label; −, not used; ALE, alemtuzumab; Anti-CD20: CD20 antibodies; AZA, azathioprine; CLAD: cladribine; DMF, dimethyl fumarate; FDA, US Food and Drug Administration; IBD, inflammatory bowel disease; IFN, interferon; MS, multiple sclerosis; MTX, methotrexate; NAT, natalizumab; PsO, psoriasis; RA, rheumatoid arthritis; S1PR, sphingosin-1-phosphate receptor modulator; SLE, systemic lupus erythematosus; TER teriflunomide; UC: ulcerative colitis.

6. MS and Autoimmune Comorbidities: Unmet Needs

The association between MS and different comorbidities is of rising interest as a factor that could explain the heterogeneity of outcomes.

Over the past decades, insufficient attention has been paid to comorbid conditions in MS, as it was considered mainly a disease of young adults with a limited comorbidity burden. Recent studies demonstrated the importance of investigating comorbidities in PwMS, as several comorbid conditions influence the course of the disease, disability progression, worsened QoL, treatment choices and mortality [84–86]. Another important point is the diagnostic delay in patients with comorbidities [13]. An explanation could be that the clinician could erroneously attribute MS symptoms to a preexisting condition, increasing the time from symptom onset to diagnosis and subsequently disability at diagnosis [13].

However, studies investigating comorbidities in PwMS have been performed mainly in small cohorts, and a global view of comorbidity in MS is lacking. A clear understanding of the risk of developing these pathologies and their prevalence in PwMS is necessary [6].

Furthermore, patients with comorbidities are usually excluded in clinical trials of disease-modifying therapies in MS. Thus, trial results may not be representative of a real-world scenario and cannot be applied to a standard clinic population with general comorbidities [87]. However, comorbidities may influence disease activity, disability progression and treatment choices [12].

Marrie et al. reviewed the published results of nine placebo-controlled trials of DMTs for MS. Of the nine trials evaluated, five excluded individuals with different comorbidities, and the explanation of the exclusions for comorbidities was unclear in four trials. None of the trials reported the comorbidity status of patients at enrollment [87].

To address this knowledge gap, the International Workshop on Comorbidity in Multiple Sclerosis produced several recommendations for future clinical trials, indicating relaxing restrictions on the inclusion of patients with comorbidity. The comorbidities prioritized were depression, anxiety, autoimmune disease, diabetes, cancer, hypertension, and migraine [87].

In this scenario, autoimmune comorbidities play a cardinal role for several reasons. First, the coexistence of autoimmune disorders may reflect common underlying pathobiology. Second, as the treatment landscape continues to expand with the development of new DMTs, concomitant pathologies may present a unique therapeutic target.

For example, MS and type 1 DM share T-cell-mediated autoimmunity [88]. Type 1 DM is caused by T lymphocytes' destruction of the insulin-producing β-cells in pancreatic islets once activated by particular insulin epitopes on antigen-presenting cells. In vitro, T-cells from type 1 DM patients reacted to pancreatic islet and CNS antigens [89]. Children with concomitant CNS demyelination and type 1 DM exhibited heightened T-cell reactivities to self-antigens, and these responses were not strictly limited to the disease target organs [90].

The presence of shared pathways and the involvement of similar cell types in the pathogenesis of autoimmune conditions may underlie the risk of both MS and other autoimmune diseases [91]. Furthermore, genetic loci shared between multiple autoimmune diseases are involved in a wide range of immune pathways (i.e., T-cell activation, B-cell activation, cytokine signaling), helping to explain common pathogenic features [92]. Apart from loci within the major histocompatibility complex (MHC) associated with the greater risk, multiple non-MHC genes mainly involved in the regulation of immune response may also explain the susceptibility of multiple autoimmune conditions [92].

However, the impact of an autoimmune comorbid condition on MS remains yet to be fully elucidated. The coexistence of another autoimmune disease may reduce the possibility of disability progression due to unknown changes in immunological pathways, increasing the tolerance against autoantigens. Alternatively, regulatory T-cells can prevent the inflammatory activity of Th1 and Th17 cells following the secretion of anti-inflammatory cytokines. The activation of more regulatory T-cells and the production of more anti-inflammatory cytokines could be another hypothesis for a milder course in patients with two concomitant autoimmune diseases. Further studies with larger sample sizes are required to confirm the finding of brain atrophy associated with autoimmune comorbidities.

7. Conclusions

In this comprehensive review, we summarized currently available data on the incidence and prevalence of autoimmune diseases comorbidity in PwMS, the possible effect of this association on clinical and neuroradiological MS course and its impact on treatment choice. We also highlighted the unmet needs in this field. The currently available data make it difficult to draw firm conclusions concerning the extent of coexistence of other autoimmune diseases in PwMS and all other topics discussed herein. Further studies in this area are warranted considering the implication on MS pathogenesis, the possible role of autoimmune co-morbidity on MS diagnosis delay, disease activity and disability progression, brain atrophy, and treatment choice.

Author Contributions: Writing—original draft preparation, V.N. and M.R.; writing—review and editing, V.N. and M.R.; visualization, V.N.; supervision, V.N.; project administration, V.N. All authors have read and agreed to the published version of the manuscript.

Funding: This research received no external funding.

Institutional Review Board Statement: Not applicable.

Informed Consent Statement: Not applicable.

Data Availability Statement: Not applicable.

Conflicts of Interest: The authors declare no conflict of interest related to this manuscript.

References

1. Dobson, R.; Giovannoni, G. Multiple sclerosis—A review. *Eur. J. Neurol.* **2019**, *26*, 27–40. [CrossRef]
2. Kalincik, T.; Diouf, I.; Sharmin, S.; Malpas, C.; Spelman, T.; Horakova, D.; Havrdova, E.K.; Trojano, M.; Izquierdo, G.; Lugaresi, A.; et al. Effect of Disease-Modifying Therapy on Disability in Relapsing-Remitting Multiple Sclerosis over 15 Years. *Neurology* **2021**, *96*, e783–e797. [CrossRef] [PubMed]
3. Naegele, M.; Martin, R. The good and the bad of neuroinflammation in multiple sclerosis. *Handb. Clin. Neurol.* **2014**, *122*, 59–87. [CrossRef] [PubMed]
4. Magyari, M.; Sorensen, P.S. Comorbidity in Multiple Sclerosis. *Front. Neurol.* **2020**, *11*, 851. [CrossRef]
5. Valderas, J.M.; Starfield, B.; Sibbald, B.; Salisbury, C.; Roland, M. Defining comorbidity: Implications for understanding health and health services. *Ann. Fam. Med.* **2009**, *7*, 357–363. [CrossRef] [PubMed]
6. Hauer, L.; Perneczky, J.; Sellner, J. A global view of comorbidity in multiple sclerosis: A systematic review with a focus on regional differences, methodology, and clinical implications. *J. Neurol.* **2021**, *268*, 4066–4077. [CrossRef]
7. Marrie, R.A.; Reingold, S.; Cohen, J.; Stuve, O.; Trojano, M.; Sorensen, P.S.; Cutter, G.; Reider, N. The incidence and prevalence of psychiatric disorders in multiple sclerosis: A systematic review. *Mult. Scler. J.* **2015**, *21*, 305–317. [CrossRef] [PubMed]
8. Jadidi, E.; Mohammadi, M.; Moradi, T. High risk of cardiovascular diseases after diagnosis of multiple sclerosis. *Mult. Scler.* **2013**, *19*, 1336–1340. [CrossRef]
9. Kaplan, T.B.; Berkowitz, A.L.; Samuels, M.A. Cardiovascular Dysfunction in Multiple Sclerosis. *Neurologist* **2015**, *20*, 108–114. [CrossRef]
10. Lo, L.M.P.; Taylor, B.V.; Winzenberg, T.; Palmer, A.J.; Blizzard, L.; van der Mei, I. Change and onset-type differences in the prevalence of comorbidities in people with multiple sclerosis. *J. Neurol.* **2021**, *268*, 602–612. [CrossRef] [PubMed]
11. Koch-Henriksen, N.; Laursen, B.; Stenager, E.; Magyari, M. Excess mortality among patients with multiple sclerosis in Denmark has dropped significantly over the past six decades: A population based study. *J. Neurol. Neurosurg. Psychiatry* **2017**, *88*, 626–631. [CrossRef] [PubMed]
12. Marrie, R.A. Comorbidity in multiple sclerosis: Implications for patient care. *Nat. Rev. Neurol.* **2017**, *13*, 375–382. [CrossRef] [PubMed]
13. Marrie, R.A.; Horwitz, R.; Cutter, G.; Tyry, T.; Campagnolo, D.; Vollmer, T. Comorbidity delays diagnosis and increases disability at diagnosis in MS. *Neurology* **2009**, *72*, 117–124. [CrossRef] [PubMed]
14. Marrie, R.A.; Reider, N.; Cohen, J.; Trojano, M.; Sorensen, P.S.; Cutter, G.; Reingold, S.; Stuve, O. A systematic review of the incidence and prevalence of sleep disorders and seizure disorders in multiple sclerosis. *Mult. Scler.* **2015**, *21*, 342–349. [CrossRef] [PubMed]
15. Koch, M.; Uyttenboogaart, M.; Polman, S.; De Keyser, J. Seizures in multiple sclerosis. *Epilepsia* **2008**, *49*, 948–953. [CrossRef]
16. Gambardella, A.; Valentino, P.; Labate, A.; Sibilia, G.; Ruscica, F.; Colosimo, E.; Nisticò, R.; Messina, D.; Zappia, M.; Quattrone, A. Temporal lobe epilepsy as a unique manifestation of multiple sclerosis. *Can. J. Neurol. Sci.* **2003**, *30*, 228–232. [CrossRef]
17. Sokic, D.V.; Stojsavljevic, N.; Drulovic, J.; Dujmovic, I.; Mesaros, S.; Ercegovac, M.; Peric, V.; Dragutinovic, G.; Levic, Z. Seizures in multiple sclerosis. *Epilepsia* **2001**, *42*, 72–79. [CrossRef]
18. Gasparini, S.; Ferlazzo, E.; Ascoli, M.; Sueri, C.; Cianci, V.; Russo, C.; Pisani, L.R.; Striano, P.; Elia, M.; Beghi, E.; et al. Risk factors for unprovoked epileptic seizures in multiple sclerosis: A systematic review and meta-analysis. *Neurol. Sci.* **2017**, *38*, 399–406. [CrossRef]
19. Kinnunen, E.; Wikström, J. Prevalence and prognosis of epilepsy in patients with multiple sclerosis. *Epilepsia* **1986**, *27*, 729–733. [CrossRef]
20. Nicoletti, A.; Sofia, V.; Biondi, R.; Lo Fermo, S.; Reggio, E.; Patti, F.; Reggio, A. Epilepsy and multiple sclerosis in Sicily: A population-based study. *Epilepsia* **2003**, *44*, 1445–1448. [CrossRef]

21. Calabrese, M.; Grossi, P.; Favaretto, A.; Romualdi, C.; Atzori, M.; Rinaldi, F.; Perini, P.; Saladini, M.; Gallo, P. Cortical pathology in multiple sclerosis patients with epilepsy: A 3 year longitudinal study. *J. Neurol. Neurosurg. Psychiatry* **2012**, *83*, 49–54. [CrossRef] [PubMed]
22. Martínez-Lapiscina, E.H.; Ayuso, T.; Lacruz, F.; Gurtubay, I.G.; Soriano, G.; Otano, M.; Bujanda, M.; Bacaicoa, M.C. Cortico-juxtacortical involvement increases risk of epileptic seizures in multiple sclerosis. *Acta Neurol. Scand.* **2013**, *128*, 24–31. [CrossRef]
23. Pack, A. Is There a Relationship Between Multiple Sclerosis and Epilepsy? If So What Does It Tell Us About Epileptogenesis? *Epilepsy. Curr.* **2018**, *18*, 95–96. [CrossRef] [PubMed]
24. Wang, L.; Zhang, J.; Deng, Z.R.; Zu, M.D.; Wang, Y. The epidemiology of primary headaches in patients with multiple sclerosis. *Brain Behav.* **2021**, *11*, e01830. [CrossRef] [PubMed]
25. Applebee, A. The clinical overlap of multiple sclerosis and headache. *Headache* **2012**, *52* (Suppl. S2), 111–116. [CrossRef]
26. Mrabet, S.; Wafa, M.; Giovannoni, G. Multiple sclerosis and migraine: Links, management and implications. *Mult. Scler. Relat. Disord.* **2022**, *68*, 104152. [CrossRef]
27. Graziano, E.; Hagemeier, J.; Weinstock-Guttman, B.; Ramasamy, D.P.; Zivadinov, R. Increased contrast enhancing lesion activity in relapsing–remitting multiple sclerosis migraine patients. *NeuroImage Clin.* **2015**, *9*, 110–116. [CrossRef]
28. D'Amico, D.; La Mantia, L.; Rigamonti, A.; Usai, S.; Mascoli, N.; Milanese, C.; Bussone, G. Prevalence of primary headaches in people with multiple sclerosis. *Cephalalgia* **2004**, *24*, 980–984. [CrossRef]
29. Gee, J.R.; Chang, J.; Dublin, A.B.; Vijayan, N. The association of brainstem lesions with migraine-like headache: An imaging study of multiple sclerosis. *Headache* **2005**, *45*, 670–677. [CrossRef]
30. Tortorella, P.; Rocca, M.A.; Colombo, B.; Annovazzi, P.; Comi, G.; Filippi, M. Assessment of MRI abnormalities of the brainstem from patients with migraine and multiple sclerosis. *J. Neurol. Sci.* **2006**, *244*, 137–141. [CrossRef]
31. Eikermann-Haerter, K.; Ayata, C. Cortical Spreading Depression and Migraine. *Curr. Neurol. Neurosci. Rep.* **2010**, *10*, 167–173. [CrossRef] [PubMed]
32. Merkler, D.; Klinker, F.; Jürgens, T.; Glaser, R.; Paulus, W.; Brinkmann, B.G.; Sereda, M.W.; Stadelmann-Nessler, C.; Guedes, R.C.; Brück, W.; et al. Propagation of spreading depression inversely correlates with cortical myelin content. *Ann. Neurol.* **2009**, *66*, 355–365. [CrossRef] [PubMed]
33. Gursoy-Ozdemir, Y.; Qiu, J.; Matsuoka, N.; Bolay, H.; Bermpohl, D.; Jin, H.; Wang, X.; Rosenberg, G.A.; Lo, E.H.; Moskowitz, M.A. Cortical spreading depression activates and upregulates MMP-9. *J. Clin. Investig.* **2004**, *113*, 1447–1455. [CrossRef] [PubMed]
34. Villani, V.; Prosperini, L.; Pozzilli, C.; Salvetti, M.; Sette, G. Quality of life of multiple sclerosis patients with comorbid migraine. *Neurol. Sci.* **2011**, *32*, 149. [CrossRef] [PubMed]
35. Marrie, R.A.; Horwitz, R.I.; Cutter, G.; Tyry, T.; Vollmer, T. Smokers with multiple sclerosis are more likely to report comorbid autoimmune diseases. *Neuroepidemiology* **2011**, *36*, 85–90. [CrossRef] [PubMed]
36. Langer-Gould, A.; Albers, K.B.; Van Den Eeden, S.K.; Nelson, L.M. Autoimmune diseases prior to the diagnosis of multiple sclerosis: A population-based case-control study. *Mult. Scler.* **2010**, *16*, 855–861. [CrossRef]
37. Nielsen, N.M.; Frisch, M.; Rostgaard, K.; Wohlfahrt, J.; Hjalgrim, H.; Koch-Henriksen, N.; Melbye, M.; Westergaard, T. Autoimmune diseases in patients with multiple sclerosis and their first-degree relatives: A nationwide cohort study in Denmark. *Mult. Scler.* **2008**, *14*, 823–829. [CrossRef]
38. Ramagopalan, S.V.; Dyment, D.A.; Valdar, W.; Herrera, B.M.; Criscuoli, M.; Yee, I.M.; Sadovnick, A.D.; Ebers, G.C. Autoimmune disease in families with multiple sclerosis: A population-based study. *Lancet Neurol.* **2007**, *6*, 604–610. [CrossRef]
39. Marrie, R.A.; Reider, N.; Cohen, J.; Stuve, O.; Sorensen, P.S.; Cutter, G.; Reingold, S.C.; Trojano, M. A systematic review of the incidence and prevalence of autoimmune disease in multiple sclerosis. *Mult. Scler.* **2015**, *21*, 282–293. [CrossRef]
40. Dobson, R.; Giovannoni, G. Autoimmune disease in people with multiple sclerosis and their relatives: A systematic review and meta-analysis. *J. Neurol.* **2013**, *260*, 1272–1285. [CrossRef]
41. Nielsen, N.M.; Westergaard, T.; Frisch, M.; Rostgaard, K.; Wohlfahrt, J.; Koch-Henriksen, N.; Melbye, M.; Hjalgrim, H. Type 1 diabetes and multiple sclerosis: A Danish population-based cohort study. *Arch. Neurol.* **2006**, *63*, 1001–1004. [CrossRef] [PubMed]
42. Marrosu, M.G.; Cocco, E.; Lai, M.; Spinicci, G.; Pischedda, M.P.; Contu, P. Patients with multiple sclerosis and risk of type 1 diabetes mellitus in Sardinia, Italy: A cohort study. *Lancet* **2002**, *359*, 1461–1465. [CrossRef]
43. Bechtold, S.; Blaschek, A.; Raile, K.; Dost, A.; Freiberg, C.; Askenas, M.; Fröhlich-Reiterer, E.; Molz, E.; Holl, R.W. Higher relative risk for multiple sclerosis in a pediatric and adolescent diabetic population: Analysis from DPV database. *Diabetes Care* **2014**, *37*, 96–101. [CrossRef] [PubMed]
44. Karni, A.; Abramsky, O. Association of MS with thyroid disorders. *Neurology* **1999**, *53*, 883–885. [CrossRef] [PubMed]
45. Niederwieser, G.; Buchinger, W.; Bonelli, R.M.; Berghold, A.; Reisecker, F.; Költringer, P.; Archelos, J.J. Prevalence of autoimmune thyroiditis and non-immune thyroid disease in multiple sclerosis. *J. Neurol.* **2003**, *250*, 672–675. [CrossRef] [PubMed]
46. Munteis, E.; Cano, J.F.; Flores, J.A.; Martinez-Rodriguez, J.E.; Miret, M.; Roquer, J. Prevalence of autoimmune thyroid disorders in a Spanish multiple sclerosis cohort. *Eur. J. Neurol.* **2007**, *14*, 1048–1052. [CrossRef]
47. Roshanisefat, H.; Bahmanyar, S.; Hillert, J.; Olsson, T.; Montgomery, S. Shared genetic factors may not explain the raised risk of comorbid inflammatory diseases in multiple sclerosis. *Mult. Scler.* **2012**, *18*, 1430–1436. [CrossRef]
48. Edwards, L.J.; Constantinescu, C.S. A prospective study of conditions associated with multiple sclerosis in a cohort of 658 consecutive outpatients attending a multiple sclerosis clinic. *Mult. Scler. J.* **2004**, *10*, 575–581. [CrossRef]

49. Kosmidou, M.; Katsanos, A.H.; Katsanos, K.H.; Kyritsis, A.P.; Tsivgoulis, G.; Christodoulou, D.; Giannopoulos, S. Multiple sclerosis and inflammatory bowel diseases: A systematic review and meta-analysis. *J. Neurol.* **2017**, *264*, 254–259. [CrossRef]
50. Marrie, R.A.; Patten, S.B.; Tremlett, H.; Wolfson, C.; Leung, S.; Fisk, J.D. Increased incidence and prevalence of psoriasis in multiple sclerosis. *Mult. Scler. Relat. Disord.* **2017**, *13*, 81–86. [CrossRef]
51. Fellner, A.; Dano, M.; Regev, K.; Mosek, A.; Karni, A. Multiple sclerosis is associated with psoriasis. A case-control study. *J. Neurol. Sci.* **2014**, *338*, 226–228. [CrossRef] [PubMed]
52. Farez, M.F.; Balbuena Aguirre, M.E.; Varela, F.; Köhler, A.A.; Correale, J. Autoimmune Disease Prevalence in a Multiple Sclerosis Cohort in Argentina. *Mult. Scler. Int.* **2014**, *2014*, 828162. [CrossRef] [PubMed]
53. Silfvast-Kaiser, A.S.; Homan, K.B.; Mansouri, B. A narrative review of psoriasis and multiple sclerosis: Links and risks. *Psoriasis* **2019**, *9*, 81–90. [CrossRef]
54. Costelloe, L.; Jones, J.; Coles, A. Secondary autoimmune diseases following alemtuzumab therapy for multiple sclerosis. *Expert Rev. Neurother.* **2012**, *12*, 335–341. [CrossRef]
55. Caraccio, N.; Dardano, A.; Manfredonia, F.; Manca, L.; Pasquali, L.; Iudice, A.; Murri, L.; Ferrannini, E.; Monzani, F. Long-term follow-up of 106 multiple sclerosis patients undergoing interferon-beta 1a or 1b therapy: Predictive factors of thyroid disease development and duration. *J. Clin. Endocrinol. Metab.* **2005**, *90*, 4133–4137. [CrossRef] [PubMed]
56. Sahraian, M.A.; Owji, M.; Naser Moghadasi, A. Concomitant multiple sclerosis and another autoimmune disease: Does the clinical course change? *Clin. Neurol. Neurosurg.* **2016**, *150*, 92–95. [CrossRef]
57. Fanouriakis, A.; Mastorodemos, V.; Pamfil, C.; Papadaki, E.; Sidiropoulos, P.; Plaitakis, A.; Amoiridis, G.; Bertsias, G.; Boumpas, D.T. Coexistence of systemic lupus erythematosus and multiple sclerosis: Prevalence, clinical characteristics, and natural history. *Semin. Arthritis Rheum.* **2014**, *43*, 751–758. [CrossRef]
58. Zéphir, H.; Gower-Rousseau, C.; Salleron, J.; Simon, O.; Debouverie, M.; Le Page, E.; Bouhnik, Y.; Lebrun-Frenay, C.; Papeix, C.; Vigneron, B.; et al. Milder multiple sclerosis course in patients with concomitant inflammatory bowel disease. *Mult. Scler.* **2014**, *20*, 1135–1139. [CrossRef]
59. Zivadinov, R.; Raj, B.; Ramanathan, M.; Teter, B.; Durfee, J.; Dwyer, M.G.; Bergsland, N.; Kolb, C.; Hojnacki, D.; Benedict, R.H.; et al. Autoimmune Comorbidities Are Associated with Brain Injury in Multiple Sclerosis. *AJNR Am. J. Neuroradiol.* **2016**, *37*, 1010–1016. [CrossRef]
60. Lorefice, L.; Fenu, G.; Pitzalis, R.; Scalas, G.; Frau, J.; Coghe, G.; Musu, L.; Sechi, V.; Barracciu, M.A.; Marrosu, M.G.; et al. Autoimmune comorbidities in multiple sclerosis: What is the influence on brain volumes? A case-control MRI study. *J. Neurol.* **2018**, *265*, 1096–1101. [CrossRef]
61. Barnett, M.; Bergsland, N.; Weinstock-Guttman, B.; Butzkueven, H.; Kalincik, T.; Desmond, P.; Gaillard, F.; van Pesch, V.; Ozakbas, S.; Rojas, J.I.; et al. Brain atrophy and lesion burden are associated with disability progression in a multiple sclerosis real-world dataset using only T2-FLAIR: The NeuroSTREAM MSBase study. *Neuroimage Clin.* **2021**, *32*, 102802. [CrossRef] [PubMed]
62. Losseff, N.A.; Webb, S.L.; O'Riordan, J.I.; Page, R.; Wang, L.; Barker, G.J.; Tofts, P.S.; McDonald, W.I.; Miller, D.H.; Thompson, A.J. Spinal cord atrophy and disability in multiple sclerosis. A new reproducible and sensitive MRI method with potential to monitor disease progression. *Brain* **1996**, *119*, 701–708. [CrossRef] [PubMed]
63. Kunchok, A.; Aksamit, A.J., Jr.; Davis, J.M., 3rd; Kantarci, O.H.; Keegan, B.M.; Pittock, S.J.; Weinshenker, B.G.; McKeon, A. Association Between Tumor Necrosis Factor Inhibitor Exposure and Inflammatory Central Nervous System Events. *JAMA Neurol.* **2020**, *77*, 937–946. [CrossRef] [PubMed]
64. La Mantia, L.; Capsoni, F. Psoriasis during interferon beta treatment for multiple sclerosis. *Neurol. Sci.* **2010**, *31*, 337–339. [CrossRef] [PubMed]
65. Hojjati, S.M.; Heidari, B.; Babaei, M. Development of rheumatoid arthritis during treatment of multiple sclerosis with interferon beta 1-a. Coincidence of two conditions or a complication of treatment: A case report. *J. Adv. Res.* **2016**, *7*, 611–613. [CrossRef] [PubMed]
66. Schott, E.; Paul, F.; Wuerfel, J.T.; Zipp, F.; Rudolph, B.; Wiedenmann, B.; Baumgart, D.C. Development of ulcerative colitis in a patient with multiple sclerosis following treatment with interferon beta 1a. *World J. Gastroenterol.* **2007**, *13*, 3638–3640. [CrossRef] [PubMed]
67. Killestein, J.; van Oosten, B. Emerging safety issues in alemtuzumab-treated MS patients. *Mult. Scler.* **2019**, *25*, 1206–1208. [CrossRef]
68. Lambrianides, S.; Kinnis, E.; Leonidou, E.; Pantzaris, M. Does Natalizumab Induce or Aggravate Psoriasis? A Case Study and Review of the Literature. *Case Rep. Neurol.* **2018**, *10*, 286–291. [CrossRef]
69. Su, E.; Novic, J.; Han, M.H. Emergence of rheumatoid arthritis following exposure to natalizumab. *Mult. Scler. Relat. Disord.* **2020**, *40*, 101936. [CrossRef]
70. Khoy, K.; Mariotte, D.; Defer, G.; Petit, G.; Toutirais, O.; Le Mauff, B. Natalizumab in Multiple Sclerosis Treatment: From Biological Effects to Immune Monitoring. *Front. Immunol.* **2020**, *11*, 549842. [CrossRef]
71. MacDonald, J.K.; McDonald, J.W. Natalizumab for induction of remission in Crohn's disease. *Cochrane Database Syst. Rev.* **2007**, CD006097. [CrossRef] [PubMed]

72. Targan, S.R.; Feagan, B.G.; Fedorak, R.N.; Lashner, B.A.; Panaccione, R.; Present, D.H.; Spehlmann, M.E.; Rutgeerts, P.J.; Tulassay, Z.; Volfova, M.; et al. Natalizumab for the treatment of active Crohn's disease: Results of the ENCORE Trial. *Gastroenterology* **2007**, *132*, 1672–1683. [CrossRef] [PubMed]
73. Abushouk, A.I.; Ahmed, H.; Ismail, A.; Elmaraezy, A.; Badr, A.S.; Gadelkarim, M.; Elnenny, M. Safety and efficacy of ocrelizumab in rheumatoid arthritis patients with an inadequate response to methotrexate or tumor necrosis factor inhibitors: A systematic review and meta-analysis. *Rheumatol. Int.* **2017**, *37*, 1053–1064. [CrossRef] [PubMed]
74. Kristjánsson, V.B.; Lund, S.H.; Gröndal, G.; Sveinsdóttir, S.V.; Agnarsson, H.R.; Jónasson, J.G.; Björnsson, E.S. Increased risk of inflammatory bowel disease among patients treated with rituximab in Iceland from 2001 to 2018. *Scand. J. Gastroenterol.* **2021**, *56*, 46–52. [CrossRef] [PubMed]
75. Darwin, E.; Romanelli, P.; Lev-Tov, H. Ocrelizumab-induced psoriasiform dermatitis in a patient with multiple sclerosis. *Dermatol. Online J.* **2018**, *24*. [CrossRef]
76. Sunjaya, D.B.; Taborda, C.; Obeng, R.; Dhere, T. First Case of Refractory Colitis Caused by Ocrelizumab. *Inflamm. Bowel. Dis.* **2020**, *26*, e49. [CrossRef]
77. Schaper, K.; Dickhaut, J.; Japtok, L.; Kietzmann, M.; Mischke, R.; Kleuser, B.; Bäumer, W. Sphingosine-1-phosphate exhibits anti-proliferative and anti-inflammatory effects in mouse models of psoriasis. *J. Dermatol. Sci.* **2013**, *71*, 29–36. [CrossRef]
78. Deguchi, Y.; Andoh, A.; Yagi, Y.; Bamba, S.; Inatomi, O.; Tsujikawa, T.; Fujiyama, Y. The S1P receptor modulator FTY720 prevents the development of experimental colitis in mice. *Oncol. Rep.* **2006**, *16*, 699–703. [CrossRef]
79. Ladrón Abia, P.; Alcalá Vicente, C.; Martínez Delgado, S.; Bastida Paz, G. Fingolimod-induced remission in a patient with ulcerative colitis and multiple sclerosis. *Gastroenterol. Hepatol.* **2021**, *44*, 156–157. [CrossRef]
80. Sandborn, W.J.; Feagan, B.G. Ozanimod Treatment for Ulcerative Colitis. *N. Engl. J. Med.* **2016**, *375*, e17. [CrossRef]
81. Altmeyer, P.J.; Matthes, U.; Pawlak, F.; Hoffmann, K.; Frosch, P.J.; Ruppert, P.; Wassilew, S.W.; Horn, T.; Kreysel, H.W.; Lutz, G.; et al. Antipsoriatic effect of fumaric acid derivatives. Results of a multicenter double-blind study in 100 patients. *J. Am. Acad. Dermatol.* **1994**, *30*, 977–981. [CrossRef]
82. Ghoreschi, K.; Brück, J.; Kellerer, C.; Deng, C.; Peng, H.; Rothfuss, O.; Hussain, R.Z.; Gocke, A.R.; Respa, A.; Glocova, I.; et al. Fumarates improve psoriasis and multiple sclerosis by inducing type II dendritic cells. *J. Exp. Med.* **2011**, *208*, 2291–2303. [CrossRef] [PubMed]
83. Brummer, T.; Ruck, T.; Meuth, S.G.; Zipp, F.; Bittner, S. Treatment approaches to patients with multiple sclerosis and coexisting autoimmune disorders. *Ther. Adv. Neurol. Disord.* **2021**, *14*, 17562864211035542. [CrossRef] [PubMed]
84. Berrigan, L.I.; Fisk, J.D.; Patten, S.B.; Tremlett, H.; Wolfson, C.; Warren, S.; Fiest, K.M.; McKay, K.A.; Marrie, R.A. Health-related quality of life in multiple sclerosis: Direct and indirect effects of comorbidity. *Neurology* **2016**, *86*, 1417–1424. [CrossRef]
85. Salter, A.; Tyry, T.; Wang, G.; Fox, R.J.; Cutter, G.; Marrie, R.A. Examining the joint effect of disability, health behaviors, and comorbidity on mortality in MS. *Neurol. Clin. Pract.* **2016**, *6*, 397–408. [CrossRef]
86. Newland, P.K.; Lorenz, R.; Budhathoki, C.; Jensen, M.P. The Presence of Symptoms With Comorbid Conditions in Individuals With Multiple Sclerosis (MS). *Clin. Nurs. Res.* **2016**, *25*, 532–548. [CrossRef]
87. Marrie, R.A.; Miller, A.; Sormani, M.P.; Thompson, A.; Waubant, E.; Trojano, M.; O'Connor, P.; Reingold, S.; Cohen, J.A. The challenge of comorbidity in clinical trials for multiple sclerosis. *Neurology* **2016**, *86*, 1437–1445. [CrossRef]
88. Handel, A.E.; Handunnetthi, L.; Ebers, G.C.; Ramagopalan, S.V. Type 1 diabetes mellitus and multiple sclerosis: Common etiological features. *Nat. Rev. Endocrinol.* **2009**, *5*, 655–664. [CrossRef]
89. Winer, S.; Astsaturov, I.; Cheung, R.; Gunaratnam, L.; Kubiak, V.; Cortez, M.A.; Moscarello, M.; O'Connor, P.W.; McKerlie, C.; Becker, D.J.; et al. Type I diabetes and multiple sclerosis patients target islet plus central nervous system autoantigens; nonimmunized nonobese diabetic mice can develop autoimmune encephalitis. *J. Immunol.* **2001**, *166*, 2831–2841. [CrossRef]
90. Banwell, B.; Bar-Or, A.; Cheung, R.; Kennedy, J.; Krupp, L.B.; Becker, D.J.; Dosch, H.M. Abnormal T-cell reactivities in childhood inflammatory demyelinating disease and type 1 diabetes. *Ann. Neurol.* **2008**, *63*, 98–111. [CrossRef]
91. Baranzini, S.E. The genetics of autoimmune diseases: A networked perspective. *Curr. Opin. Immunol.* **2009**, *21*, 596–605. [CrossRef] [PubMed]
92. Richard-Miceli, C.; Criswell, L.A. Emerging patterns of genetic overlap across autoimmune disorders. *Genome Med.* **2012**, *4*, 6. [CrossRef] [PubMed]

Journal of *Personalized Medicine*

MDPI

Article

DNA Methylation Analysis Reveals Distinct Patterns in Satellite Cell–Derived Myogenic Progenitor Cells of Subjects with Spastic Cerebral Palsy

Karyn G. Robinson [1], Adam G. Marsh [2], Stephanie K. Lee [1], Jonathan Hicks [2], Brigette Romero [3], Mona Batish [3], Erin L. Crowgey [1], M. Wade Shrader [4] and Robert E. Akins [1,*]

1 Nemours Children's Research, Nemours Children's Health System, Wilmington, DE 19803, USA
2 Center for Bioinformatics and Computational Biology, University of Delaware, Newark, DE 19716, USA
3 Medical and Molecular Sciences, University of Delaware, Newark, DE 19716, USA
4 Department of Orthopedics, Nemours Children's Hospital Delaware, Wilmington, DE 19803, USA
* Correspondence: robert.akins@nemours.org; Tel.: +1-302-651-6779

Abstract: Spastic type cerebral palsy (CP) is a complex neuromuscular disorder that involves altered skeletal muscle microanatomy and growth, but little is known about the mechanisms contributing to muscle pathophysiology and dysfunction. Traditional genomic approaches have provided limited insight regarding disease onset and severity, but recent epigenomic studies indicate that DNA methylation patterns can be altered in CP. Here, we examined whether a diagnosis of spastic CP is associated with intrinsic DNA methylation differences in myoblasts and myotubes derived from muscle resident stem cell populations (satellite cells; SCs). Twelve subjects were enrolled (6 CP; 6 control) with informed consent/assent. Skeletal muscle biopsies were obtained during orthopedic surgeries, and SCs were isolated and cultured to establish patient–specific myoblast cell lines capable of proliferation and differentiation in culture. DNA methylation analyses indicated significant differences at 525 individual CpG sites in proliferating SC–derived myoblasts (MB) and 1774 CpG sites in differentiating SC–derived myotubes (MT). Of these, 79 CpG sites were common in both culture types. The distribution of differentially methylated 1 Mbp chromosomal segments indicated distinct regional hypo– and hyper–methylation patterns, and significant enrichment of differentially methylated sites on chromosomes 12, 13, 14, 15, 18, and 20. Average methylation load across 2000 bp regions flanking transcriptional start sites was significantly different in 3 genes in MBs, and 10 genes in MTs. SC derived MBs isolated from study participants with spastic CP exhibited fundamental differences in DNA methylation compared to controls at multiple levels of organization that may reveal new targets for studies of mechanisms contributing to muscle dysregulation in spastic CP.

Keywords: cerebral palsy; muscle spasticity; primary cell culture; satellite cells; skeletal muscle; muscle; skeletal; humans; epigenomics; DNA methylation; regulatory non-coding RNAs

Citation: Robinson, K.G.; Marsh, A.G.; Lee, S.K.; Hicks, J.; Romero, B.; Batish, M.; Crowgey, E.L.; Shrader, M.W.; Akins, R.E. DNA Methylation Analysis Reveals Distinct Patterns in Satellite Cell–Derived Myogenic Progenitor Cells of Subjects with Spastic Cerebral Palsy. *J. Pers. Med.* **2022**, *12*, 1978. https://doi.org/10.3390/jpm12121978

Academic Editor: Anne-Marie Caminade

Received: 17 November 2022
Accepted: 25 November 2022
Published: 30 November 2022

Publisher's Note: MDPI stays neutral with regard to jurisdictional claims in published maps and institutional affiliations.

1. Introduction

Cerebral palsy (CP) is the most common cause of physical disability in childhood, with a prevalence of 2–4 per 1000 live births [1–3]. It is a heterogeneous set of movement disorders associated with a static encephalopathy that occurs during fetal development or early postnatal life [4]. There are three fundamental types of CP: ataxic, dyskinetic, and spastic [5], with spastic CP accounting for about 80% of cases [6]. Spastic CP is characterized by hypertonia, exaggerated reflexes, and poor muscle growth associated with progressive musculoskeletal deformities that often require surgical correction [5–13].

Individuals with spastic CP have difficulties with movement, movement control, and muscle function [2,3]. Ultrasound and MRI studies have demonstrated that subjects with spastic CP have decreased muscle length [14], cross–sectional area [15], and volume [16,17], leading to diminished force generation, reduced range of motion, and weakness [18–21].

Additionally, studies of muscle indicate that patients with CP exhibit increased sarcomere length [22], disorganized neuromuscular junctions [23–25], extracellular matrix abnormalities [22], tissue–level differences in gene expression profiles [26,27], and limited myogenic potential [28].

While all individuals with spastic CP have some movement dysfunction, there is a high degree of variability in phenotype between individuals. A more thorough understanding of the mechanisms associated with dysfunction in the peripheral neuromotor system is needed in order to develop more targeted and enhanced therapeutics addressing the major challenges facing individuals with spastic CP. Importantly, spastic CP is associated with alterations in the muscle–resident stem cell populations (satellite cells; SC) responsible for skeletal muscle growth and repair. Surgical patients with CP have a reduced SC population, which may account for aspects of their impaired muscle growth and decreased ability to strengthen muscle [28–30]. Isolated SC–derived myoblasts (MB) from CP donors appear to have altered phenotypes in culture compared to control MBs [28,31,32], and an RNASeq study from our group showed differential expression of mRNAs and miRNAs in spastic CP muscle [26]. These data suggest that MBs from CP patients may retain intrinsic differences through the cell isolation and culture process.

Genetic alterations may account for intrinsic differences in isolated MB populations. A number of potentially causative genetic variants have been identified in CP [33], but not all types of CP are easily detected or characterized by genomic data. Several rare copy number variants and mutations have been identified, but there is considerable genetic heterogeneity in patients with CP [33,34]. While some CP cases may be associated with certain types of genetic abnormality [35], the conventional view of CP remains that environmental factors affecting neuromotor maturation are responsible for most cases, especially among individuals with spastic CP [36,37].

Epigenetic modification may also account for some retained phenotypic differences in isolated MB behavior. In recent work, DNA methylation pattern differences were identified in peripheral blood cells from subjects with CP [38–41], and some early studies indicate that muscle in CP may be similarly altered, with differential DNA methylation in CP resulting in a decreased capacity for MBs to fuse and differentiate into MTs [32]. It has been demonstrated in several disease states that patterns of altered DNA methylation may uncover molecular etiologies and reveal potential therapeutic targets [42–51], and such may be the case in spastic CP as well [39]. DNA methylation changes may serve as markers for diagnosis, prognosis, tailoring the best treatment for a subclass of disease, monitoring treatment efficacy, and identifying genes to be examined for the development of genetically or epigenetically targeted therapies [52]. In the current study, DNA methylomes were analyzed to provide insight into individual CpG site differences and altered DNA methylation patterns in chromosomal segments and near transcription start site (TSS) in spastic CP SCs compared to controls.

2. Materials and Methods

2.1. Subject Enrollment

Six subjects with a diagnosis of spastic CP and 6 control subjects were enrolled in an IRB–approved study at Nemours Children's Hospital, Delaware, after informed consent/assent. The control cohort comprised children with an idiopathic condition or an injury. Subjects with a chromosomal disorder, degenerative neurological disease, or muscular dystrophy were excluded.

2.2. Satellite Cell Isolation

Skeletal muscle biopsies collected during orthopedic surgeries were enzymatically digested and a double–immunomagnetic isolation approach was used to collect a population of mononuclear Cells positive for the surface markers neural cell adhesion molecule 1 and C–X–C motif chemokine receptor 4 (anti–NCAM1 and anti–CXCR4, both at 2.5 ng/mL, Miltenyi, San Diego, CA, USA) as previously described [26]. Previous studies have demonstrated that CXCR4 marks human SCs and that selection using the combination of NCAM1 (CD56) and CXCR4 more effectively removes non–satellite cells than using either marker alone [53,54]. This method resulted in a nearly pure SC population as verified by positive PAX7 immunofluorescence signal obtained after 24–48 h in culture (Anti–PAX7 from hybridoma cells deposited to the DSHB by Kawakami, A., Developmental Studies Hybridoma Bank, Iowa City, IA, USA). Cell populations that were at least 90% positive for PAX7 expression were utilized for experiments (Supplemental Table S1).

Cells were seeded at passage 3–5 (Supplemental Table S1) and proliferated in medium consisting of Zenbio Skeletal Muscle Growth Medium (Zenbio, Triangle Park, NC, USA) supplemented to a final concentration of 20% Qualified FBS (Thermo Fisher Scientific, Philadelphia, PA, USA), 4 g/L of D Glucose (Thermo Fisher Scientific, Philadelphia, PA, USA), 1 ng/mL of human bFGF (PeproTech, Rocky Hill, NJ, USA), and 1% penicillin–streptomycin, (Thermo Fisher Scientific, Philadelphia, PA, USA) was exchanged every other day until cells reached confluence. Proliferating MBs were collected at 50% confluence (2–5 days after seeding; Supplemental Table S1). For differentiation, cultures were switched to low–serum medium consisting of high glucose DMEM (Thermo Fisher Scientific, Philadelphia, PA, USA) supplemented with 2% horse serum (Thermo Fisher Scientific, Philadelphia, PA, USA), 2% human insulin (Sigma, St. Louis, MO, USA) and 1% penicillin–streptomycin upon reaching 90–100% confluence. Cells were differentiated for 24 **h** to initiate MB fusion into MTs [28] and collected.

2.3. DNA Extraction, Library Preparation, and Sequencing

Genomic DNA was isolated using Gentra Puregene kits (Qiagen, Germantown, MD, USA). A previously published DNA methylation assay [39,55,56] was utilized. Briefly, DNA libraries for next generation sequencing (NGS) were prepared by digesting genomic DNA with methyl–sensitive restriction endonuclease HpaII, which recognizes CCGG sites. A standard sequencing protocol was then performed including randomized shearing (Covaris, Woburn, MA, USA) and synthesis of a gDNA fragment library using Illumina TruSeq Nano library synthesis kits (San Diego, CA, USA). NGS was performed on an Illumina ×10 platform by Psomagen (Rockville, MD, USA). The protocol generated single end reads (150 bp) with >20× coverage of the regions captured. FASTQ data files were processed to calculate the probability of methylation at individual CpG sites through a commercial bioinformatics pipeline and software platform (Genome Profiling LLC, Newark, DE, USA). For convenience, the term "CpG" in this paper refers to "C(CpG)G" HpaII restriction sites.

2.4. Methylation Analysis

FASTQ files were aligned to human reference genome hg19 using BWA (Burrows–Wheeler Aligner algorithm [57]. To reduce false discovery associated with the inclusion of both male and female participants in the study, sites on the X and Y chromosomes were excluded from analyses. For each CpG per sample, a methylation score was calculated proportional to the probability that a specific CpG site was methylated. These scores were then compared between cohorts using a set of analytic modules from the R–packages edgeR [58,59] and limma [60] to compare pairwise for each CpG site. The response scale of the methylation data sets used here is within the operational boundaries of log–scaled gene expression data for which edgeR and limma were designed, and the well–developed false–discovery rate calculations in these R packages are ideal for methylation score data distributions [61,62].

Statistical significance at the level of individual CpG sites and was evaluated using a Likelihood Ratio Test with a one–way ANOVA contrast (LRT–ANOVA). Potentially informative CpG sites were selected for the PCA plots by filtering the LRT–ANOVA p-values with an appropriate cutoff (<0.01). Gene annotations were derived from the Ensembl GRCh37 database based on chromosomal locus (www.ensmbl.org; last accessed on 15 November 2022 for validation). Enhancer regions were derived from Ensembl's GRCh37 BioMart tool. The ontology terms Muscle Organ Development, Muscle System Processes, and Skeletal

Muscle Cell Differentiation (GO Enrichment Analysis; amigo.geneontology.org) were used to determine 575 unique genes annotated to be involved in muscle physiology. Fisher's exact test was utilized to determine enrichment of significant CpGs within chromosomes.

Because methylation of the region around the TSS of a gene is thought to be highly informative of gene expression [63], the individual CpG methylation scores were averaged per TSS using 1000 bp upstream of the TSS and 1000 bp downstream [64]. TSS were identified from Ensembl's GRCh37 BioMart data mining tool (release 106) [65]. If one or fewer CpGs were found within this range, the TSS was excluded from further analysis. For each sample, the mean methylation load in this 2000 bp range was calculated and a likelihood ratio test performed on the methylation loads. No more than one transcript for each gene was included in the statistical analysis.

3. Results

Samples from 12 unique study participants were included in the study; demographic data are summarized in Table 1. Briefly, the control group consisted of $n = 6$ subjects (males = 3, and females = 3) with an average age of 13.9 ± 1.7 years; the CP group consisted of $n = 6$ subjects (males = 3, and females = 3) with an average age of 15.5 ± 3.0 years. Biopsies from different muscles were included based on the availability of viable muscle tissue suitable for cell isolation and to help identify methylation signals generally associated with spastic CP rather than a specific muscle: SCs were isolated from spinalis or semitendinosus muscle for controls and from spinalis, rectus femoris, adductor longus, or vastus lateralis for the CP group. Although differences likely exist between muscle types, previous studies of MTs derived from different muscles found low inter-muscle variability in RNA–sequencing data [26].

Table 1. Demographic information for subjects in the study *.

Sample	Diagnosis	Age	Sex	GMFCS	Tissue Source
CN1	Spondylolysis	16.6	M	N/A	Spinalis
CN2	Torn ACL	12.6	M	N/A	Semitendinosus
CN3	Idiopathic scoliosis	12.1	F	N/A	Spinalis
CN4	Torn ACL	12.7	F	N/A	Semitendinosus
CN5	Idiopathic scoliosis	15.1	M	N/A	Spinalis
CN6	Idiopathic scoliosis	14.3	F	N/A	Spinalis
CP1	Spastic CP	15.6	M	5	Vastus lateralis
CP2	Spastic CP	19.1	M	5	Adductor longus
CP3	Spastic CP	12.6	M	4	Rectus femoris
CP4	Spastic CP	13.8	F	2	Rectus femoris
CP5	Spastic CP	19.0	F	5	Spinalis
CP6	Spastic CP	12.8	F	5	Spinalis

* CN = control; CP = cerebral palsy; ACL = anterior cruciate ligament; M = male; F = female; GMFCS = Gross Motor Function Classification System.

To evaluate differences in DNA methylation patterns associated with affected muscle of subjects with spastic CP, whole genome methylation patterns were determined from proliferating MBs and differentiating MTs. Cells derived from subjects with CP appeared morphologically similar to those derived from controls (Supplemental Figure S1) and exhibited no significant differences in proliferation rates (Supplemental Table S1). NGS was performed after methylation sensitive restriction endonuclease (HpaII) digestion. The hg19 reference genome assembly from the University of California Santa Cruz [66] includes 2.29×10^6 HpaII target CCGG motifs, which represent ~15% of the 14×10^6 CpG sites in the haploid hg19 genome [39]. Alignment of the HpaII restricted sites in our 12 samples yielded

1,483,038 sites were in common across all subjects for MBs and MTs. DNA methylation patterns were analyzed at the individual CpG site level using dimensionality reduction by principal component analysis (PCA) to assess the degree of discrimination between CP and non–CP cohorts. All potentially informative CpG sites (n = 20,254 for MBs and 27,834 for MTs), were integrated as one pattern and demonstrated strong discrimination based on diagnosis (Figure 1, Supplemental Figure S2).

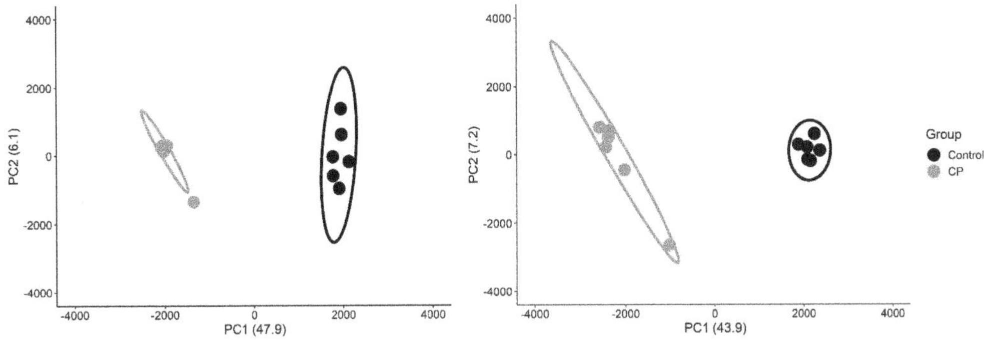

(a) (b)

Figure 1. Principal component analysis to identify discriminating methylation patterns between CP and non–CP cohorts. For MBs (**a**) and MTs (**b**), the first two component axes (PC1, PC2) were plotted with % variance explained in parenthesis. Each point represents the similarity position of a subject based on all potentially informative CpG sites ($p < 0.01$). CP subjects are represented in gray and control subjects in black. Ellipses represent 90% confidence intervals. The complete segregation of the two cohorts indicates that DNA methylation patterns fundamentally differ between cohorts.

In MBs, 525 CpG sites were found to have differential methylation load scores (FDR < 0.05; heatmap in Figure 2a, volcano plot in Supplemental Figure S3A, list of significant CpGs in Supplemental Table S2). Of these, 11 were within genes known to be involved in muscle physiology and 21 were within known gene enhancer regions. 1774 CpG sites were found to have differential methylation load scores in MTs between the CP and control cohorts (heatmap in Figure 2b, volcano plot in Supplemental Figure S3B, list of significant CpGs in Supplemental Table S3), of which 43 CpG sites were within genes known to be involved in muscle physiology and 97 were within gene enhancer regions. The differentially methylated CpGs included 79 CpG sites that were significantly different under both cell conditions. Of these, 36 were significantly hypermethylated and 43 were significantly hypomethylated in the CP cohort compared to the control cohort (Table 2).

To determine the methylation differences over larger regions of the genome, the chromosomal distribution of significant CpGs was visualized (Supplemental Figure S4) and Fisher's exact test was employed to analyze enrichment of differentially methylated CpGs on individual chromosomes. For both MB and MT cell populations, significant enrichment was found on chromosomes 12, 13, 14, 15, 18, and 20. Interestingly, the same chromosomes were identified as having significant CpG site enrichment in previous studies of both muscle tissue and peripheral blood cells from subjects with CP (Table 3). To further assess regional differences, methylation load scores were calculated across 1 Mbp chromosomal segments. There was a strong correlation between the 1 Mbp methylation loads of MBs and MTs (Figure 3a), indicating stability of the methylome. When these 1 Mpb regions were mapped to the chromosomes, regions of accentuated differential methylation load were noted on all chromosomes except 1 and 17 (Figure 3b,c). These large–scale changes in DNA methylation could affect higher–order chromatin structure and regulation of gene expression [52].

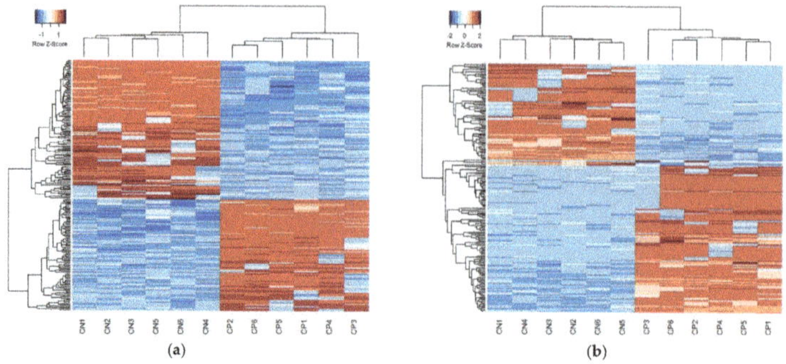

Figure 2. Heatmap clustering of the top 200 CpG sites. Of the common CpG sites, there was significant differential methylation in 525 distinct CpGs for MBs (a) and 1774 for MTs (b) (FDR-corrected p-value < 0.05). Heatmaps based on the 200 CpG sites with the lowest FDR—corrected p-values were generated using Euclidean distances and complete linkage clustering. Each row represents the score for a single CpG site across all subjects with blue indicating hypomethylation and red indicating hypermethylation. Quantitative differences in CpG site methylation by diagnosis were apparent.

Table 2. CpG sites that were differentially methylated in both MBs and MTs *.

Position	MB LogFC	MB FDR Corrected p-Value	MT LogFC	MT FDR Corrected p-Value	Gene
chr2.0003882321	0.88	2.29×10^{-2}	0.90	2.40×10^{-2}	
chr2.0029850455	1.24	6.74×10^{-4}	1.32	1.11×10^{-5}	ALK
chr2.0033057636	−0.84	3.40×10^{-3}	−0.85	1.94×10^{-2}	LINC00486
chr2.0035092870	−1.09	1.00×10^{-2}	−1.19	8.26×10^{-5}	AC012593.1
chr2.0056193463	1.32	1.79×10^{-3}	1.55	1.22×10^{-6}	RP11—481J13.1, AC011306.2
chr2.0223166989	0.87	4.07×10^{-2}	0.90	4.80×10^{-2}	CCDC140
chr2.0235215325	−1.05	6.08×10^{-4}	−1.08	6.49×10^{-4}	
chr3.0053784559	0.91	1.58×10^{-2}	1.14	8.96×10^{-4}	CACNA1D
chr3.0060919598	−0.85	1.89×10^{-3}	−0.86	3.09×10^{-2}	FHIT
chr3.0119863345	1.28	2.04×10^{-3}	1.48	1.19×10^{-5}	GPR156
chr3.0119990864	−1.14	2.05×10^{-3}	−1.01	4.01×10^{-2}	GPR156
chr3.0127606140	−1.40	1.26×10^{-4}	−1.12	6.36×10^{-3}	
chr3.0182124231	−1.14	3.82×10^{-2}	−1.15	5.10×10^{-3}	
chr3.0189791239	−1.34	1.56×10^{-2}	−1.13	5.00×10^{-2}	LEPREL1
chr3.0196595774	−1.37	3.54×10^{-2}	−1.33	1.44×10^{-2}	SENP5
chr4.0101719592	−1.04	4.74×10^{-5}	−1.15	1.77×10^{-7}	EMCN
chr5.0011534641	0.90	2.48×10^{-2}	1.05	7.82×10^{-3}	CTNND2
chr5.0039219698	1.22	3.91×10^{-2}	1.37	2.24×10^{-2}	FYB
chr5.0164483805	−0.92	2.51×10^{-2}	−1.11	1.05×10^{-3}	CTC—340A15.2

Table 2. *Cont.*

Position	MB LogFC	MB FDR Corrected *p*-Value	MT LogFC	MT FDR Corrected *p*-Value	Gene
chr5.0166472226	−1.03	1.03×10^{-2}	−1.16	4.86×10^{-4}	
chr6.0008948266	1.16	2.27×10^{-2}	1.24	2.52×10^{-3}	
chr6.0016145414	−1.17	2.94×10^{-4}	−1.24	9.54×10^{-5}	MYLIP
chr6.0019413218	0.81	2.95×10^{-3}	0.86	1.70×10^{-2}	
chr6.0031008851	0.96	2.29×10^{-2}	1.01	1.56×10^{-2}	RASSF3
chr6.0154640863	1.37	9.68×10^{-3}	1.34	3.64×10^{-3}	IPCEF1
chr6.0161063597	−2.29	3.71×10^{-3}	−1.62	2.31×10^{-2}	LPA
chr7.0016768868	−0.75	1.18×10^{-2}	−1.03	1.03×10^{-5}	
chr7.0044621160	0.91	1.71×10^{-2}	0.97	4.26×10^{-2}	TMED4
chr7.0147581299	−0.68	3.30×10^{-2}	−0.83	3.24×10^{-2}	CNTNAP2
chr11.0123045794	−1.36	1.13×10^{-3}	−1.54	2.08×10^{-6}	CLMP
chr11.0129565594	1.28	1.59×10^{-2}	1.68	1.23×10^{-4}	
chr12.0003241735	1.19	4.26×10^{-3}	1.04	3.25×10^{-2}	TSPAN9
chr12.0026672531	1.00	4.67×10^{-2}	1.28	1.68×10^{-2}	ITPR2
chr12.0048360477	−1.69	3.13×10^{-4}	−1.10	4.25×10^{-2}	TMEM106C
chr12.0054366343	0.87	4.94×10^{-2}	1.07	3.53×10^{-2}	HOTAIR
chr12.0055783991	1.19	4.36×10^{-2}	1.29	2.11×10^{-2}	
chr12.0083436417	1.62	4.94×10^{-3}	2.14	6.72×10^{-5}	TMTC2
chr12.0114887843	1.42	1.84×10^{-4}	0.62	4.91×10^{-2}	
chr12.0116068191	−1.32	1.42×10^{-2}	−1.58	5.07×10^{-6}	RP11—1028N23.4
chr12.0128167651	1.13	3.28×10^{-2}	1.57	8.87×1014	
chr12.0131689822	1.29	2.19×10^{-2}	1.74	5.09×10^{-5}	RP11—638F5.1
chr13.0021286449	−1.10	4.87×10^{-2}	−1.33	2.22×10^{-2}	IL17D
chr13.0027424109	1.51	1.23×10^{-2}	1.56	5.47×10^{-4}	
chr13.0033220266	−1.24	5.87×10^{-3}	−1.31	2.18×10^{-3}	PDS5B
chr13.0047191668	−1.30	6.51×10^{-3}	−1.34	6.36×10^{-3}	LRCH1
chr13.0093896533	1.50	2.96×10^{-2}	2.41	2.86×10^{-5}	GPC6
chr13.0099687193	1.01	3.54×10^{-2}	1.06	4.85×10^{-2}	DOCK9
chr13.0107176083	−1.69	7.74×10^{-3}	−1.68	1.29×10^{-2}	EFNB2
chr13.0109856377	−1.48	4.00×10^{-2}	−1.52	6.19×10^{-3}	MYO16
chr14.0021177142	−1.22	3.13×10^{-4}	−1.24	7.38×10^{-5}	
chr14.0021316565	−1.29	2.23×10^{-2}	−1.56	2.05×10^{-4}	
chr14.0025947530	0.91	1.48×10^{-2}	1.05	7.13×10^{-3}	
chr14.0080449863	−1.84	5.11×10^{-5}	−1.74	2.05×10^{-4}	
chr14.0085404000	−1.38	4.50×10^{-3}	−1.38	3.07×10^{-3}	
chr14.0104190006	−1.48	3.71×10^{-3}	−1.75	2.66×10^{-4}	ZFYVE21
chr15.0046178808	−0.97	1.71×10^{-2}	−0.70	4.78×10^{-2}	RP11—718O11.1
chr15.0069824154	1.42	1.25×10^{-4}	1.55	4.82×10^{-6}	RP11—279F6.1

Table 2. *Cont.*

Position	MB LogFC	MB FDR Corrected *p*-Value	MT LogFC	MT FDR Corrected *p*-Value	Gene
chr15.0092982723	1.45	3.91×10^{-5}	1.73	9.39×10^{-9}	ST8SIA2
chr16.0004815786	−0.89	8.94×10^{-4}	−0.92	9.24×10^{-3}	ZNF500
chr16.0077912976	−1.23	8.49×10^{-3}	−1.28	2.22×10^{-3}	VAT1L
chr16.0079468883	1.11	4.82×10^{-3}	1.28	3.78×10^{-5}	
chr17.0018941025	−1.85	1.91×10^{-3}	−1.96	2.09×10^{-4}	GRAP
chr17.0019045779	−1.55	7.74×10^{-3}	−1.56	8.05×10^{-4}	GRAPL, CTC—457L16.2
chr17.0028803808	−1.20	1.47×10^{-2}	−1.24	4.72×10^{-3}	
chr17.0070499160	1.04	1.42×10^{-4}	1.06	1.48×10^{-3}	LINC00511
chr17.0074566299	0.87	3.28×10^{-2}	1.14	3.76×10^{-4}	ST6GALNAC2, RP11—666A8.9
chr18.0043923940	−1.66	5.63×10^{-6}	−1.73	7.53×10^{-7}	RNF165
chr18.0045011716	1.10	2.73×10^{-2}	1.41	4.57×10^{-5}	CTD—2130O13.1
chr18.0047177650	−1.15	3.29×10^{-4}	−0.64	8.52×10^{-3}	
chr18.0047230566	−1.39	1.98×10^{-2}	−1.30	3.48×10^{-3}	
chr18.0072250823	1.12	7.11×10^{-3}	1.05	2.58×10^{-2}	CNDP1
chr19.0002867898	1.36	5.19×10^{-9}	1.12	1.72×10^{-2}	ZNF556
chr19.0041126191	−0.80	1.99×10^{-2}	−0.92	4.84×10^{-3}	LTBP4
chr20.0031210733	1.20	2.29×10^{-2}	1.36	2.68×10^{-2}	
chr20.0052825772	−1.35	1.91×10^{-3}	−1.31	3.11×10^{-3}	PFDN4
chr20.0055369320	−1.24	9.46×10^{-4}	−1.68	1.37×10^{-4}	
chr20.0060501154	2.02	6.08×10^{-4}	1.87	9.34×10^{-4}	CDH4
chr21.0030689317	−0.78	4.49×10^{-3}	−0.86	1.16×10^{-3}	BACH1
chr22.0050332646	−1.23	5.83×10^{-3}	−1.35	1.32×10^{-3}	

* LogFC = $\log_2$ fold change; positive logFC = hypermethylated in CP; negative logFC = hypomethylated in CP.

Since DNA methylation in promoter regions of genes has been associated with regulation of expression, the individual CpG methylation scores were averaged across TSS flanking regions using 1000 bp upstream of the TSS and 1000 bp downstream. The promoter methylation data revealed distinct patterns between the control and CP cohorts for both MBs (Table 4) and MTs (Table 5). Of 31,844 unique promoters identified, there were 3 promoters with statistically different methylation loads between CP and control subjects in the MBs and 10 promoters in MTs (FDR < 0.05). The majority of the differentially methylated promoters were in non-coding genes, with only one of the MB promoters and two of the MT promoters associated with protein coding transcripts. To explore the relationship between methylation loads and RNA expression, the correlation of methylation load in protein coding genes was analyzed against previously published RNA-seq count data per gene, but no correlation was found (data not shown).

Table 3. Chromosome enrichment analysis *.

Chromosome	MB		MT		Muscle		Blood	
	Significant CpGs	Enrichment *p*-Value	Significant CpGs	Enrichment *p*-Value	Significant CpGs	Enrichment *p*-Value	Significant CpGs	Enrichment *p*-Value
1	0	1.000	1	1.000	1	1.000	10	1.000
2	26	0.997	103	1.000	84	0.361	312	1.000
3	45	4.81×10^{-3}	112	0.146	77	9.94×10^{-3}	650	2.20×10^{-16}
4	3	1.000	9	1.000	7	1.000	21	1.000
5	14	0.998	82	0.863	64	0.082	222	1.000
6	29	0.348	98	0.209	109	1.22×10^{-12}	544	2.20×10^{-16}
7	8	1.000	23	1.000	23	1.000	136	1.000
8	9	1.000	26	1.000	28	0.999	169	1.000
9	18	0.921	54	1.000	22	1.000	208	1.000
10	13	0.999	35	1.000	4	1.000	76	1.000
11	33	0.103	101	0.092	58	0.196	618	2.20×10^{-16}
12	76	2.20×10^{-16}	204	2.20×10^{-16}	67	7.43×10^{-3}	519	2.20×10^{-16}
13	21	3.73×10^{-2}	104	7.37×10^{-14}	61	9.60×10^{-9}	344	2.20×10^{-16}
14	41	3.31×10^{-7}	137	2.20×10^{-16}	68	6.01×10^{-8}	448	2.20×10^{-16}
15	31	1.63×10^{-3}	130	2.20×10^{-16}	72	6.96×10^{-9}	298	3.44×10^{-9}
16	20	0.835	93	0.111	68	3.04×10^{-3}	369	1.37×10^{-5}
17	26	0.564	86	0.659	56	0.329	300	0.912
18	35	4.65×10^{-8}	107	2.20×10^{-16}	47	1.42×10^{-5}	341	2.20×10^{-16}
19	9	1.000	47	1.000	25	1.000	170	1.000
20	47	4.59×10^{-11}	153	2.20×10^{-16}	64	4.93×10^{-8}	484	2.20×10^{-16}
21	2	0.995	6	1.000	2	1.000	9	1.000
22	19	0.101	63	1.09×10^{-2}	31	0.257	293	2.20×10^{-16}
Total	525		1774		1038		6541	

* Fisher's exact test was used to determine chromosomes that contained an enrichment of differentially methylated CpGs. Significant *p*-values ($p < 0.05$) are indicated in **green**. Both cell conditions contained an enrichment of differentially methylated CpGs on the chromosomes indicated in **orange**. Significant CpG sites were enriched on the same chromosomes when analyzing muscle tissue and blood cells from subjects with CP.

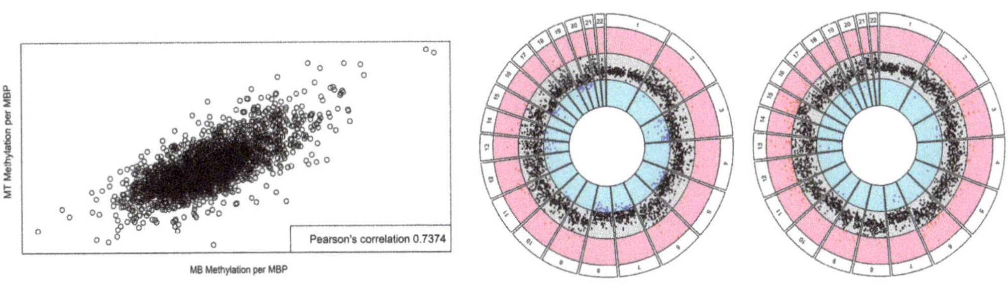

(a) (b) (c)

Figure 3. Chromosome–based circos plots. Mean logFC values for CpG methylation load (CP over control) were calculated for sequential 1 Mbp intervals. There was a strong correlation between differential methylation in MBs and MTs (**a**). Scatterplots of values for each chromosome are shown for MBs (**b**) and MTs (**c**); values outside the 95% confidence interval of the average logFCs across the whole genome are red indicating significant hypermethylation in CP or blue for significant hypomethylation in CP. The gray ring shows points within the 95% CI for the overall data.

Table 4. Significant promoter regions (+/− 1000 bp from TSS) in MBs *.

TSS	LogFC	FDR Corrected p-Value	Gene	Class
chr16:51277965	−0.85	3.82×10^{-4}	AC137527.2	Pseudogene
chr13:115039303	0.20	1.97×10^{-3}	MIR4502	miRNA
chr17:34397734	0.39	4.41×10^{-2}	CCL18	Protein coding

* TSS = transcription start site, logFC = $\log_2$ fold change; positive logFC = hypermethylated in CP; negative logFC = hypomethylated in CP.

Table 5. Significant promoter regions (+/− 1000 bp from TSS) in MTs *.

TSS	LogFC	FDR Corrected p-Value	Gene	Class
chr17:73070401	0.59	9.50×10^{-6}	AC111186.1	Pseudogene
chr17:75148756	0.36	4.01×10^{-4}	RNU4–47P	snRNA
chr19:48673949	0.60	8.84×10^{-4}	ZSWIM9	Protein coding
chr11:46134769	0.55	1.51×10^{-3}	AC024475.1	miRNA
chr4:111866955	0.30	1.81×10^{-3}	LYPLA1P2	Pseudogene
chr12:7072409	0.25	6.45×10^{-3}	U47924.27	lincRNA
chr1:242187356	−0.14	6.93×10^{-3}	RNU6–1139P	snRNA
chr12:7072408	0.25	7.37×10^{-3}	EMG1	Protein coding
chr11:93971316	1.04	1.67×10^{-2}	RP11–680H20.2	lincRNA
chr2:47454056	−0.67	4.96×10^{-2}	AC106869.2	lincRNA

* TSS = transcription start site, logFC = $\log_2$ fold change; positive logFC = hypermethylated in CP; negative logFC = hypomethylated in CP.

4. Discussion

This study found that DNA methylation patterns in skeletal muscle SCs grown in culture differed significantly between a cohort of study participants with spastic CP and non–CP controls. DNA methylation is a common and widespread chemical modification involving the addition of a methyl group to the 5–carbon position of cytosine, predominantly within CpG dinucleotides [67]. DNA methylation patterns can change during normal developmental processes, and it has been shown that altered DNA methylation can be passed to daughter cells and sustained later in life [68–75]. Specific DNA methylation changes can modify gene expression, and DNA methylation is well known to be involved in X–chromosome inactivation, gene imprinting, and the silencing of transposable elements [76]. Changes in DNA methylation patterns can also occur as a result of pathophysiologic processes or acute exposures to environmental or physiologic stress [77–79]. Altered DNA methylation has been linked to a number of risk factors and potential causes for CP including prematurity, hypoxia–ischemia, and infection [80–83]. In a prior study, we found that DNA methylation patterns in peripheral blood cells of spastic CP patients varied significantly from controls [39], raising the possibility of methylome alterations in both hematopoietic stem cell and myogenic stem/SC lineages in spastic CP. Furthermore, epigenetic patterns from adolescents were able to be used to predict diagnosis of much younger patients [39], suggesting that at least some methylation pattern differences are associated with the onset of CP and are preserved over time.

The present study examined methylation pattern differences but did not look directly at differences in cell phenotype or behavior. Studies show that individuals with CP have significantly reduced numbers of SCs and that MBs derived from these SCs have a decreased capacity to fuse and differentiate into MTs in culture [29,84]. A recent report indicated that SC–derived MB progenitors from contractured muscle in CP have globally hypermethylated DNA and gene expression patterns that favor proliferation over quiescence and differentiation; in that study, a 24 h treatment with a hypomethylating agent reduced DNA methylation to control levels and promoted an exit from mitosis [32]. While previous studies of DNA methylation in CP SC–derived MBs have either examined DNA methylation of a specific CpG island [28] or used Infinium Human MethylationEPIC Beadchip arrays of 850,000 targeted CpGs to identify hypermethylated regions [32], the current study used a different sequencing technology and computational pipeline to examine over 1.4 million CpGs distributed throughout the genome in both MBs and MTs. In addition, the current study was able to take advantage of more closely age–matched samples than had been possible previously. This combination of more closely matched controls and a broader technology platform allowed for identification of both hypermethylated and hypomethylated regions as well as individual CpGs, which provide higher likelihood candidates for biomarker platforms.

The identification of differentially methylated CpG sites and regions in common across cells from the CP cohort suggests that fundamental molecular alterations associated with diagnosis were sustained after the cells were removed from the structural, biomechanical, and humoral environments of the muscle tissue. Such an effect has been described as muscle epigenetic memory wherein DNA methylation is stably altered by prior events like biomechanical loading or acute early life exposure to inflammatory cytokines [85]. Here, we identified 525 CpGs in MBs and 1774 in MTs that were differentially methylated in spastic CP versus controls. Interestingly, MBs demonstrated similar proliferation rates and RNA–seq profiles between cohorts in a previous study [26], indicating that while these 525 CpGs may be biomarkers for CP, they may not be associated with functional changes within the cells. Further work is needed to elucidate fully specific linkages between DNA methylation and the regulation of protein levels and cell activities. The larger number of significant sites in MTs was consistent with a larger number of differentially expressed genes in RNA–seq and may indicate that a mixture of cells at different stages of myogenic differentiation was present at the time of DNA isolation. In addition, reports indicate that native SC populations actually comprise multiple different subpopulations that may differentially contribute to variability; studies focused on clonal cells rather than heterogenous populations may be needed. Interestingly, in both MBs and MTs, several significant CpG sites were within genes known to be involved in muscle physiology, including skeletal muscle differentiation (HLF, NOTCH1), muscle organ development (BMP2, COL6A3, DCN, FZD2, HEG1, HLF, ITGA11, ITGA7, LAMA5, LARGE1, MAPK14, MYBPC1, MYH6, MYLK, NOTCH1, NRG1, PKP2, SGCD, SMAD7, TBX1, TCF12, TEAD4, WNT5A, ZFHX3, ZNF609), and muscle system process (ACTN3, ATP8A2, CACNA1C, CACNA1D, DTNA, DYSF, EDN3, GNAO1, HCN4, ITGB5, KCNQ1, LTB4R, MYBPC1, MYH6, MYLK, NEDD4L, PDLIM5, PKP2, PLA2G6, ROCK1, SGCD, TRDN). Of the differentially methylated CpGs identified in MBs and MTs, 79 were in common, with 36 being significantly hypermethylated and 43 significantly hypomethylated in the CP cohort under both conditions (Table 2). Alterations in DNA methylation can be sustained long–term and previous studies indicated that the majority of the DNA methylome remained relatively preserved through myogenesis, from SC to MT formation [76,86]. These 79 sites may therefore represent stable methylation signals indicative of CP; however, more studies are needed to determine the implication(s) of these differentially methylated sites and their roles in muscle impairment in CP.

Chromosome enrichment analysis determined that there was an enrichment of significant CpGs on chromosomes 12, 13, 14, 15, 18, and 20 for both MBs and MTs (Table 3). Interestingly, significant CpGs were also enriched on the same chromosomes in the CP cohort in skeletal muscle tissue and blood cells. An analysis of differential methylation over 1 Mpb regions in MBs demonstrated that chromosomes 6, 9, 11, and 21 contained regions of hypermethylation in the CP cohort, while chromosomes 5, 7, 8, 16, 19, and 20 contained regions of hypomethylation, and chromosomes 2, 3, 4, 10, 12, 13, 14, 15, and 18 contained regions of both hypermethylation and hypomethylation. MTs contained more hypermethylated regions and fewer hypomethylated regions in the CP cohort than MBs.

Chromosomes 4, 8, 9, 11, 12, 14, 16, 21, and 22 contained regions of hypermethylation, while chromosome 7 contained regions of hypomethylation, and chromosomes 2, 3, 5, 6, 10, 13, 15, 18, and 20 contained regions of both hypermethylation and hypomethylation (Figure 3). Overall, the differential methylation levels over 1 Mbp regions were well correlated between MBs and MTs, again suggesting stability of the methylome during the course of myogenesis. The large regions of differential methylation between the CP and control cohorts throughout the genome suggest differences in chromatin structure within the CP cohort as various chromatin states based on histone modifications and nucleosome positioning can determine DNA methylation patterning [67]. There are complex mechanisms underlying the molecular crosstalk between DNA and histone methylation [87], and additional studies are needed to investigate specific histone modifications in CP to understand this complex relationship. While we are in the early stages of unraveling how these alterations are relevant to CP, several key chromosomes were identified as potential targets for future investigation in spastic CP.

We also identified statistically significant differences in methylation of the promoter regions of genes and assessed the relationship between these differences and an RNA–seq study of the same samples [26] to investigate the effect of differential DNA methylation on gene expression. Of particular note, specific associations between protein coding DNA methylation patterns and RNA expression were not readily resolved in our study. However, methylation/expression relationships are difficult to resolve in general and several studies have demonstrated that relationships between methylation status and gene expression can be complex [88,89]. Additionally, there is no current gold standard method for rolling up methylation scores across groups of individual CpG sites into a relevant burden for individual genes. Furthermore, approximately 95% of CpG island promoter regions are unmethylated independent of gene activity and recent studies suggest that methylation of promoter CpG islands is not the primary determinant of gene activity [52,90]. Many CpG islands occur in gene bodies, intergenic regions, or enhancers and may be relevant to gene expression [52]. In fact, recent studies have suggested that altered methylation in enhancer regions rather than promoter regions may be more indicative of changes in gene expression [52,91–93]. Enhancers can regulate the transcription of one or more genes, regardless of orientation or relative distance to the target promoter [94]. While 21 differentially methylated CpGs were identified within annotated enhancer regions in MBs and 97 in MTs, enhancers are difficult to map experimentally [52], enhancer activity is context and stimulus–dependent, and there are no genome–wide enhancer sets linked to specific promoters [94]. It was not possible to investigate the effects on gene expression due to these limitations in the annotation of enhancer elements. Therefore, it will be essential to continue investigating genome–wide analysis approaches that can accurately associate high–throughput expression data with methylation signatures.

Of note, the majority of the differentially methylated promotors identified in MBs and MTs were for regulatory RNAs; a result that may be indicative of differences in RNA processing within cells isolated from CP tissue (Tables 4 and 5). These regulatory RNAs comprise the majority of the transcriptome and play critical roles in maintaining gene expression regulation. The most well–known regulatory RNAs include micro RNAs and long non-coding RNAs. MicroRNAs (miRNAs) are small noncoding RNAs (~22 nucleotides in length) that play important roles in developmental processes such as myogenesis and neurogenesis [95]. Long non-coding RNAs (lncRNAs) may mediate chromatin remodeling and modification, interact with transcription factors for gene regulation, interact with mRNAs to regulate post–transcriptional processes [96], and interact with miRNAs to facilitate myogenesis [97]. Additional studies are warranted to investigate the complex interplay between DNA methylation, histone modifiers, and non-coding RNAs in order to provide a comprehensive understanding of these epigenetic modulations on SC physiology and myogenesis in spastic CP.

The findings of our study compellingly support the idea that spastic CP is associated with altered epigenetic pathways, but our studies are limited by our reliance on the ability to obtain biopsies from individuals presenting for surgery. Because of this, our number of samples is small, SCs must be derived from different muscles in order to age–match samples, and our CP population presenting for surgery consists of mostly severely affected individuals with a high level of motor impairment and inactivity (GMFCS V); therefore, extrapolation of our results to a larger CP community may require additional research. Additionally, the technology used interrogates 2.29×10^6 CCGG motifs, which represent ~15% of the 14×10^6 CpG sites in the genome. The majority of the CpG sites in our study do not overlap with those selected for inclusion in the Infinium MethylationEPIC technology making comparisons to other studies using Infinium data challenging.

5. Conclusions

In this report, an innovative DNA methylation analysis was employed on SC–derived MBs and MTs collected from individuals with and without CP. We identified differential methylation in the CP cohort at the levels of individual CpGs, 1 Mpb regions, and promoters. The work presented here leverages our novel methylation approach with ex vivo cell studies to elucidate aberrant methylation signatures.

Supplementary Materials: The following supporting information can be downloaded at https://www.mdpi.com/article/10.3390/jpm12121978/s1, Figure S1: Cell morphology, Figure S2: Scree plots for explained variance, Figure S3: Volcano plots, Figure S4: Manhattan plots, Table S1: Cell culture information, Table S2: Significant CpGs in MBs, Table S3: Significant CpGs in MTs.

Author Contributions: Conceptualization, K.G.R., A.G.M., E.L.C., M.W.S., R.E.A.; methodology, A.G.M., S.K.L., K.G.R., R.E.A.; software, K.G.R., A.G.M., J.H.; validation, J.H.; formal analysis, K.G.R., A.G.M., J.H.; investigation, K.G.R., S.K.L.; resources, M.W.S.; data curation, K.G.R., A.G.M., S.K.L., J.H.; writing—original draft preparation, K.G.R.; writing—review and editing, K.G.R., A.G.M., S.K.L., J.H., B.R., M.B., E.L.C., M.W.S., R.E.A.; visualization, K.G.R., A.G.M., S.K.L., J.H., B.R.; supervision, E.L.C., M.B., R.E.A.; project administration, R.E.A.; funding acquisition, R.E.A. All authors have read and agreed to the published version of the manuscript.

Funding: This work was supported by the Delaware Bioscience Center for Advanced Technology; an American Academy for Cerebral Palsy and Developmental Medicine Pedal with Pete Foundation award to R.E.A.; the Delaware CTR ACCEL Program [U54-GM104941] to E.L.C.; and US National Science Foundation awards [0944557, 1316055] to A.G.M. The authors would like to thank the Nemours Foundation, Nemours Biomedical Research, and the Department of Pediatrics for institutional support of E.L.C., K.G.R., S.K.L. and R.E.A., as well as The Swank Foundation for their support to R.E.A. allowing the establishment of the neuro–orthopedic tissue repository at Nemours that provided samples for the work.

Institutional Review Board Statement: The study was conduction according to the guidelines of the Declaration of Helsinki, and approved by the Institutional Review Board of Nemours (protocol number 687629).

Informed Consent Statement: Informed consent was obtained from all subjects involved in the study.

Data Availability Statement: Methylomic data will be made publicly accessible via the controlled access system provided by the NIH database of Genotypes and Phenotypes (dbGaP) or can be made available upon a direct request to the corresponding author.

Conflicts of Interest: The authors declare no conflict of interest.

References

1. Christensen, D.; Braun, K.V.N.; Doernberg, N.S.; Maenner, M.J.; Arneson, C.L.; Durkin, M.S.; Benedict, R.E.; Kirby, R.S.; Wingate, M.S.; Fitzgerald, R.; et al. Prevalence of cerebral palsy, co–occurring autism spectrum disorders, and motor functioning—Autism and Developmental Disabilities Monitoring Network, USA, 2008. *Dev. Med. Child Neurol.* **2013**, *56*, 59–65. [CrossRef] [PubMed]
2. Graham, H.K.; Rosenbaum, P.; Paneth, N.; Dan, B.; Lin, J.P.; Damiano, D.L.; Becher, J.G.; Gaebler–Spira, D.; Colver, A.; Lieber, R.L.; et al. Cerebral palsy. *Nat. Rev. Dis. Prim.* **2016**, *2*, 15082. [CrossRef] [PubMed]
3. Mandaleson, A.; Lee, Y.; Kerr, C.; Graham, H.K. Classifying cerebral palsy: Are we nearly there? *J. Pediatr. Orthop.* **2015**, *35*, 162–166. [CrossRef] [PubMed]

4. Oskoui, M.; Coutinho, F.; Dykeman, J.; Jetté, N.; Pringsheim, T. An update on the prevalence of cerebral palsy: A systematic review and meta–analysis. *Dev. Med. Child Neurol.* **2013**, *55*, 509–519. [CrossRef]
5. NICHD. What Are the Types of Cerebral Palsy? Available online: https://www.nichd.nih.gov/health/topics/cerebral--palsy/conditioninfo/types (accessed on 14 April 2022).
6. CDC. Data and Statistics for Cerebral Palsy. Available online: https://www.cdc.gov/ncbddd/cp/data.html (accessed on 14 April 2022).
7. Accardo, P. *Neurodevelopmental Disabilities in Infancy and Childhood*, 3rd ed.; Volume I: Neurodevelopmental Diagnosis and Treatment; Paul H. Brookes Publishing Co.: Baltimore, MD, USA, 2007.
8. Lieber, R.L.; Friden, J. Spasticity causes a fundamental rearrangement of muscle–joint interaction. *Muscle Nerve* **2002**, *25*, 265–270. [CrossRef]
9. Maenner, M.J.; Blumberg, S.J.; Kogan, M.D.; Christensen, D.; Yeargin–Allsopp, M.; Schieve, L.A. Prevalence of cerebral palsy and intellectual disability among children identified in two U.S. National Surveys, 2011–2013. *Ann. Epidemiol.* **2016**, *26*, 222–226. [CrossRef]
10. Mockford, M.; Caulton, J.M. The Pathophysiological Basis of Weakness in Children With Cerebral Palsy. *Pediatr. Phys. Ther.* **2010**, *22*, 222–233. [CrossRef]
11. Sankar, C.; Mundkur, N. Cerebral palsy–definition, classification, etiology and early diagnosis. *Indian J. Pediatr.* **2005**, *72*, 865–868. [CrossRef]
12. Sharan, D. *Cerebral Palsy—Challenges for the Future*; InTech: Vienna, Austria, 2014.
13. Von Walden, F.; Gantelius, S.; Liu, C.; Borgström, H.; Björk, L.; Gremark, O.; Stål, P.; Nader, G.A.; Pontén, E. Muscle contractures in patients with cerebral palsy and acquired brain injury are associated with extracellular matrix expansion, pro–inflammatory gene expression, and reduced rRNA synthesis. *Muscle Nerve* **2018**, *58*, 277–285. [CrossRef]
14. Oberhofer, K.; Stott, N.; Mithraratne, K.; Anderson, I. Subject–specific modelling of lower limb muscles in children with cerebral palsy. *Clin. Biomech.* **2010**, *25*, 88–94. [CrossRef]
15. Bandholm, T.; Magnusson, P.; Jensen, B.R.; Sonne–Holm, S. Dorsiflexor muscle–group thickness in children with cerebral palsy: Relation to cross–sectional area. *NeuroRehabilitation* **2009**, *24*, 299–306. [CrossRef] [PubMed]
16. Fry, N.R.; Gough, M.; McNee, A.E.; Shortland, A.P. Changes in the Volume and Length of the Medial Gastrocnemius After Surgical Recession in Children With Spastic Diplegic Cerebral Palsy. *J. Pediatr. Orthop.* **2007**, *27*, 769–774. [CrossRef] [PubMed]
17. Lampe, R.; Grassl, S.; Mitternacht, J.; Gerdesmeyer, L.; Gradinger, R. MRT–measurements of muscle volumes of the lower extremities of youths with spastic hemiplegia caused by cerebral palsy. *Brain Dev.* **2006**, *28*, 500–506. [CrossRef] [PubMed]
18. Barber, L.; Hastings–Ison, T.; Baker, R.; Barrett, R.; Lichtwark, G. Medial gastrocnemius muscle volume and fascicle length in children aged 2 to 5 years with cerebral palsy. *Dev. Med. Child Neurol.* **2011**, *53*, 543–548. [CrossRef]
19. de Bruin, M.; Smeulders, M.J.; Kreulen, M. Why is joint range of motion limited in patients with cerebral palsy? *J. Hand Surg. (Eur. Vol.)* **2013**, *38*, 8–13. [CrossRef] [PubMed]
20. Elder, G.C.B.; Kirk, J.; Stewart, G.; Cook, K.; Weir, D.; Marshall, A.; Leahey, L. Contributing factors to muscle weakness in children with cerebral palsy. *Dev. Med. Child Neurol.* **2003**, *45*, 542–550. [CrossRef]
21. Hussain, A.W.; Onambélé, G.L.; Williams, A.G.; Morse, C.I. Muscle size, activation, and coactivation in adults with cerebral palsy. *Muscle Nerve* **2013**, *49*, 76–83. [CrossRef]
22. Smith, L.R.; Lee, K.S.; Ward, S.R.; Chambers, H.G.; Lieber, R.L. Hamstring contractures in children with spastic cerebral palsy result from a stiffer extracellular matrix and increased in vivo sarcomere length. *J. Physiol.* **2011**, *589 Pt 10*, 2625–2639. [CrossRef]
23. Robinson, K.G.; Mendonca, J.L.; Militar, J.L.; Theroux, M.C.; Dabney, K.W.; Shah, S.A.; Miller, F.; Akins, R.E. Disruption of Basal Lamina Components in Neuromotor Synapses of Children with Spastic Quadriplegic Cerebral Palsy. *PLoS ONE* **2013**, *8*, e70288. [CrossRef]
24. Theroux, M.C.; Akins, R.E.; Barone, C.; Boyce, B.; Miller, F.; Dabney, K.W. Neuromuscular junctions in cerebral palsy: Presence of extrajunctional acetylcholine receptors. *Anesthesiology* **2002**, *96*, 330–335. [CrossRef]
25. Theroux, M.C.; Oberman, K.G.; Lahaye, J.; Boyce, B.A.; DuHadaway, D.; Miller, F.; Akins, R.E. Dysmorphic neuromuscular junctions associated with motor ability in cerebral palsy. *Muscle Nerve* **2005**, *32*, 626–632. [CrossRef] [PubMed]
26. Robinson, K.G.; Crowgey, E.L.; Lee, S.K.; Akins, R.E. Transcriptional analysis of muscle tissue and isolated satellite cells in spastic cerebral palsy. *Dev. Med. Child Neurol.* **2021**, *63*, 1213–1220. [CrossRef] [PubMed]
27. Smith, L.R.; Pontén, E.; Hedström, Y.; Ward, S.R.; Chambers, H.G.; Subramaniam, S.; Lieber, R.L. Novel transcriptional profile in wrist muscles from cerebral palsy patients. *BMC Med. Genom.* **2009**, *2*, 44. [CrossRef] [PubMed]
28. Domenighetti, A.A.; Mathewson, M.A.; Pichika, R.; Sibley, L.A.; Zhao, L.; Chambers, H.G.; Lieber, R.L. Loss of myogenic potential and fusion capacity of muscle stem cells isolated from contractured muscle in children with cerebral palsy. *Am. J. Physiol. Physiol.* **2018**, *315*, C247–C257. [CrossRef]
29. Dayanidhi, S.; Dykstra, P.B.; Lyubasyuk, V.; McKay, B.R.; Chambers, H.G.; Lieber, R.L. Reduced satellite cell number in situ in muscular contractures from children with cerebral palsy. *J. Orthop. Res.* **2015**, *33*, 1039–1045. [CrossRef]
30. Smith, L.R.; Chambers, H.G.; Subramaniam, S.; Lieber, R.L. Transcriptional Abnormalities of Hamstring Muscle Contractures in Children with Cerebral Palsy. *PLoS ONE* **2012**, *7*, e40686. [CrossRef]

31. Corvelyn, M.; De Beukelaer, N.; Duelen, R.; Deschrevel, J.; Van Campenhout, A.; Prinsen, S.; Gayan-Ramirez, G.; Maes, K.; Weide, G.; Desloovere, K.; et al. Muscle Microbiopsy to Delineate Stem Cell Involvement in Young Patients: A Novel Approach for Children With Cerebral Palsy. *Front. Physiol.* **2020**, *11*, 945. [CrossRef]
32. Sibley, L.A.; Broda, N.; Gross, W.R.; Menezes, A.F.; Embry, R.B.; Swaroop, V.T.; Chambers, H.G.; Schipma, M.J.; Lieber, R.L.; Domenighetti, A.A. Differential DNA methylation and transcriptional signatures characterize impairment of muscle stem cells in pediatric human muscle contractures after brain injury. *FASEB J.* **2021**, *35*, e21928. [CrossRef]
33. Fahey, M.C.; Maclennan, A.H.; Kretzschmar, D.; Gecz, J.; Kruer, M.C. The genetic basis of cerebral palsy. *Dev. Med. Child Neurol.* **2017**, *59*, 462–469. [CrossRef]
34. McMichael, G.L.; Bainbridge, M.N.; Haan, E.; Corbett, M.; Gardner, A.; Thompson, S.E.; Van Bon, B.W.M.; van Eyk, C.; Broadbent, J.; Reynolds, C.A.; et al. Whole–exome sequencing points to considerable genetic heterogeneity of cerebral palsy. *Mol. Psychiatry* **2015**, *20*, 176–182. [CrossRef]
35. Kubota, N.; Yokoyama, T.; Hoshi, N.; Suyama, M. Identification of a candidate enhancer for DMRT3 involved in spastic cerebral palsy pathogenesis. *Biochem. Biophys. Res. Commun.* **2018**, *496*, 133–139. [CrossRef] [PubMed]
36. Gulati, S.; Sondhi, V. Cerebral Palsy: An Overview. *Indian J. Pediatr.* **2017**, *85*, 1006–1016. [CrossRef] [PubMed]
37. Sewell, M.D.; Eastwood, D.M.; Wimalasundera, N. Managing common symptoms of cerebral palsy in children. *BMJ* **2014**, *349*, g5474. [CrossRef] [PubMed]
38. Bahado–Singh, R.O.; Vishweswaraiah, S.; Aydas, B.; Mishra, N.K.; Guda, C.; Radhakrishna, U. Deep Learning/Artificial Intelligence and Blood–Based DNA Epigenomic Prediction of Cerebral Palsy. *Int. J. Mol. Sci.* **2019**, *20*, 2075. [CrossRef] [PubMed]
39. Crowgey, E.L.; Marsh, A.G.; Robinson, K.G.; Yeager, S.K.; Akins, R.E. Epigenetic machine learning: Utilizing DNA methylation patterns to predict spastic cerebral palsy. *BMC Bioinform.* **2018**, *19*, 225. [CrossRef]
40. Mohandas, N.; Bass-Stringer, S.; Maksimovic, J.; Crompton, K.; Loke, Y.J.; Walstab, J.; Reid, S.M.; Amor, D.J.; Reddihough, D.; Craig, J.M.; et al. Epigenome–wide analysis in newborn blood spots from monozygotic twins discordant for cerebral palsy reveals consistent regional differences in DNA methylation. *Clin. Epigenetics* **2018**, *10*, 25. [CrossRef]
41. Yuan, Y. Study of global DNA methylation in monozygotic twins with cerebral palsy. *Pak. J. Pharm. Sci.* **2017**, *30* (Suppl. S4), 1467–1473.
42. Alag, A. Machine learning approach yields epigenetic biomarkers of food allergy: A novel 13–gene signature to diagnose clinical reactivity. *PLoS ONE* **2019**, *14*, e0218253. [CrossRef]
43. Aref-Eshghi, E.; Rodenhiser, D.I.; Schenkel, L.C.; Lin, H.; Skinner, C.; Ainsworth, P.; Paré, G.; Hood, R.L.; Bulman, D.E.; Kernohan, K.D.; et al. Genomic DNA Methylation Signatures Enable Concurrent Diagnosis and Clinical Genetic Variant Classification in Neurodevelopmental Syndromes. *Am. J. Hum. Genet.* **2018**, *102*, 156–174. [CrossRef]
44. Capper, D.; Jones, D.T.W.; Sill, M.; Hovestadt, V.; Schrimpf, D.; Sturm, D.; Koelsche, C.; Sahm, F.; Chavez, L.; Reuss, D.E.; et al. DNA methylation–based classification of central nervous system tumours. *Nature* **2018**, *555*, 469–474. [CrossRef]
45. De Bellis, M.; Camerino, D.C.; Desaphy, J.-F. Toward precision medicine in myotonic syndromes. *Oncotarget* **2017**, *8*, 14279–14280. [CrossRef] [PubMed]
46. Kulis, M.; Heath, S.; Bibikova, M.; Queirós, A.C.; Navarro, A.; Clot, G.; Martínez-Trillos, A.; Castellano, G.; Brun-Heath, I.; Pinyol, M.; et al. Epigenomic analysis detects widespread gene–body DNA hypomethylation in chronic lymphocytic leukemia. *Nat. Genet.* **2012**, *44*, 1236–1242. [CrossRef] [PubMed]
47. Liu, B.; Liu, Y.; Pan, X.; Li, M.; Yang, S.; Li, S.C. DNA methylation markers for Pan–Cancer prediction by deep learning. *Genes* **2019**, *10*, 778. [CrossRef]
48. Orozco, J.I.J.; Knijnenburg, T.A.; Manughian–Peter, A.O.; Salomon, M.P.; Barkhoudarian, G.; Jalas, J.R.; Wilmott, J.S.; Hothi, P.; Wang, X.; Takasumi, Y.; et al. Epigenetic profiling for the molecular classification of metastatic brain tumors. *Nat. Commun.* **2018**, *9*, 4627. [CrossRef]
49. Queiros, A.; Villamor, N.; Clot, G.; Martineztrillos, A.; Kulis, M.; Navarro, A.; Penas, E.M.M.; Jayne, S.; Majid, A.M.S.A.; Richter, J.A.; et al. A B–cell epigenetic signature defines three biologic subgroups of chronic lymphocytic leukemia with clinical impact. *Leukemia* **2014**, *29*, 598–605. [CrossRef] [PubMed]
50. Robinson, D.R.; Wu, Y.-M.; Lonigro, R.J.; Vats, P.; Cobain, E.; Everett, J.; Cao, X.; Rabban, E.; Kumar-Sinha, C.; Raymond, V.; et al. Integrative clinical genomics of metastatic cancer. *Nature* **2017**, *548*, 297–303. [CrossRef] [PubMed]
51. Zhang, X.; Hu, Y.; Aouizerat, B.E.; Peng, G.; Marconi, V.C.; Corley, M.J.; Hulgan, T.; Bryant, K.J.; Zhao, H.; Krystal, J.H.; et al. Machine learning selected smoking–associated DNA methylation signatures that predict HIV prognosis and mortality. *Clin. Epigenetics* **2018**, *10*, 155. [CrossRef]
52. Ehrlich, M. DNA hypermethylation in disease: Mechanisms and clinical relevance. *Epigenetics* **2019**, *14*, 1141–1163. [CrossRef]
53. Bareja, A.; Holt, J.A.; Luo, G.; Chang, C.; Lin, J.; Hinken, A.C.; Freudenberg, J.; Kraus, W.E.; Evans, W.J.; Billin, A.N. Human and Mouse Skeletal Muscle Stem Cells: Convergent and Divergent Mechanisms of Myogenesis. *PLoS ONE* **2014**, *9*, e90398. [CrossRef]
54. Garcia, S.M.; Tamaki, S.; Lee, S.; Wong, A.; Jose, A.; Dreux, J.; Kouklis, G.; Sbitany, H.; Seth, R.; Knott, P.D.; et al. High–Yield Purification, Preservation, and Serial Transplantation of Human Satellite Cells. *Stem Cell Rep.* **2018**, *10*, 1160–1174. [CrossRef]
55. Marsh, A.G.; Pasqualone, A.A. DNA methylation and temperature stress in an Antarctic polychaete, Spiophanes tcherniai. *Front. Physiol.* **2014**, *5*, 173. [CrossRef] [PubMed]
56. Rambo, I.M.; Marsh, A.; Biddle, J.F. Cytosine Methylation within Marine Sediment Microbial Communities: Potential Epigenetic Adaptation to the Environment. *Front. Microbiol.* **2019**, *10*, 1291. [CrossRef] [PubMed]

57. Li, H. Toward better understanding of artifacts in variant calling from high-coverage samples. *Bioinformatics* **2014**, *30*, 2843–2851. [CrossRef] [PubMed]
58. McCarthy, D.J.; Chen, Y.; Smyth, G.K. Differential expression analysis of multifactor RNA–Seq experiments with respect to biological variation. *Nucleic Acids Res.* **2012**, *40*, 4288–4297. [CrossRef] [PubMed]
59. Robinson, M.D.; McCarthy, D.J.; Smyth, G.K. EdgeR: A Bioconductor package for differential expression analysis of digital gene expression data. *Bioinformatics* **2010**, *26*, 139–140. [CrossRef]
60. Ritchie, M.E.; Belinda, P.; Wu, D.; Hu, Y.; Law, C.W.; Shi, W.; Smyth, G.K. limma powers differential expression analyses for RNA–sequencing and microarray studies. *Nucleic Acids Res.* **2015**, *43*, e47. [CrossRef]
61. Li, R.; Yang, Y.-E.; Yin, Y.-H.; Zhang, M.-Y.; Li, H.; Qu, Y.-Q. Methylation and transcriptome analysis reveal lung adenocarcinoma–specific diagnostic biomarkers. *J. Transl. Med.* **2019**, *17*, 324. [CrossRef]
62. Maksimovic, J.; Phipson, B.; Oshlack, A. A cross–package Bioconductor workflow for analysing methylation array data. *F1000Research* **2016**, *5*, 1281. [CrossRef]
63. Jones, P.A.; Baylin, S.B. The Epigenomics of Cancer. *Cell* **2007**, *128*, 683–692. [CrossRef]
64. Contreras-Romero, C.; Pérez-Yépez, E.-A.; Martinez-Gutierrez, A.D.; Campos-Parra, A.; Zentella-Dehesa, A.; Jacobo-Herrera, N.; López-Camarillo, C.; Corredor-Alonso, G.; Martínez-Coronel, J.; Rodríguez-Dorantes, M.; et al. Gene Promoter–Methylation Signature as Biomarker to Predict Cisplatin–Radiotherapy Sensitivity in Locally Advanced Cervical Cancer. *Front. Oncol.* **2022**, *12*, 773438. [CrossRef]
65. Cunningham, F.; Allen, J.E.; Allen, J.; Alvarez-Jarreta, J.; Amode, M.R.; Armean, I.M.; Austine-Orimoloye, O.; Azov, A.G.; Barnes, I.; Flicek, P.; et al. Ensembl 2022. *Nucleic Acids Res.* **2022**, *50*, D988–D995. [CrossRef] [PubMed]
66. University of California Santa Cruz (UCSC) Genome Browser Gateway. Available online: https://genome.ucsc.edu/ (accessed on 15 November 2022). (accessed multiple times in 2021 and 2022 with final access date for validation).
67. Robertson, K.D. DNA methylation and chromatin—Unraveling the tangled web. *Oncogene* **2002**, *21*, 5361–5379. [CrossRef] [PubMed]
68. Blaze, J.; Roth, T.L. Evidence from clinical and animal model studies of the long–term and transgenerational impact of stress on DNA methylation. *Semin. Cell Dev. Biol.* **2015**, *43*, 76–84. [CrossRef] [PubMed]
69. Hartley, I.; Elkhoury, F.F.; Shin, J.H.; Xie, B.; Gu, X.; Gao, Y.; Zhou, D.; Haddad, G.G. Long–Lasting Changes in DNA Methylation Following Short–Term Hypoxic Exposure in Primary Hippocampal Neuronal Cultures. *PLoS ONE* **2013**, *8*, e77859. [CrossRef] [PubMed]
70. Houtepen, L.C.; Vinkers, C.H.; Carrillo–Roa, T.; Hiemstra, M.; Van Lier, P.A.; Meeus, W.; Branje, S.; Heim, C.; Nemeroff, C.B.; Mill, J.; et al. Genome–wide DNA methylation levels and altered cortisol stress reactivity following childhood trauma in humans. *Nat. Commun.* **2016**, *7*, 10967. [CrossRef]
71. Labonté, B.; Suderman, M.; Maussion, G.; Navaro, L.; Yerko, V.; Mahar, I.; Bureau, A.; Mechawar, N.; Szyf, M.; Meaney, M.J.; et al. Genome–wide Epigenetic Regulation by Early–Life Trauma. *Arch. Gen. Psychiatry* **2012**, *69*, 722–731. [CrossRef]
72. Pacis, A.; Tailleux, L.; Morin, A.M.; Lambourne, J.; MacIsaac, J.L.; Yotova, V.; Dumaine, A.; Danckaert, A.; Luca, F.; Grenier, J.-C.; et al. Bacterial infection remodels the DNA methylation landscape of human dendritic cells. *Genome Res.* **2015**, *25*, 1801–1811. [CrossRef]
73. Rask–Andersen, M.; Martinsson, D.; Ahsan, M.; Enroth, S.; Ek, W.E.; Gyllensten, U.; Johansson, Å. Epigenome–wide association study reveals differential DNA methylation in individuals with a history of myocardial infarction. *Hum. Mol. Genet.* **2016**, *25*, 4739–4748. [CrossRef]
74. Stirzaker, C.; Zotenko, E.; Song, J.Z.; Qu, W.; Nair, S.S.; Locke, W.; Stone, A.; Armstong, N.J.; Robinson, M.D.; Dobrovic, A.; et al. Methylome sequencing in triple–negative breast cancer reveals distinct methylation clusters with prognostic value. *Nat. Commun.* **2015**, *6*, 5899. [CrossRef]
75. Xiao, X.; Zhao, Y.; Jin, R.; Chen, J.; Wang, X.; Baccarelli, A.; Zhang, Y. Fetal growth restriction and methylation of growth–related genes in the placenta. *Epigenomics* **2016**, *8*, 33–42. [CrossRef]
76. Carrió, E.; Lois, S.; Mallona, I.; Cases, I.; Forn, M.; Peinado, M.A.; Suelves, M.; Díez–Villanueva, A. Deconstruction of DNA Methylation Patterns During Myogenesis Reveals Specific Epigenetic Events in the Establishment of the Skeletal Muscle Lineage. *Stem Cells* **2015**, *33*, 2025–2036. [CrossRef] [PubMed]
77. Martin, E.M.; Fry, R.C. Environmental Influences on the Epigenome: Exposure– Associated DNA Methylation in Human Populations. *Annu. Rev. Public Health* **2018**, *39*, 309–333. [CrossRef] [PubMed]
78. Sosnowski, D.W.; Booth, C.; York, T.P.; Amstadter, A.B.; Kliewer, W. Maternal prenatal stress and infant DNA methylation: A systematic review. *Dev. Psychobiol.* **2018**, *60*, 127–139. [CrossRef] [PubMed]
79. Watamura, S.E.; Roth, T.L. Looking back and moving forward: Evaluating and advancing translation from animal models to human studies of early life stress and DNA methylation. *Dev. Psychobiol.* **2018**, *61*, 323–340. [CrossRef] [PubMed]
80. Bustelo, M.; Barkhuizen, M.; Hove, D.L.A.V.D.; Steinbusch, H.W.M.; Bruno, M.A.; Loidl, C.F.; Gavilanes, A.W.D. Clinical Implications of Epigenetic Dysregulation in Perinatal Hypoxic–Ischemic Brain Damage. *Front. Neurol.* **2020**, *11*, 483. [CrossRef]
81. Menon, R.; Conneely, K.N.; Smith, A.K. DNA Methylation: An Epigenetic Risk Factor in Preterm Birth. *Reprod. Sci.* **2012**, *19*, 6–13. [CrossRef]
82. Richetto, J.; Massart, R.; Weber–Stadlbauer, U.; Szyf, M.; Riva, M.A.; Meyer, U. Genome–wide DNA Methylation Changes in a Mouse Model of Infection–Mediated Neurodevelopmental Disorders. *Biol. Psychiatry* **2016**, *81*, 265–276. [CrossRef]

83. Vaiserman, A.M. Epigenetic programming by early–life stress: Evidence from human populations. *Dev. Dyn.* **2014**, *244*, 254–265. [CrossRef]
84. Smith, L.; Chambers, H.G.; Lieber, R.L. Reduced satellite cell population may lead to contractures in children with cerebral palsy. *Dev. Med. Child Neurol.* **2012**, *55*, 264–270. [CrossRef]
85. Sharples, A.P.; Polydorou, I.; Hughes, D.; Owens, D.J.; Hughes, T.M.; Stewart, C. Skeletal muscle cells possess a 'memory' of acute early life TNF–α exposure: Role of epigenetic adaptation. *Biogerontology* **2015**, *17*, 603–617. [CrossRef]
86. Tsumagari, K.; Baribault, C.; Terragni, J.; Varley, K.; Gertz, J.; Pradhan, S.; Badoo, M.; Crain, C.M.; Song, L.; Crawford, G.E.; et al. Early de novo DNA methylation and prolonged demethylation in the muscle lineage. *Epigenetics* **2013**, *8*, 317–332. [CrossRef] [PubMed]
87. Li, Y.; Chen, X.; Lu, C. The interplay between DNA and histone methylation: Molecular mechanisms and disease implications. *EMBO Rep.* **2021**, *22*, e51803. [CrossRef] [PubMed]
88. Lea, A.J.; Vockley, C.M.; Johnston, R.A.; Del Carpio, C.A.; Barreiro, L.; Reddy, T.E.; Tung, J. Genome–wide quantification of the effects of DNA methylation on human gene regulation. *eLife* **2018**, *7*, e37513. [CrossRef] [PubMed]
89. Wagner, J.R.; Busche, S.; Ge, B.; Kwan, T.; Pastinen, T.; Blanchette, M. The relationship between DNA methylation, genetic and expression inter–individual variation in untransformed human fibroblasts. *Genome Biol.* **2014**, *15*, R37. [CrossRef] [PubMed]
90. Angeloni, A.; Bogdanovic, O. Enhancer DNA methylation: Implications for gene regulation. *Essays Biochem.* **2019**, *63*, 707–715. [CrossRef]
91. Bae, M.G.; Kim, J.Y.; Choi, J.K. Frequent hypermethylation of orphan CpG islands with enhancer activity in cancer. *BMC Med. Genom.* **2016**, *9* (Suppl. S1), 38. [CrossRef]
92. Glass, J.L.; Hassane, D.; Wouters, B.J.; Kunimoto, H.; Avellino, R.; Garrett–Bakelman, F.E.; Guryanova, O.A.; Bowman, R.; Redlich, S.; Intlekofer, A.M.; et al. Epigenetic Identity in AML Depends on Disruption of Nonpromoter Regulatory Elements and Is Affected by Antagonistic Effects of Mutations in Epigenetic Modifiers. *Cancer Discov.* **2017**, *7*, 868–883. [CrossRef]
93. Wang, Y.; Hao, D.-P.; Li, J.-J.; Wang, L.; Di, L.-J. Genome–wide methylome and chromatin interactome identify abnormal enhancer to be risk factor of breast cancer. *Oncotarget* **2017**, *8*, 44705–44719. [CrossRef]
94. Benton, M.L.; Talipineni, S.C.; Kostka, D.; Capra, J.A. Genome–wide enhancer annotations differ significantly in genomic distribution, evolution, and function. *BMC Genom.* **2019**, *20*, 511. [CrossRef]
95. Romero, B.; Robinson, K.G.; Batish, M.; Akins, R.E. An Emerging Role for Epigenetics in Cerebral Palsy. *J. Pers. Med.* **2021**, *11*, 1187. [CrossRef]
96. Mbadhi, M.N.; Tang, J.-M.; Zhang, J.-X. Histone Lysine Methylation and Long Non-coding RNA: The New Target Players in Skeletal Muscle Cell Regeneration. *Front. Cell Dev. Biol.* **2021**, *9*, 759237. [CrossRef] [PubMed]
97. Luo, H.; Lv, W.; Tong, Q.; Jin, J.; Xu, Z.; Zuo, B. Functional Non-coding RNA During Embryonic Myogenesis and Postnatal Muscle Development and Disease. *Front. Cell Dev. Biol.* **2021**, *9*, 628339. [CrossRef] [PubMed]

Journal of
*Personalized
Medicine*

MDPI

Article

The Effect of Lumbar Belts with Different Extensibilities on Kinematic, Kinetic, and Muscle Activity of Sit-to-Stand Motions in Patients with Nonspecific Low Back Pain

Sang-Cheol Im , Seong-Wook Seo, Na-Yeon Kang, Hoon Jo and Kyoung Kim *

Department of Physical Therapy, College of Rehabilitation Science, Daegu University, Gyeongsan 38453, Korea
* Correspondence: kykim257@hanmail.net

Abstract: Although lumbar belts can be used for the treatment and prevention of low back pain, the role of the lumbar belt remains unclear without clear guidelines. This study aimed to investigate the effect of lumbar belts with different extensibilities on the kinematics, kinetics, and muscle activity of sit-to-stand motions in terms of motor control in patients with nonspecific low back pain. A total of 30 subjects participated in the study: 15 patients with nonspecific low back pain and 15 healthy adults. Participants performed the sit-to-stand motion in random order of three conditions: no lumbar belt, wearing an extensible lumbar belt, and wearing a non-extensible lumbar belt. The sit-to-stand motion's kinematic, kinetic, and muscle activity variables in each condition were measured using a three-dimensional motion analysis device, force plate, and surface electromyography. An interaction effect was found for the time taken, anterior pelvic tilt angle, and muscle activity of the vastus lateralis and biceps femoris. The two lumbar belts with different extensibilities had a positive effect on motor control in patients with nonspecific low back pain. Therefore, both types of extensible lumbar belts can be useful in the sit-to-stand motion, which is an important functional activity for patients with nonspecific low back pain.

Keywords: electromyography; kinematic; kinetic; low back pain; lumbar belt; sit-to-stand

Citation: Im, S.-C.; Seo, S.-W.; Kang, N.-Y.; Jo, H.; Kim, K. The Effect of Lumbar Belts with Different Extensibilities on Kinematic, Kinetic, and Muscle Activity of Sit-to-Stand Motions in Patients with Nonspecific Low Back Pain. *J. Pers. Med.* **2022**, *12*, 1678. https://doi.org/10.3390/jpm12101678

Academic Editor: Anne-Marie Caminade

Received: 19 September 2022
Accepted: 6 October 2022
Published: 9 October 2022

1. Introduction

Low back pain is one of the most prevalent disorders today [1], and people with low back pain move differently than healthy people do [2]. In particular, changes in motion variability may indicate an altered adaptability of motor control systems [3,4]. As reported in previous studies, a decrease in motion variability during running [5] and bending [6] activities in patients with low back pain is a protective behavior when pain is present or when pain is expected [7], and repeated tissue stress can occur over time [8]. In addition, when the motion variability increases, as observed in reaching [9] and walking [10], in patients with low back pain, errors or noises in proprioceptive sensory feedback and motor commands may occur, thereby reducing the robustness of the system [11]. Thus, low back pain can cause significant functional loss and affect the performance of dynamic tasks and static postural control [12]. The sit-to-stand motion is a common and functionally important task in daily life activities [13]. Because it is associated with worsening low back pain symptoms, studies on the sit-to-stand motion are essential in back pain-related research [14].

Although there is hope that individualized treatment will be effective for people suffering from low back pain [15], most studies claiming to be effective have been shown to be flawed [16,17]. In addition, claims of treatment efficacy in subgroups of individuals with low back pain are reliable when there is a biological basis, but this is difficult in nonspecific low back pain [18]. Because nonspecific low back pain has no known pathological cause, treatment focuses on reducing pain and consists of education, analgesics, and psychotherapy, but many patients do not require medical treatment [19]. Strategies for preventing low back pain include ergonomic control at the workplace where repetitive movements and tasks are performed [20], exercises, such as training in waist mechanics and lifting techniques [21], and the use of lumbar belts, which can also be used to treat low back pain [22].

To explain the clinical benefits of lumbar belts, psychological, neuromuscular, and biomechanical mechanisms have been proposed, but these remain unproven [23]. Psychological benefits may be related to the perception of mechanical support generated by the lumbar belt, and neuromuscular benefits may include mechanisms that affect lumbar stability, such as lumbar proprioception, trunk muscle feedforward, and reflex activity [24,25]. Biomechanical benefits are related to the mechanical stiffness of the lumbar belt, such as a reduction in lumbar range of motion, a reduction in the stress of the posterior lumbar passive tissue, and a potential reduction in the compressive load of the lumbar spine [26,27]. However, the potential role of lumbar belts in treating low back pain remains unclear [28], and low-quality studies and insufficient evidence imply that there are no clear guidelines for using lumbar belts in patients with low back pain [29]. For this reason, it is necessary to study the effect of lumbar belts on the biomechanics and motor control mechanism of patients with low back pain.

Most previous studies on lumbar belt have analyzed proprioception or balance, and there is an insufficient number of studies on the effect of wearing a lumbar belt while the patient with low back pain performs a functional task. In particular, the effect of sit-to-stand motion, an important functional activity for modern people who spend a lot of time sitting and for patients with low back pain, has not been studied so far. Therefore, this study sought to investigate the effects of wearing two lumbar belts with different extensibilities on the kinematics, kinetics, and muscle activity in terms of motor control of the sit-to-stand motion in patients with nonspecific low back pain and healthy adults. From this research, we intend to present the evidence for the effective use of lumbar belts for patients with nonspecific low back pain and provide basic data necessary for the prevention and treatment of low back pain. For the purpose of this study, we hypothesized that the effects of the sit-to-stand motion on kinematics, kinetics, and muscle activity would differ depending on the presence or absence of back pain and the extensibility of the lumbar belt.

2. Methods

2.1. Participants

Participants in this study were recruited from outpatients and caregivers between the ages of 20 and 50 years who visited a spine specialty hospital located in Daegu, Korea, from April 2020 to May 2020. Of the 30 participants, 15 patients with nonspecific low back pain were assigned to the experimental group, and 15 normal adults were assigned to the control group. Participants in the control group were matched to those in the experimental group according to age, sex, and body mass index (BMI). We used G*Power software program (version 3.1.9.4, Franz Faul, Düsseldorf, Germany) to determine the appropriate number of subjects, and we referred to a previous study with the same design (two groups × three conditions) as this study [30]. We used the effect size of the lumbar total range of motion variable of the manual material-handling task in the previous study. Based on an effect size of 0.31, a significance level of 0.05, and a power of 80%, we calculated that a total of 30 subjects were required [31].

The selection criterion for the experimental group was lumbar or lumbar spine pain for at least 4 weeks without radiating pain below the knee (nonacute phase), and the exclusion criteria were spinal surgery, specific lumbar pathology, scoliosis, systemic or degenerative disease, BMI > 30 kg/m^2, hypertension, neurological history unrelated to back pain, and medications that could affect neuronal excitability [30]. The exclusion criterion for the control group was a history of low back pain lasting more than 1 week during the past year [32]. Additional exclusion criteria for both groups were a neurological or respiratory disease, trauma, or pregnancy that could affect the sit-to-stand motion [33]. The orthopedic surgeon verified the qualifications of these participants through medical records and image evaluations and interviews.

After explaining the purpose of the study, the contents of the experiment, and the degree of exposure during the experiment to all participants in advance, written consent for voluntary participation was obtained. All participants' sex, age, and anthropometric data were collected, and pain (visual analog scale) and disability (Oswestry Disability Index) data were additionally investigated for patients with nonspecific low back pain in the experimental group. This study was approved by the Bioethics Committee of Daegu University (1040621-201911-HR-025-02) and was conducted in accordance with the 1964 Declaration of Helsinki and its later amendments. In addition, this study was registered with the Clinical Research Information Service for information sharing, ethics, and transparency of clinical research (KCT0004970).

2.2. Lumbar Belts

After consulting with an orthopedic surgeon, two models of lumbar belt were selected, taking into consideration flexibility, comfort, and economy for use in daily life and work. These two types of lumbar belts consisted of two layers of straps secured with Velcro. The initial adjustment and placement of the lumbar belt were made with the inner layer, and the final tension was adjusted with the outer layer. The extensible lumbar belt (REDIX-K210, Acetech, Seoul, Korea) was made up of elastic materials, but the nonextensible lumbar belt (REDIX-L350, Acetech) was made up of nonstretchable nylon materials only. Ready-made products in three sizes (large, medium, and small) were used based on the body type of the participants. The position of the lumbar belt to be worn was such that the lower edge did not touch the thigh when sitting but covered the anterior superior iliac spine (ASIS) of the pelvis [32]. Then, while the participant was standing still, the wearing pressure was standardized to 60 mmHg as measured from a thin force-sensing resister sensor inserted between the waist belt and the participant's right iliac crest [34].

2.3. Experimental Procedure

A comfortable environment was created so that participants did not feel uncomfortable with the temperature and surrounding environment of the laboratory. In order not to interfere with the measurement, all participants wore a sleeveless T-shirt made of lightweight material and short shorts. Before obtaining other measurements, body measurements were obtained by the same examiner to determine the participants' general characteristics.

When the electromyography (EMG) electrode placement and motion analysis reflex marker were prepared, the participants placed their bare feet at a comfortable width on two force plates installed on the floor. They sat down in height-adjustable chairs without armrests or backrests (Figure 1). The chair's height was adjusted to be 90% of the length between the knee and the floor, which is the distance between the fibular head and the floor. Participants looked forward and sat in their preferred comfortable position, with their arms folded in front of their chest to avoid interference from shaking and occlusion of markers [35].

Following verbal cues from a research assistant operating a computer connected to the data acquisition system, the participants stood up at a rate of their choosing, held a comfortable upright position for 3 s, and then sat back down on their chairs without further instructions. To help the participants understand the experimental procedure, measurements were taken after two to three practice sessions following the researcher's demonstration. Participants performed the sit-to-stand motion in the most comfortable way of their choosing, with the only limitation under which they did not change their feet positions. Paper tape was used to mark the initial position of the foot, and in the subsequent conditions, the sit-to-stand motion was performed with the same foot position.

Three experimental conditions were set: no lumbar belt-wearing condition (condition 1), extensible lumbar belt-wearing condition (condition 2), and nonextensible lumbar belt-wearing condition (condition 3). To exclude a learning effect, the experimental conditions were conducted in a random order. A card with the experimental conditions was placed in a sealed envelope, and the experiment was conducted in the order in which each participant drew the card. The participants performed the sit-to-stand motion under each condition, and the biomechanical variables were then measured (Figure 2). To increase reliability, three successful measurements were taken for each condition, and the average value was used for analysis. All participants performed the sit-to-stand motion a total of nine times. The test was considered successful if all markers were visible and the EMG signals of all muscles were recorded correctly. There was a one-minute rest period between each test and a five-minute rest period between the lumbar belt conditions.

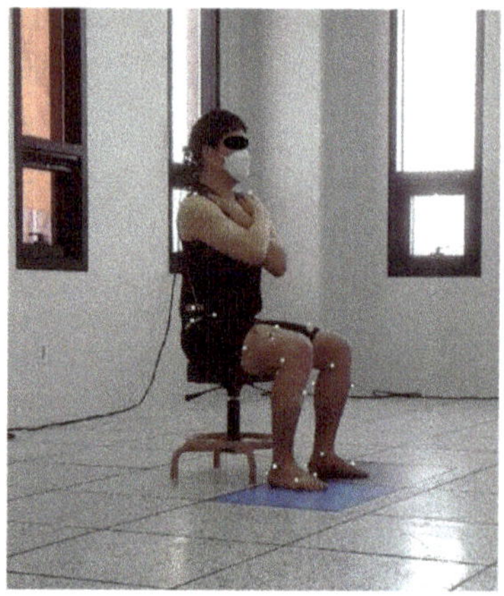

Figure 1. Experimental preparation posture.

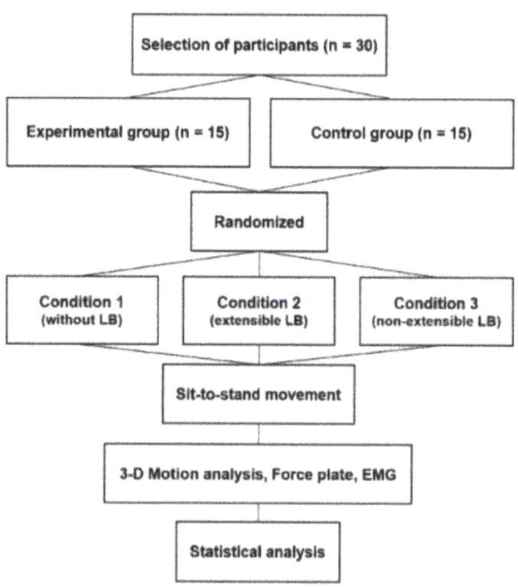

Figure 2. Study flowchart.

2.4. Measurement Method

2.4.1. Measurement of Kinematic and Kinetic Parameters

In this study, we used a three-dimensional (3D) motion analysis system (Qualisys Medical AB, Gothenburg, Sweden) that included 6 infrared cameras and a force plate (Kistler, Winterthur, Switzerland) to collect the kinematic and kinetic data of the participants' body movements when performing the sit-to-stand motion. The resolution of the Qualisys

Motion Capture System was 1280 × 1024 pixels, and the sampling frequency was set to 200 Hz. The 3D spatial coordinates were made using a nonlinear transformation method. The sampling frequency of the force plate was set to 2000 Hz. All kinematic and kinetic data were measured using Qualisys Track Manager (Qualisys Medical AB) software and were low-pass filtered using a fourth-order Butterworth filter with cutoff frequencies of 25 Hz and 10 Hz. Data were then analyzed using Visual 3D software (C-Motion, Germantown, MD, USA) [36]. Moments were normalized by the system to weight and height [37].

Prior to conducting this test, we performed calibration to correct errors that may occur in the infrared camera. One examiner modified the methods of the previous study according to the purpose of this study and attached 36 reflective markers on the landmark skin of the body [38]. A static test was performed by attaching a reflective marker and checking the position of each joint on a computer screen in the still state of standing barefoot. During the static test, which was performed before measuring each experimental condition, the participant stood still for 2 s on the force plate. All markers were fixed by banding using Kinesio tape, which does not reflect light, to prevent shaking or falling. After the static examination, the length from both lateral ankles to the markers (ASIS, PSIS, top of iliac crest) that were covered when the lumbar belt was worn while standing upright and the length between each marker were recorded with a tape measure. In the lumbar belt–wearing condition, the marker was reattached by maintaining the same position between the experimental conditions using this length.

2.4.2. Measurement of Muscle Activity

We measured the muscle activity of the vastus lateralis (VL), biceps femoris (BF), tibialis anterior (TA), and gastrocnemius (GM) muscles during the sit-to-stand motion using the Wireless Bipolar Cometa Mini Wave Plus EMG system (Cometa, Bareggio, Italy) synchronized with Visual 3D software (C-Motion). The sampling rate of the surface EMG signal was set to 2000 Hz. The raw surface EMG signal was rectified after band-pass filtering at 10–1000 Hz and low-pass filtered with a Butterworth filter [39]. As the ground electrode, a disposable anode surface electrode made of silver–silver chloride was used and attached parallel to the muscle fiber. The distance between electrodes in the surface EMG was set to 20 mm. To minimize the skin resistance and reduce the noise of muscle activity measurement, the hair on the skin was removed using a disposable razor, and after gently polishing with sandpaper, the skin was cleaned with alcohol. After the electrode attachment site was dried, two active electrodes and a reference electrode were attached. The electrodes of the four muscles of the dominant lower extremity (VL, BF, TA, GM) were placed according to the SENIAM guidelines [40] (Figure 3).

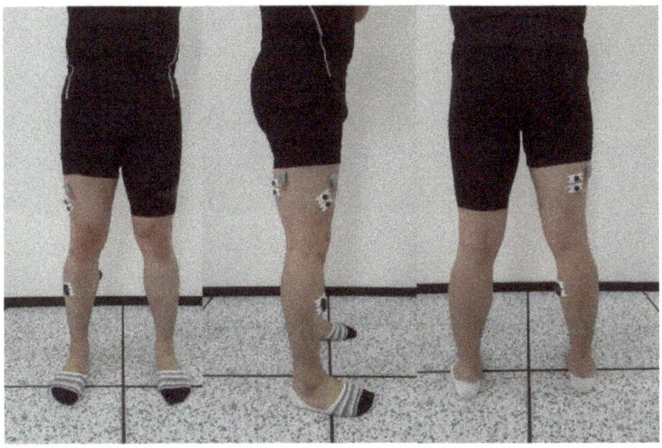

Figure 3. Attachment of EMG electrodes.

All of the measured raw EMG data were converted into root mean square, which provides a value close to the actual output value of the EMG signal. To analyze the activity of each muscle, the maximal voluntary isometric contraction (MVIC) of VL, BF, TA, and GM was measured, and the EMG signal collected during the sit-to-stand motion under each experimental condition was calculated as the percentage of MVIC [41]. Surface EMG data were synchronized using the pulses generated at the beginning of the recording of kinematics and kinetics, and the average muscle activity measurements during the sit-to-stand motion were compared.

2.5. Data Processing

The sit-to-stand motion was divided into three points and two phases to conduct the analysis of the kinematics, kinetics, and muscle activity variables (Figure 4). The three points were determined using the ground reaction force (GRF) measured on the force plate. The initiation point [42] and the seat-off point [43] were determined based on the vertical GRF, and the termination point was determined based on the anterior–posterior GRF (A-P GRF) [44]. To detect potential indicator changes in the sit-to-stand motion, the minimum and maximum values of each GRF data were used for calculation. The initiation point was detected when a 5% threshold change occurred from the reference value of the vertical GRF in the sitting position, and the termination point was detected when a 7.5% threshold change occurred from the final value of A-P GRF in the standing position, going backward [45]. The seat-off point was set as the maximum value point of the vertical GRF [44]. The phase from the initiation point to the seat-off point was set as the flexion phase, and the phase between the seat-off point to the termination point was set as the extension phase [46]. Each sit-to-stand motion was normalized to 101 data points from the initiation point to the termination point [33]. The time from the initiation point to the termination point was measured, and the value of the seat-off point of the sit-to-stand motion was used for the angle of joints in the kinematic variable. The maximum value during the sit-to-stand motion was used for the moment in the kinetic variable. The average muscle activities of the extension and flexion phases were used for the muscle activity variable.

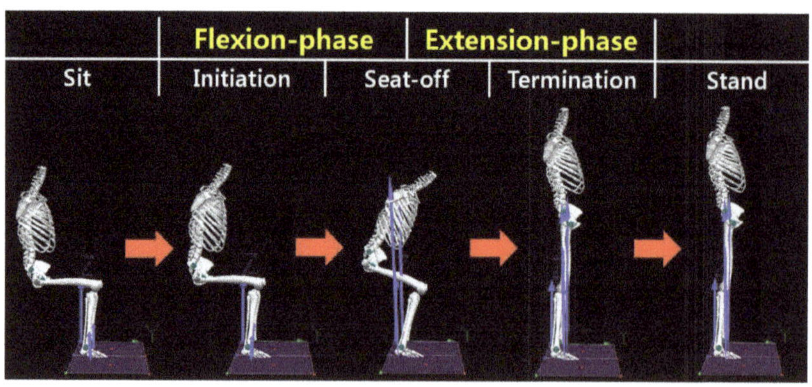

Figure 4. Indicators and classification of the sit-to-stand motion.

2.6. Statistical Analysis

For statistical processing of the data collected in this study, we used SPSS 22.0 for Windows (IBM Corp., Armonk, NY, USA), and all data are reported as mean ± standard deviation. The normality test was performed using the Shapiro–Wilk test. The participants' general characteristics were analyzed using descriptive statistics, and an independent t test and a chi-square test were used to test the homogeneity of the two groups. Two-way repeated analysis of variance (ANOVA) was performed to confirm the interaction between the presence or absence of back pain and the change in lumbar belt-wearing conditions. If the interaction was significant, the Bonferroni method was used for post hoc analysis. The least square difference was used for the main effect comparison. When both the interaction and the main effect were significant, we analyzed the interaction only. The statistical significance level (α) was set to 0.05.

Cohen's d formula was used for the effect size corresponding to the detected effect for the interaction between the experimental group and the control group, the main effect, and the comparison between groups [47]. The effect size was calculated using the G-power 3.1.9.4 program. In ANOVA, 0.10 was interpreted as "small," 0.25 as "medium," and 0.40 as "large" for effect size f [48].

3. Results

3.1. Analysis of The General Characteristics of Experimental and Control Groups

Table 1 shows the general characteristics of the experimental group and the control group. There was no significant difference between the two groups in terms of participants' general characteristics ($p > 0.05$).

Table 1. General characteristics of the experimental and control groups.

Variable	CG (n = 15)	EG (n = 15)	p
Sex (Male, %)	9 (60.0)	8 (53.3)	0.500 †
Age (year)	34.53 (4.40)	35.60 (5.19)	0.549 ‡
Height (cm)	171.40 (9.83)	170.20 (8.08)	0.718 ‡
Weight (kg)	70.60 (12.50)	67.40 (14.60)	0.525 ‡
BMI (kg/m^2)	23.85 (2.26)	23.02 (3.38)	0.437 ‡
Visual analog scale (score)		3.80 (0.86)	
Oswestry Disability Index (%)		19.60 (4.42)	
Onset (month)		43.06 (33.35)	

† Chi-squared test, ‡ Independent t-test, CG: Control group, EG: Experimental group.

3.2. Analysis of the Time Taken, Kinematics, and Kinetics Variables According to Lumbar Belt Wearing Conditions of the Experimental Group and Control Group

There was an interaction effect in the required time ($p < 0.05, f = 0.36$; Table 2). As a result of the post hoc analysis, there was a significant difference between the condition in which the experimental group did not wear a lumbar belt (condition 1) and the conditions in which the lumbar belt was worn (conditions 2 and 3). There was an interaction effect in the pelvic anterior tilt angle ($p < 0.05, f = 0.50$; Table 2). Post hoc analysis showed a significant difference between the control group not wearing the lumbar belt (condition 1) and the lumbar belt–wearing conditions (conditions 2 and 3), and there was a significant difference between the control group and experimental group in the condition of not wearing the lumbar belt (condition 1).

There were main effects among lumbar belt–wearing conditions in trunk flexion angle, hip flexion angle, knee joint flexion angle, hip flexion–extension moment, and knee joint flexion–extension moment ($p < 0.05, f = 0.36$–0.75; Table 2). Post hoc analysis showed a significant difference between the condition in which the lumbar belt was not worn (condition 1) and the conditions in which the lumbar belt was worn (conditions 2 and 3). There was a main effect between groups in hip flexion–extension moment ($p < 0.05, f = 0.65$; Table 2). Post hoc analysis showed a significant difference between the control group and experimental group.

Table 2. Analysis of the time taken, kinematics, and kinetics variables for the sit-to-stand motion according to the experimental conditions.

Variable	Condition 1 (No LB)		Condition 2 (Extensible LB)		Condition 3 (Nonextensible LB)		*p* (*f*)			Post Hoc
	CG	EG	CG	EG	CG	EG	"Group" (G)	"Condition" (C)	G × C	
Time taken (s)	1.76 (0.19)	1.90 (0.28)	1.76 (0.22)	1.78 (0.20)	1.75 (0.22)	1.77 (0.20)	0.401	0.018 * (0.39)	0.033 * (0.36)	EG: C1 > C2, C1 > C3
Trunk flexion angle (°)	45.70 (10.42)	40.79 (6.41)	45.96 (10.07)	45.17 (6.12)	46.89 (10.28)	44.09 (5.65)	0.336	0.030 * (0.38)	0.102	C1 < C2, C1 < C3
Pelvic anterior tilt angle (°)	34.14 (5.94)	42.05 (6.16)	45.08 (8.30)	45.46 (8.89)	45.55 (8.57)	44.74 (6.31)	0.290	<0.001 * (0.87)	0.002 * (0.50)	CG: C1 < C2, C1 < C3 C1: CG < EG
Hip flexion angle (°)	100.16 (8.82)	104.30 (11.45)	108.07 (13.42)	110.73 (10.79)	111.97 (11.86)	110.24 (9.72)	0.640	<0.001 * (0.75)	0.196	C1 < C2, C1 < C3
Knee flexion angle (°)	73.13 (9.09)	77.05 (9.34)	72.13 (7.82)	74.77 (9.29)	71.77 (8.94)	75.08 (9.13)	0.308	0.028 * (0.37)	0.655	C1 > C2, C1 > C3
Ankle flexion angle (°)	−6.74 (4.28)	−5.73 (3.71)	−6.14 (4.91)	−6.61 (3.85)	−6.41 (4.44)	−6.57 (5.92)	0.933	0.908	0.422	
Hip flexion–extension moment (nm/kg)	−1082.71 (272.10)	−875.20 (171.62)	−1163.58 (213.83)	−968.21 (112.71)	−1193.42 (222.59)	−948.38 (151.12)	0.002 * (0.65)	0.005 * (0.45)	0.699	CG > EG C1 < C2, C1 < C3
Hip adduction–abduction moment (nm/kg)	202.97 (164.75)	178.82 (139.13)	215.26 (117.89)	181.27 (103.57)	203.81 (94.11)	186.31 (88.26)	0.532	0.848	0.821	
Hip Int. Rot–Ext. Rot moment (nm/kg)	176.89 (117.37)	170.56 (103.88)	177.02 (163.95)	159.37 (112.66)	185.39 (186.78)	188.46 (109.36)	0.883	0.365	0.675	
Knee flexion-extension moment (nm/kg)	−593.82 (145.11)	−702.46 (168.38)	−563.46 (153.38)	−655.45 (161.79)	−549.03 (134.16)	−661.88 (155.61)	0.055	0.033 * (0.36)	0.820	C1 > C2, C1 > C3
Knee varus–valgus moment (nm/kg)	−70.03 (32.95)	−70.54 (57.46)	−70.41 (38.22)	−73.85 (56.54)	−60.15 (43.77)	−61.66 (57.35)	0.912	0.158	0.971	
Knee Int. Rot–Ext. Rot moment (nm/kg)	−108.82 (62.90)	−108.31 (67.98)	−110.77 (64.01)	−107.18 (76.34)	−102.12 (55.16)	−111.63 (68.13)	0.934	0.972	0.763	
Ankle flexion–extension moment (nm/kg)	107.93 (43.72)	104.49 (34.09)	113.31 (34.75)	96.59 (37.56)	108.47 (52.49)	93.34 (30.89)	0.322	0.756	0.621	
Ankle inversion–eversion moment (nm/kg)	29.02 (25.26)	30.87 (21.77)	32.93 (26.19)	34.26 (25.48)	30.96 (24.01)	36.06 (22.46)	0.750	0.091	0.551	
Ankle adduction–abduction moment (nm/kg)	−1.54 (6.61)	−4.37 (7.25)	−2.36 (9.06)	−3.12 (4.34)	−3.76 (7.41)	−2.69 (6.91)	0.708	0.912	0.243	

* $p < 0.05$, f: effect size "f", CG: Control group, EG: Experimental group, C1: Condition 1, C2: Condition 2, C3: Condition 3.

3.3. Analysis of Muscle Activity According to Lumbar Belt Wearing Conditions of the Experimental Group and Control Group

An interaction effect was observed in the VL flexion phase ($p < 0.05, f = 0.52$; Table 3). Post hoc analysis showed a significant difference between the control group and experimental group in the condition in which the lumbar belt was not worn (condition 1). There was an interaction effect observed in the BF flexion phase ($p < 0.05, f = 0.46$; Table 3). Post hoc analysis showed a significant difference between the condition in which the lumbar belt was not worn (condition 1) and the conditions in which the lumbar belt was worn (conditions 2 and 3) in the experimental group.

Table 3. Analysis of muscle activity variables of the sit-to-stand motion according to the experimental conditions (%MVIC).

Variable	Condition 1 (No LB)		Condition 2 (Extensible LB)		Condition 3 (Nonextensible LB)		*p* (*f*)			Post Hoc
	CG	EG	CG	EG	CG	EG	"Group" (G)	"Condition" (C)	G × C	
Vastus lateralis flexion phase	18.16 (6.48)	31.71 (10.52)	21.00 (7.14)	28.28 (9.04)	20.39 (7.97)	28.35 (9.86)	0.003 * (0.61)	0.762	0.003 * (0.52)	C1: CG < EG
Vastus lateralis extension phase	28.50 (12.09)	40.89 (13.92)	27.41 (12.94)	40.39 (14.48)	29.59 (11.53)	41.04 (13.94)	0.014 * (0.49)	0.342	0.728	CG < EG
Biceps femoris flexion phase	3.59 (1.44)	5.05 (2.87)	3.96 (1.41)	4.47 (2.38)	3.85 (1.37)	4.23 (2.87)	0.312	0.277	0.004 * (0.46)	EG: C1 > C2, C1 > C3
Biceps femoris extension phase	9.37 (4.39)	8.09 (4.63)	9.94 (4.40)	10.11 (4.92)	10.28 (5.25)	10.68 (6.69)	0.893	0.007 * (0.47)	0.219	C1 < C2, C1 < C3
Tibialis anterior flexion phase	19.91 (13.06)	24.47 (14.29)	17.97 (8.95)	21.58 (11.68)	15.49 (8.77)	22.19 (13.59)	0.246	0.007 * (0.44)	0.321	C1 > C2, C1 > C3
Tibialis anterior extension phase	5.31 (3.17)	9.19 (4.92)	5.06 (3.37)	8.14 (6.86)	5.91 (3.44)	7.78 (6.22)	0.092	0.423	0.136	
Gastrocnemius flexion phase	3.08 (1.42)	4.26 (1.75)	2.75 (1.39)	3.83 (1.48)	2.24 (1.03)	4.06 (1.56)	0.010 * (0.52)	0.011 * (0.42)	0.076	CG < EG C1 > C2, C1 > C3
Gastrocnemius extension phase	5.25 (2.32)	6.65 (2.67)	4.76 (1.84)	5.68 (1.76)	3.91 (1.80)	5.59 (2.55)	0.063	0.005 * (0.45)	0.565	C1 > C2, C1 > C3

* $p < 0.05$, f: effect size "f", CG: Control group, EG: Experimental group, C1: Condition 1, C2: Condition 2, C3: Condition 3.

There was a main effect between the lumbar belt–wearing conditions in the BF extension phase, TA flexion phase, GM flexion phase, and GM extension phase ($p < 0.05$, $f = 0.42$–0.47; Table 3). Post hoc analysis showed a significant difference between the condition in which the lumbar belt was not worn (condition 1) and the conditions in which the lumbar belt was worn (conditions 2 and 3). There was a main effect between groups in VL extension phase and GM flexion phase ($p < 0.05, f = 0.49–0.52$; Table 3). The results of the post hoc analysis suggested there was a significant difference between the control group and experimental group.

4. Discussion

The purpose of this study was to investigate the effect of the presence or absence of back pain and the difference in the extensibility of the lumbar belt on kinematics, kinetics, and muscle activity in terms of motion control in the sit-to-stand motion. The results of this study indicated that there was an interaction effect according to the presence or absence of back pain and the difference in the extensibility of the lumbar belt, and the two types of lumbar belts with different extensibility had a similar effect overall.

4.1. Analysis of The Interaction between the Presence or Absence of Low Back Pain and the Extensibility of the Lumbar Belt

The interaction effect was found in the time taken to perform the sit-to-stand motion. The post hoc analysis indicated no difference between the lumbar belt–wearing conditions in the healthy adult group, but there was a decrease in the time taken in the patients with nonspecific low back pain in the conditions in which the lumbar belt was worn (conditions 2 and 3) compared with the condition in which the lumbar belt was not worn (condition 1). The lumbar belt had effects such as improving proprioception [49], increasing mechanical stiffness [23], and relieving pain [30]. Although the mechanical stiffness increased in both the nonspecific low back pain group and the healthy adult group, wearing a lumbar belt in this study reduced the required time only in the group with nonspecific low back pain. It has been reported that proprioceptive sensation is impaired in patients with low back pain [50,51], and impaired proprioceptive sensation is associated with motor control disorders [52]. Therefore, we believe that the required time was reduced because the lumbar belt improved the proprioceptive sense of the patients with nonspecific low back pain and had a positive effect on motor control. Considering the normal effect size, we believe that the lumbar belt was partially helpful in improving the sit-to-stand motion by reducing the time taken for the patients with nonspecific low back pain.

In the kinematics of the sit-to-stand motion, an interactive effect was found on the angle of pelvic anterior tilt at the seat-off point. Results of the post hoc analysis showed that the experimental group had a greater pelvic anterior tilt angle than the control group in the condition in which the lumbar belt was not worn (condition 1). In conjunction with the angle of trunk flexion in this experiment, this difference in the angle of anterior pelvic tilt is thought to be a protection strategy for patients with low back pain to reduce the angle of trunk flexion and maintain the lumbar region in a rigid state. In the control group, the angle of the anterior pelvic tilt was greater in the conditions in which the lumbar belt was worn (conditions 2 and 3) than in the condition in which the lumbar belt was not worn (condition 1). In a previous study analyzing the effect of the lumbar belt on the range of motion of the lumbar spine and the lumbar–pelvic rhythm, it was reported that wearing a lumbar belt restricts the range of motion of the lumbar spine and also changes the lumbar–pelvic rhythm [23]. In this study, the lumbar belt was worn so that the lower edge of the lumbar belt covered the ASIS and the iliac crest to include the pelvis, which may have had a stiffening effect on the lumbar belt, as in the previous study on the spine–pelvic connection. The forward-driving force that accelerates the body's center of mass (COM) in the sit-to-stand motion is generated by trunk flexion [53]. In the experimental group, it is thought that the front driving force required to perform the sit-to-stand motion was generated as the trunk flexion angle increased by wearing the lumbar belt. In the control group, there was no change in the trunk flexion angle because the lumbar belt was worn, and the anterior pelvic tilt angle may have increased as a result of the stiffening effect of the lumbar belt. The interaction of the pelvic anterior tilt angle at the seat-off point showed a large effect size. Therefore, the difference in the angle of the anterior pelvic tilt is a characteristic that shows that the nonspecific low back pain patient group and the normal adult group performed the sit-to-stand motion differently, and the stiffening effect of the lumbar belt acting on the lumbar–pelvic connection is considered to be sufficient.

An interaction effect was found on the average muscle activity of VL and BF during the sit-to-stand motion. There was an interaction effect on the average muscle activity of the VL flexion phase. As a result of the post hoc analysis, the mean VL muscle activity of the nonspecific low back pain patient group was significantly higher than that of the healthy adult group under the condition in which the lumbar belt was not worn (condition 1). In this study, the nonspecific low back pain patient group showed a tendency to perform the sit-to-stand motion using the knee joint, increasing the angle and extension moment of the knee joint while keeping the lumbar–pelvic region in a rigid state. Thus, we believe that VL was involved in knee joint control in the flexion phase, and muscle activity was high in the patients with nonspecific low back pain. A previous study also reported that the increase in the knee joint extension moment was due to the high activity of the knee joint extensors, which supports the results of this study [54]. There was an interaction effect on the average muscle activity of the BF flexion phase. As a result of the post hoc analysis, there was no difference between the lumbar belt–wearing conditions in the healthy adult group; however, as compared with the condition without the lumbar belt (condition 1), the average muscle activity was significantly decreased in the nonspecific low back pain patient group in the condition with the lumbar belt (conditions 2 and 3). According to a study using EMG, the compensatory mechanism of muscle activation in the painful state could occur in antagonist muscles and induce changes in the pattern of muscle recruitment [55]. Because the sit-to-stand motion is an action that causes the body to stand up against gravity, the knee joint is regulated by the contraction of VL in the flexion phase. Therefore, the high muscle activity of the BF, the antagonist of the VL, in the condition that the lumbar belt is not worn (condition 1) is a change in the muscle recruitment pattern and is considered to be a compensatory mechanism in patients with nonspecific low back pain. According to the results of this study, the muscle activity of the VL and BF was reduced as the compensatory mechanism was reduced because the trunk flexion angle of the nonspecific low back pain patient group was increased by wearing a lumbar belt, which secured the trunk flexion angle necessary for the sit-to-stand motion. We found a large effect size of the interaction in the muscle activity of the VL and BF. From this large effect size, the lumbar belt is considered to have an advantage in terms of motor control.

4.2. Analysis of Main Effects between Patients with Nonspecific Low Back Pain and Healthy Adults

During the sit-to-stand motion, there was a difference between the nonspecific low back pain patient group and the healthy adult group. The extension moment of the healthy adult group was significantly greater in the flexion–extension moment of the hip joint. Although the extension moment of the nonspecific low back pain patient group was not statistically significant in the flexion–extension moment of the knee joint, it tended to be larger. Healthy individuals complete the process of standing upright by simultaneously flexing the spine and hip and then simultaneously extending the spine and hip when performing a sit-to-stand motion [56]. In this study, the healthy adult group also correctly performed the sit-to-stand motion using the hip extension moment. In contrast, in the nonspecific low back pain patient group, the sit-to-stand motion was performed using the knee joint rather than the hip joint. The surface EMG results of the lower extremity muscles showed a tendency for the muscle activity to be high in the nonspecific low back pain patient group. In the flexion phase, the VL and GM muscle activity of the nonspecific low back pain patient group was high, and in the extension phase, the VL muscle activity of the nonspecific low back pain patient group was high. In a previous study, the authors claimed there was a causal relationship between motor outcome and neuromuscular activity to create a protective mechanism [2]. In the flexion phase, the VL and GM play a role in regulating the knee and ankle joints, and in the extension phase, the VL more greatly extends the flexed knee joint; thus, it is considered that the muscle activity of the nonspecific low back pain patient group was high.

Previous studies have discussed the various mechanisms for nonspecific low back pain [57], and motor control disorders have been reported to possibly be one of the causes [58]. Because of the link between neuromuscular activity and biomechanical consequences, changes in the neuromuscular system resulting from motor control disorders lead to slower movement and affect the movement of higher muscle activity [59]. In this study, the nonspecific low back pain patient group performed the sit-to-stand motion with the lumbar spine rigid by increasing the angle of pelvic anterior tilt and decreasing the angle of trunk flexion. This could be due to a short-term protective strategy to reduce the load on the posterior structure of the lumbar spine and avoid further stress on the painful area. In addition, to compensate for reduced body flexion angle, the knee joint flexion angle was increased when performing the sit-to-stand motion. This is consistent with the results of the nonspecific low back pain patient group, which had a small hip extension moment and a large knee joint extension moment, as compared with the healthy adult group. Previous studies also reported that in patients with low back pain, the lower extremities were involved as a possible compensatory exercise for pain [60]. As such, we confirmed that movement control disorders in patients with nonspecific low back pain had an effect on daily activities, such as performing the sit-to-stand motion.

4.3. Analysis of The Main Effects between Lumbar Belt Wearing Conditions

The trunk flexion angle, hip flexion angle, and hip joint extension moment at the seat-off point of the sit-to-stand motion significantly increased in the conditions with the lumbar belt (conditions 2 and 3) compared with the condition without the lumbar belt (condition 1). In the knee joint, flexion angle and extension moment were significantly decreased. The lumbar belt can reduce the stress on the posterior viscoelastic structure of the lumbar vertebrae or the compression load of the lumbar vertebrae [26,27]. This is an effect of the limited lumbar movement due to the increased mechanical rigidity when the lumbar belt is worn, and it can prevent the load of a specific vertebral structure [61]. In this study, we found that the protection mechanism for reducing the angle of trunk flexion was reduced by wearing a lumbar belt to avoid stress on the painful area and thus the angle of trunk flexion was increased. Trunk flexion in the sit-to-stand motion creates a forward-driving force that accelerates the body's COM [53]. We believe that the compensatory action occurring at the knee joint was reduced because the angle of trunk flexion and hip flexion increased by wearing the lumbar belt.

The EMG results showed that by wearing the lumbar belt, the muscle activity in the flexion phase significantly decreased in the TA, and the muscle activity in the flexion phase and extension phase in the GM significantly decreased. The TA and GM are involved in the control of knee and ankle joints in the sit-to-stand motion. Previous studies also reported that the TA prepares for anterior displacement of the body's COM by moving the tibia forward [62]. It is thought that the muscle activity of the TA and GM decreased as the knee joint flexion angle decreased as a result of lumbar belt wearing. There was a significant increase in the muscle activity in the extension phase of the BF. Compensation is a motor strategy used to restore postural balance by sensory feedback [63]. In a previous study, a relatively small activity of the hip extensor was reported because small trunk flexion required less forward speed of the body's COM [54]. In this study, we consider that the muscle activity of the BF acting as a hip extensor increased because the body flexion angle increased by wearing a lumbar belt, which required greater forward speed of the body's COM.

4.4. Limitations

Because of the relatively small number of subjects in this study, it is difficult to generalize the results. We investigated only the immediate effect of the lumbar belt as opposed to the long-term effect. Because the spine was investigated as a single body, the movement of each vertebral segment and the compensatory action of the thoracic vertebrae were not considered. To focus on the effect of the lumbar belt on the sit-to-stand motion, there were restrictions on the height of the chair and the position of the feet. Therefore, different results may be obtained if the chair height and foot position are freely set. In addition, because the types and designs of the lumbar belts are diverse, different results could be obtained if there is a difference in the extensibility and design of the lumbar belt used. Finally, the muscle activity of the trunk muscles could not be measured because it could be disturbed by wearing a lumbar belt. Therefore, further studies using more diverse lumbar belt designs, including study designs using more subjects and a placebo effect, and lumbar belts equipped with biofeedback devices are needed in the future.

5. Conclusions

Patients with nonspecific low back pain performed the sit-to-stand motion in a pattern different from that of healthy adults, resulting in impaired motor control and decreased ability. The two lumbar belts with different extensibilities had a positive effect on motor control and improved performance in patients with nonspecific low back pain. Therefore, both types of lumbar belts with different extensibilities can help with the sit-to-stand motion, which is an important functional activity for patients with low back pain.

Author Contributions: Conceptualization, S.-C.I., S.-W.S. and K.K.; data curation, S.-C.I. and K.K.; formal analysis, S.-W.S., H.J. and K.K.; investigation, S.-C.I. and N.-Y.K.; methodology, S.-C.I., H.J. and K.K.; supervision, K.K.; validation, S.-C.I. and K.K.; visualization, S.-W.S., N.-Y.K. and H.J.; writing—original draft preparation, S.-C.I. and S.-W.S.; writing—review and editing, N.-Y.K. and K.K. All authors have read and agreed to the published version of the manuscript.

Funding: This research was supported by a Daegu University research grant in 2019 (2019-0415).

Institutional Review Board Statement: The study was conducted according to the guidelines of the Declaration of Helsinki and approved by the Institutional Review Board of the Daegu University (IRB no. 1040621-201911-HR-025-02).

Informed Consent Statement: Informed consent was obtained from all participants involved in this study. Written informed consent was obtained from the participants to publish this paper.

Data Availability Statement: The data presented in this study are available on request from the corresponding author. The data are not publicly available due to ethical restrictions.

Acknowledgments: The authors are grateful to all volunteer participants for their co-operation and participation in this study.

Conflicts of Interest: The authors declare no conflict of interest.

References

1. Hoy, D.; Bain, C.; Williams, G.; March, L.; Brooks, P.; Blyth, F.; Woolf, A.; Vos, T.; Buchbinder, R. A systematic review of the global prevalence of low back pain. *Arthritis Rheum.* **2012**, *64*, 2028–2037. [CrossRef] [PubMed]
2. Hodges, P.W.; Tucker, K. Moving differently in pain: A new theory to explain the adaptation to pain. *Pain* **2011**, *152*, S90–S98. [CrossRef] [PubMed]
3. Bartlett, R.; Wheat, J.; Robins, M. Is movement variability important for sports biomechanists? *Sports Biomech.* **2007**, *6*, 224–243. [CrossRef]
4. Stergiou, N.; Decker, L.M. Human movement variability, nonlinear dynamics, and pathology: Is there a connection? *Hum. Mov. Sci.* **2011**, *30*, 869–888. [CrossRef] [PubMed]
5. Seay, J.F.; Van Emmerik, R.E.; Hamill, J. Low back pain status affects pelvis-trunk coordination and variability during walking and running. *Clin. Biomech.* **2011**, *26*, 572–578. [CrossRef] [PubMed]
6. Mokhtarinia, H.R.; Sanjari, M.A.; Chehrehrazi, M.; Kahrizi, S.; Parnianpour, M. Trunk coordination in healthy and chronic nonspecific low back pain subjects during repetitive flexion-extension tasks: Effects of movement asymmetry, velocity and load. *Hum. Mov. Sci.* **2016**, *45*, 182–192. [CrossRef]
7. Moseley, G.L.; Hodges, P.W. Reduced variability of postural strategy prevents normalization of motor changes induced by back pain: A risk factor for chronic trouble? *Behav. Neurosci.* **2006**, *120*, 474–476. [CrossRef] [PubMed]

8. Hamill, J.; van Emmerik, R.E.; Heiderscheit, B.C.; Li, L. A dynamical systems approach to lower extremity running injuries. *Clin. Biomech.* **1999**, *14*, 297–308. [CrossRef]
9. Silfies, S.P.; Bhattacharya, A.; Biely, S.; Smith, S.S.; Giszter, S. Trunk control during standing reach: A dynamical system analysis of movement strategies in patients with mechanical low back pain. *Gait Posture* **2009**, *29*, 370–376. [CrossRef] [PubMed]
10. Lamoth, C.J.; Meijer, O.G.; Daffertshofer, A.; Wuisman, P.I.; Beek, P.J. Effects of chronic low back pain on trunk coordination and back muscle activity during walking: Changes in motor control. *Eur. Spine. J.* **2006**, *15*, 23–40. [CrossRef] [PubMed]
11. Reeves, N.P.; Narendra, K.S.; Cholewicki, J. Spine stability: Lessons from balancing a stick. *Clin. Biomech.* **2011**, *26*, 325–330. [CrossRef] [PubMed]
12. Claeys, K.; Dankaerts, W.; Janssens, L.; Brumagne, S. Altered preparatory pelvic control during the sit-to-stance-to-sit movement in people with non-specific low back pain. *J. Electromyogr. Kinesiol.* **2012**, *22*, 821–828. [CrossRef] [PubMed]
13. Shum, G.L.; Crosbie, J.; Lee, R.Y. Effect of low back pain on the kinematics and joint coordination of the lumbar spine and hip during sit-to-stand and stand-to-sit. *Spine* **2005**, *30*, 1998–2004. [CrossRef] [PubMed]
14. Sedrez, J.A.; Mesquita, P.V.; Gelain, G.M.; Candotti, C.T. Kinematic Characteristics of Sit-to-Stand Movements in Patients with Low Back Pain: A Systematic Review. *J. Manip. Physiol. Ther.* **2019**, *42*, 532–540. [CrossRef]
15. Bishop, F.L.; Dima, A.L.; Ngui, J.; Little, P.; Moss-Morris, R.; Foster, N.E.; Lewith, G.T. "Lovely Pie in the Sky Plans": A Qualitative Study of Clinicians' Perspectives on Guidelines for Managing Low Back Pain in Primary Care in England. *Spine* **2015**, *40*, 1842–1850. [CrossRef]
16. Hancock, M.J.; Kjaer, P.; Korsholm, L.; Kent, P. Interpretation of subgroup effects in published trials. *Phys. Ther.* **2013**, *93*, 852–859. [CrossRef]
17. Mistry, D.; Patel, S.; Hee, S.W.; Stallard, N.; Underwood, M. Evaluating the quality of subgroup analyses in randomized controlled trials of therapist-delivered interventions for nonspecific low back pain: A systematic review. *Spine* **2014**, *39*, 618–629. [CrossRef]
18. Sun, X.; Briel, M.; Walter, S.D.; Guyatt, G.H. Is a subgroup effect believable? Updating criteria to evaluate the credibility of subgroup analyses. *BMJ* **2010**, *340*, c117. [CrossRef]
19. Maher, C.; Underwood, M.; Buchbinder, R. Non-specific low back pain. *Lancet* **2017**, *389*, 736–747. [CrossRef]
20. Gallagher, K.M.; Callaghan, J.P. Standing on a declining surface reduces transient prolonged standing induced low back pain development. *Appl. Ergon.* **2016**, *56*, 76–83. [CrossRef]
21. Lahad, A.; Malter, A.D.; Berg, A.O.; Deyo, R.A. The effectiveness of four interventions for the prevention of low back pain. *JAMA* **1994**, *272*, 1286–1291. [CrossRef] [PubMed]
22. Ammendolia, C.; Kerr, M.S.; Bombardier, C. Back belt use for prevention of occupational low back pain: A systematic review. *J. Manip. Physiol. Ther.* **2005**, *28*, 128–134. [CrossRef] [PubMed]
23. Larivière, C.; Caron, J.M.; Preuss, R.; Mecheri, H. The effect of different lumbar belt designs on the lumbopelvic rhythm in healthy subjects. *BMC Musculoskelet. Disord.* **2014**, *15*, 307. [CrossRef]
24. Preuss, R.; Fung, J. Can acute low back pain result from segmental spinal buckling during sub-maximal activities? A review of the current literature. *Man. Ther.* **2005**, *10*, 14–20. [CrossRef] [PubMed]
25. Panjabi, M.M. Clinical spinal instability and low back pain. *J. Electromyogr. Kinesiol.* **2003**, *13*, 371–379. [CrossRef]
26. McGill, S.M.; Kippers, V. Transfer of loads between lumbar tissues during the flexion-relaxation phenomenon. *Spine* **1994**, *19*, 2190–2196. [CrossRef]
27. Katsuhira, J.; Sasaki, H.; Asahara, S.; Ikegami, T.; Ishihara, H.; Kikuchi, T.; Hirai, Y.; Yamasaki, Y.; Wada, T.; Maruyama, H. Comparison of low back joint moment using a dynamic 3D biomechanical model in different transferring tasks wearing low back belt. *Gait Posture* **2008**, *28*, 258–264. [CrossRef]
28. Foster, N.E.; Anema, J.R.; Cherkin, D.; Chou, R.; Cohen, S.P.; Gross, D.P.; Ferreira, P.H.; Fritz, J.M.; Koes, B.W.; Peul, W.; et al. Prevention and treatment of low back pain: Evidence, challenges, and promising directions. *Lancet* **2018**, *391*, 2368–2383. [CrossRef]
29. Van Duijvenbode, I.C.; Jellema, P.; van Poppel, M.N.; van Tulder, M.W. Lumbar supports for prevention and treatment of low back pain. *Cochrane Database Syst. Rev.* **2008**, *2008*, CD001823. [CrossRef]
30. Shahvarpour, A.; Preuss, R.; Sullivan, M.J.L.; Negrini, A.; Larivière, C. The effect of wearing a lumbar belt on biomechanical and psychological outcomes related to maximal flexion-extension motion and manual material handling. *Appl. Ergon.* **2018**, *69*, 17–24. [CrossRef]
31. Faul, F.; Erdfelder, E.; Buchner, A.; Lang, A.G. Statistical power analyses using G*Power 3.1: Tests for correlation and regression analyses. *Behav. Res. Methods* **2009**, *41*, 1149–1160. [CrossRef]
32. Boucher, J.A.; Roy, N.; Preuss, R.; Larivière, C. The effect of two lumbar belt designs on trunk repositioning sense in people with and without low back pain. *Ann. Phys. Rehabil. Med.* **2017**, *60*, 306–311. [CrossRef] [PubMed]
33. Ippersiel, P.; Robbins, S.; Preuss, R. Movement variability in adults with low back pain during sit-to-stand-to-sit. *Clin. Biomech.* **2018**, *58*, 90–95. [CrossRef] [PubMed]
34. Ludvig, D.; Preuss, R.; Larivière, C. The effect of extensible and non-extensible lumbar belts on trunk muscle activity and lumbar stiffness in subjects with and without low-back pain. *Clin. Biomech.* **2019**, *67*, 45–51. [CrossRef] [PubMed]
35. Orakifar, N.; Shaterzadeh-Yazdi, M.J.; Salehi, R.; Mehravar, M.; Namnik, N. Muscle Activity Pattern Dysfunction During Sit to Stand and Stand to Sit in the Movement System Impairment Subgroups of Low Back Pain. *Arch. Phys. Med. Rehabil.* **2019**, *100*, 851–858. [CrossRef] [PubMed]

36. Sinclair, J.; Atkins, S.; Richards, J.; Vincent, H. Modelling of Muscle Force Distributions During Barefoot and Shod Running. *J. Hum. Kinet.* **2015**, *47*, 9–17. [CrossRef] [PubMed]
37. Costigan, P.A.; Wyss, U.P.; Deluzio, K.J.; Li, J. Semiautomatic three-dimensional knee motion assessment system. *Med. Biol. Eng. Comput.* **1992**, *30*, 343–350. [CrossRef] [PubMed]
38. Cappozzo, A.; Catani, F.; Croce, U.D.; Leardini, A. Position and orientation in space of bones during movement: Anatomical frame definition and determination. *Clin. Biomech.* **1995**, *10*, 171–178. [CrossRef]
39. Koopman, A.S.; Kingma, I.; Faber, G.S.; de Looze, M.P.; van Dieën, J.H. Effects of a passive exoskeleton on the mechanical loading of the low back in static holding tasks. *J. Biomech.* **2019**, *83*, 97–103. [CrossRef] [PubMed]
40. Hermens, H.J.; Freriks, B.; Disselhorst-Klug, C.; Rau, G. Development of recommendations for SEMG sensors and sensor placement procedures. *J. Electromyogr. Kinesiol.* **2000**, *10*, 361–374. [CrossRef]
41. Farahpour, N.; Jafarnezhadgero, A.; Allard, P.; Majlesi, M. Muscle activity and kinetics of lower limbs during walking in pronated feet individuals with and without low back pain. *J. Electromyogr. Kinesiol.* **2018**, *39*, 35–41. [CrossRef] [PubMed]
42. Etnyre, B.; Thomas, D.Q. Event standardization of sit-to-stand movements. *Phys. Ther.* **2007**, *87*, 1651–1666. [CrossRef] [PubMed]
43. Schenkman, M.; Riley, P.O.; Pieper, C. Sit to stand from progressively lower seat heights—alterations in angular velocity. *Clin. Biomech.* **1996**, *11*, 153–158. [CrossRef]
44. Stevermer, C.A.; Gillette, J.C. Kinematic and Kinetic Indicators of Sit-to-Stand. *J. Appl. Biomech.* **2016**, *32*, 7–15. [CrossRef] [PubMed]
45. Pai, Y.C.; Rogers, M.W. Control of body mass transfer as a function of speed of ascent in sit-to-stand. *Med. Sci. Sports Exerc.* **1990**, *22*, 378–384. [CrossRef]
46. Nuzik, S.; Lamb, R.; VanSant, A.; Hirt, S. Sit-to-stand movement pattern. A kinematic study. *Phys. Ther.* **1986**, *66*, 1708–1713. [CrossRef]
47. Lakens, D. Calculating and reporting effect sizes to facilitate cumulative science: A practical primer for *t*-tests and ANOVAs. *Front. Psychol.* **2013**, *4*, 863. [CrossRef] [PubMed]
48. Fritz, C.O.; Morris, P.E.; Richler, J.J. Effect size estimates: Current use, calculations, and interpretation. *J. Exp. Psychol. Gen.* **2012**, *141*, 2–18. [CrossRef] [PubMed]
49. Calmels, P.; Fayolle-Minon, I. An update on orthotic devices for the lumbar spine based on a review of the literature. *Rev. Rhum. Engl. Ed* **1996**, *63*, 285–291. [PubMed]
50. Boucher, J.A.; Abboud, J.; Nougarou, F.; Normand, M.C.; Descarreaux, M. The Effects of Vibration and Muscle Fatigue on Trunk Sensorimotor Control in Low Back Pain Patients. *PLoS ONE* **2015**, *10*, e0135838. [CrossRef] [PubMed]
51. Willigenburg, N.W.; Kingma, I.; Hoozemans, M.J.; van Dieën, J.H. Precision control of trunk movement in low back pain patients. *Hum. Mov. Sci.* **2013**, *32*, 228–239. [CrossRef] [PubMed]
52. Koch, C.; Hänsel, F. Chronic Non-specific Low Back Pain and Motor Control During Gait. *Front. Psychol.* **2018**, *9*, 2236. [CrossRef] [PubMed]
53. Roebroeck, M.E.; Doorenbosch, C.A.; Harlaar, J.; Jacobs, R.; Lankhorst, G.J. Biomechanics and muscular activity during sit-to-stand transfer. *Clin. Biomech.* **1994**, *9*, 235–244. [CrossRef]
54. Jeon, W.; Jensen, J.L.; Griffin, L. Muscle activity and balance control during sit-to-stand across symmetric and asymmetric initial foot positions in healthy adults. *Gait Posture* **2019**, *71*, 138–144. [CrossRef] [PubMed]
55. Svendsen, J.H.; Svarrer, H.; Laessoe, U.; Vollenbroek-Hutten, M.; Madeleine, P. Standardized activities of daily living in presence of sub-acute low-back pain: A pilot study. *J. Electromyogr. Kinesiol.* **2013**, *23*, 159–165. [CrossRef] [PubMed]
56. Tully, E.A.; Fotoohabadi, M.R.; Galea, M.P. Sagittal spine and lower limb movement during sit-to-stand in healthy young subjects. *Gait Posture* **2005**, *22*, 338–345. [CrossRef]
57. Saragiotto, B.T.; Maher, G.C.; Yamato, T.P.; Costa, L.O.; Menezes Costa, L.C.; Ostelo, R.W.; Macedo, L.G. Motor control exercise for chronic non-specific low-back pain. *Cochrane Database Syst. Rev.* **2016**, *2016*, CD012004. [CrossRef] [PubMed]
58. Götze, M.; Ernst, M.; Koch, M.; Blickhan, R. Influence of chronic back pain on kinematic reactions to unpredictable arm pulls. *Clin. Biomech.* **2015**, *30*, 290–295. [CrossRef]
59. Van Dieën, J.H.; Selen, L.P.; Cholewicki, J. Trunk muscle activation in low-back pain patients, an analysis of the literature. *J. Electromyogr. Kinesiol.* **2003**, *13*, 333–351. [CrossRef]
60. Sung, P.S.; Danial, P. Analysis of relative kinematic index with normalized standing time between subjects with and without recurrent low back pain. *Eur. Spine J.* **2017**, *26*, 518–527. [CrossRef]
61. Van Poppel, M.N.; de Looze, M.P.; Koes, B.W.; Smid, T.; Bouter, L.M. Mechanisms of action of lumbar supports: A systematic review. *Spine* **2000**, *25*, 2103–2113. [CrossRef]
62. Khemlani, M.M.; Carr, J.H.; Crosbie, W.J. Muscle synergies and joint linkages in sit-to-stand under two initial foot positions. *Clin. Biomech.* **1999**, *14*, 236–246. [CrossRef]
63. Haddas, R.; Satin, A.; Lieberman, I. What is actually happening inside the "cone of economy": Compensatory mechanisms during a dynamic balance test. *Eur. Spine J.* **2020**, *29*, 2319–2328. [CrossRef] [PubMed]

Journal of *Personalized Medicine*

MDPI

Article

Postoperative Inpatient Rehabilitation Does Not Increase Knee Function after Primary Total Knee Arthroplasty

Dominik Rak [1,*], Alexander J. Nedopil [1], Eric C. Sayre [2], Bassam A. Masri [3] and Maximilian Rudert [1]

1 Orthopädische Klinik König-Ludwig-Haus, Lehrstuhl für Orthopädie der Universität Würzburg, 97074 Würzburg, Germany
2 Arthritis Research Canada, Vancouver, BC V5Y 3P2, Canada
3 Department of Orthopedic Surgery, University of British Columbia, Vancouver, BC V5Z 1M9, Canada
* Correspondence: d-rak.klh@uni-wuerzburg.de

Abstract: Inpatient rehabilitation (IR) is a common postoperative protocol after total knee replacement (TKA). Because IR is expensive and should therefore be justified, this study determined the difference in knee function one year after TKA in patients treated with IR or outpatient rehabilitation, fast-track rehabilitation (FTR) in particular, which also entails a reduced hospital length of stay. A total of 205 patients were included in this multi-center prospective cohort study. Of the patients, 104 had primary TKA at a German university hospital and received IR, while 101 had primary TKA at a Canadian university hospital and received FTR. Patients receiving IR or FTR were matched by pre-operative demographics and knee function. Oxford Knee Score (OKS), Western Ontario and McMaster Universities Arthritis Index (WOMAC), and EuroQol visual analogue scale (EQ-VAS) determined knee function one year after surgery. Patients receiving IR had a 2.8-point lower improvement in OKS ($p = 0.001$), a 6.7-point lower improvement in WOMAC ($p = 0.063$), and a 12.3-point higher improvement in EQ-VAS ($p = 0.281$) than patients receiving FTR. IR does not provide long-term benefits to patient recovery after primary uncomplicated TKA under the current rehabilitation regime.

Keywords: total knee arthroplasty; fast track rehabilitation; inpatient rehabilitation; postoperative rehabilitation; patient reported outcome measures

Citation: Rak, D.; Nedopil, A.J.; Sayre, E.C.; Masri, B.A.; Rudert, M. Postoperative Inpatient Rehabilitation Does Not Increase Knee Function after Primary Total Knee Arthroplasty. *J. Pers. Med.* **2022**, *12*, 1934. https://doi.org/10.3390/jpm12111934

Academic Editor: Anne-Marie Caminade

Received: 14 October 2022
Accepted: 17 November 2022
Published: 21 November 2022

1. Introduction

Over the last two decades, the number of total joint replacement surgeries has consistently increased due to ageing in Western societies and growing numbers of people with overweight or obesity [1]. Due to the expected increase in patients needing a total knee arthroplasty (TKA), alternative rehabilitation pathways besides inpatient rehabilitation (IR) have been explored with the aim to provide comparable functional outcomes while reducing the time a patient is required to stay away from home.

IR after TKA is common practice in Western European countries. IR usually includes 3 weeks of medical rehabilitation at a rehabilitation center to reintroduce patients to independent living and social integration. Patients receive daily physical therapy, occupational therapy, and functional training. In addition, patients are provided accommodation and daily meals. After discharge from IR, outpatient physical therapy is continued for four more weeks.

Post-TKA fast-track rehabilitation (FTR) was developed with the goal of reducing costs and providing patients with a faster return to home while maintaining comparable functional outcomes. FTR entails a reduced length of stay in the hospital with discharge to home and not to IR. Outpatient physical therapy or inhouse visits by a physical therapist help patients to regain their knee function and return to an independent lifestyle, while forgoing IR can save a patient $3450 [2].

Because of the potential cost reduction, multiple countries have steered away from utilizing IR after uncomplicated TKAs [3]. One of these countries is Canada. In Germany,

IR is still the standard postoperative rehabilitation regime after uncomplicated TKA. As insured patients in Germany would be reluctant to be randomized to a therapy other than the one to which they feel entitled, it is difficult to conduct a randomized controlled trial that compares a resource-intensive pathway (=IR) with a less resource-intensive alternative (=FTR). Consequently, the similarities between the Canadian and German healthcare systems offer an opportunity to compare functional improvements after these two different rehabilitation regimes [4].

Multiple studies—mainly performed outside Germany—have demonstrated equal or even improved patient joint function when FTR after TKA or total hip arthroplasty was compared to IR [2,5–7]. While German patients remain unwilling to forfeit their right to IR, a plausible method to compare the influence of IR and FTR on the functional improvement after TKA is to analyze PROMs between German patients receiving IR and patients from another country receiving FTR. To minimize the inherent bias when comparing patients from two different countries, the patients' general health and pre-operative knee function and the surgical technique and implant design should be comparable.

Accordingly, the purpose of the study was to compare the functional improvement after TKA among patients receiving IR in Germany with patients receiving FTR in Canada. The hypothesis was that FTR results in comparable functional improvement one year after TKA. A follow-up period of one year seemed reasonable, since knee function reaches a plateau within the first postoperative year, which remains stable in the following years [8–11]. A cost analysis, however, was not performed because of the countries' different reimbursement policies.

2. Materials and Methods

The study was planned and conducted as a transnational evaluation of prospectively collected data, in which the one-year clinical improvement after the Canadian FTR and the German IR following primary TKA was compared. The study was a mutual project of the orthopedic departments of University of British Columbia, Vancouver, Canada, and Julius Maximilian Universität, Würzburg, Germany. IR for the German patients took place at a preselected rehabilitation center. Canadian patients instead conducted outpatient physical therapy with corresponding home exercises or through home visits by a physical therapist.

Upon receipt of approval from both the Canadian (The University of British Columbia, Office of Research Ethics, Study ID: H18-02307, 05.09.2018) and German (Ethik-Kommission der Universität Würzburg, Number: 20210925 01, 26.20.2021) institutional ethics boards, we identified all patients between December 2019 and February 2020 who fulfilled the criteria for medical need for TKA treatment. Included were all patients with primary osteoarthritis and radiographic evidence of Kellgren–Lawrence grades II-IV osteoarthritic changes who underwent an uncomplicated unilateral TKA using the Stryker Triathlon TKA System. Patients receiving simultaneous bilateral TKAs, patients with previous fracture of the affected lower extremity, patients with metabolic or inflammatory joint disease (e.g., rheumatoid arthritis or osteonecrosis), and patients not following the standardized postoperative rehabilitation regime were not eligible to participate in the study. Patients lost to follow-up were excluded from the final data analysis.

During the above-mentioned time-period, a total of 117 TKAs were performed at the German hospital, of which the Stryker Triathlon System was used in 113. Nine patients were lost to follow-up, leaving 104 patients for final analysis. In Canada, 115 TKAs were performed in the above-mentioned time-period, all utilizing the Stryker Triathlon System; 14 patients were lost to follow-up, leaving 101 patients for final analysis (Table 1).

All TKAs were performed using a standard medial parapatellar approach, mechanical alignment principles, and a cruciate retaining femoral component with a cruciate retaining (CR) or cruciate-substituting polyethylene (CR-CS) Triathlon TKA system (Stryker, Kalamazoo, MI, USA).

The target hospital length-of-stay for the IR patients was 4–5 days before they were transferred to IR. IR was scheduled for three weeks and included daily physical therapy, occupational therapy, and functional training. In the FTR group, the target hospital length-of-stay was 1–2 days before home discharge and onset of outpatient rehabilitation. Outpatient rehabilitation included physical therapy twice per week for 6 weeks with instruction to perform independent exercises at home.

Table 1. Patient pre-operative demographics.

Rehabilitation Protocol		FTR-Canada (n = 101)	IR-Germany (n = 104)	p-Value
Age (years) mean ± SD		66 ± 8	67 ± 10	0.703
Sex	**Female (n) (%)**	65 (64%)	57 (55%)	0.205
	Male (n) (%)	36 (36%)	47 (45%)	
BMI mean ± SD		30.1 ± 6.4	31.6 ± 6.0	0.029
ASA				0.189
	I (n) (%)	9 (9%)	4 (4%)	
	II (n) (%)	71 (70%)	70 (67%)	
	III (n) (%)	21 (21%)	30 (29%)	

BMI, body mass index; ASA, American Society of Anesthesiologists Physical Status; SD, standard deviation; FTR, fast-track rehabilitation; IR, inhouse rehabilitation.

To compare the baseline health status between the two groups the pre-operative age, sex, body mass index (BMI), and American Society of Anesthesiologists (ASA) Physical Status score were obtained.

To evaluate and assess knee function and the functional improvement after TKA the following patient-reported outcome measures (PROMs) were conducted. The Oxford Knee Score (OKS) is scored from 0 (worst) to 48 (best) and assesses knee function before and after TKA. It is self-conducted by the patient and comprises 12 questions that are divided into two subscales: pain and physical function. The Western Ontario and McMaster Universities Arthritis Index (WOMAC) is a self-administered questionnaire to assess pain, function, and stiffness of the knee. The descriptors range from no difficulty (0 point) to extreme difficulties (4 points). The WOMAC Score is a commonly used standardized questionnaire to evaluate the condition of patients with osteoarthritis of the knee and hip. The WOMAC is scored from 0 (best) to 96 (worst). Furthermore, the EuroQol visual analogue scale (EQ-VAS), which measures health-related quality of life, was completed by patients. It is scored from 0 (worst to 100 (best) and quantifies the patient's perception of their health state. It is self-conducted by the patient and has 5 different dimensions: mobility, self-care, usual activities, pain/discomfort, and anxiety/depression.

The difference between the OKS, WOMAC, and EQ-VAS obtained pre-operatively and at the one-year follow-up visit defined clinical improvement. The minimum clinically important difference (MCID) for OKS was 5.0, for WOMAC, 10, and for EQ-VAS, 8.0 [12–14].

Statistical Analysis

In unadjusted analyses, IR and FTR were compared with pre-operative age, sex, BMI, ASA, OKS, WOMAC, and EQ-5D VAS. Binary or categorical comparisons were made via exact chi-squared tests, while continuous comparisons were made via the Wilcoxon rank-sum test. Adjusted multivariable linear regression models were fitted to predict improvements from pre-operative to follow-up in OKS, WOMAC, and EQ-5D VAS vs. rehabilitation protocol, adjusting for age, sex, BMI, ASA, and the pre-operative value of the PROM being analyzed. Fit of linear regression models were assessed via normal quantile–quantile plots of the standardized residuals. Fit of logistic regression models were assessed with the Hosmer and Lemeshow goodness-of-fit test at alpha = 0.05 [15]. SAS software (version 9.4, SAS Institute Inc., Cary, NC, USA) was used for statistical analyses.

3. Results

A total of 205 patients participated in the study with available PROMs at 1-year follow-up. Patient pre-operative demographics and knee function were comparable between the FTR and IR groups (Tables 1 and 2).

Table 2. Clinical scores in both groups.

	Rehabilitation Protocol	Pre-Operative Mean ± SD	*p*-Value	1-Year Follow-Up Mean ± SD	*p*-Value	Clinical Improvement Δ Mean ± SD	*p*-Value
OKS	FTR	21.9 ± 8.4	0.111	38.7 ± 8.8	<0.001	16.8 ± 10.6	0.001
	IR	19.9 ± 6.6		33.9 ± 9.9		14.0 ± 10.7	
WOMAC	FTR	47.7 ± 20.1	0.177	14.7 ± 15.2	0.571	33.2 ± 20.6	0.063
	IR	44.8 ± 19.8		18.3 ± 18.7		26.5 ± 25.5	
EQ-VAS	FTR	66.4 ± 19.6	<0.001	78.6 ± 17.6	0.103	11.9 ± 24.2	0.281
	IR	48.4 ± 17.3		72.6 ± 22.1		24.2 ±26.9	

OKS, Oxford Knee Score; WOMAC, Western Ontario and McMaster Universities Arthritis Index; EQ-VAS, EuroQol visual analogue scale; FTR, fast-track rehabilitation; IR, inhouse rehabiliation; *p*-values for clinical improvements are from multivariable models; improvements in WOMAC are shown in the positive direction for comparison.

Patients receiving IR had a significantly lower mean improvement in OKS (14.0 ± 10.7) and a trend of lower improvement in WOMAC (−26.5 ± 25.5) than patients receiving FTR (OKS 16.8 ± 10.6; WOMAC −33.2 ± 20.6), resulting in adjusted models for *p*-values of 0.001 and 0.063, respectively. The EQ-VAS tended to improve more after IR (24.2 ± 26.9) than after FTR (11.9 ± 24.2) (p = 0.281) (Table 2 and Figure 1). All differences in follow-up OKS, WOMAC, and EQ-VAS between the FTR and IR groups were below the MCID.

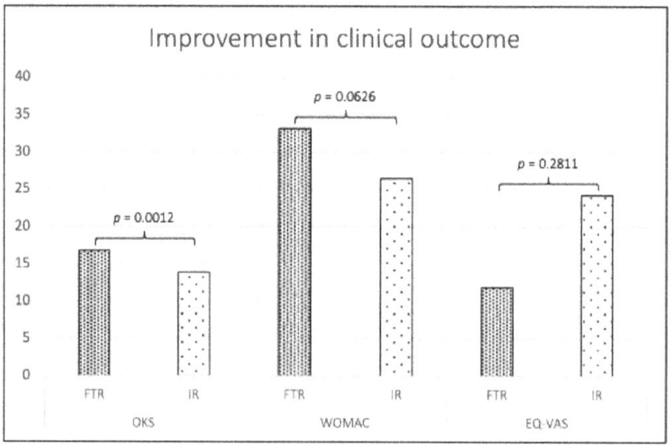

Figure 1. Improvement in clinical outcome. OKS, Oxford Knee Score; WOMAC, Western Ontario and McMaster Universities Arthritis Index; EQ-VAS, EuroQol visual analogue scale; FTR, fast-track rehabilitation; IR, inhouse rehabiliation; *p*-values for clinical improvements are from multivariable models; improvements in WOMAC are shown in the positive direction for comparison.

4. Discussion

The most important finding from the present study was that FTR after TKA did not result in inferior functional improvement when compared to IR. Indeed, patients receiving FTR showed higher improvement in PROMs than patients receiving IR. While outcomes scores at one year were higher in the FTR group regardless of the degree of improvement, this difference was not clinically significant. These results may encourage insurance companies to recommend FTR to their patients, as FTR does not compromise patient outcome.

These findings agree with multiple international studies comparing outpatient and inpatient rehabilitation [4,7]. Prospective and retrospective studies, mainly from outside Europe, demonstrated comparable or even improved function upon utilizing outpatient rehabilitation after TKA, with reduced treatment costs [2,4,16–19]. As the risk of complications seemed not to increase, the conclusion of these studies was predominantly to not recommend IR to patients undergoing uncomplicated TKA. A dissertation conducted at the University of Rostock/Germany comparing outpatient and inpatient rehabilitation was in agreement with these findings [5]. Unfortunately, the dissertation has not yet been published in the medical literature. Because the improvements in PROMs in the present study were adjusted for age, sex, ASA, and the pre-operative value of the corresponding PROM, the observed differences in improvements being influenced by these variables can be ruled out. I.e., irrespective of the patients' age, sex, ASA, and pre-operatively assessed PROM, the improvement in PROMs is comparable or slightly higher after FTR than IR. While comparison of clinical improvement after FTR or IR can be confounded by a selection bias, i.e., patients receiving FTR could be fitter, healthier, and more motivated to improve than patients receiving IR, the results from this study are not biased by a selection process because both patient groups followed the standard postoperative protocol of the treating hospital [17].

In addition to providing comparable functional recovery, FTR has also been shown to reduce costs [2]. These cost reductions include costs associated with the necessary clinical wound check and staple removal follow-up of patients not receiving IR. Obviously, not providing patients with accommodation, food, and sanitary needs reduces costs. Outpatient rehabilitation also reduces readmission rate after TKA and the risk of peri-prosthetic complications [7,18]. The number of diagnostic tests can be reduced with no negative effect on patient outcome [19]. Because FTR includes a reduced length of hospital stay, it can be expected that costs can be further reduced using this rehabilitation protocol [20].

Because this study favors FTR due to comparable or slightly higher functional improvement one year after TKA, it could motivate carriers of IR facilities to reorient the focus of their rehabilitation regime. One potential reason that patients after FTR achieve higher improvement than after IR is the self-guidance and patient responsibility to regain their independence and function. The self-guidance potentially improves function in two ways. (1) It fosters salutogenesis by focusing the patient towards solving problems after TKA, which include regaining range of motion, mobility, and strength [21]. (2) It prevents externalization of the rehabilitation process, i.e., success or failure to improve is not the physical therapist's responsibility, it is the patient's responsibility.

There are some limitations to this study. First, the difference in clinical improvement could be influenced by country-specific pain awareness of patients and their interpretation of functional limitations. Because the pre-operative OKS and WOMAC were not different between the two countries and because the comparison of clinical improvement was normalized to the pre-operative PROM, the country-specific influence on clinical improvement should be minimal. Second, because patients were treated by two different orthopedic surgeon teams in different countries, multiple variables can influence patient clinical improvement besides the rehabilitation protocol. To minimize the influence of other variables, we ensured that surgical technique, component alignment target, and component design were identical between patient groups. Consequently, only a single hospital participated in Germany and in Canada, which could limit the generalizability of the results. Still, these were large teaching hospitals with high volumes, and as such were representative for the comparison of the two different rehabilitation protocols. Third, a cost analysis and comparison between both rehabilitation protocols could not be made, since profoundly different reimbursement policies were used. However, the gained knowledge from this study that functional gains after TKA are not compromised by omitting IR could motivate German insurance carriers, hospital systems, and healthcare authorities to conduct prospective studies analyzing the financial impact of omitting IR after uncomplicated TKA. Fourth, shorter follow-up intervals within the first year after TKA could have provided a more detailed comparison between FTR and IR [10]. Ideally, this comparison can be performed in a prospective study within Germany, which will also allow an accurate cost analysis that includes the financial impact of patient sick leave. Finally, the results from this study only apply to patients undergoing primary TKA with an ASA of I to III.

5. Conclusions

Among adults undergoing primary TKA, the use of IR compared with FTR does not yield a higher improvement in knee function. These findings do not support IR for this group of patients with the current rehabilitation regime. This study can provide a thought-provoking impulse to IR carriers to reorient their focus on rehabilitation after TKA by fostering patient salutogenesis.

Author Contributions: Conceptualization, D.R., A.J.N., B.A.M. and M.R.; methodology, D.R., A.J.N., B.A.M. and M.R.; software, A.J.N. and E.C.S.; validation, D.R., A.J.N., B.A.M., E.C.S. and M.R.; formal analysis, A.J.N. and E.C.S.; investigation, D.R., A.J.N. and E.C.S.; data curation, D.R., A.J.N., E.C.S., B.A.M. and M.R.; writing—original draft preparation, D.R. and A.J.N.; writing—review and editing, D.R., A.J.N., E.C.S., B.A.M. and M.R.; visualization, D.R. and A.J.N.; supervision, B.A.M. and M.R. All authors have read and agreed to the published version of the manuscript.

Funding: This research received no external funding.

Institutional Review Board Statement: The study was conducted in accordance with the Declaration of Helsinki, and approved by both the Canadian (The University of British Columbia, Office of Research Ethics, Study ID: H18-02307, 05.09.2018) and German (Ethik-Kommission der Universität Würzburg, Number: 20210925 01, 26.20.2021) Institutional Review Board.

Informed Consent Statement: Informed consent was obtained from all subjects involved in the study.

Data Availability Statement: The data sets to support the findings of this study are included within the article, including figures and tables. Any other data used to support the findings of this study are available from the corresponding authors upon request.

Acknowledgments: The authors thank BaCaTeC for supporting open access publication and Raman Johal at UBC for her help in research coordination.

Conflicts of Interest: The authors declare no conflict of interest.

References

1. Jiang, L.; Rong, J.; Wang, Y.; Hu, F.; Bao, C.; Li, X.; Zhao, Y. The relationship between body mass index and hip osteoarthritis: A systematic review and meta-analysis. *Jt. Bone Spine Rev. Du Rhum.* **2011**, *78*, 150. [CrossRef] [PubMed]
2. Mahomed, N.N.; Davis, A.M.; Hawker, G.; Badley, E.; Davey, J.R.; Syed, K.A.; Coyte, P.C.; Gandhi, R.; Wright, J.G. Inpatient compared with home-based rehabilitation following primary unilateral total hip or knee replacement: A randomized controlled trial. *J. Bone Jt. Surg. Am.* **2008**, *90*, 1673. [CrossRef] [PubMed]
3. Burnett, R.A., III; Serino, J.; Yang, J.; Della Valle, C.J.; Courtney, P.M. National trends in post-acute care costs following total knee arthroplasty from 2007 to 2016. *J. Arthroplast.* **2021**, *36*, 2268. [CrossRef]
4. Naylor, J.M.; Hart, A.; Mittal, R.; Harris, I.; Xuan, W. The value of inpatient rehabilitation after uncomplicated knee arthroplasty: A propensity score analysis. *Med. J. Aust.* **2017**, *207*, 250. [CrossRef] [PubMed]
5. El-Aarid, N. *Effektivität von Stationärer und Ambulanter Rehabilitation bei Patienten Nach Knieendoprothesen-Implantation*; Universität Rostock: Rostock, Germany, 2019.
6. Fussenich, W.; Gerhardt, D.M.; Pauly, T.; Lorenz, F.; Olieslagers, M.; Braun, C.; van Susante, J.L. A comparative health care inventory for primary hip arthroplasty between Germany versus the Netherlands. Is there a downside effect to fast-track surgery with regard to patient satisfaction and functional outcome? *Hip. Int.* **2020**, *30*, 423. [CrossRef] [PubMed]
7. Onggo, J.R.; Onggo, J.D.; De Steiger, R.; Hau, R. The Efficacy and Safety of Inpatient Rehabilitation Compared with Home Discharge After Hip or Knee Arthroplasty: A Meta-Analysis and Systematic Review. *J. Arthroplast.* **2019**, *34*, 1823. [CrossRef]
8. Scott, C.E.H.; Bell, K.R.; Ng, R.T.; MacDonald, D.J.; Patton, J.T.; Burnett, R. Excellent 10-year patient-reported outcomes and survival in a single-radius, cruciate-retaining total knee arthroplasty. *Knee Surg. Sports Traumatol. Arthrosc.* **2019**, *27*, 1106. [CrossRef]
9. Cosendey, K.; Eudier, A.; Fleury, N.; Pereira, L.C.; Favre, J.; Jolles, B.M. Ten-year follow-up of a total knee prosthesis combining multi-radius, ultra-congruency, posterior-stabilization and mobile-bearing insert shows long-lasting clinically relevant improvements in pain, stiffness, function and stability. *Knee Surg. Sports Traumatol. Arthrosc.* 2022; *online ahead of print*.

10. Fransen, B.L.; Hoozemans, M.J.; Argelo, K.D.; Keijser, L.; Burger, B.J. Fast-track total knee arthroplasty improved clinical and functional outcome in the first 7 days after surgery: A randomized controlled pilot study with 5-year follow-up. *Arch. Orthop. Trauma Surg.* **2018**, *138*, 1305. [CrossRef] [PubMed]
11. Seetharam, A.; Deckard, E.R.; Ziemba-Davis, M.; Meneghini, R.M. The AAHKS clinical research award: Are minimum two-year patient-reported outcome measures necessary for accurate assessment of patient outcomes after primary total knee arthroplasty? *J. Arthroplast.* **2022**, *37*, S716–S720. [CrossRef] [PubMed]
12. Clement, N.; MacDonald, D.; Simpson, A. The minimal clinically important difference in the Oxford knee score and Short Form 12 score after total knee arthroplasty. *Knee Surg. Sports Traumatol. Arthrosc.* **2014**, *22*, 1933. [CrossRef] [PubMed]
13. Clement, N.D.; Bardgett, M.; Weir, D.; Holland, J.; Gerrand, C.; Deehan, D.J. What is the Minimum Clinically Important Difference for the WOMAC Index After TKA? *Clin. Orthop. Relat. Res.* **2018**, *476*, 2005. [CrossRef] [PubMed]
14. Tsai, A.P.Y.; Hur, S.A.; Wong, A.; Safavi, M.; Assayag, D.; Johannson, K.A.; Morisset, J.; Fell, C.; Fisher, J.H.; Manganas, H.; et al. Minimum important difference of the EQ-5D-5L and EQ-VAS in fibrotic interstitial lung disease. *Thorax* **2021**, *76*, 37. [CrossRef] [PubMed]
15. Hosmer, D.W., Jr.; Lemeshow, S.; Sturdivant, R.X. *Applied Logistic Regression*; John Wiley & Sons: Hoboken, NJ, USA, 2013.
16. Buhagiar, M.A.; Naylor, J.M.; Harris, I.A.; Xuan, W.; Kohler, F.; Wright, R.; Fortunato, R. Effect of Inpatient Rehabilitation vs a Monitored Home-Based Program on Mobility in Patients With Total Knee Arthroplasty: The HIHO Randomized Clinical Trial. *JAMA* **2017**, *317*, 1037. [CrossRef] [PubMed]
17. Chan, H.Y.; Sultana, R.; Yeo, S.J.; Chia, S.L.; Pang, H.N.; Lo, N.N. Comparison of outcome measures from different pathways following total knee arthroplasty. *Singap. Med. J.* **2018**, *59*, 476.
18. Jorgenson, E.S.; Richardson, D.M.; Thomasson, A.M.; Nelson, C.L.; Ibrahim, S.A. Race, Rehabilitation, and 30-Day Readmission After Elective Total Knee Arthroplasty. *Geriatr. Orthop. Surg. Rehabil.* **2015**, *6*, 303. [CrossRef] [PubMed]
19. White, P.B.; Carli, A.V.; Meftah, M.; Ghazi, N.; Alexiades, M.M.; Windsor, R.E.; Ranawat, A.S. Patients Discharged to Inpatient Rehabilitation Facilities Undergo More Diagnostic Interventions with No Improvement in Outcomes. *Orthopedics* **2018**, *41*, E841. [CrossRef] [PubMed]
20. Nöth, U.; Rackwitz, L.; Clarius, M. Herausforderungen der Fast-Track-Endoprothetik in Deutschland. *Der Orthopäde* **2020**, *49*, 334. [CrossRef] [PubMed]
21. Lindström, B.; Eriksson, M. Salutogenesis. *J. Epidemiol. Community Health* **2005**, *59*, 440. [CrossRef] [PubMed]

Journal of *Personalized Medicine*

MDPI

Article

Individualized 3D-Printed Bone-Anchored Maxillary Protraction Device for Growth Modification in Skeletal Class III Malocclusion

Minji Kim [1,2], Jingwen Li [1], Sehyang Kim [1], Wonho Kim [2], Sun-Hyun Kim [1], Sung-Min Lee [3], Young Long Park [3], Sook Yang [4] and Jin-Woo Kim [1,3,*]

1. Graduate School of Clinical Dentistry, Ewha Womans University, Seoul 07985, Korea; minjikim@ewha.ac.kr (M.K.); jingwen7883@gmail.com (J.L.); shkim0363@gmail.com (S.K.); dhsh0828@gmail.com (S.-H.K.)
2. Department of Orthodontics, School of Medicine, Ewha Womans University, Seoul 07985, Korea; kimwonho311@gmail.com
3. Department of Oral and Maxillofacial Surgery, School of Medicine, Ewha Womans University, Seoul 07985, Korea; sm930513@dent.dku.edu (S.-M.L.); 21628@eumc.ac.kr (Y.L.P.)
4. Cusmedi Co., Ltd., Suwon-si 400815, Korea; syang@cusmedi.com

* Correspondence: jwkim84@ewha.ac.kr; Tel.: +82-2-2650-2720; Fax: +82-2-2650-2754

Abstract: Bone-anchored maxillary protraction (BAMP) is effective for skeletal Class III malocclusion. However, infection, screw and plate loosening, and device failures occur with conventional plates. This pilot prospective study analyzed the feasibility of individualized BAMP using preoperative simulation and 3D titanium printing in patients referred by the orthodontic department for four BAMP miniplates. Preoperative cone beam computed tomography data were analyzed using CAD/CAM software to fabricate the individualized 3D-printed BAMP device. The customized plates were printed using selective laser sintering and inserted onto the bone through an adjunct transfer jig. The accuracy of preoperative simulation and actual placement of the BAMP device were tested by superimposing simulated positioned digital images and postoperative computed tomography data. The growth modification effect depended on superimposition of lateral cephalograms and comparative changes in SNA, SNB, ANB, and Wits. Two male patients were finally included in the study. BAMP decreased the ANB difference (−4.56 to −1.09) and Wits appraisal (−7.52 to −3.26) after 2 years. Normal measurement indices for sagittal and vertical growth indicated successful growth modification. The mean accuracy between preoperative simulation and actual surgery was 0.1081 ± 0.5074 mm. This treatment modality involving preoperative simulation and 3D titanium printing for fabricating and placing customized BAMP devices precisely at planned locations is effective for treating skeletal Class III malocclusion.

Keywords: bone-anchored maxillary protraction; skeletal class III malocclusion; 3D printing; growth modification

Citation: Kim, M.; Li, J.; Kim, S.; Kim, W.; Kim, S.-H.; Lee, S.-M.; Park, Y.L.; Yang, S.; Kim, J.-W. Individualized 3D-Printed Bone-Anchored Maxillary Protraction Device for Growth Modification in Skeletal Class III Malocclusion. *J. Pers. Med.* **2021**, *11*, 1087. https://doi.org/10.3390/jpm11111087

Academic Editor: Anne-Marie Caminade

Received: 27 September 2021
Accepted: 23 October 2021
Published: 26 October 2021

1. Introduction

Skeletal Class III malocclusion is a common orthodontic deformity involving a single or a combination of protrusive mandibles, deficient or retrusive maxilla, protrusive mandibular dentition, and retrusive maxillary dentition, which not only affect a patient's masticatory function but also their esthetic appearance and mental health [1,2]. Prevalence of Skeletal Class III was reported as 7.8% [3], and Class III presents the highest prevalence among orthognathic cases [4].

In growing patients, early intervention for correcting the craniofacial relationship has been used for several decades with significant outcomes [5–7].

It is better to use extraoral anchorage than intraoral anchorage for skeletal growth control in the early intervention for Class III malocclusion to prevent undesirable dental changes from anchored teeth force application [8]. Among the different therapies, bone-anchored maxillary protraction (BAMP) is considered one of the most reliable methods for Class III malocclusion; it eliminates dentoalveolar effects and facilitates patient full-time wear compliance by using only intraoral skeletal anchorage. BAMP with miniplates is used in growing patients, with multiple miniscrews to ensure anchor stability and avoid any possible damage to the tooth germ [9]. A recent systematic review and clinical study showed reliable results for the treatment of skeletal Class III malocclusion, in which BAMP was reported to enhance maxillary growth and inhibit mandibular growth [10].

Despite its advantages, there are limitations to BAMP treatment, such as the need for additional surgery, possible tooth germ injury, and irritation of the adjacent tissues by elastics or miniplates. The miniplates can also loosen due to insufficient bone density at early ages. Therefore, the precise application of ready-made miniplates of sufficient bone quality following the contour of the cortical surface while avoiding the tooth germ is often surgically challenging [11,12].

Recent developments in 3D printing technology and the introduction of digital dentistry have made it possible to fabricate patient-customized simulations and individualized dental devices. In addition, metal 3D printing using selective laser sintering (SLS) allows the fabrication of individualized bone fixation plates and bone reconstruction material. BAMP is an effective treatment modality for skeletal Class III treatment; however, it is challenging to apply conventional plates onto the correct location. Therefore, the clinical feasibility of individualized BAMP plates using preoperative simulation and 3D titanium printing was tested in this pilot study.

2. Methods

2.1. Study Sample

This pilot prospective study was performed from 2019 to 2021 in departments of orthodontics and oral and maxillofacial surgery at Ewha Womans University Mokdong Hospital. This study was approved by the institutional ethics committees (EUMC 2019-06-014). Patients who were referred by the orthodontics department for four BAMP miniplates were included in the study. Any syndromic patients and cleft patients were excluded.

2.2. Data Acquisition

Cone beam computed tomography (CBCT) and intraoral scanning were performed to collect essential pre-surgical information to fabricate the individualized 3D-printed BAMP device. Since the BAMP device itself is placed on the maxillary and mandibular bone surface, it can be manufactured solely with CBCT, without digital information of the dentition. However, we previously experienced inaccurate placing of the BAMP device without dentitional information during the actual surgery. Therefore, an additional tooth-guided jig that can deliver the BAMP device to the precise simulated position was manufactured, and intraoral scanning was performed for an accurate tooth-guided transfer jig for properly positioning the plate.

Dicom data of CBCT were extracted into STL format and merged with intraoral scanning STL data. The CBCT dataset obtained 2 weeks before surgery was surface-rendered in the 3D model (STL format) of the bone. Intraoral Scanning with Trios3 (3 shape, Copenhagen, Denmark) started with the most distal tooth in the third quadrant and continuing to the anterior teeth. Next, the fourth quadrant was scanned, again beginning with the most distal tooth. Scanning of the maxilla started with the most distal tooth in the second quadrant and ended at the central incisor. The first quadrant was recorded starting with the most distal tooth. The camera was positioned at 45 degrees (or as close as possible to the axis of the tooth) to the buccal and lingual scans. The scanning device works by means of confocal microscopy, with a fast scanning time; the light source provides an illumination pattern to cause light oscillation on the object. As for the superimposition and merger of DICOM+STL (including software info), 3D Slicer(extension slicer RT, ver.4.11, open source) was used to create a 3D bone model file (stl format), and Meshmixer (ver.3.5, Autodesk) was used to edit the surface model created by 3D slicer.

2.3. Preoperative Simulation

Using Ondemand® CAD/CAM software (Cybermed Co., Seoul, Korea) and Doctor Check software® (Cusmedi Co., Seoul, Korea), virtual miniplates and an additional transfer jig were designed with respect to the patient's bone contour, location of the hook, and shape of the surgical site. The maxillary plates for BAMP are preferably positioned on the zygomatico-maxillary junction and maxillary first molar. Positioning as close to the teeth as possible is more advantageous for surgical convenience and minimal surgical invasion. When located in the alveolar bone vertically too close to the teeth, there is a risk of tooth germ injury and screw loosening due to relatively more active bone remodeling and the low density of the maxillary bone during treatment. When positioned higher, near the zygomatic bone, permanent tooth germ injury can be avoided, and fixation strength can be obtained due to the greater bone density. The proper position, considering bone density for fixational support and avoiding tooth germ injury, could be determined preoperatively on the CBCT.

2.4. Surgical Procedure

The BAMP device consisted of a closed subperiosteal part for bone fixation and an open intraoral part for applying the elastics to deliver orthopedic force. The conventional flat-design bone fixation miniplate does not follow the bone surface contour nor does it form a tight contact with the bone, leaving small gaps that could cause biological complications. To minimize these complications, the BAMP plate is designed to be positioned to emerge intraorally at the mucogingival junction 5 mm above the gingival margin so that it can be placed along the anatomical contour when it is opened intraorally in the bone fixation part. This design minimizes the biological complications of the device maintained in the oral cavity for at least 6 months and up to 3 years. The elastics can be applied between the maxillary premolar and the first molar to maximize the patient's treatment adherence.

Accurate application of the simulated BAMP device to the determined location during actual surgery is a challenge. An additional transfer jig can be fabricated using precise intraoral scanning data to minimize the error between simulation and actual surgery. The transfer jig is suitable for being referenced to adjacent maxillary teeth and is manufactured to carry and deliver the BAMP device. The transfer jig is manufactured using resin or titanium metal printing (SLA type, EP-3500, Shinin). The BAMP device to be placed on the mandibular anterior teeth is designed in the same manner. The mandibular device is placed on the mandibular symphysis; therefore, tooth germ injury and bone density are less of a concern than with the maxillary device.

The design of the customized BAMP device through preoperative simulation is exported as an STL file from a CAD software program (Magics, Materialise, Leuven, Belgium) and submitted for 3D printing (Metalsys150, Winforsys Co., Seoul, Korea; laser power 120–200 W) using titanium alloy (Ti6Al4V ELI, medical grade in accordance with ASTM F136, AP&C, Quebec, QC, Canada) following the SLS additive manufacturing technique. Titanium powder was melted into thin layers, which were used in the 3D printer. Post-processing of the plate involved removal of the loose power and the support structures and polishing to improve the surface quality and mechanical properties and reduce errors during CBCT and scanning data acquisition.

A rapid prototyping model of the patient was manufactured before surgery, and a virtual surgery was performed through the process of pre-fitting. In the actual surgery, customized plates were installed at the surgical site using the transfer jig as a guide. An incision (1.5–2.5 cm) was made above the bone where the fixation plate had to be placed, and minimal dissection was performed. Then, the jig was precisely matched to the reference point, indicating accurate installation of the BAMP device. After the correct position of the BAMP device was confirmed, fixation was performed using 1.5 mm screws. After confirming that the BAMP device was sufficiently fixed to the bone, the incision was sutured using vicryl 4-0 and 5-0.

Postoperatively, patients were instructed to use a chlorhexidine mouth rinse three times a day for 7 days. Analgesics and non-steroidal anti-inflammatory drugs were prescribed. After 7–10 days of soft tissue healing, Class III elastics were applied with an initial force of ~70 g on each side. The force of the elastics was increased to 150 g after 1 month of traction and again increased to 200–250 g after 3 months. Patients were instructed to wear the elastics full-time, except for meals and for brushing. They had to replace the elastics daily, with regular follow-up. Patients were also instructed to follow-up once a month postoperatively.

2.5. Cephalometric and Superimposition Analysis

Simulated positioned digital images and postoperative CT data were superimposed for comparison to evaluate the accuracy of preoperative simulation and actual application of the fabricated BAMP device. In addition, lateral cephalometric analysis was performed before BAMP treatment and 2 years postoperatively to evaluate the treatment effects of BAMP's skeletal Class III correction. The superimposition of lateral cephalograms was analyzed, and changes in SNA, SNB, ANB, and Wits were compared to determine the effect of growth modification.

The 3D deviation between the position of the BAMP designed with CBCT and the position of the BAMP after surgery was investigated to evaluate the accuracy between virtual simulation and actual surgery. Twenty points were randomly selected from BAMP and used as reference points to investigate the single linear deviation, reported in mm, along the x-, y-, and z-axis. The x-axis is the transverse (lateral/medial) direction, the y-axis is the sagittal (anterior/posterior) direction, and the z-axis is the axial (cranial/caudal) direction. The deviation in the cranial, anterior, and lateral directions of the BAMP was considered positive (+), and the deviation in the contralateral direction was considered negative (−). The 3D average distance is expressed in mm, and the difference is indicated by color (Supplementary Figure S1). The 3D geometrical deviation was assessed after color coding high or low geometrical deviation using Geomagic software (Freeform Plus, 3D Systems, Morrisville, NC, USA) (Supplementary Figure S2).

3. Results

Four patients were screened, and two were included in this study for BAMP therapy. The two included patients were both male, 9 and 10 years old. All operations were performed under local anesthesia, and the average time from incision to the upper bone of the BAMP device through the transfer jig, from screw fixation to sutures, was 8.3 min per site, on average. Unlike for conventional plate surgery, the time to consider tooth germ injury and sufficiently dense bone and application work through screw bending were not relevant; therefore, the surgery time was dramatically reduced. In addition, since the plate fixing position was accurately known in advance, minimally invasive surgery, including incision and dissection, was possible.

During the follow-up, the customized miniplates showed no loosening or inflammation. The miniplates were well-retained for over a year, had sufficient retention force, and maintained good oral hygiene in the surrounding tissue.

Skeletal measurements revealed that BAMP decreased ANB difference (−4.56 to −1.09) and Wits appraisal (−7.52 to −3.26). Other measurements indices for sagittal and vertical growth were also in the normal range, indicating successful growth modification with customized BAMP therapy for skeletal Class III malocclusion (Table 1).

The accuracy between simulation and actual surgery was satisfactory. In the measurement of the linear deviation between preoperative simulation and actual surgery, the average difference was −0.0154 mm on the x-axis, −0.0946 mm on the y-axis, and 0.0579 mm on the z-axis, with an average distance of 0.1081 mm (Table 2).

Table 1. Skeletal measurements: preoperative and postoperative (24 months) *.

Measurement	Reference Value	Preoperative	Postoperative 24 Months
Maxillo-mandibular measurement			
SNA	81.6 (3.1)	77.98	84.63
SNB	79.1 (3.0)	82.54	85.72
ANB difference	2.4 (1.8)	−4.56	−1.09
Wits appraisal	−2.7 (2.4)	−7.52	−3.26
Angular measurement			
FMA	29.63 (5.66)	30.34	28.98
SN-GoMe	36 (4)	37.1	32.72
A point—N Perpend	0.4 (2.3)	−5.37	−1.86
Facial height measurement			
Facial Height Ratio	65 (9)	61.7	66.37
Posterior Facial Height	85 (5.5)	76.84	90.76
Anterior Facial Height	127.4 (5.6)	124.52	136.74
Dentoalveolar measurement			
U1 to SN	107 (6)	123.45	120.62
IMPA	95.9 (6.3)	78.2	78.3
Interincisal angle	124 (8.3)	121.2	128.32
Soft tissue measurement			
Upper Lip E-plane	0.1 (2)	0.44	−1.58
Lower Lip E-plane	0.1 (2)	2.69	0.26
Nasolabial angle	84.9 (5)	93.86	101.9

* Values are presented as mean (standard deviation).

Table 2. Linear deviations in the *x*-, *y*-, and *z*-axis at 20 selected reference points.

	Linear Deviation			
	x-Axis (Lateral/Medial)	*y*-Axis (Anterior/Posterior)	*z*-Axis (Cranial/Caudal)	Distance Difference
N	20	20	20	20
Median	−0.0154	−0.0946	0.0579	0.1081
Range	−0.1729–0.2448	−1.1643–1.5076	−0.9189–0.7126	0.0211–1.3772
SD	0.1463	0.9012	0.5515	0.5074

4. Case Presentation

A 10-year-old boy presented with maxillary retrognathism, mandible prognathism, and dentofacial deformity and was diagnosed with skeletal Class III malocclusion (Figures 1 and 2). The patient had no systemic diseases and no signs of temporomandibular joint dysfunction. Cephalometric analysis showed a concave facial profile and hyperdivergent growth pattern. Considering the patient's age and growth patterns, treatment with customized BAMP miniplates was considered as the most optimal therapy.

For the left maxilla, a 'Y' shaped plate was used and fixed at the zygomatic process to avoid possible disruption of the molar roots. An 'L' shaped plate was designed for the mandible and fixed at the symphysis of mandible. After preoperative simulation and fabrication, BAMP plates were used for the individualized rapid prototyping model (Figure 3). Under local anesthesia, customized BAMP plates were installed at the simulated location using the transfer jig as a guide. The jig was precisely matched with the reference point for accurate installation of the miniplates (Figure 4). The mean surgical time from incision to suture was 6.4 min for each side.

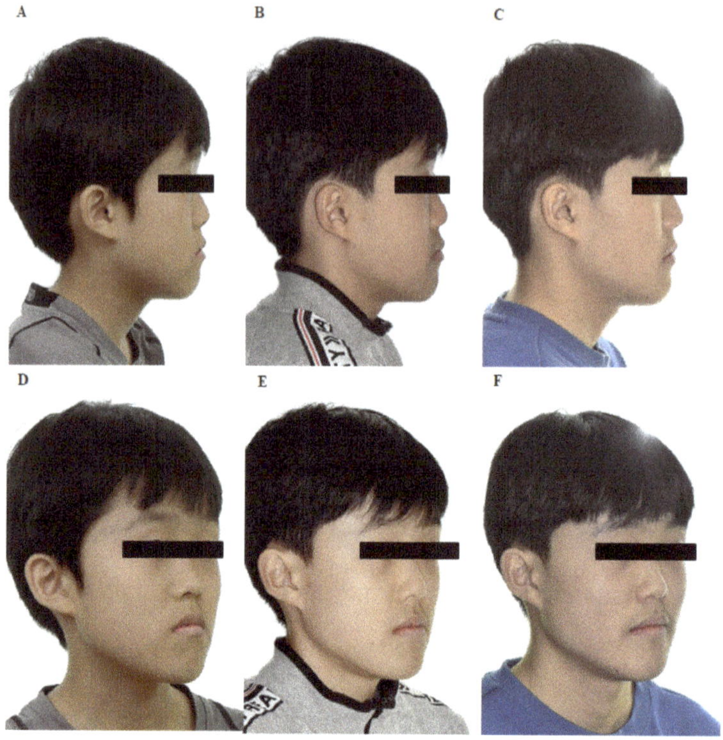

Figure 1. Facial photographs showing lateral and 45 degree profiles (**A**,**D**) at initial, (**B**,**E**) 12 months after customized bone-anchored maxillary protraction (BAMP) therapy, and (**C**,**F**) 24 months after customized BAMP therapy.

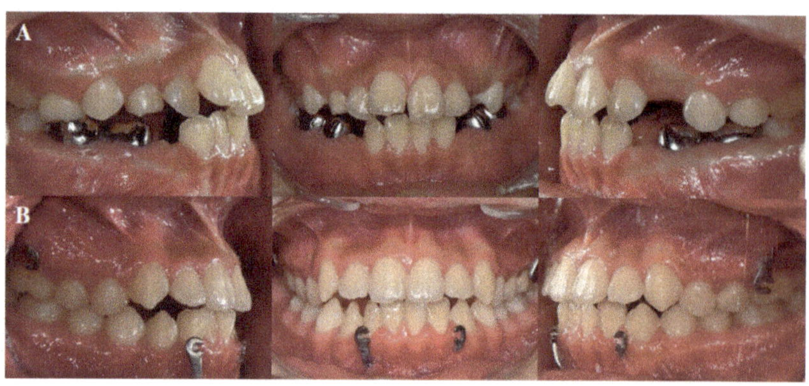

Figure 2. (**A**) Initial intraoral photographs. (**B**) Intraoral photographs 24 months after customized bone anchored maxillary protraction therapy showing Class I molar key.

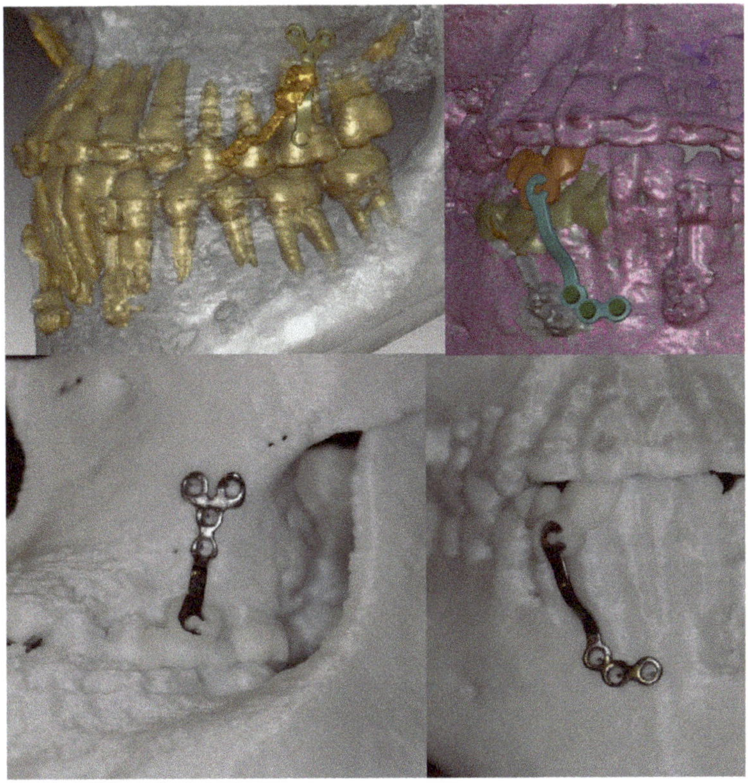

Figure 3. Preoperative simulation and application of fabricated customized plates and transfer jig on the individualized rapid prototyping model.

The patient was advised to maintain good oral hygiene after the operation. After soft tissue healing, Class III elastics (5/8 3.50 oz) were applied with an initial force of ~70 g on each side. The force of elastics (left: 5/16 3.50 oz; right: 3/16 3.50 oz) was increased to 150 g after 1 month of traction and again increased to 250 g after 3 months. The patient was instructed to wear the elastics full-time, except during meals and brushing. The patient also had to replace the elastics daily, with regular follow-up (Figure 4). During the follow-up, the customized miniplates showed no loosening or inflammation. The miniplates were well-retained for over two years, showing sufficient retention force, and maintained good oral hygiene in the surrounding tissue. Superimposition of lateral cephalograms indicated that BAMP therapy successfully protracted the maxilla and prevented mandibular growth and hyperdivergent vertical growth (Figure 5).

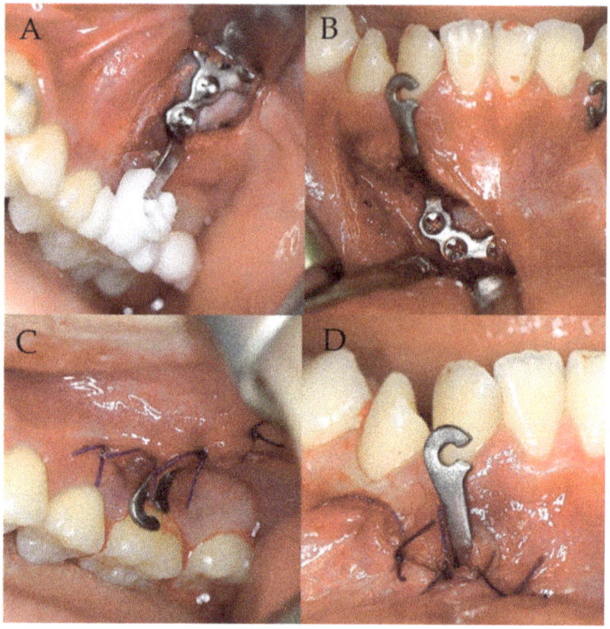

Figure 4. Surgical photographs. (**A**,**B**) Customized BAMP miniplates were transferred to the pre-simulated location using a transfer jig. (**C**,**D**) Postoperative photographs.

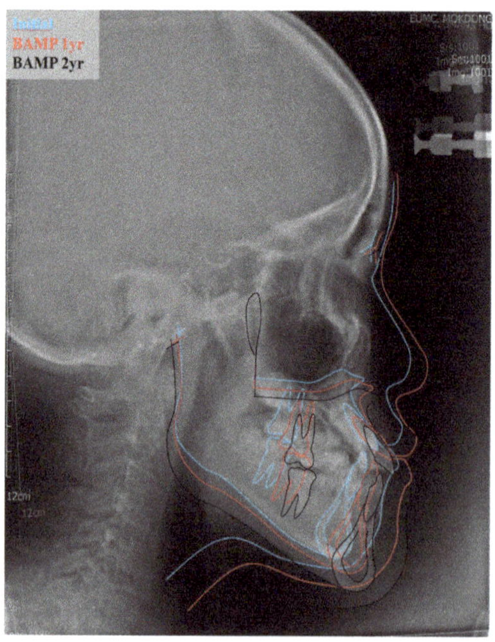

Figure 5. Superimposition of lateral cephalograms at initial stage, 12 months after BAMP therapy, and 24 months after BAMP therapy. The superimposition was based on the cranial base.

5. Discussion

Recently, bone-anchored maxillary protraction has been regarded as an effective treatment method for growth modification in growing skeletal Class III patients; however, an individualized approach, especially using 3D metal plate printing, has not been previously reported. The authors believe sharing our experience in the study will contribute to the acceleration of the medical use of individualized metal printing to the field of dentistry.

Skeletal Class III malocclusion correction in growing patients is considered one of the most challenging treatments in orthodontics. It is conducted using either an extraoral or intraoral device. The use of a facemask has been one of the most common approaches for maxillary traction, but it may result in undesired dentoalveolar outcomes, such as upper incisors or extrusion of the upper molars. A chin cup is effective for mandibular development restriction, but it may cause anterior teeth inclination in the mandible due to the pressure applied on the bone and soft tissue [13]. Wearing an extraoral appliance may be esthetically unpleasing and may cause social discomfort in growing patients, oftentimes leading to noncompliance [14]. Miniscrews or miniplates have replaced the use of extraoral devices for such orthopedic force transfer. Intraoral miniscrews and miniplates have shown sufficient biocompatibility and durability for the loading of orthodontic force, significantly influencing the advancement of maxillary structures [9,11].

However, the use of bone anchorages has been increasing, and various factors relating to the failure of implants are being studied. The location of the implant, diameter of the screw, shape of the miniplate, inflammation, and keratinization of peri-implant tissues affect the stability of bone anchorage [15,16]. Age is also a crucial factor in adolescent patients. The stability of BAMP therapy can be attributed to mechanical retention and the density and thickness of the cortical bone. In our case, the plate on the left maxilla and right mandible were found mobile after 5 months of installation and eventually failed. As a child, the patient's developing bone may have not been strong enough to endure the force applied for anchorage. Applied orthopedic force for bone correction, which is relatively higher than that applied for tooth movement, may also have influenced the stability of bone anchorage. Bone remodeling occurs quickly and plays a significant role in young patients during their growing period. When excess force is applied on the anchorage device, its loading can significantly influence the density and turnover of the alveolar bone, loosening the BAMP correction device and leading to eventual failure.

Oral hygiene and plaque accumulation may also affect the success of plate insertion. Microbial flora at the interface between the artificial material and mucosal epithelium influence plate fixation. Anaerobic bacteria at the insertion site may cause inflammation. In our case, only mild infection was observed in the right mandible. Keeping the emergence point within the attached gingiva is essential to avoid bacterial invasion [17].

A customized titanium miniplate was manufactured after the failure of anchorage on the left maxilla and right mandible. The transfer jig created for each surgical site allowed clinicians to precisely determine the final location and delivery of the plate during the surgery [18,19]. CBCT and oral scan data were used to evaluate and design the customized miniplates accordingly. In the second operation, the customized miniplates were to be fixed at the zygomatic process and symphyseal to avoid the previous insertion site and any possible damage to the surrounding structures, such as the root and tooth germ. Bending and adjusting the plate contour not only alters the mechanical strength of the plate but also causes its inevitable deformation. Customized plates manufactured according to the skeletal contour of the patient eliminate the need for adjustment during the surgery. The use of customized devices allows clinicians to make smaller but more accurate incisions and deliver the plates in their ideal locations within a shorter period. There was a report of a minimally invasive approach for BAMP placement without surface incisions in the gingival margins or papillae. This potentially minimizes the gingival recession that sometimes accompanies flap surgery [20]. A combination of maxillary protraction, expansion and contraction would be beneficial in this case.

A BAMP device ideally fixes to the jawbone and directly contacts the oral surface through the open mucosa. There are inherent risks of plaque accumulation and chronic inflammation, but BAMP therapy is shorter and more efficient than extraoral orthopedic devices such as a chin cup.

6. Conclusions

Preoperative simulation and a 3D titanium printing technique allowed the fabrication of customized BAMP devices and their precise application at the planned locations. The clinical results demonstrated the stability and effectiveness of this modality for the treatment of skeletal Class III malocclusion.

Supplementary Materials: The following are available online at https://www.mdpi.com/article/10.3390/jpm11111087/s1, Figure S1: Distance measurement of 20 selected reference points, Figure S2: A color-coded deviation figure that visualizes the high or low geometrical deviation.

Author Contributions: S.K., S.-H.K., M.K.—Conception and design of the work, analysis, interpretation of data, critical revision of draft; S.-M.L., Y.L.P., S.Y., J.L.—Conception and design of the work, analysis, interpretation of data, original draft preparation; W.K.—Analysis, interpretation of data, critical revision of draft; J.-W.K.—Conception and design of the work, analysis, interpretation of data, critical revision of draft. All authors have read and agreed to the published version of the manuscript.

Funding: This research was supported by the KIAT (Korea Institute for Advancement of Technology) grant funded by the Korea Government (MOTIE:Ministry of Trade Industry and Energy). (P0008799), and the Korea Medical Device Development Fund grant from the Korea Government (the Ministry of Science and ICT, the Ministry of Trade, Industry and Energy, the Ministry of Health & Welfare, the Ministry of Food and Drug Safety) (Project Number: 9991006713, KMDF_PR_20200901_0040).

Institutional Review Board Statement: The study was conducted according to the guidelines of the Declaration of Helsinki and approved by the Institutional Review Board (or Ethics Committee) of Ewha Womans Unviersity Mok-Dong Hospital (EUMC 2019-06-14).

Informed Consent Statement: Informed consent was obtained from all subjects involved in the study.

Data Availability Statement: Data is available with permission of corressponding author.

Conflicts of Interest: The authors declare no conflict of interest. The funders had no role in the design of the study; in the collection, analyses, or interpretation of data; in the writing of the manuscript, or in the decision to publish the results.

References

1. Proffit, W.R.; Fields, H.W., Jr.; Sarver, D.M. *Contemporary Orthodontics*; Elsevier Health Sciences: Amsterdam, The Netherlands, 2006.
2. Litton, S.F.; Ackermann, L.V.; Isaacson, R.J.; Shapiro, B.L. A genetic study of Class 3 malocclusion. *Am. J. Orthod.* **1970**, *58*, 565–577. [CrossRef]
3. Borzabadi-Farahani, A.; Borzabadi-Farahani, A.; Eslamipour, F. Malocclusion and occlusal traits in an urban Iranian population. An epidemiological study of 11- to 14-year-old children. *Eur. J. Orthod.* **2009**, *31*, 477–484. [CrossRef] [PubMed]
4. Eslamian, L.; Borzabadi-Farahani, A.; Badiee, M.R.; Le, B.T. An Objective Assessment of Orthognathic Surgery Patients. *J. Craniofac. Surg.* **2019**, *30*, 2479–2482. [CrossRef]
5. White, L. Early orthodontic intervention. *Am. J. Orthod. Dentofac. Orthop.* **1998**, *113*, 24–28. [CrossRef]
6. Zhang, H.; Deng, F.; Wang, H.; Huang, Q.; Zhang, Y. Early orthodontic intervention followed by fixed appliance therapy in a patient with a severe Class III malocclusion and cleft lip and palate. *Am. J. Orthod. Dentofac. Orthop.* **2013**, *144*, 726–736. [CrossRef] [PubMed]
7. Azamian, Z.; Shirban, F. Treatment Options for Class III Malocclusion in Growing Patients with Emphasis on Maxillary Protraction. *Sci. Cairo* **2016**, *2016*. [CrossRef] [PubMed]
8. Sugawara, J.; Baik, U.B.; Umemori, M.; Takahashi, I.; Nagasaka, H.; Kawamura, H.; Mitani, H. Treatment and posttreatment dentoalveolar changes following intrusion of mandibular molars with application of a skeletal anchorage system (SAS) for open bite correction. *Int. J. Adult Orthodon. Orthognath. Surg.* **2002**, *17*, 243–253. [PubMed]
9. De Clerck, H.; Cevidanes, L.; Baccetti, T. Dentofacial effects of bone-anchored maxillary protraction: A controlled study of consecutively treated Class III patients. *Am. J. Orthod. Dentofac. Orthop.* **2010**, *138*, 577–581. [CrossRef]
10. Rodríguez de Guzmán-Barrera, J.; Sáez Martínez, C.; Boronat-Catalá, M.; Montiel-Company, J.M.; Paredes-Gallardo, V.; Gandía-Franco, J.L.; Almerich-Silla, J.M.; Bellot-Arcís, C. Effectiveness of interceptive treatment of class III malocclusions with skeletal anchorage: A systematic review and meta-analysis. *PLoS ONE* **2017**, *12*, e0173875. [CrossRef] [PubMed]

11. Cevidanes, L.; Baccetti, T.; Franchi, L.; McNamara, J.A., Jr.; De Clerck, H. Comparison of two protocols for maxillary protraction: Bone anchors versus face mask with rapid maxillary expansion. *Angle Orthod.* **2010**, *80*, 799–806. [CrossRef] [PubMed]
12. Cornelis, M.A.; Scheffler, N.R.; Mahy, P.; Siciliano, S.; De Clerck, H.J.; Tulloch, J.F. Modified miniplates for temporary skeletal anchorage in orthodontics: Placement and removal surgeries. *J. Oral Maxillofac. Surg.* **2008**, *66*, 1439–1445. [CrossRef] [PubMed]
13. Kim, J.H.; Viana, M.A.; Graber, T.M.; Omerza, F.F.; BeGole, E.A. The effectiveness of protraction face mask therapy: A meta-analysis. *Am. J. Orthod. Dentofac. Orthop.* **1999**, *115*, 675–685. [CrossRef]
14. Sergl, H.G.; Klages, U.; Zentner, A. Functional and social discomfort during orthodontic treatment—Effects on compliance and prediction of patients' adaptation by personality variables. *Eur. J. Orthod.* **2000**, *22*, 307–315. [CrossRef] [PubMed]
15. Miyawaki, S.; Koyama, I.; Inoue, M.; Mishima, K.; Sugahara, T.; Takano-Yamamoto, T. Factors associated with the stability of titanium screws placed in the posterior region for orthodontic anchorage. *Am. J. Orthod. Dentofac. Orthop.* **2003**, *124*, 373–378. [CrossRef]
16. Cheng, S.J.; Tseng, I.Y.; Lee, J.J.; Kok, S.H. A prospective study of the risk factors associated with failure of mini-implants used for orthodontic anchorage. *Int. J. Oral Maxillofac. Implant.* **2004**, *19*, 100–106.
17. Murakami, S.; Mealey, B.L.; Mariotti, A.; Chapple, I.L.C. Dental plaque-induced gingival conditions. *J. Clin. Periodontol.* **2018**, *45*, S17–S27. [CrossRef] [PubMed]
18. Kim, S.H.; Choi, Y.S.; Hwang, E.H.; Chung, K.R.; Kook, Y.A.; Nelson, G. Surgical positioning of orthodontic mini-implants with guides fabricated on models replicated with cone-beam computed tomography. *Am. J. Orthod. Dentofac. Orthop.* **2007**, *131*, S82–S89. [CrossRef] [PubMed]
19. Yu, J.J.; Kim, G.T.; Choi, Y.S.; Hwang, E.H.; Paek, J.; Kim, S.H.; Huang, J.C. Accuracy of a cone beam computed tomography-guided surgical stent for orthodontic mini-implant placement. *Angle Orthod.* **2012**, *82*, 275–283. [CrossRef] [PubMed]
20. Zadeh, H.H.; Borzabadi-Farahani, A.; Fotovat, M.; Kim, S.H. Vestibular Incision Subperiosteal Tunnel Access (VISTA) for Surgically Facilitated Orthodontic Therapy (SFOT). *Contemp. Clin. Dent.* **2019**, *10*, 548. [PubMed]

Journal of *Personalized Medicine*

MDPI

Article

A Pan-Cancer Atlas of Differentially Interacting Hallmarks of Cancer Proteins

Medi Kori [1,†], Gullu Elif Ozdemir [1,†], Kazim Yalcin Arga [1,2,*] and Raghu Sinha [3,*]

1 Department of Bioengineering, Marmara University, Istanbul 34854, Turkey
2 Genetic and Metabolic Diseases Research and Investigation Center, Marmara University, Istanbul 34854, Turkey
3 Department of Biochemistry and Molecular Biology, Penn State College of Medicine, Hershey, PA 17033, USA
* Correspondence: kazim.arga@marmara.edu.tr (K.Y.A.); rus15@psu.edu (R.S.)
† These authors contributed equally to this work.

Abstract: Cancer hallmark genes and proteins orchestrate and drive carcinogenesis to a large extent, therefore, it is important to study these features in different cancer types to understand the process of tumorigenesis and discover measurable indicators. We performed a pan-cancer analysis to map differentially interacting hallmarks of cancer proteins (DIHCP). The TCGA transcriptome data associated with 12 common cancers were analyzed and the differential interactome algorithm was applied to determine DIHCPs and DIHCP-centric modules (i.e., DIHCPs and their interacting partners) that exhibit significant changes in their interaction patterns between the tumor and control phenotypes. The diagnostic and prognostic capabilities of the identified modules were assessed to determine the ability of the modules to function as system biomarkers. In addition, the druggability of the prognostic and diagnostic DIHCPs was investigated. As a result, we found a total of 30 DIHCP-centric modules that showed high diagnostic or prognostic performance in any of the 12 cancer types. Furthermore, from the 16 DIHCP-centric modules examined, 29% of these were druggable. Our study presents candidate systems' biomarkers that may be valuable for understanding the process of tumorigenesis and improving personalized treatment strategies for various cancers, with a focus on their ten hallmark characteristics.

Keywords: hallmarks of cancer; differential interactome; system biomarkers; druggability; personalized treatments

Citation: Kori, M.; Ozdemir, G.E.; Arga, K.Y.; Sinha, R. A Pan-Cancer Atlas of Differentially Interacting Hallmarks of Cancer Proteins. *J. Pers. Med.* **2022**, *12*, 1919. https://doi.org/10.3390/jpm12111919

Academic Editor: Anne-Marie Caminade

Received: 23 October 2022
Accepted: 15 November 2022
Published: 17 November 2022

1. Introduction

Cancer is one of the leading causes of death in almost every country in the world. According to GLOBOCAN, there were about 19.3 million new cancer cases and 10 million cancer deaths in 2020 [1], and these numbers make the fateful picture clear. Cancer is an uncontrolled cell-division process that leads to cell transformation through the occurrence of various and successive genetic alterations. This transformation is the result of a complex process, and this complexity makes the disease an enigma shrouded in mystery and incurable [2].

In 2000, Hanahan and Weinberg wondered, as there were a variety of cancers, how many regulatory mechanisms were affected, or were the same regulatory mechanisms destroyed in the cell to become cancerous? Additionally, were there common signals? After asking these questions, they proposed that cancer cells exhibited six biological phenomena, which they called "hallmarks of cancer". They proposed that these six hallmarks play a critical role in cancer development and are likely to occur in all cancers [3]. In the coming years, Hanahan and Weinberg added four more cancer features to the list of characteristics with the ongoing studies. Subsequently, with this update, the ten hallmarks of cancer approved today (eight trait characteristics and two enabling characteristics) were formed.

These hallmarks include: (i) activating invasion and metastasis; (ii) enabling replicative immortality; (iii) evading growth suppressors; (iv) avoiding immune destruction; (v) genome instability and mutation; (vi) inducing angiogenesis; (vii) deregulating cellular energetics; (viii) resisting cell death; (ix) sustaining proliferative signaling; and (x) tumor-promoting inflammation [4]. Hanahan suggested, in his latest article, that potential features such as cellular plasticity, non-mutative epigenetic reprogramming, polymorphic microbiomes, and impaired differentiation could be added to the list in the future. However, he believes that the definition of these conceptual features needs to be discussed and experimentally validated with cancer biology studies [5].

Since the hallmarks of cancer genes and proteins orchestrate and drive carcinogenesis to a large extent, it is noteworthy to study these features in different cancers to understand the process of tumorigenesis and discover measurable indicators of altered biological states such as biomarkers. A study which was conducted by Nagy et al. [6] performed a pan-cancer survival analysis of cancer hallmark genes by not considering the associations among them. However, the development of cancer is generally induced by a combination of genes, proteins, metabolites, and other factors. In 2019, Yu and co-workers, by considering the association between genes, identified cancer hallmarks based on the gene co-expression networks for seven cancers [7]. Nevertheless, deciphering the changes at the protein level (protein interactome) is essential to understand tumorigenesis at the systemic level. Therefore, deciphering the hallmarks of different cancers using protein–protein interactions (PPIs) within the protein interactome is a promising approach to understand the mechanisms of cancer and to propose diagnostic, prognostic or therapeutic biomarker targets.

In our study, we performed a pan-cancer analysis and aimed to map differential hallmarks of cancer-associated protein–protein interactions (Figure 1). We examined the Cancer Genome Atlas (TCGA) transcriptome data from 12 different cancers: breast invasive carcinoma (BRCA); colon adenocarcinoma (COAD); head and neck squamous cell carcinoma (HNSC); kidney renal clear cell carcinoma (KIRC); kidney renal papillary cell carcinoma (KIRP); liver hepatocellular carcinoma (LIHC); lung adenocarcinoma (LUAD); lung squamous cell carcinoma (LUSC); prostate adenocarcinoma (PRAD); stomach adenocarcinoma (STAD); thyroid carcinoma (THCA); and uterine corpus endometrial carcinoma (UCEC), having sufficient samples in both tumor and control groups ($n > 30$) and applied a differential protein interactome algorithm [8], to determine differentially interacting hallmark of cancer proteins (DIHCPs) and DIHCP-centered modules (i.e., DIHCPs and their interacting partners) that represent significant changes in their interaction patterns between the tumor and control phenotypes. The diagnostic and prognostic capabilities of the identified DIHCP-centered modules were assessed in order to identify the modules' potential ability to function as system biomarkers. In addition, the druggability of the prognostic and diagnostic DIHCPs were investigated. Ultimately, this study presents candidate system biomarkers that may be useful for understanding tumorigenesis, developing novel diagnostic tools, and improving personalized treatment strategies for various cancers, with a focus on the hallmarks of cancer.

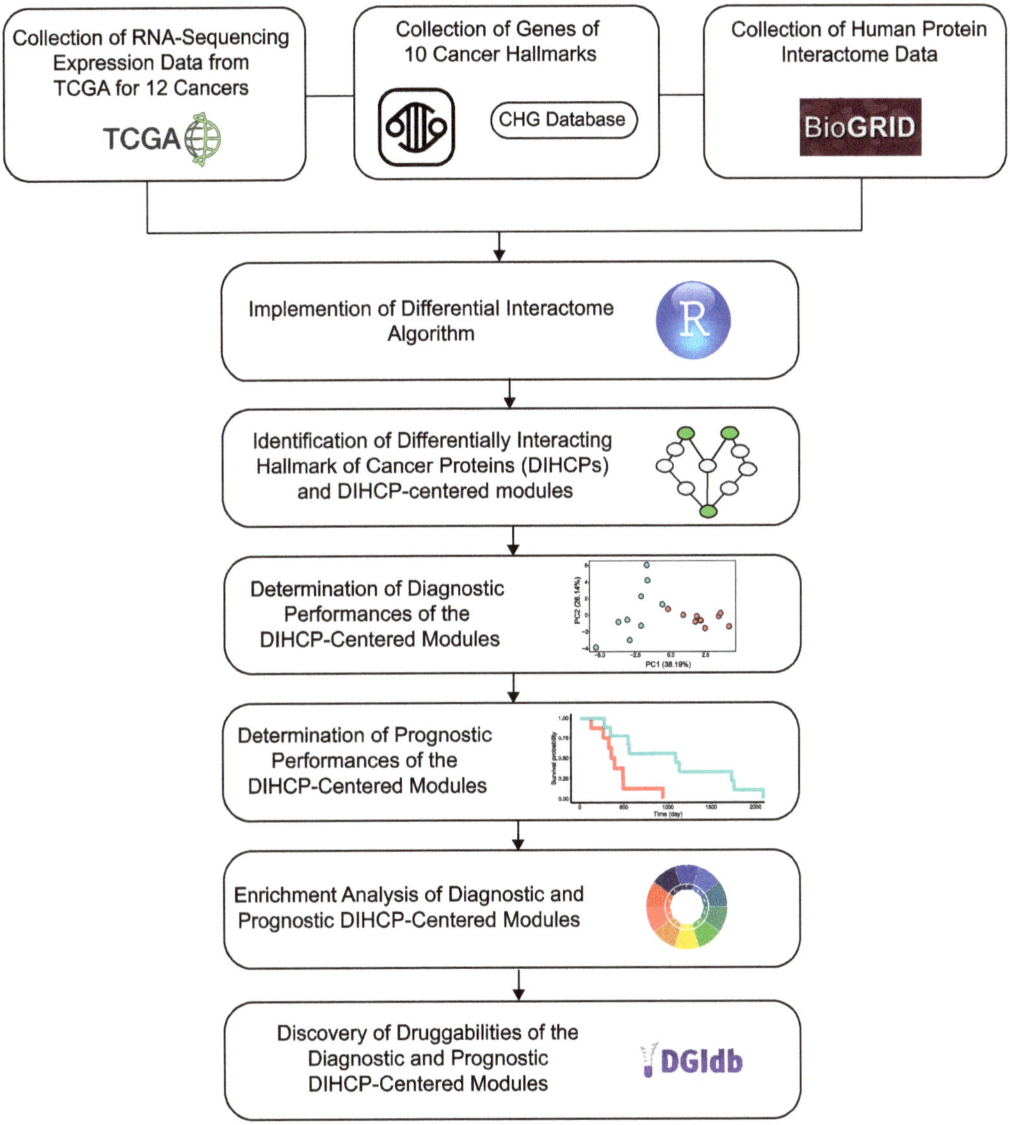

Figure 1. The systematic approach for the study. The steps of the applied systems' biology methodology for the study. After collecting the data from different repositories, the differential interactome algorithm was applied. This algorithm was used to identify differentially interacting hallmark of cancer proteins (DIHCPs) and DIHCPs-centered modules. The prognostic and diagnostic performance of the DIHCPs-centered modules was evaluated. The enrichment analyses were performed and finally the components of the modules' druggability were discovered.

2. Materials and Methods

2.1. Collecting of Gene Expression Data

RNA-sequencing (RNA-seq) fragments per kilobase of transcript per million fragments mapped (FPKM) normalized gene expression data were collected from the Cancer Genome Atlas (TCGA) [9]. For this study, gene expression data of 12 cancers with at least 30 normal and tumor samples of 33 different cancers were collected from the TCGA. In total, gene expression data were collected from 6239 tumor and 637 matched normal tissue samples. The 12 cancer types studied and their sample numbers are represented in Figure 2.

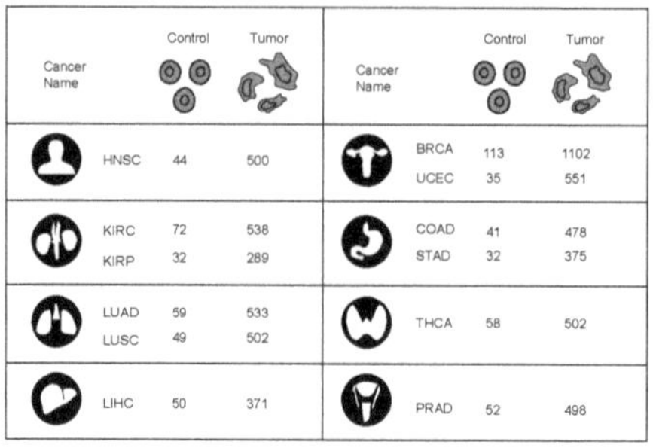

Figure 2. Distribution of samples from 12 human cancers. The diagram shows control and tumor samples from 12 human cancers analyzed in the study.

2.2. Collection of Cancer Hallmark Genes

Genes representing all the ten hallmarks of cancer [4] were collected from two publicly available biological repositories which are Cancer Hallmark Genes (CHG) da-tabase [10] and the Catalogue of Somatic Mutations in Cancer (COSMIC) database [11] (v.95). A total of 1906 different cancer hallmark genes were collected from these two repositories. The GeneCards: The Human Gene Database [12] was used to define proteins encoded by the cancer hallmark genes.

2.3. Collection of Human Protein Interactome Data

Human PPI data were extracted from the BioGRID database [13] (MV-Physical-4.2.191), which contains 51,745 physical and experimentally detected PPIs among 10,177 human proteins. Integration of the obtained PPI interaction data with proteins encoded by genes for which gene expression data are available in TCGA and with proteins encoded by cancer hallmark genes resulted in a network consisting of 7422 PPIs among 1906 proteins.

2.4. Identification of Differentially Interacting Hallmark of Cancer Proteins

The differential protein interactome algorithm [8] was applied to the gene expression profiles of all 12 cancer types using the R platform [14] (version 4.0.2). Briefly, differential hallmarks of cancer-associated PPIs (dHCPPIs) were determined to detect changes in PPI patterns between tumors and controls. To this end, the algorithm uses gene expression profiling to predict the relative frequency of observation (q-value) for each PPI to represent the dHCPPIs.

The following criteria were chosen to determine dHCPPIs: (i) if the predicted q value for PPI is less than 0.10, the interaction is significantly suppressed in the tumor state; (ii) if the predicted q value for PPI is greater than 0.90, the interaction is significantly activated in the tumor state; (iii) the normalized frequency of observation in the tumor or normal phenotype is greater than 20%.

Application of the differential protein interactome algorithm yielded DIHCPs representing significant changes in their interaction patterns during the transition between the normal and tumor phenotypes. The DIHCPs were classified into 2 groups according to their interaction partners: (i) proteins with suppressed interactions in the tumor state and (ii) proteins with activated interactions in the tumor state. DIHCPs (the hub hallmark of cancer proteins) together with their interacting protein partners were designated as DIHCP-centered modules (modular structures around hubs). Further analyses were performed using the DIHCP-centered modules with at least 10 proteins. The dHCPPIs and DIHCP-centered modules were visualized using Cytoscape (v3.5.0) [15].

2.5. Diagnostic Performance Analyses of Differentially Interacting Hallmark of Cancer Protein Modules

Principal component analyses (PCA) were performed based on TCGA-derived gene expression profiles of the genes encoding DIHCPs in each DIHCP-centered module. Each simulation was performed with at least 30 randomly selected normal and 30 tumor samples, and the first 3 principal components representing the highest variances (at least 80% of the total variance) were considered in determining the sensitivity (the proportion of positive test results among diseased individuals) and specificity (the proportion of negative test results among healthy individuals) metrics. Simulations were repeated until the robustness of the average value of the sensitivity and specificity metrics was ensured. The DIHCP-centered modules with at least 90% of the sensitivity and specificity values were considered statistically significant in this study and were accepted as a diagnostic DIHCP-centered module.

2.6. Prognostic Performance Analyses of Differentially Interacting Hallmarks of Cancer Protein Modules

To evaluate the prognostic performance of the DIHCP-centered modules, clinical information on 12 cancer types was collected from TCGA and used in the prognostic performance analyses. Prognostic abilities were evaluated by Kaplan–Meier plots and the log-rank test. All analyses were performed using the Survival package in R [14] (version 4.0.2). Samples were analyzed according to the prognostic index (PI), which is the linear component of the Cox model (PI = β1 × 1 + β2x2 + ... + βpxp, where xi is the expression value of each gene, βi is the coefficient obtained from the Cox fit). The hazard ratio (HR = (O1/E1)/(O2/E2)) was calculated using the ratio between the relative mortality rate in group 1 and the relative mortality rate in group 2, where O and E are the observed and expected number of deaths, respectively. DIHCP-centered modules with a log-rank p-value < 0.01 were considered as statistically significant and accepted as a prognostic DIHCP-centered module in this study.

2.7. Enrichment Analyses of Diagnostic and Prognostic Modules

To obtain clues about the biological characteristics of the diagnostic and prognostic modules, over-representation analyses were performed using the bioinformatics tool Database for Annotation, Visualization and Integrated Discovery (DAVID) [16] to identify functional annotations (i.e., biological pathways) significantly associated with DIHCPs. Pathway p-values were determined using Fisher's exact test, and the Benjamini–Hochberg correction was used as a correction technique for multiple testing. Pathways with adjusted $p < 0.01$ were considered statistically significant.

Moreover, although the DIHCPs were already associated with hallmarks of cancer and to determine which cancer hallmarks are prominent and to further examine the distribution of the hallmarks, the components of the DIHCP-centered modules (i.e., DIHCPs) were subjected to enrichment analysis. For the analyses, we used the obtained cancer hallmark genes from the two repositories [10,11] and integrated them with the diagnostic and prognostic DIHCPs.

2.8. Druggability of the Diagnostic and Prognostic Differentially Interacting Hallmark of Cancer Protein Modules

The DIHCP-centered modules that were (i) accepted as diagnostic in this study, (ii) accepted as prognostic in this study, and (iii) all dHCPPIs that were suppressed or activated in the tumor state were included in the druggability analysis. Components of the DIHCP-centered modules that met the established criteria were screened for druggability using the Drug Gene Interaction Database (DGIdb v4.2.0) [17]. Only FDA-approved drugs were considered throughout the screening process. The types of protein–drug interaction (activator, suppressor, inhibitor, etc.) were also considered. Namely, if all dHCPPIs in the modules were suppressed, the drug candidates with activator activity were considered, or if all dHCPPIs in the modules were activated, the drug candidates with inhibitor activity were considered.

3. Results

3.1. Interpreting Differential Protein Interactome Algorithm in Human Cancers

To identify dHCPPIs between tumor and normal tissue samples, the differential protein interaction algorithm [8] was independently applied to the gene expression profiles of the 12 cancer types and integrated with the reconstructed human protein interactome (7422 PPIs between 1906 proteins). Application of the algorithm yielded the hub proteins representing significant changes in the interaction patterns between "tumor phenotype" and "normal phenotype", which we named "DIHCPs".

We identified a total of 4405 dHCPPIs among 832 different hallmarks of cancer proteins (i.e., DIHCPs) for 12 cancer types. While HNSC had the highest number of dHCPPIs, KIRP had the lowest dHCPPIs. The tumor specificity of dHCPPIs varied by cancer type. Of all the dHCPPIs found, 786 (18.5%) of the dHCPPIs were specific to a cancer type. COAD had the highest number of specific interactions (230 specific dHCPPIs), whereas there was one specific dHCPPI for LUAD (0.2%) (Figure 3A). In addition, HNSC had the highest number of nonspecific dHCPPIs. On the other hand, none of the dHCPPIs was the same for all cancer types studied, and five interactions were the same in at most seven different cancers (i.e., CAV1-CTNNB1, COL1A1-IGFBP3, LDHA-LDHB, PCNA-GAPDH, and PSMB7-PSMB3). Comparative analysis of the interaction type (i.e., activated or suppressed) of these dHCPPIs revealed that the dHCPPIs were generally activated in the tumor state (88% activated).

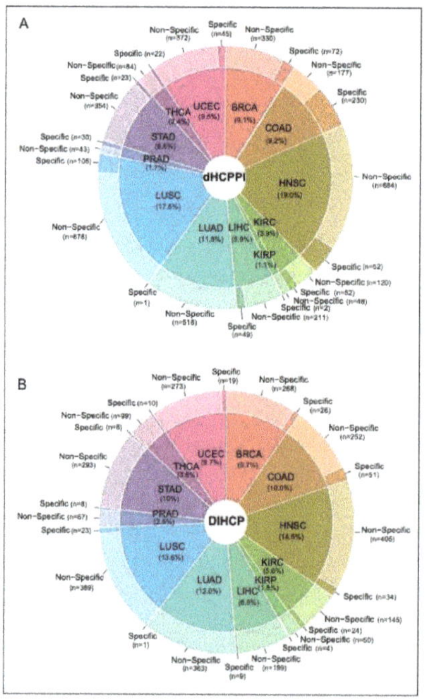

Figure 3. Distribution of differential hallmark of cancer-associated protein–protein interactions and differentially interacting hallmark of cancer proteins. (**A**) The pie-donut chart showing the percentage of specific and non-specific differential hallmark of cancer-associated protein–protein interaction (dHCPPIs) between cancer types. (**B**) The pie-donut chart showing the percentage of specific and non-specific differentially interacting hallmark of cancer proteins (DIHCPs) between cancer types.

3.2. Differentially Interacting Hallmark Proteins and Modules in Human Cancers

Of the 832 DIHCPs identified, 26% of the DIHCPs were specific for one cancer type. There were two common proteins (EGFR and ESR1) that had DIHCP features in all cancers studied. COAD had the highest number of specific DIHCPs. Namely, 16.9% of the DIHCPs of COAD were COAD-specific (Figure 3B).

The differential interactome showed a network topology with modular organization. Therefore, the DIHCPs (hub cancer hallmark proteins) together with their interacting protein partners, were referred to as modules, which we called "DIHCP-centered modules". The DIHCP-centered modules with at least 10 components were considered, and each DIHCP-centered module was named after the name of the hub protein of the module.

In the comparative analysis of DIHCP-centered modules, we found that modules whose hubs belong to the protease or 14-3-3 protein complex shared the vast majority of DIHCPs. Therefore, DIHCP-centered modules sharing at least 70% of the common proteins were pooled based on cancer type to increase the robustness of DIHCP-centered modules. These pooled modules were designated as mPSMcomp or mYWHAcomp. For BRCA, four proteasome complex-associated hub modules were pooled. For COAD, two proteasome and two 14-3-3 protein complex-associated hub modules were pooled separately. In HNSC, 25 proteasome-associated and 3 of the 14-3-3 protein complex-associated hub modules were pooled. In LUAD and UCEC, five protease complex-associated hub modules were pooled independently. In LUSC, 31 proteasome complex-associated hub modules were pooled.

As a result, a total of 111 DIHCP-centered modules were identified (Table S1 (Supplementary Materials)). HNSC provided the highest number of DIHCP-centered modules (21 DIHCP-centered modules), while KIRP provided only 2 DIHCP-centered modules (Figure 4).

3.3. *Diagnostic and Prognostic Power of Differentially Interacting Hallmarks of Cancer Protein-Centered Modules*

In this study, the DIHCP-centered modules have the ability to be potential systems' biomarkers because we believe that the potential disease differences depending on diseased and control status are mainly due to the coordinated action of a group of biological entities. We further hypothesize that the DIHCP-centered module, as a systems' biomarker, would have high diagnostic and prognostic capabilities.

The diagnostic feature of each module was analyzed by PCA. Considering the most significant principal components (accounting for at least 80% of the total variance), sensitivity and specificity metrics were calculated, and modules with at least 90% of the sensitivity and specificity values were considered as diagnostic DIHCP-centered modules. Of the 111 DIHCP-centered modules, 39 had significantly high diagnostic performance (sensitivity $\geq$ 90% and specificity $\geq$ 90%), and some of these are shown in Figure 5. No diagnostic DIHCP-centered module could be observed for PRAD and THCA (Table S2).

The prognostic performance of the modules was assessed using Kaplan–Meier survival plots. The log-rank p-value and hazard ratios were considered to determine whether the DIHCP-centered modules had a high impact on overall patient survival. A total of 88 DIHCP-centered modules had a high impact on patients' overall survival (log-rank p-value < 0.01) in different cancer types (Figure 6), except for PRAD that had no effect on the prognostic DIHCP-centered modules (Table S2).

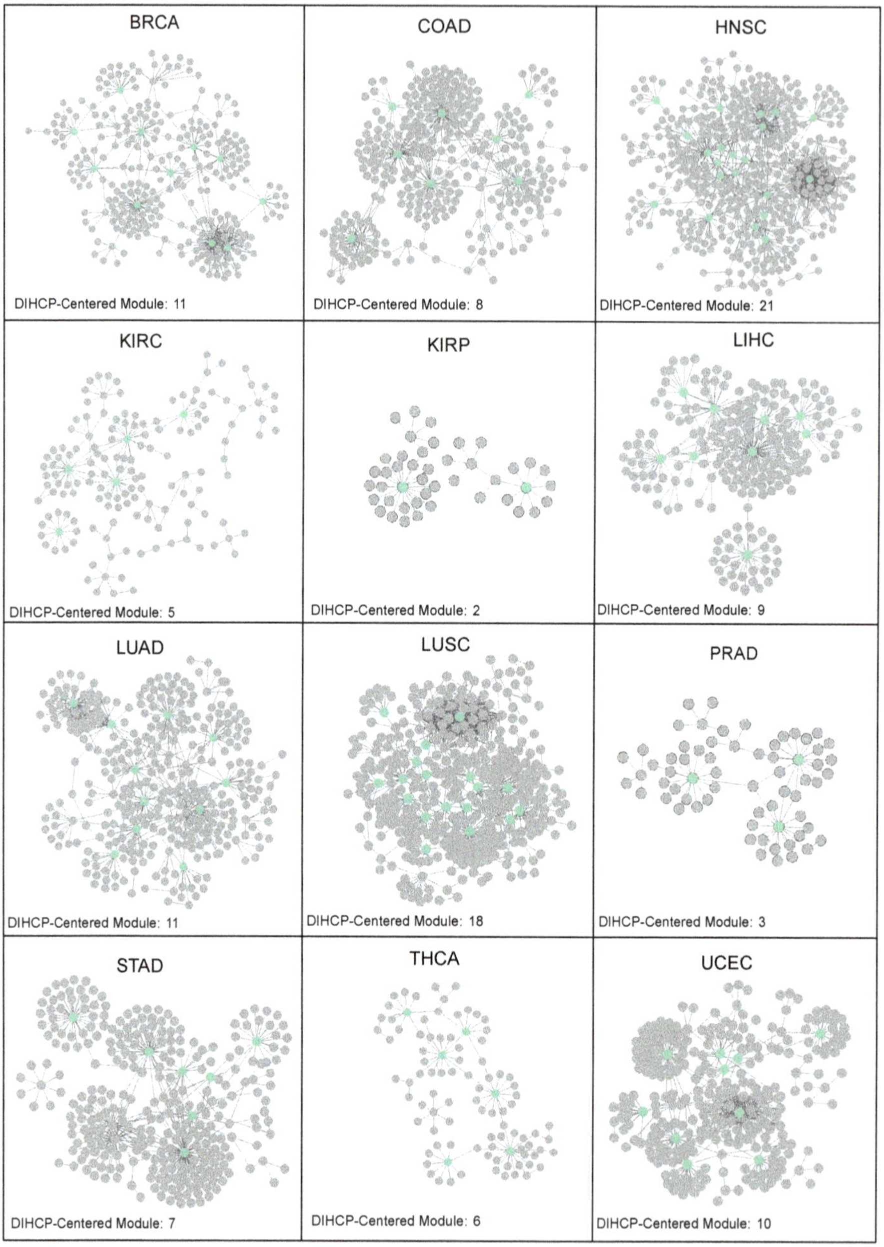

Figure 4. Differential interactome networks in 12 human cancers. For each cancer type, a differential interactome network was constructed around a differential hallmark of cancer-associated PPIs (dHCPPIs). The number of differentially interacting hallmarks of cancer proteins (DIHCPs) for each cancer type was presented.

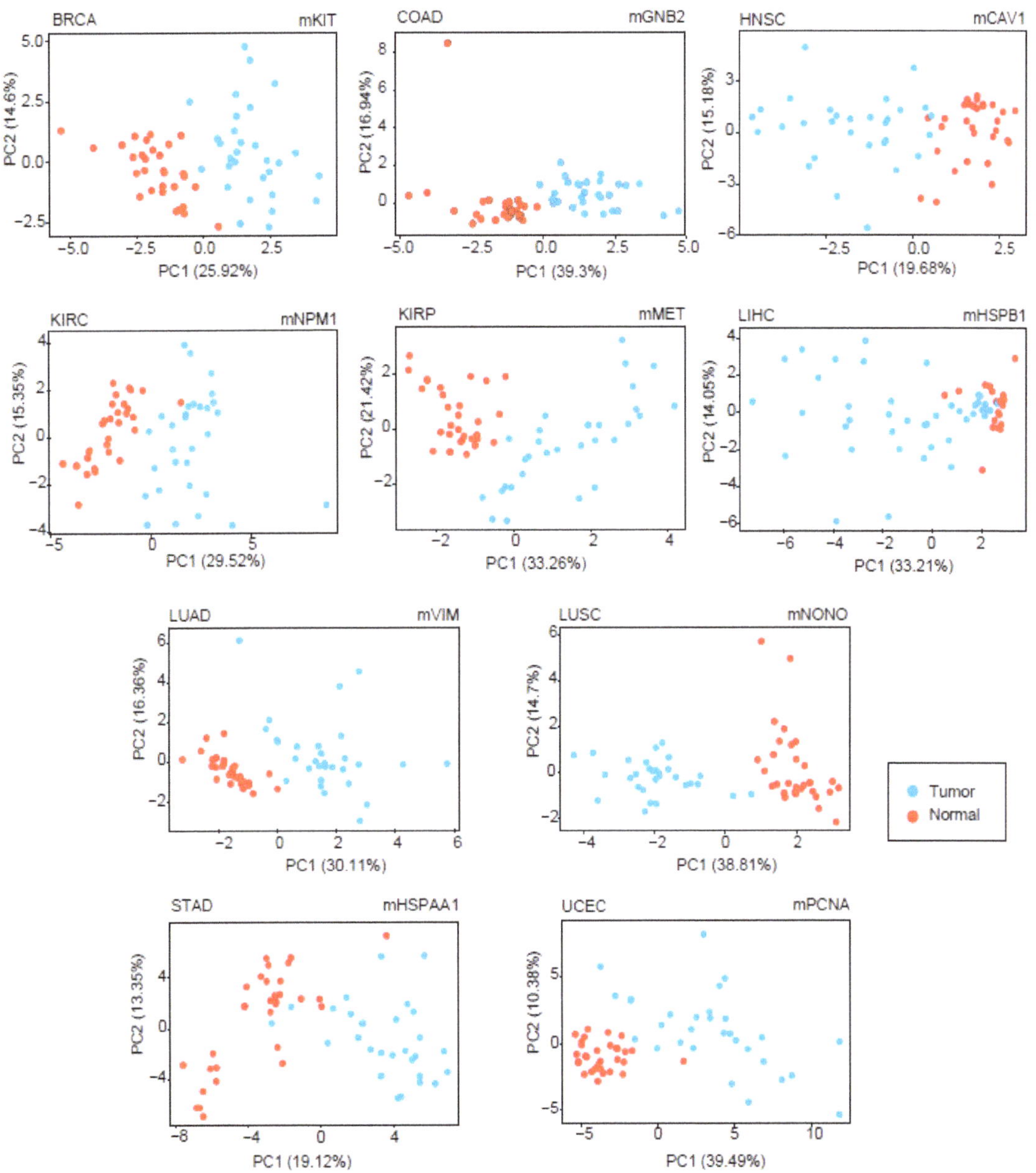

Figure 5. Principal component analyses for different cancer types. PCA plots showing individual differences in protein expression profiles between cancer types comprising at least 30 individuals in each type.

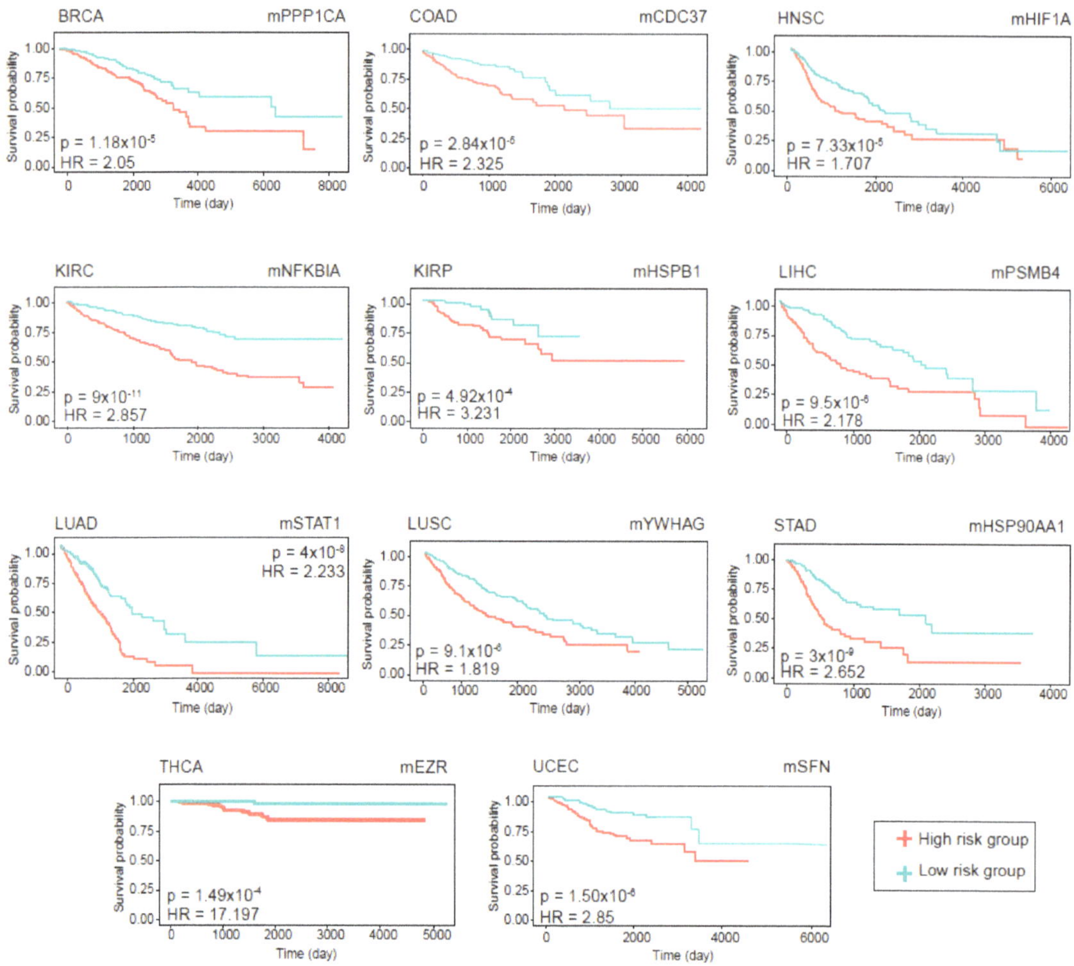

Figure 6. Prognostic analyses for different types of cancer. Kaplan–Meier Plots estimating patients' survival for 11 cancers with p-value and hazard ratio given for each curve.

Among the DIHCP-centered modules, there were a total of 30 DIHCP-centered modules that showed a high diagnostic or prognostic performance in any of the cancer types: one module in BRCA (mPPP1CA), KIRC (mCCND1) and STAD (mHSP90AA1); two modules in HNSC (mCAV1 and mEGFR), KIRP (mHSPB1 and mMET) and LIHC (mHSP90AA1 and mHSPB1; three modules in UCEC (mGNB1, mPCNA and mSFN); four modules in LUAD (mNPM1, mPPP1CA, mSTAT1, and mVIM); seven modules in COAD (mCDC37, mCTNNB1, mGNB2, mJUN, mMYC, mNONO, and mYWHAcomp); and LUSC (mCAV1, mCDH1, mNONO, mPCNA, mPPP1CA, mPSMcomp, and mYWHAG) indicated both diagnostic and prognostic properties.

3.4. Enrichment Analyses of Diagnostic and Prognostic Protein Modules

Pathway and hallmark enrichment analyses were performed to obtain further biological characteristics of the diagnostic and prognostic 30 modules. The DIHCPs pathway over-representation analysis of modules based on annotations stored in the KEGG database revealed various pathways (Figure 7). For instance, cancer-associated pathways such as bladder cancer, chronic myeloid leukemia, gastric cancer, glioma, and melanoma come into prominence. Signaling pathways such as Hippo, JAK-STAT, MAPK, PI3K-AKT, and WNT, which are known to be associated with cancer development and progression, were found to be statistically significant. Interestingly, viral carcinogenesis and viral infections associated with cancer (i.e., HBV, HPV, HTLV1, and KSHV) [18] were remarkable pathways associated with DIHCPs. Moreover, the hallmarks of cancer comprise ten distinct biological capabilities derived during the multi-staged development of tumors; however, to reveal the most fundamental trait of cancer cells, hallmark enrichment analyses were performed in diagnostic and prognostic DIHCP-centered modules (Figure 8A). Among the ten hallmarks of cancer, DIHCPs were most enriched in sustaining proliferative signaling, presumably due to the fact that unlimited replication is essential in cancer cells for tumor development. In addition, genome instability and mutation and enabling replicative immortality hallmarks remain in the background when compared to others.

Figure 7. The pathway enrichment analyses results. The bubble plot indicates the diagnostic and prognostic 30 modules differentially interacting hallmark of cancer protein's (DIHCP's) pathway enrichment results according to q-value ($-\log_{10}$) significance.

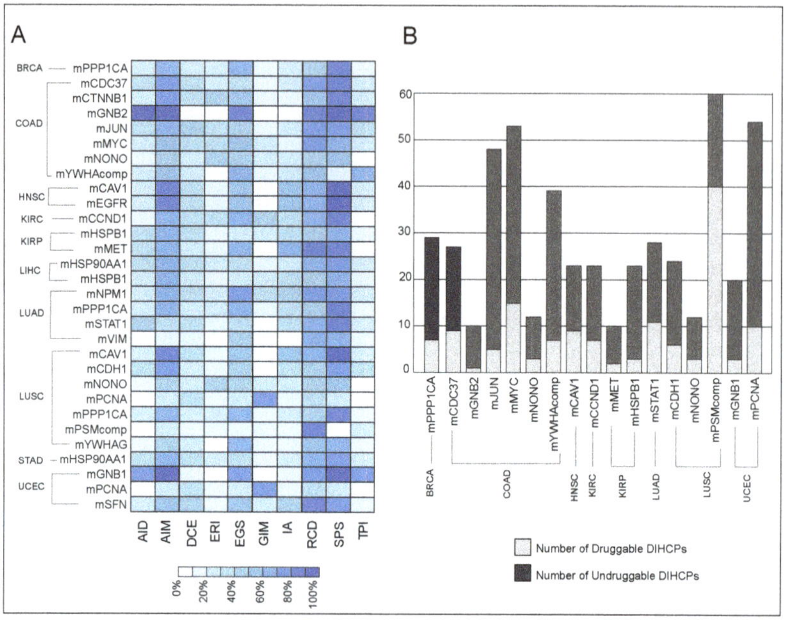

Figure 8. The hallmark enrichment and druggability analyses results. (**A**) The heat-map represents the hallmark enrichment results of the module's differentially interacting hallmark of cancer proteins (DIHCPs). AID: avoiding immune destruction; AIM: activating invasion and metastasis; DCE: deregulating cellular energetics; ERI: enabling replicative immortality; EGS: evading growth suppressors; GIM: genome instability and mutation; IA: inducing angiogenesis; RCD: resisting cell death; SPS: sustaining proliferative signaling; TPI: tumor-promoting inflammation. (**B**) The druggability distribution of module's DIHCPs. The bar graph indicates the number of DIHCPs, which are druggable (light grey bars) and undruggable (dark grey bars).

3.5. Druggability of Differentially Interacting Hallmarks of Cancer Proteins

The DIHCPs were promising therapeutic targets because they represent significant changes in the interaction patterns between tumor and normal phenotype. For this reason, we investigated the druggability of the modules. Of the 30 diagnostic and prognostic DIHCP-centered modules, the DIHCPs of 17 modules were examined according to the established third criteria, because 13 of the dHCPPIs modules had both suppressed and activated interactions that did not meet our third criteria (see Materials and Methods).

We found that of the 17 DIHCP-centered modules, 28% of the DIHCPs were druggable. The mPSMcomp belonging to LUSC had the highest percentage (66%) of druggable proteins, whereas the lowest number of druggable DIHCPs among all cancers analyzed belonged to mGNB2 (10%) of COAD (Figure 8B).

A total of 200 different drugs targeting DIHCPs were found. A total of 54% of the drugs were specific for only one cancer type. None of the drugs were specific for all cancers studied, and 25 of the drugs were specific for no more than 6 different cancers (Table S3).

4. Discussion

Complex diseases such as cancer usually arise as a result of interactions between biological entities. The biological molecules that trigger the development of a cancer or biological process never function alone; they work together in a complex, interconnected network to carry out these biological functions. The state of interconnectedness of molecules establishes the phenomenon of the "biological network". A biological network is an advanced concept that allows researchers to model, characterize, and decipher complex interactions between different biologically relevant entities in a biological system [19]. Biological networks thus change our perspective and shed light on the internal organization of the cell, allowing us to understand the mechanisms underlying complex diseases. Moreover, predictive, preventive, and personalized medicine is a holistic healthcare strategy that aims to predict individual predisposition, provide targeted prevention, and provide personalized treatment [20]. To advance personalized medicine strategies, the discovery of new treatments is essential. However, with today's financial resources, it is nearly impossible to discover drugs from scratch using traditional methods. Using biological network analysis, researchers can also discover and/or predict anti-cancer drugs [21]. We believe that network-based systems biology is a new approach for discovering treatment and prevention strategies, especially for cancer, which is currently a major burden worldwide, and researchers are embracing this concept. For example, researchers have integrated gene expression profiling with multiple networks to identify new biomarkers and drug candidates for breast cancer [22], pancreatic cancer [23], cervical cancer [24], acute myeloid leukemia [25], thyroid cancer [26], and so on. Accordingly, these holistic analyses enable the discovery of potential biomarkers for diagnosis, prognosis, or therapeutic purposes. Therefore, biological networks hold great potential for the present and the future, especially for complex diseases such as cancer [27].

The protein–protein interactome represents a biological network that encompasses all physical protein interactions within the cell. These protein–protein interactions play a crucial role in the living organism. These are essential for almost all cellular functions, involved in the physiology and biochemistry of the organism. These can also influence cancer cell growth, transmit oncogenic signals, and even cause the development of typical cancer features. Targeting such interactions is a promising strategy for new drug development [28,29]. Thus, it is strongly suggested that significant changes in protein–protein interactions occur depending on the phenotype of the individual (i.e., diseased versus healthy) [30].

We applied our constructed differential interactome algorithm, which had proven to be a powerful tool in other studies [8,31,32], for the protein–protein interactome data on hallmarks of cancer. Our results revealed DIHCPs showing remarkable changes in the interaction patterns of patients during the transition from the "normal" phenotype to the "cancer" phenotype. As mentioned earlier, cancers occur and spread usually due to the coordinated action of a group of biological molecules. In this study, DIHCPs and the modular structures around these proteins (DIHCP-centered modules) were considered as potential systems' biomarkers. However, we believe that a valid biomarker must provide prognostic or diagnostic information about the specific cancer. For this reason, to provide confidence in the precise diagnostic and prognostic capacity of the systems' biomarkers we clearly demonstrated their diagnostic and prognostic performances by PCA and survival analyses and demonstrated high biomarker performances in multiple tumors (30 diagnostic and prognostic DIHCP-centered modules).

In this study, we only used hallmark of cancer proteins, while we integrated proteomics data to apply the differential interactome algorithm. Hallmarks of cancer proteins are considered as driver proteins of tumorigenesis. These proteins are responsible for the most basic phenotypic features of tumor initiation and progression [3]. Therefore, we hypothesized that we can highlight the principles and mechanisms of tumorigenesis by focusing only on the proteins essential for carcinogenesis (hallmarks of cancer proteins), we can reduce the complexity of the disease, and typify the process by characteristic and complementary features. In addition, if the potential system biomarker components (i.e., DIHCPs) presented in the study are all among the hallmark characteristics of human cancers, we believe we will identify highly robust cancer biomarkers. Moreover, since these hallmarks represent features of cancer cells, recent therapeutic approaches have been increasingly aiming at targeting hallmark proteins [33]. Therefore, the systems' biomarkers identified in this study will have remarkable clinical value. For this reason, this study also evaluated the druggability of the diagnostic and prognostic systems' biomarkers.

The hubs of the diagnostic and prognostic systems' biomarkers are mostly associated with the sustaining proliferative signaling involved as an important feature in the hallmarks of cancer. It is known that cancer cells can stimulate their own proliferation indefinitely and this cancer hallmark characteristic is chronically activated in all cancers. The well-known mechanism for sustaining proliferative signaling is to activate oncogenes [34]. Oncogenes encode proteins that stimulate cell proliferation and programmed cell death. An oncogene arises when a proto-oncogene is altered genetically (i.e., mutated) and over-expressed. For instance, in cancers, the most noticeable oncogenes are BRAF and RAS [35]. Moreover, resisting the cell-death hallmark feature comes into prominence for diagnostic and prognostic hubs. Resistance to cell death is a natural barrier to cancer development and is mostly regulated by programmed cell death. The best-known form of programmed cell death is apoptosis, in which cells are destined to die. In addition to apoptosis, necroptosis is referred to as the second important process for inducing cell death. Apoptosis and necroptosis cause different immune responses. Necroptosis releases molecules that promote inflammation, while apoptosis triggers silent immunological responses [36]. Autophagy is another process that leads to cell death [34]. The autophagy process allows cells to remove unnecessary or dysfunctional components. In cancer, autophagy may play a tumor-suppressor or oncogene role under certain conditions and at certain stages of carcinogenesis [37]. A recent review systematically described the hallmarks of different cell death modes [38].

The major limitation of the study is the lack of experimental validations of the diagnostic and prognostic modules with relevant tissue samples or cell lines. Therefore, the most important aspect of this study is to translate these computational findings into experimental approaches. For example, future in vitro studies need to be performed to investigate the impact of the identified modules in terms of their response to cell viability, cell migration and disease progression. In addition, the accuracy, consistency, reproducibility, and reliability of the biomarkers presented in this study should be experimentally validated if they prove to be clinically useful. In addition, the mechanism of action of these modules needs to be experimentally evaluated to clearly elucidate their effects on the hallmarks of cancer characteristics. We believe that computational analysis is an important and first step in biomarker development. However, to address the broad medical and scientific audience, the need for experimental validation is inevitable.

In conclusion, the knowledge of specific proteins being impacted within a given hallmark of the patient's cancer type could offer the opportunity to personalize treatments. One such approach would be to target proteins from the proteins in the enrichment analysis that match the patient's specific tumor type and prioritize repurposing the drugs proposed from our analysis. Furthermore, our investigation on the hallmarks of cancer protein provides valuable data for further experimental and clinical efforts in a variety of cancer types since the proposed systems' biomarkers have the potential to be diagnostic and/or prognostic. Moreover, the protein–protein interactions could be utilized as therapeutic targets.

Supplementary Materials: The following supporting information can be downloaded at: https://www.mdpi.com/article/10.3390/jpm12111919/s1, Table S1: Differentially interacting hallmark of cancer protein (DIHCP)-centered modules; Table S2: Diagnostic and prognostic analysis of differentially interacting hallmark of cancer protein (DIHCP)-centered modules; Table S3: Drugs targeting module's differentially interacting hallmark of cancer proteins (DIHCPs).

Author Contributions: Conceptualization, K.Y.A. and R.S.; Methodology, K.Y.A. and R.S.; Formal Analysis, M.K. and G.E.O.; Investigation, M.K. and G.E.O.; Writing—Original Draft Preparation, M.K.; Writing—Review and Editing, M.K., G.E.O., K.Y.A. and R.S.; Supervision, K.Y.A. and R.S. All authors have read and agreed to the published version of the manuscript.

Funding: This research received no external funding.

Institutional Review Board Statement: Not applicable.

Informed Consent Statement: Not applicable.

Data Availability Statement: Publicly available datasets were analyzed in this study. The datasets analyzed during the current study are available in The Genome Cancer Atlas (https://portal.gdc.cancer.gov/, accessed on 1 March 2022). Protein interactome data are available in Biological General Repository for Interaction Datasets (https://thebiogrid.org, accessed on 30 March 2022). The cancer hallmark gene data are available in the Catalogue of Somatic Mutations in Cancer (https://cancer.sanger.ac.uk/cosmic, accessed on 23 May 2022) and Cancer Hallmark Genes (http://www.bio-bigdata.com/CHG/index.html, accessed on 23 May 2022). Source codes for the differential interactome algorithm (implemented in R, version 4.0.2) are freely available at http://sysbio.bioe.eng.marmara.edu.tr/diff-int-ome, accessed on 25 July 2022.

Acknowledgments: The author(s) received no financial support for the research, authorship, and/or publication of this article.

Conflicts of Interest: The authors declare no conflict of interest.

References

1. Sung, H.; Ferlay, J.; Siegel, R.L.; Laversanne, M.; Soerjomataram, I.; Jemal, A.; Bray, F. Global Cancer Statistics 2020: GLOBOCAN Estimates of Incidence and Mortality Worldwide for 36 Cancers in 185 Countries. *CA Cancer J. Clin.* **2021**, *71*, 209–249. [CrossRef] [PubMed]
2. Ravi, S.; Alencar, A.M.; Arakelyan, J.; Xu, W.; Stauber, R.; Wang, C.I.; Papyan, R.; Ghazaryan, N.; Pereira, R.M. An Update to Hallmarks of Cancer. *Cureus* **2022**, *14*, e24803. [CrossRef] [PubMed]
3. Hanahan, D.; Weinberg, R.A. The hallmarks of cancer. *Cell* **2000**, *100*, 57–70. [CrossRef]
4. Hanahan, D.; Weinberg, R.A. Hallmarks of cancer: The next generation. *Cell* **2011**, *144*, 646–674. [CrossRef]
5. Hanahan, D. Hallmarks of Cancer: New Dimensions. *Cancer Discov.* **2022**, *12*, 31–46. [CrossRef] [PubMed]
6. Nagy, Á.; Munkácsy, G.; Győrffy, B. Pancancer survival analysis of cancer hallmark genes. *Sci. Rep.* **2021**, *11*, 6047. [CrossRef] [PubMed]
7. Yu, L.H.; Huang, Q.W.; Zhou, X.H. Identification of Cancer Hallmarks Based on the Gene Co-expression Networks of Seven Cancers. *Front. Genet.* **2019**, *10*, 99. [CrossRef]
8. Gulfidan, G.; Turanli, B.; Beklen, H.; Sinha, R.; Arga, K.Y. Pan-cancer mapping of differential protein-protein interactions. *Sci. Rep.* **2020**, *10*, 3272. [CrossRef]
9. Tomczak, K.; Czerwińska, P.; Wiznerowicz, M. The Cancer Genome Atlas (TCGA): An immeasurable source of knowledge. *Contemp. Oncol. (Pozn)* **2015**, *19*, A68–A77. [CrossRef]
10. Zhang, D.; Huo, D.; Xie, H.; Wu, L.; Zhang, J.; Liu, L.; Jin, Q.; Chen, X. CHG: A Systematically Integrated Database of Cancer Hallmark Genes. *Front. Genet.* **2020**, *11*, 29. [CrossRef]
11. Tate, J.G.; Bamford, S.; Jubb, H.C.; Sondka, Z.; Beare, D.M.; Bindal, N.; Boutselakis, H.; Cole, C.G.; Creatore, C.; Dawson, E.; et al. COSMIC: The Catalogue of Somatic Mutations in Cancer. *Nucleic Acids Res.* **2019**, *47*, D941–D947. [CrossRef] [PubMed]
12. Stelzer, G.; Rosen, N.; Plaschkes, I.; Zimmerman, S.; Twik, M.; Fishilevich, S.; Stein, T.I.; Nudel, R.; Lieder, I.; Mazor, Y.; et al. The GeneCards Suite: From Gene Data Mining to Disease Genome Sequence Analyses. *Curr. Protoc. Bioinform.* **2016**, *54*, 1.30.1–1.30.33. [CrossRef]
13. Oughtred, R.; Stark, C.; Breitkreutz, B.J.; Rust, J.; Boucher, L.; Chang, C.; Kolas, N.; O'Donnell, L.; Leung, G.; McAdam, R.; et al. The BioGRID interaction database: 2019 update. *Nucleic Acids Res.* **2019**, *47*, D529–D541. [CrossRef]
14. Huber, W.; Carey, V.J.; Gentleman, R.; Anders, S.; Carlson, M.; Carvalho, B.S.; Bravo, H.C.; Davis, S.; Gatto, L.; Girke, T.; et al. Orchestrating high-throughput genomic analysis with Bioconductor. *Nat. Methods* **2015**, *12*, 115–121. [CrossRef] [PubMed]
15. Shannon, P.; Markiel, A.; Ozier, O.; Baliga, N.S.; Wang, J.T.; Ramage, D.; Amin, N.; Schwikowski, B.; Ideker, T. Cytoscape: A software environment for integrated models of biomolecular interaction networks. *Genome Res.* **2003**, *13*, 2498–2504. [CrossRef] [PubMed]
16. Sherman, B.T.; Hao, M.; Qiu, J.; Jiao, X.; Baseler, M.W.; Lane, H.C.; Imamichi, T.; Chang, W. DAVID: A web server for functional enrichment analysis and functional annotation of gene lists (2021 update). *Nucleic Acids Res.* **2022**, *50*, W216–W221. [CrossRef]
17. Freshour, S.L.; Kiwala, S.; Cotto, K.C.; Coffman, A.C.; McMichael, J.F.; Song, J.J.; Griffith, M.; Griffith, O.L.; Wagner, A.H. Integration of the Drug-Gene Interaction Database (DGIdb 4.0) with open crowdsource efforts. *Nucleic Acids Res.* **2021**, *49*, D1144–D1151. [CrossRef]
18. Kori, M.; Arga, K.Y. Human oncogenic viruses: An overview of protein biomarkers in viral cancers and their potential use in clinics. *Expert Rev. Anticancer Ther.* **2022**, 1–14. [CrossRef]
19. Zhang, B.; Tian, Y.; Zhang, Z. Network biology in medicine and beyond. *Circ. Cardiovasc. Genet.* **2014**, *7*, 536–547. [CrossRef]
20. Goetz, L.H.; Schork, N.J. Personalized medicine: Motivation, challenges, and progress. *Fertil Steril.* **2018**, *109*, 952–963. [CrossRef]
21. Li, K.; Du, Y.; Li, L.; Wei, D.Q. Bioinformatics Approaches for Anti-cancer Drug Discovery. *Curr. Drug Targets.* **2020**, *21*, 3–17. [CrossRef] [PubMed]
22. Khan, A.; Rehman, Z.; Hashmi, H.F.; Khan, A.A.; Junaid, M.; Sayaf, A.M.; Ali, S.S.; Hassan, F.U.; Heng, W.; Wei, D.Q. An Integrated Systems Biology and Network-Based Approaches to Identify Novel Biomarkers in Breast Cancer Cell Lines Using Gene Expression Data. *Interdiscip. Sci.* **2020**, *12*, 155–168. [CrossRef] [PubMed]

23. Kaushik, A.C.; Wang, Y.J.; Wang, X.; Wei, D.Q. Irinotecan and vandetanib create synergies for treatment of pancreatic cancer patients with concomitant TP53 and KRAS mutations. *Brief Bioinform.* **2021**, *22*, bbaa149. [CrossRef] [PubMed]
24. Kori, M.; Arga, K.Y.; Mardinoglu, A.; Turanli, B. Repositioning of Anti-Inflammatory Drugs for the Treatment of Cervical Cancer Sub-Types. *Front. Pharmacol.* **2022**, *13*, 884548. [CrossRef]
25. Kelesoglu, N.; Kori, M.; Turanli, B.; Arga, K.Y.; Yilmaz, B.K.; Duru, O.A. Acute Myeloid Leukemia: New Multiomics Molecular Signatures and Implications for Systems Medicine Diagnostics and Therapeutics Innovation. *OMICS* **2022**, *26*, 392–403. [CrossRef]
26. Gulfidan, G.; Soylu, M.; Demirel, D.; Erdonmez, H.; Beklen, H.; Ozbek Sarica, P.; Arga, K.Y.; Turanli, B. Systems biomarkers for papillary thyroid cancer prognosis and treatment through multi-omics networks. *Arch Biochem. Biophys.* **2022**, *15*, 109085. [CrossRef]
27. Yan, W.; Xue, W.; Chen, J.; Hu, G. Biological Networks for Cancer Candidate Biomarkers Discovery. *Cancer Inform.* **2016**, *15*, 1–7. [CrossRef]
28. Lu, H.; Zhou, Q.; He, J.; Jiang, Z.; Peng, C.; Tong, R.; Shi, J. Recent advances in the development of protein-protein interactions modulators: Mechanisms and clinical trials. *Sig. Transduct. Target. Ther.* **2020**, *5*, 213. [CrossRef]
29. Ivanov, A.A.; Khuri, F.R.; Fu, H. Targeting protein-protein interactions as an anticancer strategy. *Trends Pharmacol. Sci.* **2013**, *34*, 393–400. [CrossRef]
30. Caldera, M.; Buphamalai, P.; Müller, F.; Menche, J. Interactome-based approaches to human disease. *Curr. Opin. Syst. Biol.* **2017**, *3*, 88–94. [CrossRef]
31. Ayyildiz, D.; Gov, E.; Sinha, R.; Arga, K.Y. Ovarian cancer differential interactome and network entropy analysis reveal new candidate biomarkers. *OMICS* **2017**, *21*, 285–294. [CrossRef] [PubMed]
32. Caliskan, A.; Gulfidan, G.; Sinha, R.; Arga, K.Y. Differential interactome proposes subtype-specific biomarkers and potential therapeutics in renal cell carcinomas. *J. Pers. Med.* **2021**, *11*, 158. [CrossRef] [PubMed]
33. Sameri, S.; Mohammadi, C.; Mehrabani, M.; Najafi, R. Targeting the hallmarks of cancer: The effects of silibinin on proliferation, cell death, angiogenesis, and migration in colorectal cancer. *BMC Complement. Med. Ther.* **2021**, *21*, 160. [CrossRef] [PubMed]
34. Hanahan, D.; Weinberg, R.A. Biological hallmarks of cancer. In *Holland-Frei Cancer Medicine*, 9th ed.; Bast, R.C., Hait, W.N., Kufe, D.W., Weichselbaum, R.R., Holland, J.F., Croce, C.M., Piccart-Gebart, M., Wang, H., et al., Eds.; Wiley-Blackwell: Hoboken, NJ, USA, 2017; pp. 7–16, ISBN 978-1-118-93469-2.
35. Botezatu, A.; Iancu, I.V.; Popa, O.; Plesa, A.; Manda, D.; Huica, I.; Vladoiu, S.; Anton, G.; Badiu, C. Mechanisms of Oncogene Activation. In *New Aspects in Molecular and Cellular Mechanisms of Human Carcinogenesis*; Bulgin, D., Ed.; IntechOpen: London, UK, 2016; pp. 1–52, ISBN 978-953-51-2253-1.
36. Heckmann, B.L.; Tummers, B.; Green, D.R. Crashing the computer: Apoptosis vs. necroptosis in neuroinflammation. *Cell Death Differ.* **2019**, *26*, 41–52. [CrossRef]
37. Li, X.; He, S.; Ma, B. Autophagy and autophagy-related proteins in cancer. *Mol. Cancer* **2020**, *19*, 12. [CrossRef]
38. Yan, G.; Elbadawi, M.; Efferth, T. Multiple cell death modalities and their key features (Review). *World Acad. Sci. J.* **2020**, *2*, 39–48. [CrossRef]

Journal of *Personalized Medicine*

MDPI

Article

Dynamic Collaborations for the Development of Immune Checkpoint Blockade Agents

Arisa Djurian [1], Tomohiro Makino [1], Yeongjoo Lim [2], Shintaro Sengoku [3] and Kota Kodama [1,4,*]

1 Graduate School of Technology Management, Ritsumeikan University, Osaka 567-8570, Japan; gr0434hi@ed.ritsumei.ac.jp (A.D.); makino.tomohiro@gmail.com (T.M.)
2 Faculty of Business Administration, Ritsumeikan University, Osaka 567-8570, Japan; lim40@fc.ritsumei.ac.jp
3 School of Environment and Society, Tokyo Institute of Technology, Tokyo 108-0023, Japan; sengoku.s.aa@m.titech.ac.jp
4 Center for Research and Education on Drug Discovery, The Graduate School of Pharmaceutical Sciences in Hokkaido University, Sapporo 060-0812, Japan
* Correspondence: kkodama@fc.ritsumei.ac.jp; Tel.: +81-726652448; Fax: +81-726652448

Citation: Djurian, A.; Makino, T.; Lim, Y.; Sengoku, S.; Kodama, K. Dynamic Collaborations for the Development of Immune Checkpoint Blockade Agents. *J. Pers. Med.* **2021**, *11*, 460. https://doi.org/10.3390/jpm11060460

Academic Editor: Anne-Marie Caminade

Received: 28 April 2021
Accepted: 18 May 2021
Published: 24 May 2021

Publisher's Note: MDPI stays neutral with regard to jurisdictional claims in published maps and institutional affiliations.

Abstract: We studied the overview of drug discovery and development to understand the recent trends and potential success factors of interorganizational collaboration by reviewing 1204 transactions performed until 2019 for 107 anticancer drugs approved by the US Food and Drug Administration (FDA) from 1999 to 2018. Immune checkpoint blockade was found to be a significantly active area in interorganizational transactions, especially the number of alliances, compared with other mechanisms of action of small molecules and biologics for cancer treatment. Furthermore, the analysis of pembrolizumab and nivolumab showed that the number of approved indications for these two drugs has been rapidly expanding since their first approval in 2014. Examination of the acquisitions and alliances regarding pembrolizumab and nivolumab showed that many combination partners were developed by US-based biotechnology or start-up companies, the majority of which were biologics. These findings suggest that immune checkpoint blockade is a paradigm for cancer treatment, resulting in huge product sales and continuous indication expansion. Additionally, interorganizational collaboration, especially trial collaboration, is a strategic approach for the development of immune checkpoint blockade agents. The translation of these empirical practices to new drug candidates is expected for the research and development of innovative drugs in the future.

Keywords: cancer treatment; immune checkpoint blockade; PD-(L)1 inhibitors; interorganizational transaction; combination therapy

1. Introduction

Business in the pharmaceutical industry is unique in terms of product universality, where a company needs to research and develop each pipeline of drugs according to strict local regulations and requirements to obtain approval from the respective regulatory authority to launch their products. Therefore, drug research and development (R&D) is an extremely costly and time-consuming process [1–4]. In addition, business success in the pharmaceutical industry requires innovative products and expansion of product values to secure high pricing, grant of reimbursement and longer market exclusivities. Recently, R&D has become more complicated as modality becomes diverse by introducing new technologies and knowledge to address existing unmet medical needs in specific target populations. Thus, companies in the pharmaceutical industry have strived to enhance sustainability by adopting various strategic approaches, such as pursing globalized or region-oriented business models [5,6], selective therapeutic areas in R&D, optimizing R&D productivity [7], and open innovation, including external collaboration and interorganizational transactions [8–11]. In particular, the importance of external collaboration in R&D has been disseminated and many pharmaceutical companies have pursued interorganizational transactions and established open innovation platforms to acquire external knowledge and pipelines across organizations [8,10–13].

In the pharmaceutical industry, cancer therapeutics have a long history in the transformation of treatment options, including chemotherapy [14,15], targeted therapy [16], biologics [17], and combination therapy [18]. The recent hot spot is immuno-oncology [19,20], in which immune checkpoint blockade is a paradigm. Currently available immune checkpoint inhibitors are cytotoxic T lymphocyte-associated antigen 4 (CTLA4) inhibitors, programmed cell death protein 1 (PD-1) inhibitors, and programmed death-ligand 1 (PD-L1) inhibitors. The first entry into the market for immune checkpoint blockade was the anti-CTLA4 antibody ipilimumab (Bristol-Myers Squibb (BMS) (New York, NY, USA), and Medarex (Princeton, NJ, USA)), approved for metastatic melanoma in 2011 by the Food and Drug Administration (FDA) in the US. Subsequently, two anti-PD-1 antibodies, pembrolizumab (Merck) and nivolumab (BMS), were approved for metastatic melanoma in 2014 [21]. A number of PD-1 or PD-L1 inhibitors have been approved or are under development [22,23], thereby expanding the target tumor types [24] not only in the US but also in Europe, Japan, and other regions.

After the isolation of PD-1 by Dr. Honjo's laboratory in Japan in 1992 [25], there was a long lag period until its first approval in 2014. The approval process involved extensive collaborations, supported by academia, biotechnology companies, pharmaceutical companies, biomarker companies, non-profit organizations, and regulatory agencies [20,26]. Chen and Han reviewed several important histories regarding anti-PD-(L)1 therapy for human cancer [26]. For example, the discovery of the PD pathway resulted from several collaborations by scientists, belonging to different laboratories, who identified and isolated at least five interacting molecules, PD-1 [25], PD-L1 (B7 homolog 1 [B7-H1]) [27,28], PD-L2 (B7-DC) [29,30], CD80 (B7-1) [31,32], and molecule family member b (RGMb) [33], and explored their functions and discovered mutual interactions [28,34–37]. Hoos suggested that collaboration among Medarex, BMS, a community organization called the Cancer Immunotherapy Consortium (CIC) of the Cancer Research Institute (New York, NY, USA), and CIC's partners would generate breakthrough in the clinical evaluation of cancer immunotherapies by establishing new criteria and evaluation method. This led to the success of ipilimumab in clinical trials and eventually contributed to the acquisition of Medalex and responsible regulatory guidance by the FDA and the European Medicine Agency (EMA) (Amsterdam, The Netherlands). During the R&D of immune checkpoint blockade agents, there were important interorganizational transactions. For example, LifeArc (London, UK), which has a technology to generate a humanized clinical candidate, engineered pembrolizumab in collaboration with Organon in 2007 [38], and Merck acquired Organon later that year. Medarex, which was a biopharmaceutical company focused on the discovery, development, and potential commercialization of fully human antibody-based therapeutics and engineered ipilimumab and nivolumab, was acquired by BMS in 2009 [39]. Additionally, the development of combination therapy in this segment is quite active before and even after product launch [10,40,41] involving the collaboration of multiple companies. Given these previous successful practices, one of the effective approaches for sustainable business in the discovery and development of anticancer drugs may consider how a company that tries to generate a potential innovative drug can be a pioneer in terms of open innovation. Our interests were to understand whether immune checkpoint blockade is the active area in terms of interorganizational collaboration in anticancer drug development, and whether there are strategic collaborations with various external partners behind the recent successes of the indication and market expansion of immune checkpoint blockade agents. However, no previous study has investigated the trends in interorganizational collaborations for the R&D of immune checkpoint blockade that potentially led to successful drug development.

Therefore, the objective of this study was to understand the recent trends and success factors, based on various experiences of active transactions, for the development of immune checkpoint blockade agents. The definition of success in our study is a positive outcome derived from drug development or interorganizational transactions which enable a company to maintain sustainable growth, specifically drug approvals, new indication approvals, and market sales. We used the number of approved indications and market sales amount as proxies for success. We also investigated the important transactions in the discovery and development of immune checkpoint blockade agents, including extensive combination therapy development, especially pembrolizumab and nivolumab, which triggered the paradigm of standard of care in cancer therapeutics. In the present study, we investigated these from the angle of the mechanism of action (MOA), which defines how a drug or other substance produces an effect in the body; for example, how it affects a specific target in a cell, such as an enzyme, or a cell function, such as cell growth. Therefore, the efficacy and safety of drugs are highly dependent on the MOA [42]. Immune checkpoint blockade is the current major MOA in cancer therapy, and we expect that comparisons between immune checkpoint blockade and other MOAs of anticancer drugs are valuable for our research objectives. As an immune checkpoint blockade agent is a monoclonal antibody, that is, a biological anticancer drug, we precisely compared immune checkpoint blockade agents with other biologics. We also added small molecules for comparison, which are the most conventional therapeutic agents used in cancer treatment [14,15]. These small molecules were divided into two groups, namely kinase inhibitors (which are the major MOAs of small molecules) and other small molecules. Our goal was to uncover the potential key success factors of R&D for generating innovative and novel drugs to fulfill unmet medical needs in the future.

2. Materials and Methods

2.1. Samples and Data Sources

Sample data on cancer drugs were collected from the FDA's New Molecular Entity (NME) list of approved small molecules and the New Biological Entity (NBE) list of approved biologics [24] from 1999 to 2018. Target drugs were determined based on CenterWatch's list [43] of new cancer drug approvals, which was cross-referenced against the FDA's NME and NBE lists. Using this approach, we selected 77 small molecules and 30 biologics as samples for this study.

The data source of approved indications for pembrolizumab and nivolumab was a package insert published by the FDA [24] from 2014 to 2019. The data source of product sales in pembrolizumab and nivolumab was Form 10-K from 2014 to 2019. The data source of company type and development phase of combination partners tested with pembrolizumab or nivolumab in interorganizational transactions was the Biomedtracker Deal Search [44].

2.2. Variables and Data Sources

Information on the number of transactions per product was collected from the Informa database's Biomedtracker Deal Search [44]. We examined 1204 transactions related to the identified 107 cancer drugs approved by the FDA. The deal types defined by Informa included "acquisitions," "alliances," and "financing." Each concept of "acquisitions," "alliances," and "financing." are defined as follows:

- An acquisition is when one company purchases all or most of another company's shares to gain control of that company.
- An alliance is an agreement between two or more companies regarding a pharmaceutical product, technology, service, etc.
- A financing involves a company raising money publicly or privately through the sale of equity, debt, or royalty monetization.

The deal characteristics defined by in the alliances consisted of "co-promotions," "includes contract," "includes equity," "includes royalties or profit split information," "intra-biotech deals," "marketing-licensing," "product or technology swaps," "R+D and marketing-licensing," "trial collaborations," and "reverse licensing." The development phase is derived from each company's source information since the information on Biomedtracker Deal Search is obtained exclusively from publicly available sources such as company press releases, medical conference presentations, etc.

2.3. Statistical Analysis

SPSS Statistics 27 (IBM, Armonk, NY, USA) and Microsoft Excel 2016 (Microsoft, Redmond, WA, USA) were used for the statistical analysis. We performed multiple comparison Tukey–Kramer tests corresponding to 95% confidence intervals using a pre-set significance level (two-sided p-value of 0.05). The Tukey–Kramer test is the extension of Tukey's test. Both Tukey's test and the Tukey–Kramer test are a parametric multiple comparisons procedure and applies simultaneous to the set of all pairwise comparisons to find means that are significantly different from each other. We chose the Tukey–Kramer test which is used in the case of unequal samples sizes while Tukey's test is used in the case of equal samples sizes. In the figures, the obtained p-values are presented as follows: $p < 0.01$ as **, $p < 0.05$ as *, and $p < 0.1$ as †.

3. Results

3.1. Interorganizational Transaction per Mechanism of Action

We initially investigated the trend of interorganizational transactions for drug discovery and development in the oncology area by reviewing 1193 transactions performed until July 2019. We performed a multiple comparison Tukey–Kramer test to compare the number of transactions among four categories of MOA: immune checkpoint blockade (which is the largest number of interorganizational transactions in biological anticancer drugs approved under Biologics License Application (BLA)), other biologics approved under BLA, kinase inhibitor (which is the largest number of interorganizational transactions in small molecule anticancer drugs under New Drug Application (NDA)), and other small molecules approved under NDA. The number of interorganizational transactions in immune checkpoint blockade was significantly higher than that in other MOAs (immune checkpoint blockade: other biologics: kinase inhibitors: other small molecules' total = 44.00, SD 31.97: 11.30, SD 16.21: 6.95, SD 7.70: 9.26, SD 7.09, $p < 0.05$; acquisition = 3.29, SD 3.30: 1.17, SD 1.95: 0.68, SD 1.12: 1.46, SD 1.77, $p < 0.05$; alliances = 35.43, SD 24.67: 8.65, SD 12.63: 5.45, SD 5.94: 6.08, SD 5.40, $p < 0.05$; financing = 5.29, SD 4.89: 1.48, SD 2.25: 0.82, SD 1.66: 1.72, SD 2.25, $p < 0.05$; Figure 1). It has also been suggested that alliances are the major category of interorganizational transactions in immune checkpoint blockade.

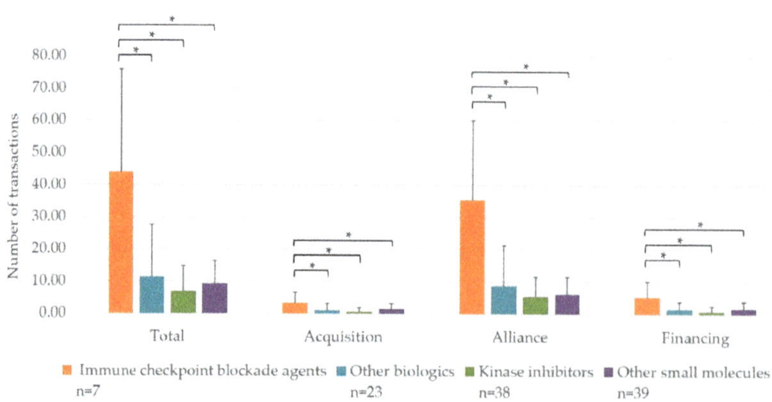

Figure 1. Multiple comparisons of the number of interorganizational transactions by the pairwise comparison procedure among immune checkpoint blockade agents, other biologics, kinase inhibitors and other small molecules. n = 77 (small molecules) and 30 (biologics). * $p < 0.05$.

We then explored the main purpose of transactions for alliances in immune checkpoint blockade. We classified the deal characteristics carefully. Some deals classified in deal types are classified two or more deal characteristics on Biomedtracker Deal Search, and we counted all deal characteristics in these cases. It was suggested that transactions for "alliance" were performed for trial collaborations most frequently, followed by R+D and marketing–licensing (Table 1).

Table 1. The average number of interorganizational transactions for seven immune checkpoint blockade agents in each deal characteristic defined in the alliances.

	Average	SD
Co-promotion	3.571	2.225
Includes Contract	0.714	0.756
Includes Equity	2.143	2.116
Includes Royalty or Profit Split Information	9.000	5.916
Intra-biotech Deal	4.714	6.396
Marketing-licensing	0.714	1.496
Product or Technology Swap	0.429	0.535
R+D and Marketing-licensing	14.000	9.678
Trial Collaborations	24.429	18.636
Reverse Licensing	0.286	0.488

3.2. Market Landscape and Interorganizational Collaborations for Combination Therapy Development of Top Two Immune Checkpoint Blockades: Pembrolizumab and Nivolumab

To understand the landscape of immune checkpoint blockade in the market, we examined the history of approved indications and annual product sales of two immune checkpoint blockades in the US, namely pembrolizumab and nivolumab, which triggered the paradigm of immuno-oncology and are the top two successful products in terms of the number of indications and product sales. Both pembrolizumab and nivolumab have achieved rapid indication expansion since they were initially approved in 2014 (Figure 2). Both products target various tumor types, such as solid and hematological tumors, including microsatellite instability-high and mismatch repair-deficient colorectal cancer. We also confirmed that product sales of both pembrolizumab and nivolumab have increased very rapidly, and the total annual sales of pembrolizumab surpassed that of nivolumab in 2018.

Next, we investigated the empirical landscape of combination therapy of pembrolizumab and nivolumab, as trial collaboration is the most common area in which active interorganizational transactions are performed in immune checkpoint blockade (Table 1), and most of them are related to combination therapy development. In this investigation, we selected only interorganizational transactions that directly influence the drug discovery and development of pembrolizumab and nivolumab, hereinafter referred to as effective interorganizational transactions. These included 67 transactions in alliance and financing for pembrolizumab until 2019 and 59 transactions in acquisition, alliance, and financing for nivolumab until 2019.

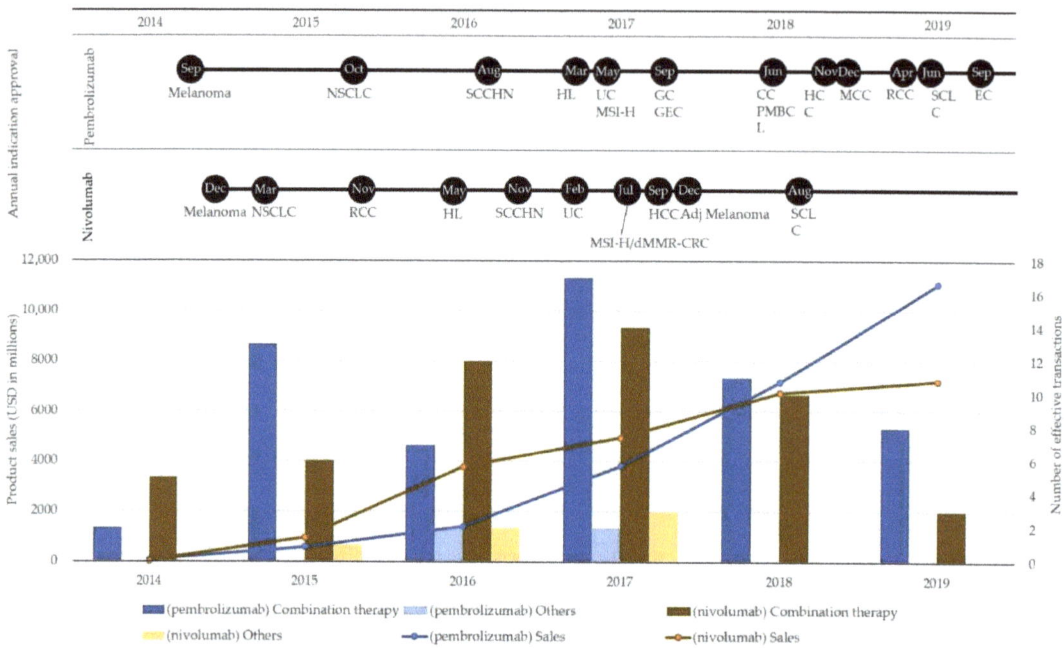

Figure 2. Annual approved indications, annual product sales (line graph), and annual number of effective interorganizational transactions (bar graph) of pembrolizumab and nivolumab from 2014 to 2019. NSCLC: non-small cell lung cancer, SCCHN: squamous cell carcinoma of the head and neck, HL: Hodgkin lymphoma, UC: urothelial carcinoma, MSI-H: microsatellite instability-high, GC: gastric cancer, GEC: gastroesophageal cancer, CC: cervical cancer, PMBCL: primary mediastinal large B-cell lymphoma, HCC: hepatocellular carcinoma, MCC: Merkel cell carcinoma, RCC: renal cell carcinoma, SCLC: small cell lung cancer, EC: endometrial carcinoma, BC: bladder cancer, TMB-H: tumor mutational burden-high, cSCC: cutaneous squamous cell carcinoma, dMMR: mismatch repair deficient, CRC: colorectal cancer, Adj Melanoma: adjuvant treatment of melanoma, ESCC: esophageal squamous cell carcinoma. Effective interorganizational transactions means interorganizational transactions that directly influenced the drug discovery and development of pembrolizumab and nivolumab. Combination therapy: interorganizational transactions for combination therapy development. Others: interorganizational transactions for other purpose.

Figure 2 shows the annual number of effective interorganizational transactions for pembrolizumab and nivolumab, sorted into transactions for combination therapy development and others from 2014 to 2019. Throughout the period, the main purpose of transactions was combination therapy development for both pembrolizumab and nivolumab. Transactions for companion diagnostics (CoDx) development have been observed for both pembrolizumab and nivolumab. Ono Pharmaceutical and Dako AS contracted for CoDx development in 2015, BMS contracted with Enterome Bioscience SA for CoDx development in 2016 and invested in GeneCentric Diagnostics for translational biomarker research in 2017 for nivolumab, and Merck contracted with NanoString Technologies for CoDx development in 2016 for pembrolizumab. Interestingly, Merck lost two cases in 2017, where Merck paid $625 million upfront and global sales royalties to BMS owing to patent infringement to obtain non-exclusive rights and paid $19.5 million to PDL Biopharma owing to patent infringement to obtain non-exclusive rights.

Outside of the period from 2014 to 2019, one of the outstanding transactions was the contract between Ono and Medalex for joint R&D in 2005, in which Medalex obtained exclusive rights in North America. Of note, the deal search based on our algorithm in the database did not detect the acquisition of Medalex by BMS in 2009.

We also examined the features of combination partners, including developing companies, which were identified in effective interorganizational transactions for combination therapy development. As transactions in financing involve collaborations not directly related to combination therapy development, we only used transactions in acquisitions and alliances; there were 46 transactions in alliances for pembrolizumab and 41 transactions, including 3 acquisitions and 38 alliances, for nivolumab. Figure 3 shows the number of products distinguished by the type of company that develops combination partners tested with pembrolizumab or nivolumab, displayed in each development phase from preclinical to market. We classified developing companies into pharmaceutical companies or biotechnology company/start-up companies, and classified development phases into preclinical, phase 1, phase 1/2 or phase 2, phase 3, initial regulatory filing, or marketed after the initial regulatory approval. Some transaction information did not indicate the development phase of the combination partners and was classified as not applicable. It was observed that the main partners were biotechnology/start-up companies for both pembrolizumab and nivolumab; 8 pharmaceutical companies and 35 biotechnology/start-up companies for pembrolizumab and 8 pharmaceutical companies and 24 biotechnology/start-up companies for nivolumab. In addition, different trends were observed, as mainly combination partners under phase 1, phase 1/2, or phase 2 statuses developed by biotechnology companies or start-up companies were tested with pembrolizumab, whereas nivolumab was combined with combination partners under broader development phases from preclinical to marketed phases.

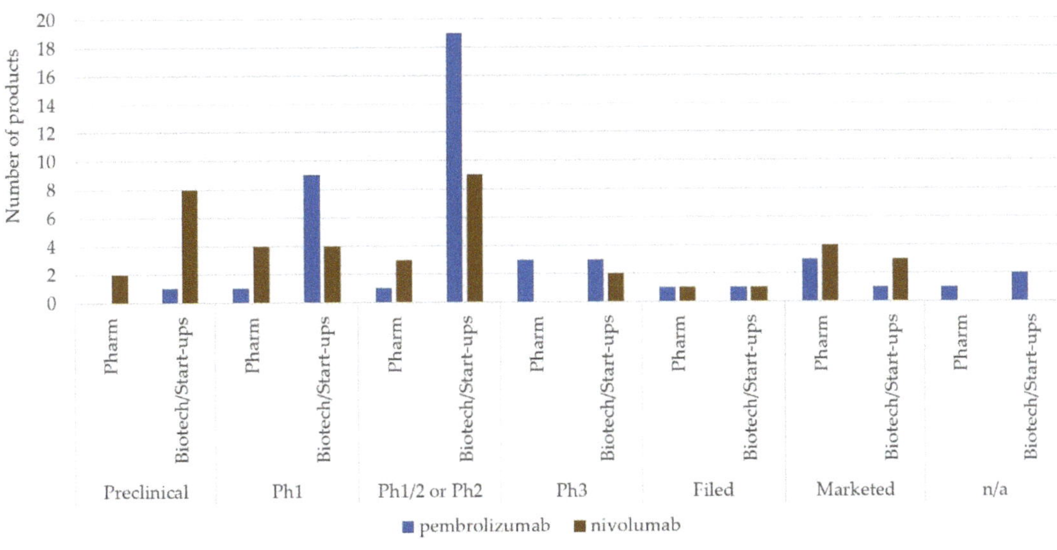

Figure 3. Number of products that were tested with pembrolizumab or nivolumab for combination therapy development. The number of products is displayed by the type of developing company as Pharm or Biotech/Start-ups. Pharm: pharmaceutical company, Biotech: biotechnology company, Start-ups: start-up company, Ph1: Phase 1, Ph1/2 or Ph2: phase 1/2 or phase 2, Ph3: phase 3, Filed: initial regulatory filing, Marketed: marketed after the initial regulatory approval, n/a: not applicable.

4. Discussion

We investigated the recent trends and success factors of interorganizational transactions in cancer therapeutics, especially immune checkpoint blockade. We also explored in depth transactions for the drug discovery and development of immune checkpoint blockade agents, including extensive combination therapy development, especially pembrolizumab and nivolumab. We collected data for each parameter from an open-source database and performed summary statistics and statistical analysis to study the trends of interorganizational transactions, combination therapies, especially with pembrolizumab and nivolumab, and the features of combination partners. Our findings suggest that immune checkpoint blockade is the most active area in terms of interorganizational collaboration in cancer drug development, where alliances related to combination therapy development are the center of practice. It was also suggested from a detailed investigation of transactions for pembrolizumab and nivolumab that different approaches were taken in indication expansion and phase of combination partners.

There were many more interorganizational transactions for drug discovery and development for immune checkpoint blockade than for other MOAs, and the main purpose of transactions was for trial collaboration, as shown in Figure 1 and Table 1, respectively. When we examined the details of interorganizational transactions for pembrolizumab and nivolumab, the majority of these were performed for combination therapy development. Based on these two observations, it was suggested that interorganizational collaborations are important in drug discovery and development of immune checkpoint blockade and alliances is the most effective means for activating combination therapy development which requires an in-licensing pipeline from another company or co-development with another company. These consequences may bring momentum to R&D activities in immune checkpoint blockade area by enabling partnering companies to achieve continuous regulatory authorization and lead business success in sales. In addition, indication expansions observed in Figure 2 may promote the potential combinations of immune checkpoint blockade with other agents for broader cancer types. Combination therapy is the key to expanding the use of pembrolizumab and nivolumab. This may be because combination therapy is expected to increase the number of responders who are non-responders with monotherapy, as previous studies argued that a combination strategy where various combination targets are tested with PD-(L)1 inhibitors is one of the most promising approaches to address treatment of these non-responders [21,40,45–47]. Although immune checkpoint blockade is a paradigm in cancer treatment, as shown by higher overall response rate, achievability of off treatment survival, and higher safety profile, there is clear challenge that only a fraction of cancer patients can derive clinical benefit from immune checkpoint blockade [21,47–50] thus much more research should be necessary for exploring further potential.

Both pembrolizumab and nivolumab are leading products in immune checkpoint blockade and both have common areas in drug discovery and development, including the same initial approval year for the same indication by the FDA. In addition, there are differences between pembrolizumab and nivolumab in terms of market landscape and interorganizational transactions. We anticipated that these differences could lead to successful drug discovery and development. Regarding the approval history and product sales transition of pembrolizumab and nivolumab, as shown in Figure 2, the indication expansion of nivolumab was initially more rapid than that of pembrolizumab from the first FDA approval in 2014 to 2017, indicating that the expansion of pembrolizumab increased from 2017 to 2019, whereas that for nivolumab has slowed down. This transition seems to link the transition of product sales, as initial product sales of nivolumab were higher than those of pembrolizumab until 2017, and this trend has been reversed since 2018. Both products are innovative and successful against various cancers. However, the difference in development strategies might have affected the result of indication expansion and product sales.

As shown in Figure 2, it was observed that the distribution of the effective transactions appears to present a "Gaussian" shape with the peak in 2017 while the sales trend keeps increasing even in 2019 [10]. Our previous study shows that that the observed distribution with the peak in 2017 is unique to pembrolizumab and nivolumab while other biologics and small molecules show different distribution of interorganizational transactions with peaks in different years. We anticipate that the reason for this is because newer immune checkpoint blockades and new drugs using new technologies such as gene therapy have been developed after the launch of pembrolizumab and nivolumab, and the interorganizational collaboration has shifted in combination with those newer agents. On the other hand, pembrolizumab and nivolumab still have solid positions with a large amount of evidence for various indications, which could be associated with an increase in sales as the previous research argued that the order of entry and strong efficacy are correlated with product sales [51,52]. In addition, we confirmed that several interorganizational transactions were performed for the R&D of CoDx and biomarkers. Although immune checkpoint blockade is a remarkably effective approach, previous researchers have argued that the overall response rate varies and is highly dependent on the microenvironment or genetic variations in patients with cancer [47,53]. Therefore, predictive biomarkers are very important [46,49,50,54] and PD-L1 expression is one of the most common biomarkers. However, it has been previously reported that a single biomarker is insufficient to predict clinical benefit or durability of the response to treatment in patients with cancer, and there are no reliable predictive biomarkers [21,50]. Therefore, our observations reflect the fact that further research in this area is necessary.

Regarding the observation of differences in the phase of combination partners between pembrolizumab and nivolumab, different strategies for partner selection can be expected. Although we did not investigate the correlation between strategic transactions and success in drug development and commercialization, the result shown in Figure 3 suggests that the selective approach of collaboration partners led to more successful indication expansion and increases in sales, as pembrolizumab was tested with many combination targets that were in phase 1, phase 1/2, or phase 2. It is also interesting to note that Merck's transactions for payment owing to patent infringement seemed to be decisive, which enabled Merck to obtain exclusive rights to develop and sell pembrolizumab globally. In addition to interorganizational transactions, different approaches can be observed in clinical development, as in the case of nivolumab, which was ahead of pembrolizumab in clinical development entry; the duration from initial investigational new drug application to initial BLA submission was 100 months, with data from 886 subjects in six clinical trials, whereas it was 56 months with data from 1577 subjects in five clinical trials for pembrolizumab, which eventually enabled BLA approval earlier than that of nivolumab. These different approaches in drug development are referable when companies in the pharmaceutical industry develop pipeline strategies.

These considerations, based on our findings, can be translated into R&D strategies when new innovative pipelines are developed. In the future study, we will perform similar analysis focused on the MOA of additional combination therapy approved in recent years in order to clarify the findings in this study which were specific for immune checkpoint blockade.

5. Conclusions

The results of this study confirmed that active research and development related to immune checkpoint blockade is extraordinary in terms of continuous indication expansion and interorganizational transactions, especially for trial collaboration after the initial success of regulatory approval. In particular, trial collaboration for testing combinations with other cancer drugs was main deal characteristic specifically in the cases of pembrolizumab and nivolumab, which also achieved continuous indication expansion and increases in market sales. In addition, we found a remarkable variety of interorganizational collaborations, from the discovery of potential molecules, such as PD-1, to the post-launch of products.

However, there are limitations in our study. We performed database research using an open-source database and investigated only anticancer drugs, especially immune checkpoint blockade. In addition, detailed investigations were performed only for pembrolizumab and nivolumab. Further research is expected in the future to investigate whether the results from our study are applicable to other MOAs or modalities or whether the same kind of strategic collaboration can be seen in newer MOAs or modalities in the future. We expect the results of future research to address the remaining unmet medical needs and sustainability of the pharmaceutical industry.

Author Contributions: Conceptualization, A.D.; methodology, A.D.; software, A.D.; validation, A.D.; formal analysis, A.D.; investigation, A.D.; resources, A.D.; data curation, A.D.; writing—original draft preparation, A.D.; writing—review and editing, T.M., Y.L., S.S. and K.K.; supervision, K.K. All authors have read and agreed to the published version of the manuscript.

Funding: This work was supported by Grants-in-Aid for Challenging Exploratory Research (grant numbers 20K20769, 26301022, and 23730336). The funding sources did not participate in the study design, data collection, analysis, interpretation, report writing, or the decision to submit this article for publication. In addition, this work was supported by the FFJ/Air Liquide Fellowship. The author gratefully acknowledges the generous support and assistance of the Fondation France-Japon (FFJ) de Ecole des Hautes Etudes en Sciences Sociales (EHESS) and Air Liquide. The authors would like to express their sincere thanks to all the respondents who participated in the survey of this study.

Institutional Review Board Statement: Not applicable.

Informed Consent Statement: Not applicable.

Acknowledgments: We are grateful to the editors and referees for their valuable comments, which helped to improve this paper.

Conflicts of Interest: Arisa Djurian is an employee of Takeda Pharmaceutical Company Limited, and Tomohiro Makino is an employee of Daiichi Sankyo Company Limited. Neither company has any direct relationship with the content of this article.

References

1. Kola, I.; Landis, J. Can the pharmaceutical industry reduce attrition rates? *Nat. Rev. Drug Discov.* **2004**, *3*, 711–715. [CrossRef] [PubMed]
2. Adams, C.P.; Van Brantner, V. Market watch: Estimating the cost of new drug development: Is it really $802 million? *Health Aff.* **2006**, *25*, 420–428. [CrossRef] [PubMed]
3. DiMasi, J.A.; Hansen, R.W.; Grabowski, H.G. The price of innovation: New estimates of drug development costs. *J. Health Econ.* **2003**, *22*, 151–185. [CrossRef]
4. Ciani, O.; Jommi, C. The role of health technology assessment bodies in shaping drug development. *Drug Des. Devel. Ther.* **2014**, *8*, 2273–2281. [CrossRef]
5. Rugman, A.M.; Verbeke, A. A perspective on regional and global strategies of multinational enterprises. *J. Int. Bus. Stud.* **2004**, *35*, 3–18. [CrossRef]
6. Teramae, F.; Makino, T.; Lim, Y.; Sengoku, S.; Kodama, K. International strategy for sustainable growth in multinational pharmaceutical companies. *Sustainability* **2020**, *12*, 867. [CrossRef]
7. Sengoku, S.; Yoda, T.; Seki, A. Assessment of Pharmaceutical Research and Development Productivity With a Novel Net Present Value–based Project Database. *Ther. Innov. Regul. Sci.* **2011**, *45*, 175–185. [CrossRef]
8. Bianchi, M.; Cavaliere, A.; Chiaroni, D.; Frattini, F.; Chiesa, V. Organisational modes for Open Innovation in the bio-pharmaceutical industry: An exploratory analysis. *Technovation* **2011**, *31*, 22–33. [CrossRef]
9. Makino, T.; Sengoku, S.; Ishida, S.; Kodama, K. Trends in interorganizational transactions in personalized medicine development. *Drug Discov. Today* **2019**, *24*, 364–370. [CrossRef]
10. Djurian, A.; Makino, T.; Lim, Y.; Sengoku, S.; Kodama, K. Trends of business-to-business transactions to develop innovative cancer drugs. *Sustainability* **2020**, *12*, 5535. [CrossRef]
11. Wang, L.; Plump, A.; Ringel, M. Racing to define pharmaceutical R&D external innovation models. *Drug Discov. Today* **2015**, *20*, 361–370. [CrossRef] [PubMed]
12. Schuhmacher, A.; Germann, P.-G.; Trill, H.; Gassmann, O. Models for open innovation in the pharmaceutical industry. *Drug Discov. Today* **2013**, *18*, 1133–1137. [CrossRef]
13. Makino, T.; Lim, Y.; Kodama, K. Strategic R&D transactions in personalized drug development. *Drug Discov. Today* **2018**, *23*, 1334–1339. [CrossRef] [PubMed]
14. DeVita, V.T.; Chu, E. A History of Cancer Chemotherapy. *Cancer Res.* **2008**, *68*, 8643–8653. [CrossRef]
15. Chabner, B.A.; Roberts, T.G. Chemotherapy and the war on cancer. *Nat. Rev. Cancer* **2005**, *5*, 65–72. [CrossRef] [PubMed]
16. Sawyers, C. Targeted cancer therapy. *Nature* **2004**, *432*, 294–297. [CrossRef] [PubMed]
17. Lord, C.J.; Ashworth, A. Biology-driven cancer drug development: Back to the future. *BMC Biol.* **2010**, *8*, 1–12. [CrossRef]
18. Al-Lazikani, B.; Banerji, U.; Workman, P. Combinatorial drug therapy for cancer in the post-genomic era. *Nat. Biotechnol.* **2012**, *30*, 679–692. [CrossRef]

19. Finn, O.J. Immuno-oncology: Understanding the function and dysfunction of the immune system in cancer. *Ann. Oncol.* **2012**, *23*, 8–11. [CrossRef]
20. Hoos, A. Development of immuno-oncology drugs-from CTLA4 to PD1 to the next generations. *Nat. Rev. Drug Discov.* **2016**, *15*, 235–247. [CrossRef] [PubMed]
21. Sharma, P.; Allison, J.P. The future of immune checkpoint therapy. *Science* **2015**, *348*, 56–61. [CrossRef] [PubMed]
22. Postow, M.A.; Callahan, M.K.; Wolchok, J.D. Immune checkpoint blockade in cancer therapy. *J. Clin. Oncol.* **2015**, *33*, 1974–1982. [CrossRef]
23. Topalian, S.L.; Drake, C.G.; Pardoll, D.M. Immune checkpoint blockade: A common denominator approach to cancer therapy. *Cancer Cell* **2015**, *27*, 450–461. [CrossRef] [PubMed]
24. Drugs@FDA: FDA-Approved Drugs. Available online: https://www.accessdata.fda.gov/scripts/cder/daf/ (accessed on 5 December 2020).
25. Ishida, Y.; Agata, Y.; Shibahara, K.; Honjo, T. Induced expression of PD-1, a novel member of the immunoglobulin gene superfamily, upon programmed cell death. *EMBO J.* **1992**, *11*, 3887–3895. [CrossRef] [PubMed]
26. Chen, L.; Han, X. Anti-PD-1/PD-L1 therapy of human cancer: Past, present, and future Find the latest version: Anti– PD-1/PD-L1 therapy of human cancer: Past, present, and future. *J. Clin. Investig.* **2015**, *125*, 3384–3391. [CrossRef]
27. Dong, H.; Zhu, G.; Tamada, K.; Chen, L. B7-H1, a third member of the B7 family, co-stimulates T-cell proliferation and interleukin-10 secretion. *Nat. Med.* **1999**, *5*, 1365–1369. [CrossRef]
28. Freeman, G.J.; Long, A.J.; Iwai, Y.; Bourque, K.; Chernova, T.; Nishimura, H.; Fitz, L.J.; Malenkovich, N.; Okazaki, T.; Byrne, M.C.; et al. Engagement of the PD-1 immunoinhibitory receptor by a novel B7 family member leads to negative regulation of lymphocyte activation. *J. Exp. Med.* **2000**, *192*, 1027–1034. [CrossRef]
29. Tseng, S.Y.; Otsuji, M.; Gorski, K.; Huang, X.; Slansky, J.E.; Pai, S.I.; Shalabi, A.; Shin, T.; Pardoll, D.M.; Tsuchiya, H. B7-DC, a new dendritic cell molecule with potent costimulatory properties for T cells. *J. Exp. Med.* **2001**, *193*, 839–845. [CrossRef]
30. Latchman, Y.; Wood, C.R.; Chernova, T.; Chaudhary, D.; Borde, M.; Chernova, I.; Iwai, Y.; Long, A.J.; Brown, J.A.; Nunes, R.; et al. PD-L2 is a second ligand for PD-1 and inhibits T cell activation. *Nat. Immunol.* **2001**, *2*, 261–268. [CrossRef]
31. Butte, M.J.; Keir, M.E.; Phamduy, T.B.; Sharpe, A.H.; Freeman, G.J. Programmed Death-1 Ligand 1 Interacts Specifically with the B7-1 Costimulatory Molecule to Inhibit T Cell Responses. *Immunity* **2007**, *27*, 111–122. [CrossRef]
32. Park, J.J.; Omiya, R.; Matsumura, Y.; Sakoda, Y.; Kuramasu, A.; Augustine, M.M.; Yao, S.; Tsushima, F.; Narazaki, H.; Anand, S.; et al. B7-H1/CD80 interaction is required for the induction and maintenance of peripheral T-cell tolerance. *Blood* **2010**, *116*, 1291–1298. [CrossRef] [PubMed]
33. Xiao, Y.; Yu, S.; Zhu, B.; Bedoret, D.; Bu, X.; Duke-Cohan, L.M.F.; Umetsu, D.T.; Sharpe, A.H.; DeKruyff, R.H.; Freeman, G.J. RGMb is a novel binding partner for PD-l2 and its engagement with PD-l2 promotes respiratory tolerance. *J. Exp. Med.* **2014**, *211*, 943–959. [CrossRef] [PubMed]
34. Nishimura, H.; Nose, M.; Hiai, H.; Minato, N.; Honjo, T. Development of lupus-like autoimmune diseases by disruption of the PD-1 gene encoding an ITIM motif-carrying immunoreceptor. *Immunity* **1999**, *11*, 141–151. [CrossRef]
35. Dong, H.; Zhu, G.; Tamada, K.; Flies, D.B.; Van Deursen, J.M.A.; Chen, L. B7-H1 determines accumulation and deletion of intrahepatic CD8+ T lymphocytes. *Immunity* **2004**, *20*, 327–336. [CrossRef]
36. Chen, L. Co-inhibitory molecules of the B7-CD28 family in the control of T-cell immunity. *Nat. Rev. Immunol.* **2004**, *4*, 336–347. [CrossRef]
37. Zou, W.; Chen, L. Inhibitory B7-family molecules in the tumour microenvironment. *Nat. Rev. Immunol.* **2008**, *8*, 467–477. [CrossRef]
38. Lifearc, A. LifeArc: Case Study-Keytruda. Available online: https://www.lifearc.org/case-studies/keytruda-new-generation-cancer-treatment/ (accessed on 13 December 2020).
39. BMS Newsroom: Bristol-Myers Squibb to Acquire Medarex. Available online: http://news.bms.com/press-release/partnering-news/bristol-myers-squibb-acquire-medarex (accessed on 13 December 2020).
40. Yu, J.X.; Hodge, J.P.; Oliva, C.; Neftelinov, S.T. Trends in clinical development for PD-1 / PD-L1 inhibitors. *Nat. Rev. Drug Discov.* **2019**, *18*, 13–14.
41. Tang, J.; Yu, J.X.; Hubbard-Lucey, V.M.; Neftelinov, S.T.; Hodge, J.P.; Lin, Y. The clinical trial landscape for PD1/PDl1 immune checkpoint inhibitors. *Nat. Rev. Drug Discov.* **2018**, *17*, 854–855. [CrossRef]
42. Definition of Mechanism of Action. Available online: https://www.cancer.gov/publications/dictionaries/cancer-terms/def/mechanism-of-action (accessed on 1 November 2020).
43. CenterWatch. Available online: https://www.centerwatch.com/directories/1067-fda-approved-drugs (accessed on 20 July 2019).
44. Biomedtracker Deal Search. Available online: https://www.biomedtracker.com/DealSearch.cfm (accessed on 26 November 2020).
45. Alsaab, H.O.; Sau, S.; Alzhrani, R.; Tatiparti, K.; Bhise, K.; Kashaw, S.K.; Iyer, A.K. PD-1 and PD-L1 checkpoint signaling inhibition for cancer immunotherapy: Mechanism, combinations, and clinical outcome. *Front. Pharmacol.* **2017**, *8*, 1–15. [CrossRef]
46. Pardoll, D.M. The blockade of immune checkpoints in cancer immunotherapy. *Nat. Rev. Cancer* **2012**, *12*, 252–264. [CrossRef]
47. Zappasodi, R.; Merghoub, T.; Wolchok, J.D. Emerging Concepts for Immune Checkpoint Blockade-Based Combination Therapies. *Cancer Cell* **2018**, *33*, 581–598. [CrossRef] [PubMed]

48. Gong, J.; Chehrazi-Raffle, A.; Reddi, S.; Salgia, R. Development of PD-1 and PD-L1 inhibitors as a form of cancer immunotherapy: A comprehensive review of registration trials and future considerations. *J. Immunother. Cancer* **2018**, *6*, 1–18. [CrossRef] [PubMed]
49. Taube, J.M.; Klein, A.; Brahmer, J.R.; Xu, H.; Pan, X.; Kim, J.H.; Chen, L.; Pardoll, D.M.; Topalian, S.L.; Anders, R.A. Association of PD-1, PD-1 ligands, and other features of the tumor immune microenvironment with response to anti-PD-1 therapy. *Clin. Cancer Res.* **2014**, *20*, 5064–5074. [CrossRef] [PubMed]
50. Meng, X.; Huang, Z.; Teng, F.; Xing, L.; Yu, J. Predictive biomarkers in PD-1/PD-L1 checkpoint blockade immunotherapy. *Cancer Treat. Rev.* **2015**, *41*, 868–876. [CrossRef]
51. Schulze, U.; Ringel, M. What matters most in commercial success: First-in-class or best-in-class. *Nat. Rev. Drug Discov.* **2013**, *12*, 419–420. [CrossRef]
52. Teramae, F.; Makino, T.; Sengoku, S.; Lim, Y.; Natori, T.; Kodama, K. Research on pharmaceutical product life cycle patterns for sustainable growth. *Sustainability* **2020**, *12*, 8938. [CrossRef]
53. Lawrence, M.S.; Stojanov, P.; Polak, P.; Kryukov, G.V.; Cibulskis, K.; Sivachenko, A.; Carter, S.L.; Stewart, C.; Mermel, C.H.; Roberts, S.A.; et al. Mutational heterogeneity in cancer and the search for new cancer-associated genes. *Nature* **2013**, *499*, 214–218. [CrossRef]
54. Patel, S.P.; Kurzrock, R. PD-L1 expression as a predictive biomarker in cancer immunotherapy. *Mol. Cancer Ther.* **2015**, *14*, 847–856. [CrossRef]

Journal of *Personalized Medicine*

MDPI

Article

A Retrospective Analysis of the De Ritis Ratio in Muscle Invasive Bladder Cancer, with Focus on Tumor Response and Long-Term Survival in Patients Receiving Neoadjuvant Chemotherapy and in Chemo Naïve Cystectomy Patients—A Study of a Clinical Multicentre Database

Victoria Eriksson [1,†], Oscar Holmkvist [1,†], Ylva Huge [2], Markus Johansson [3], Farhood Alamdari [4], Johan Svensson [5], Firas Aljabery [2,‡] and Amir Sherif [1,*,‡]

1 Department of Surgical and Perioperative Sciences, Urology and Andrology, Umeå University, 901 87 Umeå, Sweden
2 Department of Clinical and Experimental Medicine, Division of Urology, Linköping University, 581 83 Linköping, Sweden
3 Department of Surgery, Division of Urology, Sundsvall-Härnösand County Hospital, 856 43 Sundsvall, Sweden
4 Department of Urology, Västmanland Hospital, 721 89 Västerås, Sweden
5 Department of Statistics, Umeå School of Business, Economics and Statistics (USBE), Umeå University, 901 87 Umeå, Sweden
* Correspondence: amir.sherif@umu.se
† These authors contributed equally to this work.
‡ These authors contributed equally to this work.

Citation: Eriksson, V.; Holmkvist, O.; Huge, Y.; Johansson, M.; Alamdari, F.; Svensson, J.; Aljabery, F.; Sherif, A. A Retrospective Analysis of the De Ritis Ratio in Muscle Invasive Bladder Cancer, with Focus on Tumor Response and Long-Term Survival in Patients Receiving Neoadjuvant Chemotherapy and in Chemo Naïve Cystectomy Patients—A Study of a Clinical Multicentre Database. *J. Pers. Med.* **2022**, *12*, 1769. https://doi.org/10.3390/jpm12111769

Academic Editor: Anne-Marie Caminade

Received: 30 September 2022
Accepted: 25 October 2022
Published: 27 October 2022

Abstract: Background: A high pre-treatment De Ritis ratio, the aspartate transaminase/alanine aminotransferase ratio, has been suggested to be of prognostic value for mortality in muscle-invasive bladder cancer (MIBC). Our purpose was to evaluate if a high ratio was associated with mortality and downstaging. Methods: A total of 347 Swedish patients with clinically staged T2-T4aN0M0, with administered neoadjuvant chemotherapy (NAC) or eligible for NAC and undergoing radical cystectomy (RC) 2009–2021, were retrospectively evaluated with a low ratio < 1.3 vs. high ratio > 1.3, by Log Rank test, Cox regression and Mann–Whitney U-test (MWU), SPSS 27. Results: Patients with a high ratio had a decrease of up to 3 years in disease-free survival (DFS), cancer-specific survival (CSS) and overall survival (OS) ($p = 0.009$, $p = 0.004$ and $p = 0.009$) and 5 years in CSS and OS ($p = 0.019$ and $p = 0.046$). A high ratio was associated with increased risk of mortality, highest in DFS (HR, 1.909; 95% CI, 1.265–2.880; $p = 0.002$). No significant relationship between downstaging and a high ratio existed ($p = 0.564$ MWU). Conclusion: A high pre-treatment De Ritis ratio is on a population level, associated with increased mortality post-RC in endpoints DFS, CSS and OS. Associations decrease over time and require further investigations to determine how strong the associations are as meaningful prognostic markers for long-term mortality in MIBC. The ratio is not suitable for downstaging-prediction.

Keywords: urinary bladder neoplasms; neoadjuvant therapy; cystectomy; clinical decision rules; prognosis

1. Introduction

In 2020, 573,000 people were diagnosed with urinary bladder cancer (UBC), making it the 10th most common cancer worldwide and 6th most common in men [1]. The incidence varies between countries. Sweden has approximately 3200 new cases every year, with 700 deaths yearly. Men are three times more likely to be affected than women. At diagnosis, 70–75% are non-muscle-invasive bladder cancer (NMIBC; clinical stages Ta, T1 and Tis) and 20–25% are muscle-invasive bladder cancer (MIBC; clinical stages T2–T4) [2]. MIBC-patients have had a worse prognosis, with a five-year survival of less than 50% [3]. The addition of

Cisplatinum-based neoadjuvant chemotherapy (NAC) prior to radical cystectomy (RC) has improved survival. However, if patients are non-responders to NAC, this may result in delayed RC and adverse events with decreased downstaging. Downstaging is considered a surrogate marker for overall survival [4,5].

The classical staging systems of clinical tumor-node-metastasis (cTNM) from the American Joint Committee on Cancer (AJCC) and the World Health Organization (WHO) classifications of high and low tumor grade, is routine in classifying and evaluating prognosis in MIBC-patients. Still, the classifications have less significance in predicting 5-year progression-free survival (PFS) [6]. Finding pre-treatment factors with the potential to predict aggressive disease and survival post-RC, would be useful [7]. A more precise evaluation of prognosis could also affect treatment selection prior to RC [8]. No clinical markers concerning prognosis in MIBC are available from a simple blood test at present [9].

However, the De Ritis ratio has recently been described as having prognostic value in multiple individual cancer forms, such as renal cell carcinoma (RCC), myeloma, colonic, pancreatic and urothelial cancer [8,10–15]. The ratio consists of the well-known circulating enzymes aspartate aminotransferase (AST) and alanine aminotransferase (ALT) [3,10], which are retrievable by blood test. The AST/ALT ratio was initially presented by De Ritis in 1957 as a tool to asses liver disease [16]. The underlying mechanism between the AST/ALT enzymes and cancer is not fully clear, but aminotransaminases are expressed by both cancerous and noncancerous cells and found in the liver, heart, skeletal muscle, and kidney [13]. UBC is considered a glucose-dependent malignancy, and it is known that AST plays an essential role in glycolysis, which leads to a potential condition in which increased glucose metabolism induced by cancer, known as the Warburg effect, exists with an increase in AST/ALT ratio. Thus, a potential correlation between a high ratio and cancer disease has been hypothesized [7,8,13,17,18].

Retrospective studies on MIBC have reported findings in which a high ratio was prognostic for inferior PFS, lesser cancer-specific survival (CSS) and lower overall survival (OS). Similar methods to establish a cut-off value were used, including ROC curve with a high De Ritis ratio if >1.1–1.5. The studies may have made deeper reflections regarding potential weaknesses in the statistical methods that established the cut-off level [3,7,8,10,12,19]. One study was based on a population where 30% had disseminated disease and there was no mention of NAC [19]. Another study had a mixed population of MIBC and NMIBC (non-MIBC) and only 18% of the patients received NAC. One of the most important factors for assessing long-term survival, lymph node status pre-treatment, was reported in a vague way, where it is not possible to determine whether it was overt cN+ or whether postoperative pN+ was intended [3]. One study presented survival predictions solely in OS [7], which may include, but is not limited to, death by other types of cancers than MIBC, noncancer-related death and death from treatment. Thus, many events in patients that were specifically unrelated to MIBC were included [20]. Two meta-analyses from 2020 concluded that further validation of the De Ritis is required and that previous retrospective studies from up to 2020, including UBC, were often single-centre and lacking a diverse study population such as patients of African or American origin [11].

We intended to evaluate a larger cohort solely on NAC-treated and NAC-eligible patients after the exclusion of pre-treatment disseminated disease, bacillus Calmette-Guérin (BCG)-resistant UBC and liver disease, as opposed to some previous studies that included patients with known lymph-node involvement. Thus, our aspiration was a cohort with potentially less bias due to high comorbidity and disseminated disease. The aim was to investigate if an association between a high De Ritis ratio > 1.3 and increased mortality existed, by evaluating endpoints in different time periods regarding survival outcomes post-RC and evaluating the proxy marker downstaging of primary tumor. To the best of our knowledge, downstaging has not yet been evaluated in this research area.

2. Materials and Methods

Study design and patient selection; Cystectomized patients were derived from an existing clinical, multi-centre database with all cystectomized patients between the years 2009 and 2021 (n = 973). The patients were treated at either four Swedish cystectomy centres; Norrlands Universitetssjukhus, Umeå; Länssjukhuset Sundsvall, Sundsvall; Västmanlands sjukhus, Västerås and Universitetssjukhuset Linköping, Linköping. Inclusion criteria were MIBC; cT2-T4aN0M0, urothelial histopathology and NAC-administered or NAC-eligible NAC-naïve patients. All patients were discussed after diagnosis at multidisciplinary conference and assigned NAC pre-RC if clinically acceptable. Exclusion criteria were liver disease, disseminated cancer, non-urothelial histopathology, and contraindications for NAC; i.e., age > 75, reduced kidney function, Charlson Age Comorbidity Index (CACI) > 6 and hearing impairment. Patients with missing data regarding AST/ALT, due to there being no available laboratory results, were excluded. Final analysis included 347 patients, all who had undergone RC (Figure 1).

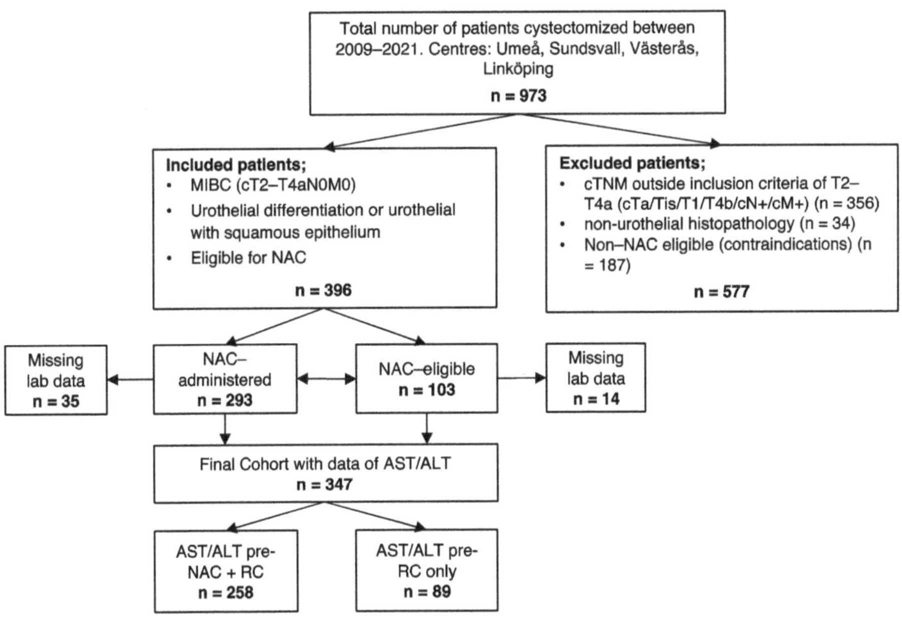

Figure 1. Flowchart displaying selection process of patient cohort, including inclusion and exclusion criteria. Muscle invasive bladder cancer (MIBC); clinical tumor node metastasis (cTNM); neoadjuvant chemotherapy (NAC); aspartate aminotransferase/alanine aminotransferase (AST/ALT); radical cystectomy (RC).

Study procedure: clinicopathological variables from patient medical records were individually collected and compiled into an existing clinical database. Laboratory data were updated from 1 January 2009 to 31 January 2022. Variables included age, Body Mass Index (BMI), American Society of Anaesthesiologists (ASA) score, CACI, cTNM, pTNM, NAC or NAC-legibility, active smoker yes/no, outcome data on cancer recurrence and disease-free survival (DFS), CSS and OS. DFS was chosen instead of PFS as the presented endpoint in this study, with the included events initially described by Punt et al. [21] and later by Birgisson et al. [20]. AST/ALT was documented within 30 days pre-NAC or pre-RC, depending on the treatment the patients were assigned. AST and ALT were routinely tested in lithium heparin plasma with upper reference level 0.75 and 1.1 μ/L, respectively. De Ritis cut-off value in analysis was set to 1.3 based on a frequent used cut-off in previous studies [3,7,19]. Patients were categorized into two groups based on high ratio ≥ 1.3 or low ratio < 1.3. Downstaging of the primary tumor was divided into four different outcomes and ranked with values 1–4 on an ordinal scale; progressive disease (PD; pN+, pT4b), stable disease (SD; pT2–4aN0M0), partial response (PR; pTa, pTis, pT1N0M0) and complete response (CR; pT0N0M0). In analysis, variables on two levels were treated as nominal, such as age > 70 or age < 70, active smoker; yes or no, NAC; yes or no, ASA and pT-stage > 0; pT0 or pTis-T4b, considering the importance of pT0 as a survival benefit, according to Rosenblatt et al. [4]. The variables BMI and CACI were treated as interval variables.

Statistical analysis: Analysis on downstaging was performed separately on NAC-receivers, and the entire cohort of NAC and NAC-eligible NAC-naïve patients, to compare if the association with downstaging and a high ratio differed depending on the different NAC levels. The association between the pre-treatment De Ritis ratio and downstaging outcomes was evaluated by the Mann–Whitney U tests. Survival was predicted up to 3, 5 and 13 years with the study endpoints DFS, CSS and OS. The influence of the pre-treatment De Ritis ratio on the study endpoints was visualized and compared with the Kaplan–Meier estimator and the Log Rank test. Crude and adjusted Cox proportional regression analyses were carried out to determine the influence of patient gender, age, histological TN, pathological T stage, tumour grade and pre-treatment De Ritis ratios on DFS, CSS and OS. HRs estimated from Cox models are shown as the HR with the corresponding 95% CI. The Cox proportional hazard assumption were visually evaluated in STATA. All the used tests are two-sided with a significance level of 5%. The data were statistically evaluated in SPSS Statistics 27.0 for Mac (IBM Corporation, Armonk, NY, USA), and STATA version 15 (Stata Corp, Houston, TX, USA).

3. Results

The final cohort consisted of 347 patients (Figure 1). A total of 258 patients with NAC had pre-treatment data on AST/ALT and 89 NAC-eligible NAC-naïve patients had pre-treatment data on AST/ALT. A total of 76% of the total cohort were men, but a larger proportion of the women in the cohort existed within the high ratio group, compared to the low ratio group (Table 1, $p = 0.044$). A total of 35% of the patients in the cohort were aged over 70. Regarding comorbidity, median level of CACI was 5 and approximately 23% of 347 patients had ASA grade III. Additionally, 75% received some sort of NAC, of which 78% within the group of high ratios were NAC-receivers, compared to 73% NAC-receivers within the low-ratio group. The majority of the cohort-patients had initial stage cT2 (61%) (Table 1).

Table 1. Patient cohort characteristics within groups of low and high De Ritis ratio and total cohort.

Variables	Low De Ritis Ratio < 1.3	High De Ritis Ratio ≥ 1.3	Total Cohort	p-Value
	n = 218	n = 129	n = 347	
		Preoperative		
Sex, n (%)				
Male	173 (79.4)	90 (69.8)	263 (75.8)	0.044 *
Female	45 (20.6)	39 (30.2)	84 (24.2)	
Age, n (%)				0.055 *
<70 years	149 (68.3)	75 (58.1)	224 (64.6)	
>70 years	69 (31.7)	54 (41.9)	123 (35.4)	
BMI (mean, SD)	26.46 (3.9)	25.3 (3.9)	26.03 (3.9)	0.008 †
ASA, n (%)				0.592 *
I	37 (17)	21 (16.3)	58 (16.7)	
II	134 (61.5)	74 (57.4)	208 (59.9)	
III	47 (21.6)	34 (26.4)	81 (23.3)	
CACI (median, IQR)	5 (4, 5)	5 (4, 5)	5 (4, 5)	0.275 ^{f}
Smoker, n (%)				0.966 *
Yes	57 (26.1)	34 (26.4)	91 (26.2)	
No	161 (73.9)	95 (73.6)	256 (73.8)	

Table 1. *Cont.*

Variables	Low De Ritis Ratio < 1.3	High De Ritis Ratio ≥ 1.3	Total Cohort	p-Value
cT-stage, n (%)				0.317 *
cT2	136 (62.4)	77 (59.7)	213 (61.4)	
cT3	73 (33.5)	41 (31.8)	114 (32.8)	
cT4a	9 (4.1)	11 (8.5)	20 (5.8)	
NAC, n (%)				0.299 *
Yes	158 (72.5)	100 (77.5)	258 (74.4)	
No	60 (27.5)	29 (22.5)	89 (25.6)	
NAC; MVAC, n (%)				0.521 *
Yes	131 (60.1)	82 (63.6)	213 (61.4)	
No	87 (39.9)	47 (36.4)	134 (38.6)	
		Postoperative		
pT-stage n (%)				0.612 *
pT0	62 (28.4)	40 (31.0)	102 (29.4)	
pTCis-T4b	156 (71.6)	89 (69.0)	245 (70.6)	

* Chi-square test; † t-test; ^{f} Mann–Whitney U-test; BMI; Body Mass Index, ASA; American Society of Anaesthesiologists, CACI; Charlson Age Comorbidity Index, cT2-4a-clinical-tumor-stage; NAC; neoadjuvant chemotherapy, pT0-T4b; pathological-tumor-stage.

Significantly increased mortality was found in patients with high De Ritis ratio; Log Rank-test, in endpoints DFS (p = 0.009), CSS (p = 0.004), and OS (p = 0.009) up to three years (36 months) post-RC (Figure 2a–c). Significantly increased mortality existed in patients with high De Ritis ratio in endpoints CSS (p = 0.019) and OS (p = 0.046) but not in DFS (p = 0.057) up to five years (60 months) post-RC (Figure 3a–c). No significant increased mortality existed in patients with high ratio, up to 13 years (160 months) post-RC in any endpoint; DFS (p = 0.090), CSS (p = 0.067), or OS (p = 0.261, Figure 4a–c).

Kaplan–Meier curves predicting survival endpoints up to 36 months.

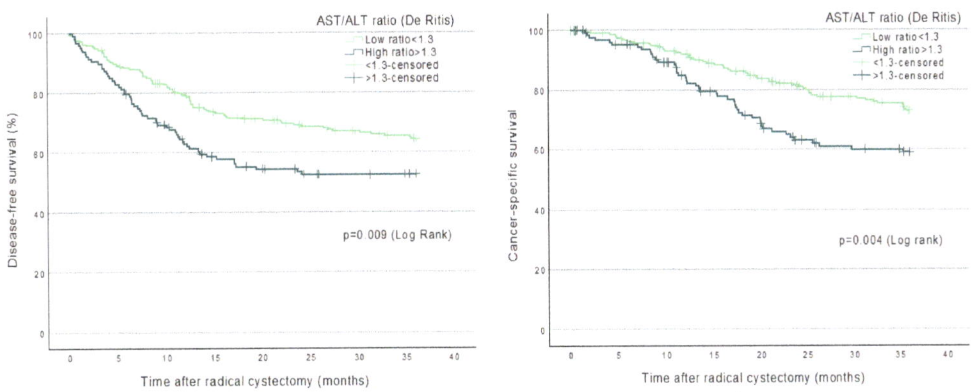

(a) (b)

Figure 2. *Cont.*

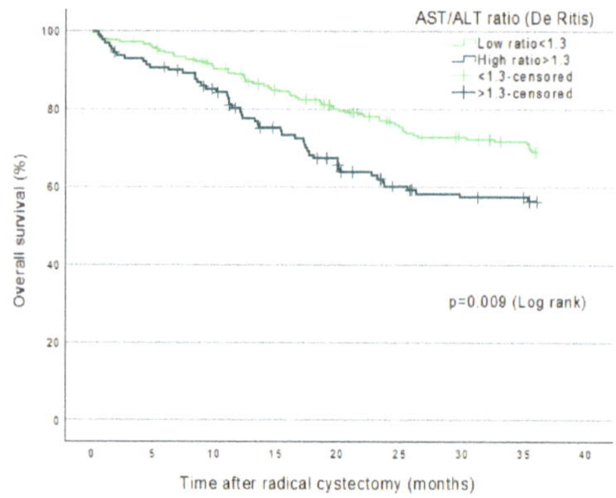

(c)

Figure 2. Kaplan–Meier curves predicting 36 (3 years) and 60 months (5 years) survival endpoints (**a**) Disease-free survival; (**b**) Cancer-specific survival; (**c**) Overall survival.

Kaplan–Meier curves predicting survival endpoints up to 60 months.

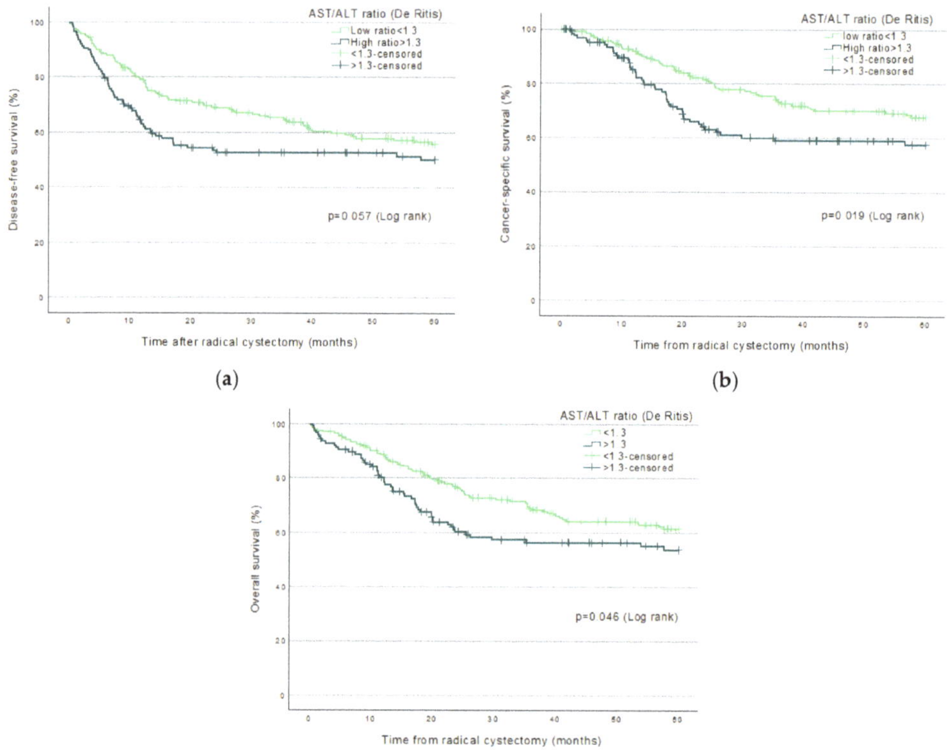

(a) (b)

(c)

Figure 3. Kaplan–Meier curves predicting 60 months (5 years) survival endpoints (**a**) Disease-free survival; (**b**) Cancer-specific survival; (**c**) Overall survival.

Kaplan–Meier curves predicting survival endpoints up to 160 months.

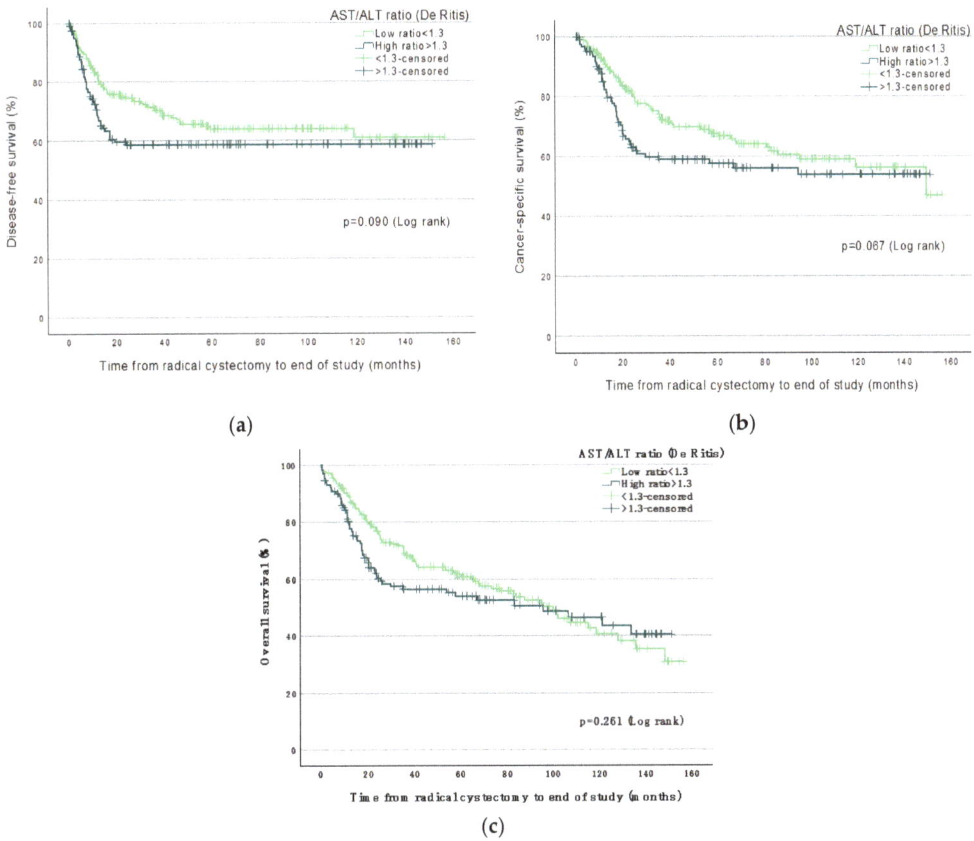

Figure 4. Kaplan–Meier curves predicting 5-13 years (160 months) post-RC survival endpoints (**a**) Disease-free survival; (**b**) Cancer-specific survival; (**c**) Overall survival.

The Cox proportional hazard assumption was visually evaluated and the hazard ratio (HR) between low and high De Ritis ratio seems constant up to three years, and then decreases somewhat. After five years, the proportional hazard assumption can no longer be considered valid. Hence, HR is only evaluated up to three and five years (Table 2). Regarding NAC, in neither crude nor adjusted Cox analysis can a significant interaction effect be visualized between a high De Ritis ratio and NAC, indicating that a high De Ritis ratio equally affects mortality in NAC-receivers and those who are NAC-eligible in our material. We used this as an argument to evaluate the effect of the De ritis ratio in the entire cohort.

A high De Ritis ratio > 1.3 was associated with increased risk of mortality in endpoints DFS, CSS and OS up to three years post-RC, up to five years in CSS and OS, both crude and adjusted for pT-stage > 0, age > 70, active smoking, ASA, CACI, BMI, and NAC. In addition, pT-stage > 0 was associated with increased risk of mortality in all endpoints up to three- and five-years post-RC when adjusted for ratio > 1.3, age > 70, active smoking, ASA, CACI, BMI, and NAC. No individual association was seen between mortality and the other variables age > 70, active smoking, ASA, BMI, and NAC, in any timeframe or endpoint (Table 2).

Table 2. Cox analysis crude and adjusted on entire cohort; NAC and NAC-eligible NAC-naïve, n = 347, up to 3- and 5-years post-RC.

		Crude Estimates [1]			Adjusted Estimates [2]		
Type of Survival	Variables	HR	CI	*p*	HR	CI	*p*
Survival up to 3 years after RC							
DFS	AST/ALT > 1.3	1.573	1.117–2.215	0.010 *	1.673	1.175–2.380	0.004 **
	pT-stage > 0				2.654	1.628–4.325	0.000 **
CSS	AST/ALT > 1.3	1.772	1.191–2.636	0.005 *	1.909	1.265–2.880	0.002 **
	pT-stage > 0				3.913	2.059–7.435	0.000 **
OS	AST/ALT > 1.3	1.625	1.125–2.348	0.010 *	1.724	1.183–2.513	0.006 **
	pT-stage > 0				2.493	1.485–4.186	0.001 **
Survival up to 5 years after RC							
DFS	AST/ALT > 1.3	1.372	0.989–1.902	0.058	1.485	1.060–2.080	0.021 **
	pT-stage > 0				2.783	1.747–4.434	0.000 **
CSS	AST/ALT > 1.3	1.575	1.075–2.308	0.020 *	1.759	1.185–2.612	0.005 **
	pT-stage > 0				4.334	2.292–8.193	0.000 **
OS	AST/ALT > 1.3	1.427	1.005–2.025	0.047 *	1.556	1.086–2.228	0.016 **
	pT-stage > 0				2.564	1.567–4.194	0.000 **

[1] Crude estimates of the De Ritis ratio < 1.3 or ratio > 1.3; [2] Adjusted estimates for variables ratio > 1.3, Age > 70, CACI; Charlson Age Comorbidity Index, ASA; American Society of Anesthesiologists, active smoking, BMI; Body Mass Index, pT-stage; pathological T-stage post-RC, NAC; neoadjuvant chemotherapy. DFS; Disease-free survival, CSS; Cancer-Specific Survival, OS; Overall survival. * Significant *p*-value in crude estimates suggest association between increased risk of mortality and a ratio > 1.3. ** Significant *p*-value in adjusted estimates age > 70, CACI, ASA active smoking, BMI, pT-stage, or NAC, suggest an association between increased risk of mortality and significant variable. Only variables with values of statistical significance any time in analysis are displayed in table.

There is no significant relationship between the downstaging of primary tumor and a high De Ritis ratio > 1.3 (p = 0.564 MWU) either in the entire cohort (Figure 5a) or in NAC-receivers only (Figure 5b) (p = 0.276 MWU). A visible difference can be seen in distribution between the downstaging grades, where NAC-receivers have a higher percentage of increased downstaging within the survival markers of CR and PR (Figure 5b), compared to the entire cohort (Figure 5a).

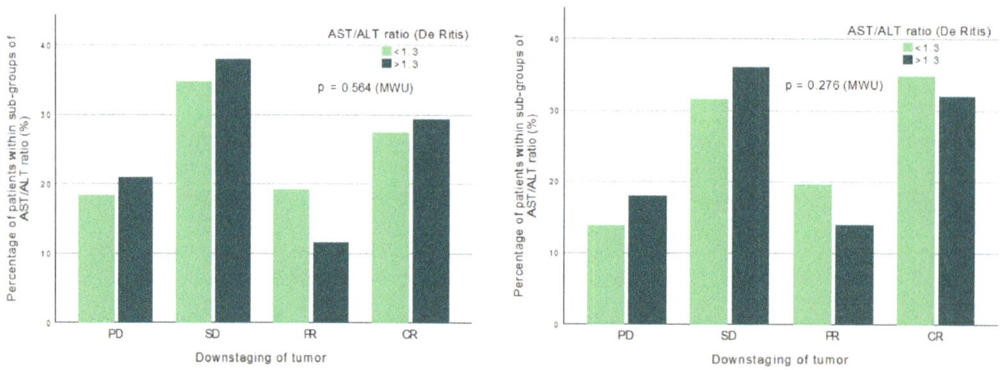

(**a**) (**b**)

Figure 5. (**a**) Downstaging of primary tumor in patient cohort NAC and NAC-naïve n = 347, by the four downstaging categories with patients (%) within sub-groups of high or low De Ritis ratio. There is no significant relationship between downstaging and a high De Ritis ratio > 1.3 (p = 0.564 MWU); (**b**) Downstaging of primary tumor in NAC patients only n = 258, by the four downstaging categories with patients (%) within sub-groups of high or low De Ritis ratio. There is no significant relationship between downstaging and a high ratio > 1.3 (p = 0.276 MWU).

Downstaging of tumor vs high AST/ALT ratio within all patients (left) and only NAC patients (right).

4. Discussion

A high pre-treatment De Ritis ratio was associated with higher mortality in both survival analysis (Figures 2a–c, 3a–c and 4a–c) and in crude and adjusted risk-analysis with increased HR if the ratio was >1.3 in patients, compared to low ratio <1.3. pT-stage >0 was also frequently associated with higher mortality in more endpoints than the De Ritis ratio (Table 2). Thus, a higher tumor stage with more severe cancer post-RC is most statistically associated with mortality than all of our other included variables, as was to be expected. We suggest that pT-stage post-RC that is of the current standard is still of higher prognostic value in predicting survival compared to a high De Ritis ratio pre-treatment. However, UBC is considered a glucose-dependent malignancy, and it is known that AST plays an essential role in glycolysis, which leads to a potential condition in which the increased glucose metabolism by cancer, known as the Warburg effect, exists with an increase in AST/ALT ratio and, therefore, possibly has a high ratio with cancer disease. However, the exact mechanism of the suggested condition has not been shown [8,13] and can only be hypothesized, which must be considered a weakness.

No significant relationship existed between downstaging of the primary tumor and a high De Ritis ratio in either the mixed cohort ($p = 0.564$ MWU) or in NAC-receivers only ($p = 0.276$ MWU) (Figure 5a,b). We, therefore, propose that the different NAC-levels did not affect the potential association between a high De Ritis ratio and downstaging, and that a high pre-treatment ratio may not be of prognostic value regarding downstaging response in either NAC- or NAC-eligible patients. To the best of our knowledge, these findings have not previously been described. The strength in this study is in the evaluation of a relatively large cohort (n = 347), considering extensive inclusion and exclusion criteria compared to other similar studies with higher heterogeneity in the study population. A selection bias could be argued regarding the selection process of patients described in the Methods, where patients with extra high comorbidity may have been excluded due to the inclusion criteria. However, our result may possibly reflect a healthier population than other, similar studies that made no such distinction between levels of disease pre-analysis. Consequently, we propose that our significant results are not primarily due to the high comorbidity in the study population, which may be considered a strength.

A potential weakness may be the HRs in Table 2. The Cox proportional hazard assumption was visually evaluated and the hazard ratio between low and high De Ritis ratio seems constant up to three years, and then decreases somewhat. After five years, the proportional hazard assumption can no longer be considered valid. Other unknown factors may influence the hazard around and after 3–5 years, which may be considered a weakness in the results displayed in Table 2. However, many recurrences were detected before three years post-RC, hence suggesting that the risk analysis utilizing De Ritis ratio has a clear interest for evaluation. Another weakness may lie in our cut-off level of a high De Ritis ratio > 1.3. This was selected from previous studies [3,7]. This may have influenced our results. However, we still had unique individual AST/ALT levels on our population, and can thus compare our results with studies that had the same cut-off level of 1.3, but different AST/ALT levels due to the different study population. We had similar results to several other studies regarding the association between higher mortality and a high ratio [7,10,19,22]. Finally, our results regarding an association between increased mortality and a high ratio lack statistical calculations regarding how strong that association may be. Therefore, we suggest that the strength in the association should be further investigated. This may provide more insight into whether said association between a high De Ritis ratio and mortality is so strong that the ratio can be considered a future prognostic factor for survival in MIBC.

5. Conclusions

A high pre-treatment De Ritis ratio is associated with an increased risk of mortality post-RC in endpoints DFS, CSS and OS. The association decreases over time and requires further investigation to determine how strong the association is as a long-term potential prognostic factor for mortality in MIBC. The De Ritis ratio is not suitable for predicting downstaging.

Author Contributions: Conceptualization, A.S. and V.E.; methodology, A.S.; software, A.S. and J.S.; validation, V.E., O.H. and A.S.; formal analysis, V.E. and A.S.; investigation, V.E.; resources, A.S., F.A. (Farhood Alamdari) and F.A. (Firas Aljabery).; data curation, V.E., O.H., M.J. and Y.H.; writing—original draft preparation, V.E., O.H. and A.S.; writing—review and editing, V.E., A.S., O.H., J.S., Y.H., M.J., F.A. (Farhood Alamdari) and F.A. (Firas Aljabery).; visualization, V.E.; supervision, A.S.; project administration, V.E.; funding acquisition, A.S. All authors have read and agreed to the published version of the manuscript.

Funding: This research was funded by the Swedish Research Council funding for clinical research in medicine (ALF) in Västerbotten, VLL, grant number Bas-ALF/VLL RV-848051" and "The APC was funded by AS".

Institutional Review Board Statement: The study was conducted in accordance with the Declaration of Helsinki and approved by the Institutional Review Board (or Ethics Committee) of Etikprövningsnämnden (EPN) Umeå, Sweden (protocol code 2013/463-31M, date of approval: 3 June 2014).

Informed Consent Statement: Patient consent was waived due to the study being retrospective and therefore no informed consent was required according to EPN.

Data Availability Statement: On reasonable request, the corresponding author can make available all codified data from the clinical database used for this study.

Conflicts of Interest: A.S., corresponding author, declares that he received Grant No. Bas-ALF/VLL RV-848051 as funding concerning this project, from The Swedish Research Council funding for clinical research in medicine (ALF) in Vaesterbotten, VLL, Sweden. All other authors declare no conflict of interest. The funders had no role in the design of the study; in the collection, analyses, or interpretation of data; in the writing of the manuscript; or in the decision to publish the results.

References

1. Bladder Cancer Statistics: World Cancer Research Fund International. [Updated 23 March 2022]. Available online: https://www.wcrf.org/cancer-trends/bladder-cancer-statistics/ (accessed on 23 March 2022).
2. Epidemiologi—RCC Kunskapsbanken: Kunskapsbanken.cancercentrum.se; On Behalf of SKR (Sveriges Kommuner och regioner). Available online: https://kunskapsbanken.cancercentrum.se/diagnoser/urinblase-och-urinvagscancer/vardprogram/bakgrund-och-orsaker/#chapter-3--1-Epidemiologi (accessed on 23 March 2022).
3. Ha, Y.S.; Kim, S.W.; Chun, S.Y.; Chung, J.W.; Choi, S.H.; Lee, J.N.; Kim, B.S.; Kim, H.T.; Yoo, E.S.; Kwon, T.G.; et al. Association between De Ritis ratio (aspartate aminotransferase/alanine aminotransferase) and oncological outcomes in bladder cancer patients after radical cystectomy. *BMC Urol.* **2019**, *19*, 10. [CrossRef] [PubMed]
4. Rosenblatt, R.; Sherif, A.; Rintala, E.; Wahlqvist, R.; Ullén, A.; Nilsson, S.; Malmström, P.U. Pathologic downstaging is a surrogate marker for efficacy and increased survival following neoadjuvant chemotherapy and radical cystectomy for muscle-invasive urothelial bladder cancer. *Eur. Urol.* **2012**, *61*, 1229–1238. [CrossRef] [PubMed]
5. Eriksson, V.; Holmlund, J.; Wiberg, E.; Johansson, M.; Huge, Y.; Alamdari, F.; Svensson, J.; Aljabery, F.; Sherif, A. Adverse events during neoadjuvant chemotherapy for muscle invasive bladder cancer—A Swedish retrospective multicentre study of a clinical database. *Transl. Androl. Urol.* **2022**, *11*, 1105–1115. [CrossRef] [PubMed]
6. Xu, X.; Wang, Y.; Zhang, S.; Zhu, Y.; Wang, J. Exploration of Prognostic Biomarkers of Muscle-Invasive Bladder Cancer (MIBC) by Bioinformatics. *Evol. Bioinform. Online* **2021**, *17*, 117693432110492702021. [CrossRef] [PubMed]
7. Ghahari, M.; Salari, A.; Yazdi, M.G.; Nowroozi, A.; Fotovat, A.; Momeni, S.A.; Nowroozi, M.R.; Amini, E. Association between Preoperative de Ritis (AST/ALT) Ratio and Oncological Outcomes Following Radical Cystectomy in Patients with Urothelial Bladder Cancer. *Clin. Genitourin Cancer* **2022**, *20*, e89–e93. [CrossRef] [PubMed]
8. Yuk, H.D.; Jeong, C.W.; Kwak, C.; Kim, H.H.; Ku, J.H. De Ritis Ratio (Aspartate Transaminase/Alanine Transaminase) as a Significant Prognostic Factor in Patients Undergoing Radical Cystectomy with Bladder Urothelial Carcinoma: A Propensity Score-Matched Study. *Dis. Markers* **2019**, *2019*, 6702964. [CrossRef] [PubMed]
9. Batista, R.; Vinagre, N.; Meireles, S.; Vinagre, J.; Prazeres, H.; Leão, R.; Máximo, V.; Soares, P. Biomarkers for Bladder Cancer Diagnosis and Surveillance: A Comprehensive Review. *Diagnostics* **2020**, *10*, 39. [CrossRef] [PubMed]

10. Bezan, A.; Mrsic, E.; Krieger, D.; Stojakovic, T.; Pummer, K.; Zigeuner, R.; Georg, C.H.; Pichler, M. The Preoperative AST/ALT (De Ritis) Ratio Represents a Poor Prognostic Factor in a Cohort of Patients with Nonmetastatic Renal Cell Carcinoma. *J. Urol.* **2015**, *194*, 30–35. [CrossRef] [PubMed]
11. Su, S.; Liu, L.; Li, C.; Zhang, J.; Li, S. Prognostic Role of Pretreatment De Ritis Ratio (Aspartate Transaminase/Alanine Transaminase Ratio) in Urological Cancers: A Systematic Review and Meta-Analysis. *Front. Oncol.* **2020**, *10*, 1650. [CrossRef] [PubMed]
12. Lee, H.; Lee, S.E.; Byun, S.S.; Kim, H.H.; Kwak, C.; Hong, S.K. De Ritis ratio (aspartate transaminase/alanine transaminase ratio) as a significant prognostic factor after surgical treatment in patients with clear-cell localized renal cell carcinoma: A propensity score-matched study. *BJU Int.* **2017**, *119*, 261–267. [CrossRef] [PubMed]
13. Cheng, X.; Zhou, X.; Yi, M.; Xu, S.; Zhang, C.; Wang, G. Preoperative aspartate transaminase/alanine transaminase ratio as a prognostic biomarker in primary non-muscle-invasive bladder cancer: A propensity score-matched study. *BMC Urol.* **2021**, *21*, 1362021. [CrossRef] [PubMed]
14. Lindmark, G.; Gerdin, B.; Påhlman, L.; Bergström, R.; Glimelius, B. Prognostic predictors in colorectal cancer. *Dis. Colon Rectum* **1994**, *37*, 1219–1227. [CrossRef] [PubMed]
15. Stocken, D.D.; Hassan, A.B.; Altman, D.G.; Billingham, L.J.; Bramhall, S.R.; Johnson, P.J.; Freemantle, N. Modelling prognostic factors in advanced pancreatic cancer. *Br. J. Cancer* **2008**, *99*, 883–893. [CrossRef]
16. De Ritis, F.; Coltorti, M.; Giusti, G. An enzymic test for the diagnosis of viral hepatitis; the transaminase serum activities. *Clin. Chim. Acta* **1957**, *2*, 70–74. [CrossRef]
17. Liberti, M.V.; Locasale, J.W. The Warburg Effect: How Does it Benefit Cancer Cells? *Trends Biochem. Sci.* **2016**, *41*, 211–218. [CrossRef] [PubMed]
18. Sookoian, S.; Pirola, C.J. Alanine and aspartate aminotransferase and glutamine-cycling pathway: Their roles in pathogenesis of metabolic syndrome. *World J. Gastroenterol.* **2012**, *18*, 3775–3781. [CrossRef] [PubMed]
19. Gorgel, S.N.; Kose, O.; Koc, E.M.; Ates, E.; Akin, Y.; Yilmaz, Y. The prognostic significance of preoperatively assessed AST/ALT (De Ritis) ratio on survival in patients underwent radical cystectomy. *Int. Urol. Nephrol.* **2017**, *49*, 1577–1583. [CrossRef] [PubMed]
20. Birgisson, H.; Wallin, U.; Holmberg, L.; Glimelius, B. Survival endpoints in colorectal cancer and the effect of second primary other cancer on disease free survival. *BMC Cancer* **2011**, *11*, 438. [CrossRef] [PubMed]
21. Punt, C.J.; Buyse, M.; Köhne, C.H.; Hohenberger, P.; Labianca, R.; Schmoll, H.J.; Påhlman, L.; Sobrero, A.; Douillard, J.Y. Endpoints in adjuvant treatment trials: A systematic review of the literature in colon cancer and proposed definitions for future trials. *J. Natl. Cancer Inst.* **2007**, *99*, 998–1003. [CrossRef] [PubMed]
22. Hu, X.; Yang, W.X.; Wang, Y.; Shao, Y.X.; Xiong, S.C.; Li, X. The prognostic value of De Ritis (AST/ALT) ratio in patients after surgery for urothelial carcinoma: A systematic review and meta-analysis. *Cancer Cell Int.* **2020**, *20*, 39. [CrossRef] [PubMed]

Journal of *Personalized Medicine*

MDPI

Article

Thromboembolic Events in Patients Undergoing Neoadjuvant Chemotherapy and Radical Cystectomy for Muscle-Invasive Bladder Cancer: A Study of Renal Impairment in Relation to Potential Thromboprophylaxis

Harriet Rydell [1,†], Anna Ericson [1,†], Victoria Eriksson [1], Markus Johansson [1,2], Johan Svensson [3], Viqar Banday [1,‡] and Amir Sherif [1,*,‡]

1 Department of Surgical and Perioperative Sciences, Urology and Andrology, Umeå University, 90187 Umeå, Sweden
2 Department of Surgery, Urology Section, Sundsvall-Härnösand Hospital, 85186 Sundsvall, Sweden
3 Department of Statistics, Umeå School of Business, Economics and Statistics, Umeå University, 90187 Umeå, Sweden
* Correspondence: amir.sherif@umu.se; Tel.: +46-705-229-104
† These authors contributed equally to this work.
‡ These authors contributed equally to this work.

Citation: Rydell, H.; Ericson, A.; Eriksson, V.; Johansson, M.; Svensson, J.; Banday, V.; Sherif, A. Thromboembolic Events in Patients Undergoing Neoadjuvant Chemotherapy and Radical Cystectomy for Muscle-Invasive Bladder Cancer: A Study of Renal Impairment in Relation to Potential Thromboprophylaxis. *J. Pers. Med.* **2022**, *12*, 1961. https://doi.org/10.3390/jpm12121961

Academic Editor: Anne-Marie Caminade

Received: 30 September 2022
Accepted: 22 November 2022
Published: 27 November 2022

Abstract: Recent studies on patients with muscle-invasive bladder cancer (MIBC) undergoing neoadjuvant chemotherapy (NAC) have shown an association between NAC and thromboembolic events (TEE) prior to radical cystectomy (RC). Recent studies suggest that central venous access catheters (CVAs) may induce TEEs, and low-molecular-weight heparin (LMWH) has been mentioned as possible prophylaxis. However, other studies have shown a high incidence of decreased renal function in these patients. The purpose of this study was to determine the portion of MIBC patients with NAC-induced TEEs who had decreased preoperative renal function for whom LMWH potentially would not be beneficial as prophylaxis. We identified 459 cystectomized MIBC patients from two Swedish medical centers from 2009 to 2021. The inclusion criterion of cT2-T4aN0M0 resulted in 220 eligible patients, who were further divided into NAC-administered (n = 187) and NAC-eligible (n = 33), the tentative control group. Values of renal function before, during, and after each NAC cycle were retrospectively collected from individual medical records. Amongst the NAC-administered patients with TEE (n = 29), 41% (95% CI 23.5–61.1%) of patients had decreased renal function. Thus, a substantial portion of NAC-administered patients who developed TEEs had reduced renal function and would have been less likely to have benefited from renal clearance-dependent LMWH prophylaxis.

Keywords: complications; cystectomy; low-molecular-weight heparin; neoadjuvant therapy; thromboembolism; urinary bladder neoplasms

1. Introduction

Urinary bladder cancer accounted for around 570,000 new cancer cases and about 200,000 deaths worldwide in 2020, making it the tenth most commonly diagnosed cancer [1]. Bladder cancer manifests either as non-muscle-invasive cancer or a muscle-invasive form (MIBC); the latter form accounts for approximately 25% of cases at diagnosis. Urothelial MIBC was, prior to the introduction of neoadjuvant chemotherapy (NAC), associated with a poor prognosis, with a 5-year overall survival (OS) post-RC of 50%. The current treatment regimen in Europe for MIBC includes NAC in medically fit patients and radical cystectomy (RC) [2]. The inclusion of NAC pre-RC aims to eradicate micro-metastatic dissemination and has been shown to significantly increase survival compared with RC only [3,4]. We previously showed that chemo-responding patients who downstaged with complete response (CR) to pT0N0M0) had an absolute risk reduction of 31% for death at the five-year median observation time (OS) [5].

The NAC regimen for MIBC patients in Sweden in most centers consists of a cisplatin-based combination of methotrexate, vinblastine, doxorubicin, and cisplatin (MVAC), while in patients who are ineligible for cisplatin, a carboplatin-gemcitabine combination is used [6]. Eligibility for NAC, according to Swedish guidelines, are age $\leq$ 75 years, renal function with eGFR > 50–60, no hearing impediments, and acceptable comorbidities [7]. NAC treatment is generally well tolerated; however, adverse events, including thromboembolic events (TEEs). have been shown to occur during and after NAC treatment [8,9]. Eriksson et al., using a multicenter clinical dataset, showed serious grade 4 TEEs in patients treated with NAC. In addition, acute kidney injury and chronic kidney disease were seen in 41% and 11% of patients, respectively [10]. Interestingly, 30% of patients receiving NAC had to terminate the treatment prematurely. Among these patients, 62% of terminations were due to acute kidney injury. There was a significant association between decreased kidney injury and increased downstaging, possibly reflecting the effect of completed intended NAC cycles on downstaging, as reported earlier [5]. Renal impairment is a fraught consequence of cisplatin-based combinations and one of the key determinants for NAC eligibility [7,11]. Mehrazin et al., in a recent study, showed a continued decline in eGFR rates in MIBC-NAC patients after discharge. In their study, the rate of decline was related to eGFR rates at discharge, with up to 43% of the patients showing a decline [12]. Deep vein thrombosis (DVT) and pulmonary embolism (PE) are two lethal complications during NAC and RC. Endothelial damage and the hypercoagulable state during cancer treatment continues post treatment [13]. Thromboprophylaxis pre- and post-surgery has become a standard in tertiary care centers and is highly encouraged as a quality-of-care indicator [14]. Low-molecular-weight heparin (LMWH) is frequently prescribed for the prevention and treatment of TEEs in patients [15]. LMWH provides a reduction in the incidence of venous thromboembolism and requires minimal monitoring [16,17]. However, combined with impaired renal function, the major mode of excretion for LMWH, there is an increased risk of supratherapeutic accumulation of LMWH with an associated risk of major bleeding [12]. To the best of our knowledge, no international or national guidelines exist advocating LMWH as prophylaxis against TEEs amongst MIBC-NAC patients during the NAC period. In this retrospective study, we investigate the incidence of NAC-induced renal impairment in our research database. The objective was to assess the incidence of patients with NAC-induced TEEs who may not have benefited from LMWH if prophylaxis had hypothetically been used.

2. Materials and Methods

2.1. Data Collection and NAC Eligibility

The study population was derived from an extensive dataset based on records of radically cystectomized patients with urothelial carcinoma in the urinary bladder in the years 2009–2021. Patients' data were retrospectively collected from individual medical records from the two participating Swedish centers: Norrlands Universitetsjukhus in Umeå and Länssjukhuset in Sundsvall. The inclusion criteria allowed for MIBC carcinomas categorized as cT2-T4bN0M0 with urothelial or mixed-type histopathology, thus excluding non-muscle-invasive cancers, disseminated disease before treatment initiation, and histopathology without any urothelial component. Hence, of the 459 radically cystectomized patients who were initially available for consideration, 220 were suitable for analysis, as 239 patients were excluded due to unfit histopathology, non-cT2-4aN0M0, or not fulfilling the criteria for NAC treatment and thus being deemed as non-NAC-eligible patients. For detailed information on patient selection, see Figure 1.

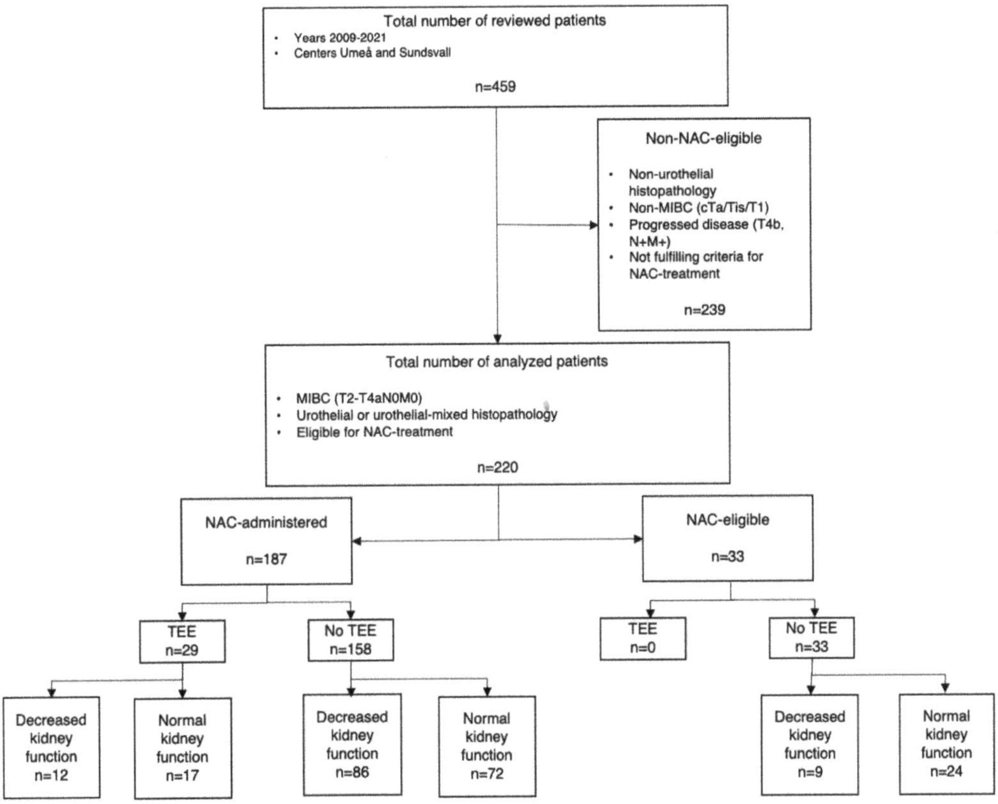

Figure 1. Flowchart of study population. MIBC; muscle-invasive bladder cancer. NAC; neoadjuvant chemotherapy. TEE; thromboembolic event.

The majority of MIBC patients in Sweden today receive 3 or 4 cycles of chemotherapy before RC, and in 2021, 71% of all MIBC patients received NAC in Umeå. On the contrary, in 2009, only 17% received NAC in Umeå, as NAC was not recommended by clinical guidelines until more recent years. Hence, a portion of all MIBC patients in Sweden did not receive NAC and was solely radically cystectomized. As chemotherapy by time was introduced into modern MIBC guidelines, the number of NAC-administrated patients rose. However, a few patients still are untreated due to other circumstances; for example, some choose to reject NAC. Thus, non-NAC but NAC-eligible patients provide us with a suitable parallel control group, as these patients can be regarded as exhibiting the same clinical status as those treated with NAC (Table 1). Consequently, this study focused on the NAC-administered patients (n = 187) and NAC-eligible non-NAC patients (n = 33), who were considered being the tentative control group. By analyzing the NAC-eligible patients, any confounding factors regarding the carcinoma and its innate effects on the cohort's kidney function could be overviewed. The basic characteristics of both NAC-treated as well as NAC-eligible patients are shown in Table 1. To establish which patients would have been eligible for NAC, medical journals and preestablished NAC criteria were used with the guidance of a senior urologist and the national guidelines. To be considered NAC-eligible, patients had to be $\leq$75 years old with an eGFR > 50 or creatinine < 100 and a Charlson Age Comorbidity Index (CACI) of 6 or less. The CACI is used to determine a standardized 10-year survival prediction and was individually calculated for each patient. If a patient fulfilled the criteria to receive NAC but the multidisciplinary team recommended abstaining from treatment, the latter judgement was regarded by the study being the final verdict regarding a patient's NAC eligibility.

Table 1. Baseline characteristics of the study cohorts.

		NAC-ADMINISTRATED n = 187			NAC-ELIGIBLE n = 33
		WITH TEE n = 29	NO TEE n = 158	TEE + NO TEE n = 187	
VARIABLE		**Mean (SD)**	**Mean (SD)**	**Mean (SD)**	**Mean (SD)**
AGE		67 (7)	68 (7)	68 (7)	67 (9)
CACI		5 (1)	5 (1)	5 (1)	5 (1)
NO. NAC CYCLES		3 (1)	3 (1)	3 (1)	0 (0)
		n (%)	n (%)	n (%)	n (%)
MEDICAL CENTER					
	UMEÅ	25 (86)	120 (76)	145 (77)	22 (67)
	SUNDSVALL	4 (14)	38 (24)	42 (22)	11 (33)
GENDER					
	WOMAN	4 (14)	41 (26)	45 (24)	5 (15)
	MAN	25 (86)	118 (75)	143 (76)	28 (85)
cT-STADIUM					
	T2	15 (51)	91 (57)	106 (57)	22 (67)
	T3	12 (41)	54 (34)	66 (35)	11 (33)
	T4 & T4a	2 (7)	13 (8)	15 (8)	0 (0)
THROMBOSIS PROPHYLAXIS					
	Anticoagulant	12 (41)	9 (6)	21 (11)	3 (9)
	Antiplatelet	28 (96)	23 (14)	51 (27)	5 (15)

NAC: neoadjuvant chemotherapy. TEE: thromboembolic event. SD: standard deviation. CACI: Charlson comorbidity index.

Laboratory values regarding renal function were gathered, adding to the database's already large collection of clinicopathological variables, such as age, gender, smoking habits, BMI, ASA, CACI, and TEE. Values for renal function are monitored in MIBC patients but are rarely measured by one single variable. Thus, four measurements were obtained for the study to obtain a conclusive portrayal of each patient's clinical kidney status: serum/plasma creatinine (S/P-creatinine), eGFR by cystatin c, eGFR by creatinine and chrome-ethylenediaminetetraacetic acid (Cr-EDTA). The cut-offs for each measurement were in accordance with Swedish clinical practice and the standardized reference intervals: S/P-creatinine > 90 for females and > 100 for males, eGFR (creatinine) < 60, eGFR (cystatin c) < 60, and Cr-EDTA < 60. The reference intervals for all analyzed measurements are presented in Table 2. Values were collected at predetermined time intervals: pre-TUR-b, before initiation of NAC, after each provided NAC cycle, and, finally, before RC. Not all patients had an available value for each measurement in every time period, and if values were inaccessible, they were considered missing data. As no conclusion can be made from unavailable measurements, the study considered missing data to be non-pathological values. No patient had missing values in all 4 extracted measurements. Utilizing each exact value for every available measurement that was collected allowed the stratification of the cohort's population into two sub-categories: patients with *or* without reduced kidney function. As one patient could potentially have either one single or several pathological values, multiple groupings regarding the type of measurement of kidney function at different time periods were considered.

Another distinction in the study cohort was made by separating the NAC-administrated who did or did not suffer a TEE. The TEE incidences, presented in Table 3, were used for analysis, with PE, DVT, TEEs anatomically related to the central venous access catheters (CVA), angina/myocardial infarction (MI), and transient ischemic attack (TIA)/stroke used as included types of events. The events were registered if the time of the TEE occurred during the period from TUR-b to RC. Four patients had a total number of 2 TEEs, and one patient had a total of 3 TEEs.

Table 2. Reference interval and cut-off values for pathology of included measurements.

MEASUREMENT		REFERENCE INTERVAL	CUT OFF VALUE
P/S-Creatinine			
	Male	60–100	>100
	Female	50–90	>90
eGFR (creatinine)			
	18–50 years	>80	<60
	>50 years	>60	<60
eGFR (cystatin c)			
	18–50 years	>80	<60
	>50 years	>60	<60
Cr-EDTA			
	20–50 years	80–125 mL/min/1.73 m^2	<60
	51–65 years	60–110 mL/min/1.73 m^2	<60
	66–80 years	50–90 mL/min/1.73 m^2	<60

P/S: plasma or serum. eGFR: estimated glomerular filtration rate. Cr-EDTA: chrome ethylenediaminetetraacetic acid.

Table 3. TEE incidences amongst NAC-administrated patients pre-cystectomy.

	NAC-ADMINISTRATED n = 187
	WITH TEE n = 29
TYPE OF TEE	
DVT	2
THROMBOPHLEBITIS	3
PE	18
FROM CVA	11
ANGINA/MI	1

TEE, thromboembolic events; NAC, neoadjuvant chemotherapy; DVT, deep vein thrombosis; PE, pulmonary embolism; FROM CVA, anatomically related to the central venous access; MI, myocardial infarction.

2.2. Statistics

Population characteristics for interval variables are described as means and standard deviation, and those for nominal variables are described as frequencies and percentages. The confidence interval for the percentage of patients with reduced kidney function among the defined groups of the cohort was calculated with the Clopper–Pearson method. The statistical analyses were performed with IBM SPSS Statistics for Windows, Version 28.0. Armonk, NY, USA: IBM Corp.

2.3. Ethics

The study was approved by the regional ethics board in Umeå: EPN-Umeå, dnr: 2013/463-31M and amendment 2016/403-32. The study conformed to the provisions of the Declaration of Helsinki (as revised in Fortaleza, Brazil, October 2013). The regional ethics board specifically decided that informed consent from the participants was to be considered redundant, especially due to the high mortality of MIBC as well as the retrospective nature of the study.

3. Results

The clinicopathological variables for each of the 220 analyzed patients were extracted from the extensive official database charting all MIBC patients who have undergone RC in Sweden and are presented in Table 1. Amongst these variables, TEE incidences were collected, enabling division into subgroups depending on whether patients suffered postulated NAC-induced thrombosis or not. Twenty-nine patients in the NAC-administrated group suffered one or more TEEs. Out of them, 18 were PEs and 11 were thromboses anatomically connected to the CVA. Three patients suffered thrombophlebitis as a TEE, two had DVTs, and, finally, one patient presented with angina. Four patients developed two individual TEEs, and one patient developed three.

The findings regarding pathological renal values before the initiation of chemotherapy are presented in Table 4, elucidating any existing kidney damage in the study population pre-NAC and thus establishing the baseline measurements in each group. Non-pathological values for P/S-creatinine and eGFR are predetermined requirements for NAC eligibility; consequently, no patients in this group had reduced kidney function. In addition, as no patient in the NAC-eligible group suffered any TEE pre-RC, only patients without TEEs are presented. The reduced kidney function in the study population during the entire period, TUR-b to RC, is presented in Table 5. In the NAC-administrated group, 98 patients (52%) out of the total number of 187 had reduced kidney function pre-RC. Pathological measurements were found in 12 patients belonging to the NAC-administrated subgroup who also suffered TEEs (n = 29), resulting in 41% (95% CI 23–61%).

Table 4. Reduced kidney function after TUR-B, before first NAC cycle.

	NAC-ADMINISTRATED (n = 187)			NAC-ELIBIGLE (n = 33)
	WITH TEE n (%)	NO TEE n (%)	TEE + NO TEE n (%)	NO TEE n (%)
	n = 29	n = 158	n = 187	n = 33
REDUCED KIDNEY FUNCTION PRE-NAC 1				
MEASUREMENT				
P/S-Creatinine	6 (21)	29 (18)	35 (19)	0 (0)
eGFR (creatinine)	3 (10)	14 (9)	17 (9)	0 (0)
eGFR (cystatin c)	1 (3)	25 (16)	26 (14)	0 (0)
Cr-EDTA	3 (10)	21 (13)	24 (13)	0 (0)
ALL MEASUREMENTS	9 (31)	50 (32)	59 (31)	0 (0)

For NAC eligibility, non-pathological renal functional values were required as predetermined by the study; thus, no NAC-eligible patients present any kidney damage at diagnosis. All measurements: one or more of the four values were pathological in the same patient. TUR-B: transurethral resection of the bladder. NAC: neoadjuvant chemotherapy. TEE: thromboembolic event. P/S: plasma or serum. eGFR: estimated glomerular filtration rate. Cr-EDTA: chrome ethylenediaminetetraacetic acid.

Table 5. Reduced kidney function at any point during the entire period; pre-TUR-B to before cystectomy.

	NAC-ADMINISTRATED n = 187			NAC-ELIBIGLE (n = 33)
	WITH TEE n (%)	NO TEE n (%)	TEE + NO TEE n (%)	NO TEE n (%)
	n = 29	n = 158	n = 187	n = 33
REDUCED KIDNEY FUNCTION PRE CE				
MEASUREMENT				
P/S-Creatinine	9 (31)	69 (44)	78 (42)	9 (27)
eGFR (creatinine)	5 (17)	37 (23)	42 (22)	1 (3)
eGFR (cystatin c)	2 (7)	31 (20)	33 (18)	0 (0)
Cr-EDTA	3 (10)	22 (14)	25 (13)	0 (0)
ALL MEASUREMENTS	12 (41)	86 (54)	98 (52)	9 (27)

All measurements: one or more of the four values were pathological in the same patient. TUR-B: Transurethral resection of the bladder. NAC: neoadjuvant chemotherapy. TEE: thromboembolic event. RC: radical cystectomy. P/S: plasma or serum. eGFR: estimated glomerular filtration rate. Cr-EDTA: chrome ethylenediaminetetraacetic acid.

Table 6 presents the portion of the study population in which at least one of the four predetermined and analyzed measurements were accessible. Out of the 220 patients, 35 (16%) had no obtainable values between TUR-b and the first NAC cycle and were thus left as missing data.

Table 6. Patients with available values at any point in the entire time period: TUR-b–pre-cystectomy.

	NAC-ADMINISTRATED n = 187			NAC-ELIGIBLE n = 33
	WITH TEE n (%)	NO TEE n (%)	TEE + NO TEE n (%)	NO TEE n (%)
	n = 29	n = 158	n = 187	n = 33
MEASUREMENT				
P/S-Creatinine	29 (100)	158 (100)	187 (100)	32 (97)
eGFR (creatinine)	20 (69)	109 (69)	129 (69)	10 (30)
eGFR (cystatin c)	0 (0)	15 (9)	15 (8)	1 (3)
Cr-EDTA	1 (3)	2 (1)	3 (2)	0 (0)

TUR-b: transurethral resection of the bladder, NAC: neoadjuvant chemotherapy. TEE: thromboembolic event. P/S: plasma or serum. eGFR: estimated glomerular filtration rate. Cr-EDTA: chrome ethylenediaminetetraacetic acid.

4. Discussion

According to our findings, a considerable portion of the NAC-administrated patients who suffered a TEE during treatment had reduced renal function, as 41% (95% CI 23–61%) had one or more pathological values throughout the entire time period, up until the day of cystectomy. Fifty-four percent of the NAC-administrated patients who underwent their chemotherapy without a thrombolytic adverse event also had reduced kidney function, presenting a similar result as the TEE group. The most frequent pathological value amongst the NAC-treated TEE patients was P/S-creatinine, with nine patients (31%) showcasing a higher value than the established reference interval. This is possibly in accordance with P/S-creatinine being the most frequently used measurement for kidney function in Sweden. No noteworthy differences were found between the NAC-administrated subgroups regarding kidney function before the initiation of chemotherapy. Hence, the two groups of TEE patients and those who did not suffer a TEE could be considered to have the same baseline renal function before the first NAC cycle.

The main objective of the study was to investigate renal function in NAC-administrated patients who had suffered a TEE and to assess their ability to potentially respond to LMWH-prophylaxis. Hence, the participants were divided into subgroups of those who did or did not develop thrombosis during the time between diagnosis and RC. NAC inherently contributes to an amplified TEE risk in MIBC patients undergoing RC, and recent findings have suggested that the choice of CVA for chemotherapy can influence the increased risk [8,18,19]. As LMWH regularly is used in TEE treatment and commonly as prophylaxis in other high-risk patient groups, suggestions have been made to also use LMWH in cases in which MIBC-NAC patients would be considered high-risk for TEE. Accordingly, Mehrazin et al. suggested recently that established reduced kidney function would render heparin-based prophylaxis sub-therapeutic by accumulation, which in turn would increase the risk of clinically significant bleeding. Our results showed that in NAC-administrated patients who did suffer a TEE, 41% would potentially not have responded to prophylactic treatment. Thus, the argument can be made to contemplate whether LMWH can be considered the single solution to the increased TEE risk in this patient group or whether instead a shift in the main focus should be made to the choice of CVA.

As the hypothetical LMWH treatment would be initiated at the start of chemotherapy (NAC), the distinction between the TEE and non-TEE patients would not be yet established. Therefore, there is relevance in also establishing the incidence of reduced renal function in NAC-administrated patients who have not suffered a TEE. In the NAC-administrated no-TEE group, 54% were found to have reduced renal function during the entire pre-RC period. Hence, this group showed similar results to the TEE-positive subgroup, indicating that LMWH could be subtherapeutic in the NAC-MIBC group as a whole. The NAC-eligible non-NAC group presented with a lower prevalence of pathological renal values than the groups (w/wo TEE) treated with NAC, possibly reflecting the negative effect that MIBC might inherently possess on kidney function.

However, because of this study's numerically limited cohort population, a larger patient selection is needed to further establish the incidence of reduced kidney function amongst NAC-administrated MIBC patients. Further important associated subjects to evaluate are how the current clinical use of LMWH contributes to reduced renal function, as well as extending the time period of analysis both pre-RC as well as post-RC. A strength of this study is that the analyzed measurements are meticulously extracted from individual medical records; thus, the presented findings could be regarded as highly reliable.

5. Conclusions

Our findings show that a substantial portion of NAC-administered patients with TEEs had reduced renal function pre-RC, and therefore, they would potentially not be suited for renal clearance-dependent LMWH prophylaxis. However, further investigations are needed to fully comprehend the prevalence of reduced renal function amongst MIBC patients.

Author Contributions: Conceptualization, A.S., H.R. and A.E.; methodology, A.S.; software, A.S and J.S.; validation, H.R., A.E., J.S. and A.S.; formal analysis, H.R., V.B., A.E. and A.S.; investigation, H.R., A.E., V.E. and A.S.; resources, A.S; data curation, V.E., H.R., A.E., M.J. and J.S.; writing—original draft preparation, H.R., A.E., V.B. and A.S.; writing—review and editing, H.R., A.E., V.B., J.S., M.J. and A.S.; visualization, A.E.; supervision, A.S.; project administration, H.R., V.B. and A.S.; funding acquisition, A.S. All authors have read and agreed to the published version of the manuscript.

Funding: This research was funded by the Swedish Research Council funding for clinical research in medicine (ALF) in Västerbotten, VLL, grant number Bas-ALF/VLL RV-848051, and the APC was funded by AS.

Institutional Review Board Statement: The study was conducted in accordance with the Declaration of Helsinki and approved by the institutional review board (or ethics committee) of Etikprövningsnämnden (EPN) Umeå, Sweden (protocol code 2013/463-31M, date of approval: 3rd of June 2014).

Informed Consent Statement: Patient consent was waived due to the study being retrospective, and therefore, no informed consent was required according to EPN.

Data Availability Statement: On reasonable request, the corresponding author can make available all codified data from the clinical database used for this study.

Conflicts of Interest: A.S., the corresponding author, declares that he received grant no. Bas-ALF/VLL RV-848051 as funding concerning this project from the Swedish Research Council funding for clinical research in medicine (ALF) in Västerbotten, VLL, Sweden. All other authors declare no conflicts of interest. The funders had no role in the design of the study; in the collection, analyses, or interpretation of data; in the writing of the manuscript; or in the decision to publish the results.

References

1. Sung, H.; Ferlay, J.; Siegel, R.L.; Laversanne, M.; Soerjomataram, I.; Jemal, A.; Bray, F. Global Cancer Statistics 2020: GLOBOCAN Estimates of Incidence and Mortality Worldwide for 36 Cancers in 185 Countries. *CA Cancer J. Clin.* **2021**, *71*, 209–249. [CrossRef] [PubMed]
2. European Association of Urology. Muscle-Invasive and Metastatic Bladder Cancer. Available online: https://uroweb.org/guidelines/muscle-invasive-and-metastatic-bladder-cancer (accessed on 30 September 2022).
3. Hamid, A.; Ridwan, F.R.; Parikesit, D.; Widia, F.; Mochtar, C.A.; Umbas, R. Meta-analysis of neoadjuvant chemotherapy compared to radical cystectomy alone in improving overall survival of muscle-invasive bladder cancer patients. *BMC Urol.* **2020**, *20*, 158. [CrossRef] [PubMed]
4. Sherif, A.; Holmberg, L.; Rintala, E.; Mestad, O.; Nilsson, J.; Nilsson, S.; Malmström, P.U.; Group, N.U.C. Neoadjuvant cisplatinum based combination chemotherapy in patients with invasive bladder cancer: A combined analysis of two Nordic studies. *Eur. Urol.* **2004**, *45*, 297–303. [CrossRef] [PubMed]

5. Rosenblatt, R.; Sherif, A.; Rintala, E.; Wahlqvist, R.; Ullén, A.; Nilsson, S.; Malmström, P.U. Pathologic downstaging is a surrogate marker for efficacy and increased survival following neoadjuvant chemotherapy and radical cystectomy for muscle-invasive urothelial bladder cancer. *Eur. Urol.* **2012**, *61*, 1229–1238. [CrossRef] [PubMed]
6. Jerlström, T.; Chen, R.; Liedberg, F.; Andrén, O.; Ströck, V.; Aljabery, F.A.S.; Hosseini, A.; Sherif, A.; Malmström, P.U.; Ullén, A.; et al. No increased risk of short-term complications after radical cystectomy for muscle-invasive bladder cancer among patients treated with preoperative chemotherapy: A nation-wide register-based study. *World J. Urol.* **2020**, *38*, 381–388. [CrossRef] [PubMed]
7. Leiberg, F. Nationellt Vårdprogram Cancer i Urinblåsa, Njurbäcken, Urinledare och Urinrör. Available online: https://cancercentrum.se/samverkan/cancerdiagnoser/urinblasa-urinvagar/vardprogram/ (accessed on 28 June 2022).
8. Ottosson, K.; Pelander, S.; Johansson, M.; Huge, Y.; Aljabery, F.; Sherif, A. The increased risk for thromboembolism pre-cystectomy in patients undergoing neoadjuvant chemotherapy for muscle-invasive urinary bladder cancer is mainly due to central venous access: A multicenter evaluation. *Int. Urol. Nephrol.* **2020**, *52*, 661–669. [CrossRef] [PubMed]
9. Ording, A.G.; Nielsen, M.E.; Smith, A.B.; Horváth-Puhó, E.; Sørensen, H.T. Venous thromboembolism and effect of comorbidity in bladder cancer: A danish nationwide cohort study of 13,809 patients diagnosed between 1995 and 2011. *Urol. Oncol.* **2016**, *34*, e291–e298. [CrossRef] [PubMed]
10. Eriksson, V.; Holmlund, J.; Wiberg, E.; Johansson, M.; Huge, Y.; Alamdari, F.; Svensson, J.; Aljabery, F.; Sherif, A. Adverse events during neoadjuvant chemotherapy for muscle invasive bladder cancer-a Swedish retrospective multicentre study of a clinical database. *Transl. Androl. Urol.* **2022**, *11*, 1105–1115. [CrossRef] [PubMed]
11. Seng, S.; Liu, Z.; Chiu, S.K.; Proverbs-Singh, T.; Sonpavde, G.; Choueiri, T.K.; Tsao, C.K.; Yu, M.; Hahn, N.M.; Oh, W.K.; et al. Risk of venous thromboembolism in patients with cancer treated with Cisplatin: A systematic review and meta-analysis. *J. Clin. Oncol.* **2012**, *30*, 4416–4426. [CrossRef] [PubMed]
12. Mehrazin, R.; Piotrowski, Z.; Egleston, B.; Parker, D.; Tomaszweski, J.J.; Smaldone, M.C.; Abbosh, P.H.; Ito, T.; Bloch, P.; Iffrig, K.; et al. Is extended pharmacologic venous thromboembolism prophylaxis uniformly safe after radical cystectomy? *Urology* **2014**, *84*, 1152–1156. [CrossRef] [PubMed]
13. Sørensen, C.; Andersen, M.; Kristiansen, V.B.; Jensen, R.; Wille-Jørgensen, P. The occurrence of late thromboembolic complications after elective abdominal surgery. *Ugeskr. Laeger* **1990**, *152*, 1586–1587. [PubMed]
14. Lyman, G.H.; Carrier, M.; Ay, C.; Di Nisio, M.; Hicks, L.K.; Khorana, A.A.; Leavitt, A.D.; Lee, A.Y.Y.; Macbeth, F.; Morgan, R.L.; et al. American Society of Hematology 2021 guidelines for management of venous thromboembolism: Prevention and treatment in patients with cancer. *Blood Adv.* **2021**, *5*, 927–974. [CrossRef] [PubMed]
15. Geerts, W.H.; Bergqvist, D.; Pineo, G.F.; Heit, J.A.; Samama, C.M.; Lassen, M.R.; Colwell, C.W. Prevention of venous thromboembolism: American College of Chest Physicians Evidence-Based Clinical Practice Guidelines (8th Edition). *Chest* **2008**, *133*, 381s–453s. [CrossRef] [PubMed]
16. ENOXACAN Study Group. Efficacy and safety of enoxaparin versus unfractionated heparin for prevention of deep vein thrombosis in elective cancer surgery: A double-blind randomized multicentre trial with venographic assessment. *Br. J. Surg.* **1997**, *84*, 1099–1103.
17. Nicolaides, A.N.; Fareed, J.; Kakkar, A.K.; Comerota, A.J.; Goldhaber, S.Z.; Hull, R.; Myers, K.; Samama, M.; Fletcher, J.; Kalodiki, E.; et al. Prevention and treatment of venous thromboembolism–International Consensus Statement. *Int. Angiol.* **2013**, *32*, 111–260. [PubMed]
18. Browne, C.; Davis, N.F.; Nolan, W.J.; MacCraith, E.D.; Lennon, G.M.; Mulvin, D.W.; Galvin, D.J.; Quinlan, D.M. Neoadjuvant Platinum-Based Chemotherapy is an Independent Predictor for Preoperative Thromboembolic Events in Bladder Cancer Patients Undergoing Radical Cystectomy. *Curr. Urol.* **2017**, *10*, 132–135. [CrossRef] [PubMed]
19. Duivenvoorden, W.C.; Daneshmand, S.; Canter, D.; Lotan, Y.; Black, P.C.; Abdi, H.; van Rhijn, B.W.; Fransen van de Putte, E.E.; Zareba, P.; Koskinen, I.; et al. Incidence, Characteristics and Implications of Thromboembolic Events in Patients with Muscle Invasive Urothelial Carcinoma of the Bladder Undergoing Neoadjuvant Chemotherapy. *J. Urol.* **2016**, *196*, 1627–1633. [CrossRef] [PubMed]

Journal of *Personalized Medicine*

MDPI

Review

COVID-19, Mucormycosis and Cancer: The Triple Threat—Hypothesis or Reality?

Ishika Mahajan [1], Aruni Ghose [2,3,4,5], Deepika Gupta [6], Manasi Manasvi [7], Saisha Bhandari [8], Aparimita Das [9], Elisabet Sanchez [4] and Stergios Boussios [4,10,11,*]

1 Department of Medical Oncology, Apollo Cancer Centre, Chennai 600001, India; ishikaishan16@gmail.com
2 Department of Medical Oncology, Barts Cancer Centre, St. Bartholomew's Hospital, Barts Health NHS Trust, London KT1 2EE, UK; aruni.ghose@nhs.net
3 Department of Medical Oncology, Mount Vernon Cancer Centre, East and North Hertfordshire NHS Trust, London KT1 2EE, UK
4 Department of Medical Oncology, Medway NHS Foundation Trust, Windmill Road, Gillingham ME7 5NY, UK; elizabet.sanchez@nhs.net
5 Division of Research, Academics and Cancer Control, Saroj Gupta Cancer Centre and Research Institute, Kolkata 700001, India
6 Microbiology, College of Medicine and Sagore Dutta Hospital, Kolkata 700001, India; guptadeepika.587@gmail.com
7 Internal Medicine, Kasturba Medical College, Mangalore 574142, India; manasimanasvi01@gmail.com
8 Internal Medicine, Post Graduate Institute of Medical Education and Research, Chandigarh 133301, India; saisha393@gmail.com
9 Faculty of Allied Health Sciences, Chettinad Academy of Research and Education, Chennai 600001, India; k.aparimita@gmail.com
10 Faculty of Life Sciences & Medicine, School of Cancer & Pharmaceutical Sciences, King's College London, London SE1 9RT, UK
11 AELIA Organization, 9th Km Thessaloniki-Thermi, 57001 Thessaloniki, Greece
* Correspondence: stergiosboussios@gmail.com or stergios.boussios@nhs.net or stergios.boussios@kcl.ac.uk

Citation: Mahajan, I.; Ghose, A.; Gupta, D.; Manasvi, M.; Bhandari, S.; Das, A.; Sanchez, E.; Boussios, S. COVID-19, Mucormycosis and Cancer: The Triple Threat—Hypothesis or Reality? *J. Pers. Med.* **2022**, *12*, 1119. https://doi.org/10.3390/jpm12071119

Academic Editor: Anne-Marie Caminade

Received: 16 June 2022
Accepted: 8 July 2022
Published: 10 July 2022

Abstract: COVID-19 has been responsible for widespread morbidity and mortality worldwide. Invasive mucormycosis has death rates scaling 80%. India, one of the countries hit worst by the pandemic, is also a hotbed with the highest death rates for mucormycosis. Cancer, a ubiquitously present menace, also contributes to higher case fatality rates. All three entities studied here are individual, massive healthcare threats. The danger of one disease predisposing to the other, the poor performance status of patients with all three diseases, the impact of therapeutics for one disease on the pathology and therapy of the others all warrant physicians having a better understanding of the interplay. This is imperative so as to effectively establish control over the individual patient and population health. It is important to understand the interactions to effectively manage all three entities together to reduce overall morbidity. In this review article, we search for an inter-relationship between the COVID-19 pandemic, emerging mucormycosis, and the global giant, cancer.

Keywords: cancer; mucormycosis; COVID-19; COVID-associated mucormycosis; diabetes; steroids; immunosuppression

1. Introduction

The Severe Acute Respiratory Syndrome Coronavirus-2 (SARS-CoV-2)-induced Coronavirus disease 2019 (COVID-19) pandemic has witnessed more than 545 million illnesses and 6.3 million fatalities worldwide. It affected all national healthcare systems at different levels [1,2]. The southeast Asian belt constitutes the world's second largest COVID-affected area. Second only to the USA, India boasts numbers larger than 43 million cases and 0.52 million deaths [2].

Bacterial and fungal co-infection alongside COVID-19 had been nil or minimal in Middle Eastern Respiratory Syndrome Coronavirus (MERS-CoV) and SARS-CoV-1 [1].

However, Huang et al. detected co-infection rates of 8 to 15% in COVID-19 patients [3]. With regard to fungi, the rate was around 3 to 6% [4]. The above invasive fungal infections (IFIs) include invasive aspergillosis, mucormycosis, candidiasis, and cryptococcosis [5].

Invasive Mucormycosis (IM), since the advent of the 21st century, has been deemed an emerging, angio-invasive fungal infection with drastically high mortality, of around 80% [6–8]. Historically speaking, India has been the principal hotbed of mucormycosis, recording a prevalence of 140 cases per million population. Globally, it is the highest and is eighty times higher than that of the affluent nations [6]. Hailing from the order *Mucorales*, Rhizopus is the most common species in the world. Apart from that, *Mucor* and *Lichtheimia* are the predominant causative agents in Europe, whereas *Apophysomyces* is the Asian, primarily Indian, counterpart [6,9]. The clinical spectrum includes five forms, primarily rhinocerebral, pulmonary, cutaneous, gastrointestinal, and disseminated disease [9]. Uncontrolled hyperglycaemia, steroid and immunomodulatory therapy, haematological malignancy (HMs) and haematopoietic stem cell transplantation (HSCT), solid organ malignancy (SOM) and transplant (SOT), persistent neutropenia, and iron overload are among the most notorious risk factors of IM [7,9].

In high-income countries (HIC) such as Europe, the USA, and Australia, HMs are the most common underlying cause of IM (38–62%) [7,9,10]. Acute myeloid leukaemia (AML) and acute lymphoblastic leukaemia (ALL), which frequently necessitate HSCT, are prominent examples [9,10]. These cases are at increased risk of IM, especially in their neutropenic state. Additionally, in these above patients, antifungal prophylaxis with azoles and echinocandins predisposes to breakthrough mucormycosis—voriconazole (52%), fluconazole (25%), caspofungin (9.8%), itraconazole (7.6%), and posaconazole (5%) [7,10].

Meanwhile, in low-and-middle-income countries (LMICs) such as the Indian subcontinent, diabetes mellitus (DM) is the most prevalent predisposing factor to IM [6,7,9]. Patel et al. identified rhino-orbito-cerebral mucormycosis (ROCM), constituting 65.7% of their IM cohort, to be the most common form [8]. The diabetic population of India is on the rise, from 60 million in 2011 to around 100 million by 2030 [11]. This risk relationship with IM is directly proportional. Prakash et al. designed a multi-centre study in India where 56.8% of IM patients were diabetics and 18% had diabetic ketoacidosis (DKA). Approximately 65% of them manifested ROCM [6].

COVID-19 patients have a higher prevalence of DM and DKA than that of the general population [12]. Compared to the other IFIs, mucormycosis in COVID-19 has proven to be a better-documented disease. A systematic review by John et al. reported the incidence of COVID-associated Mucormycosis (CAM) to be 87% [13]. Until 13 May 2021, Singh et al. reported that out of 101 CAM cases, 82 were from India and the remaining 19 were from the rest of the world [14]. From then until July, around 42,000 cases and 4000 deaths have occurred as a result of CAM in the Indian population [15].

The prevalence of COVID-19 among patients with cancer was higher in HMs (10.9%) as compared to SOMs (7.8%). Among SOMs, prostate cancer showed the highest frequency (10%). Patients receiving cancer treatment within 6 months (6.2%) were less susceptible to COVID-19 than those exceeding the above time interval (7.8%). COVID-19 patients with cancer have higher morbidity and mortality rates as compared to their negative counterparts—hospitalizations (43.8% vs. 31.5%), intensive care unit admissions (19.7% vs. 7.8%), respiratory support (7.9% vs. 1.3%), and deaths (14% vs. 3.1%). Among these COVID-19-positive cancer patients, mortality was lower in those treated with immunotherapy (7.1%) as compared to chemotherapy (14%), hormone therapy (16.2%), or even targeted therapy (14.1%) [16]. Among the 30.6% COVID-19 cancer patients enrolled in the UK Coronavirus Cancer Monitoring Project (UKCCMP) who died, fatality rates were higher in HMs as compared to SOMs. Leukaemia had the highest case fatality rate of 2.25. These HM patients who received recent chemotherapy experienced higher death rates during COVID-19 hospital admissions (odds ratio (OR) 2.09) [17].

In this review article, we search for an inter-relationship between the COVID-19 pandemic, emerging mucormycosis, and cancer.

2. CAM: Clinical Features

Rhizopus, as already mentioned, is ubiquitously present in soil, manure and as moulds on food [18]. Transmission of the infection can occur by airborne fungal spores entering via the nasal, oral, and conjunctival mucosa or by ingesting contaminated food. It may also be a commensal on the skin and mucosa, thus precipitating in opportunistic infections among the immunocompromised [19].

As discussed before, ROCM accounts for about a third of the cases of mucormycosis, with a steep rise noted in the COVID-19 pandemic [20]. The second most common type is pulmonary mucormycosis (PM) [13,14,20–22]. ROCM and PM originate in the nose and paranasal sinuses. The hallmarks of the disease are vascular invasion, thrombosis, and tissue necrosis. It results in bony destruction in the walls of the sinuses, leading to orbital and cranial spread [20,23].

Smith and Kirchner's diagnostic criteria for ROCM from 1958 still remain the gold standard for clinical diagnosis [14]. In an immunocompromised setting, a combination of the following features ought to raise suspicion of mucormycosis:

- Facial pain and blood-tinged nasal discharge on the same side;
- Soft periorbital or peri-nasal swelling with skin and mucosal discoloration progressing to induration;
- Conjunctival suffusion, eyelid ptosis, eyeball proptosis, and complete ophthalmoplegia;
- Necrotic, black turbinates appearing like clotted, crusted blood;
- Cranial nerve palsies.

The other non-specific features include fever, unilateral headache, sinus tenderness, puffiness of the face, loosening of teeth, and nasal congestion [24].

PM has non-specific symptoms such as fever, cough, chest pain, dyspnoea, and haemoptysis [24]. The symptoms appear innocuously similar to the underlying viral infection. Hence, the early signs are written off as COVID-19. Suspicion of PM should arise in COVID-19 patients when these symptoms persist, or new symptoms develop after the resolution of the viral disease. Strict monitoring for reappearance or prolonged persistence of symptoms in diabetics and those receiving steroids and immunosuppressive therapies should prompt evaluation for mucormycosis [13,14,20].

In the immunocompromised population, the invasive nature of the fungus quickly turns the disease lethal [13,14,21,25]. A mortality rate of 30–50% is seen in CAM [13,14,20,21].

3. CAM: Road to Diagnosis

The first line of investigation performed is an imaging study [20]. The modalities recommended are magnetic resonance imaging of the paranasal sinuses with cerebral contrast for ROCM and simple computerized tomography (CT) thorax for pulmonary disease.

For pathologies of the paranasal sinuses, the imaging preferred is a CT scan [26]. IM may show as a unilateral hypodense opacification of the sinuses. It has a limited role as the soft tissue changes, and critical features such as perineural spread and cavernous sinus involvement can be missed. Its use is limited to picking up bony destruction only. Magnetic resonance imaging (MRI) has higher sensitivity than CT (86% vs. 57–69%) and similar specificity (83% vs. 81%) for acute invasive fungal sinusitis [26,27]. It produces hypo- or isointense lesions in all sequences with variable enhancement on contrast. The invasion of the fungus along the soft tissue and fatty planes is delineated with an increased clarity by T1-weighted images, aiding in the staging of the disease [28]. Detection of neurological complications such as cavernous sinus thrombosis, meningitis, necrosis, and subtle perineural invasion requires the utilization of cerebral contrast material such as gadolinium [27,29].

In pulmonary disease, with concomitant COVID-19 infection, it is exceedingly difficult to differentiate due to the similarities in the lesions and the acute respiratory distress syndrome picture. CT thorax shows ground-glass opacities and infiltration. Some cases may show features of consolidation and cavitation [30]. Angio-invasion in the lungs may cause thrombosis of the pulmonary vessels, appearing as wedge-shaped infarcts.

The suspicion of mucormycosis is based on direct microscopic examination of a wet mount. Staining with comparative fluorescent fungal stain, calcofluor white, and Giemsa is also performed on the clinical specimens at the outset [20,31].

The standard practice for confirmation of the diagnosis is a biopsy from the infected area demonstrating the presence of broad, wide ribbon-like aseptate hyphae branching at right angles, on a background of necrotic debris on histopathology [19,20,32,33]. Stains commonly used are haematoxylin-eosin, periodic acid-Schiff, or Grocott–Gomori's methenamine-silver stain (GMS) [20,31].

On culture, the presence of cotton white or grey-black colonies at 28–30 °C and 35–37 °C is characteristic. The media commonly employed are Sabouraud Dextrose Agar and Brain and Heart Infusion Agar. Morphological identification from the fungal culture can be carried out by microscopic examination or by DNA sequencing on the basis of barcodes 18s, ITS, and MALDI-TOF [5,34,35]. Isolation from a fungal culture may be performed for differentiation of genus and species, and for antimicrobial susceptibility testing [5,32]. The treatment, however, is independent of the genus and species and is of epidemiological importance [36].

Serological investigations such as galactomannan assays and 1,3-β-D glucan assays support a diagnosis of fungal infection [20,35]. These tests are, however, typically negative in a pure mucor infection [20]. The utility of using the negative test to rule out the disease is questionable, as the tests are positive in mixed fungal infections [34]. While, presently, there is no serological test available for the rapid diagnosis of mucormycosis, ELISA and lateral flow immunoassay (LFIA) are in the process of development [37].

The clinical features, though characteristic of a fungal infection, are unfortunately nonspecific, and may be mistaken for aspergillosis. Further, there exists a degree of misidentification of the Mucorales for the more ubiquitous Aspergillus spp. on microscopy [20,25,38]. There is thus a challenge in obtaining a confirmation of the diagnosis without immunohistochemistry staining with monoclonal antibodies or DNA-PCR of the tissue section. Both of these modalities are not only expensive but also not readily available at the point of contact [20,39].

The utility of complete blood counts is not established, especially in the present cohort of COVID-19 patients, where the white cell picture is deranged as a result of the primary infection [30,32,36,40]. For any analysis into the association of the derangement of these parameters due to the secondary fungal infection, confounding variables of the primary disease will have to be considered. The data on this are insufficient to comment upon and require further study.

4. Treatment

4.1. Mucormycosis

As discussed above, the suspicion of IM in the immunocompromised must be treated as a medical emergency due to its lethality [20]. In the present cohort, individuals infected with SARS-CoV-2, receiving treatment with steroids and immunomodulatory drugs, fall into the aforementioned high-risk category. Hyperglycaemia, due to a primary diabetic status or as a consequence of steroid therapy, increases the risk of infection [8,13,14].

Early diagnosis can lead to early initiation of treatment for positive outcomes and decreased case fatality rates: 20% fatality in early diagnosis vs. 60% in delayed diagnosis. Treatment guidelines can be set up in a three-pronged approach involving the utilization of antifungal drugs, surgical debridement, and an effort in part to increase immunity levels [41].

Prompt diagnosis directs the initiation of antifungal therapy wherein the first-line treatment is 5–10 mg/kg of Liposomal Amphotericin B (LAmB). A higher dosage of 10 mg/kg/day is recommended for central nervous system (CNS) involvement. If LAmB is not available, amphotericin B lipid complex or amphotericin deoxycholate can be used, with caution because of their known nephrotoxicity, and should be subjected to stringent therapeutic drug monitoring [20,42,43].

Other broad-spectrum antifungals such as posaconazole and isavuconazole may be utilized in the initial induction therapy where the patient cannot tolerate amphotericin B. Although not studied in randomized trials, their efficacy has been suggested to be similar to amphotericin [44]. There is, however, a noticeable need for updating the medical guideline in terms of induction of these agents for the treatment of CAM and their role as prophylactic agents [45].

Although antifungal prophylaxis for mucormycosis has been used for diseases such as underlying haematological malignancies, it is not recommended in COVID-19 patients [20,46].

Studies on combination antifungal therapy, where amphotericin B is combined with posaconazole or other antifungals such as caspofungin, have been carried out, which have not been shown to enhance survival or be more effective than monotherapy (35% versus 39%, respectively), and their cost to benefit ratio needs to be further evaluated on a case-by-case basis [42,44,47–49].

Step-down therapy is initiated with posaconazole delayed-release tablets (300 mg every 12 h on the first day, then 300 mg once daily) or infusions, which is preferred over oral suspensions. Oral isavuconazole is also an acceptable alternative, with doses of 200 mg three times a day on day 1–2 and 200 mg once a day from day 3, which is continued for a minimum of 6 weeks [20,43,50].

Add-on aggressive surgical intervention paves the way for a favourable response, which can be seen on imaging and clinical stabilization, in a minimum of 4–6 weeks [20,42,43]. Surgical debridement is carried out with clean margins, which serves three purposes—controls the disease, obtains specimens for histopathology, and confirms the microbiological diagnosis [20]. In cases of ROCM, there is extensive surgical debridement of the affected craniofacial tissues and risk of orbital exenteration, which may be lifesaving and has been found to be helpful even in cases where the infection has spread intracranially, according to the stage of involvement. Retrobulbar injection of amphotericin B may be given before surgery, according to the stage of involvement [42,43]. Lobectomy has also been performed with success in cases of PM where the infection was localized to a single lobe. The debridement procedures might need to be repeated as and when required. Endoscopic sinus surgery with limited tissue removal has also been performed [20].

While immediate surgical debridement has been the convention for lung, Saraiya et al. found that, after response to antifungal therapy, delayed surgical debridement (10 days) showed better patient outcomes than conventional aggressive surgical management. All five patients where the new protocol was followed survived [47].

Other modalities of treatment include hyperbaric oxygen therapy, administration of cytokines concurrently with antifungals, and deferasirox, but their use is not routinely recommended in the treatment guidelines [20].

Even after aggressive therapy and immediate initiation of the treatment, the prognosis is still quite grim with multiple studies showing a 90-day mortality average of 50%, with multiple patients discontinuing treatment because of financial constraints [8,48,49].

4.2. CAM

Methylprednisolone and dexamethasone, the cornerstones of COVID-19 treatment, are known to cause immunosuppression [50]. The World Health Organization (WHO) and the National Institute of Health (NIH) have recommended judicious use of systemic corticosteroids in COVID-19, which is in cases with evidence of respiratory failure or with oxygen saturation below 93–94%. These drugs unfortunately are known to be the most common cause of drug-induced hyperglycaemia [46,51].

A widespread disruption of antibiotic stewardship programs was noted due to rampant and injudicious use of antimicrobial agents (AMAs). Increased exposure of patients to AMAs, due to fear of superadded bacterial and fungal infection on COVID pneumonia, poor infection control measures due to rapidly changing protocols and the resulting confusion, led to an upstroke of nosocomial superbug infections in the background of COVID itself [52].

Hyperglycaemia is the most common risk factor for developing mucormycosis. In CAM, there should be immediate induction of antifungal therapy along with COVID-19 treatment protocols, and control of hyperglycaemia. Diabetic ketoacidosis, if present, should be emergently managed along with the reversal of immunosuppression in these patients [13,24,43,46,49,53,54].

Tocilizumab is another immunomodulator that is being used in the treatment of the disease. Studies and guidelines have recommended against indiscriminatory use of tocilizumab, which targets the immune pathways. Its use is only recommended for severe refractory disease where inflammatory markers are increased (e.g., C-reactive protein, interleukin 6) and in the absence of active fungal or bacterial infections [20,55].

Hence, upon detection of CAM, steroid doses should be reduced and immunomodulating drugs such as tocilizumab should be discontinued for reversal of immunosuppression [43,46,51].

5. COVID-19, Mucormycosis, and Cancer: The Triple Threat

5.1. COVID and Mucormycosis

As explained above, high-risk factors for mucormycosis include DM and metabolic acidosis, steroid usage in COVID-19, neutropenia, elevated serum iron levels, immunomodulatory therapies, and concomitant chronic illness. An increased likelihood of fungal infection has also been noted with haematological malignancies, organ and bone marrow transplant recipients, steroid therapy, patients on maintenance haemodialysis and iron chelation, trauma and burn victims [22].

SARS-CoV-2-mediated immune dysregulation of the cytokine storm in the background of a compromised immune system predisposes to invasive fungal infections [56]. This hyper-inflammatory state is also noted as a consequence of some malignancies, autoimmune diseases, and immunosuppressive medications [57]. High inflammatory markers, interleukins, interferon-gamma, tumour necrosis factors and hyperferritinaemia produce a picture similar to secondary haemophagocytic lymphohistiocytosis [3,58]. This hypercytokinaemia of severe COVID-19 produces diffuse alveolar lung damage, microvascular thrombosis, hyaline membrane formation and fibrosis, leading to an acute respiratory distress picture [59–61]. This unregulated increase in inflammatory markers and acute phase reactants downregulates the CD4+ T lymphocyte count, causing lymphopenia, propagating a decrease in viral clearance [62,63]. Additionally, there is a diffuse systemic vasculitic endotheliitis and microvascular damage, causing multiorgan dysfunction [64,65].

The vascular endothelial and alveolar damage is due to angiotensinogen converting enzyme- 2 (ACE-2) receptor-mediated viral entry into these cells. Other tissues with higher expression of these receptors such as gastrointestinal mucosa and proximal tubular cells of the kidney show a greater localised inflammatory response [66,67]. Patients with already compromised systems, i.e., with cardiovascular comorbidities such as hypertension, severe dyslipidaemia, obesity, DM, and those with chronic kidney disease thus have poorer outcomes in the event of disease progression to a cytokine storm [68,69]. Cardiac infection-producing myocarditis and infarctions are similarly hypothesized [70].

Immunosuppressive therapies such as steroids are among the few interventions proven to blunt this unchecked inflammation and improve survival. The high rate of usage of steroids in the period of cytokine storm in these patients exponentially increases the risk of them acquiring the IFIs in the recovery period [13,26].

The association of steroid therapy with mucormycosis has been the most constant in the recent upstroke of cases. Steroids are one of the only therapies proven to consistently reduce mortality in COVID-19 patients [13,26]. They are considered an essential therapy for COVID-19 patients on supplemental oxygen therapy [71]. Their use, however, has a multitude of complications attached. Worsening hyperglycaemia or new-onset DM and immunosuppression are among the common mechanisms that predispose to angio-invasive mucormycosis. Corticosteroids are implicated in immune system destruction by impairing leukocyte chemotaxis and prevention of phagolysosome fusion [72]. This steroid-induced immune dysregulation is an added burden on the virus-induced lymphopenia and hypercytokinaemia, leaving the patient wide open to a host of super-added co-infections. This may be compounded by the fact that steroids are given as over-the-counter pills in India leading to rampant misuse of the drugs [14].

DM remains the maximally associated factor predisposing to mucormycosis, with a mortality of 46% [48]. SARS-CoV-2 has been seen to worsen the glycaemic profile in patients with DM through multiple mechanisms, though they are all not completely understood. Notably, the virus can infect the pancreatic islet cells causing damage to the β-cells, causing hypoinsulinaemia and worsening hyperglycaemia [22,73,74]. DM, by nature, causes a degree of endothelial dysfunction due to chronic inflammation, leading to microvascular complications. Direct viral invasion of the vascular lining, causing endothelial damage, is mediated by ACE-2 receptors. This results in apoptosis and pyroptosis of the endothelium predisposing to angio-invasive secondary infections [61].

Among diabetics, the risk multiplies in patients with diabetic ketoacidosis (DKA). The virus and the hyperimmune stress state themselves predispose to dysglycaemia and DKA, even in the absence of poorly controlled sugars [22,74]. Hyperglycaemia produces a state of secondary immune deficiency. The dysglycaemia and acidic pH in ketoacidosis results in phagocyte dysfunction and defective intracellular killing of the fungi. The low-grade chronic inflammation causes an impaired immune response—both innate and adaptive response to infections. This also paves the way for the cytokine storm. The acidotic serum pH of 6.88–7.3 in DKA is also conducive to florid fungal proliferation [75].

The endothelial dysfunction and cytokine storm are procoagulant states leading to thrombosis and tissue ischemia [59,61]. The ischemic necrosis prevents leukocyte chemotaxis and effective delivery of antifungal agents to the foci of infection. The concomitant endotheliitis that occurs as a result of oxidative damage causes adherence of the fungal elements to the vessel walls. These factors combined with the angio-invasive nature of the fungus are responsible for the haematogenous dissemination of the disease [76].

Iron is required for fungal hyphal growth and development. One of the main host defence mechanisms against mucormycosis is limiting the availability of free iron to the fungus by having it bind to forms such as transferrin, ferritin, and lactoferrin. The acidic pH in DKA displaces the iron from its bound forms in the human body, increasing its availability to the fungal elements. The presence of increased unbound iron plays a vital role in predisposing patients with DKA to developing the disease. Some Mucorales secrete high-affinity iron chelators or siderophores such as rhizoferrin to acquire the iron from the host cells. Those species without rhizoferrin utilize exogenous xenosiderophores such as deferoxamine that are administered in DKA patients to treat iron overload, in order to fulfil their iron requirements. This explains why patients who are on long-term iron chelation therapy are at elevated risk of mucormycosis. Severe COVID-19 disease is also a hyperferritinaemic state, due to interleukin-six-mediated ferritin synthesis and downregulation of iron transport. There is an intracellular accumulation of iron in the hepatocytes, eventually causing necrosis and release of the accumulated iron. The high serum iron forms a fertile ground for florid fungal proliferation [77,78].

Tocilizumab, an immunomodulatory monoclonal antibody that targets the deregulated interleukin-six pathway implicated in the COVID-19 cytokine storm, is associated with an increased risk of infections. It modulates and suppresses the immune system, leaving the body vulnerable to other pathogens such as bacteria and fungi [79,80].

Neutrophil and phagocyte activity are the main defences against fungal infection. Mononuclear and polymorphonuclear cells generate reactive oxygen species and defensin peptides that inhibit fungal growth. Patients with impaired neutrophil function (primary immune deficiencies) are thus at higher risk. These are unaffected in COVID-19 infection; thus, a primary cause–association relationship is unlikely [72].

5.2. Cancer and Mucormycosis

Invasive fungal infections are a common cause of morbidity in patients with HMs. Second only to aspergillosis, zygomycosis occurs as a result of a confluence of multiple risk factors in this population [10,81]. Among the malignancies, acute leukaemia and lymphomas are found to be the most commonly associated with mucormycosis [82,83]. Other risk factors in the subgroup included haematopoietic stem cell transplants, myelodysplastic syndromes, concomitant graft versus host disease and high-dose steroid therapy [84]. The pulmonary and disseminated forms of the disease appear to be more common with haematologic malignancies [83,84].

Management of patients with HMs involves profound immunosuppression due to highly cytotoxic drugs, immunomodulatory therapies, and long-term high-dose steroids. All of these treatments break down the entire immune system to a virtually negligible existence, leaving the patient vulnerable to a host of opportunistic pathogenic infections [85].

These patients are at substantial risk for granulocytopenia and febrile neutropenia due to the chemotherapeutic medications. With neutrophils being the primary defence against the fungal elements, reduced numbers in the circulation lower the body's defence and immunity predisposing to infection [86]. These patients show extensive angio-invasion and hyphal elements, but inflammatory infiltrates are minimal. Conversely, non-neutropenic HM patients with steroid compromised immune systems have a far less invasive disease, but more inflammatory infiltration with polymorphonuclear leukocytes [87,88].

Febrile neutropenia patients are on concomitant broad-spectrum antibiotic therapy to prevent flare-ups of bacterial infections. This in conjunction with the primary cytotoxic immunosuppressive medications leave this cohort of patients at risk for fungal infections, especially in the absence of primary antifungal prophylaxis [89].

Steroids form an essential arm in the primary therapeutic regimens of some lymphomas and myelomas. They are additionally used as adjuvant therapy for pain management [90,91], reducing inflammation and oedema, and for symptomatic relief of refractory symptoms of obstructions and dyspnoea [92]. The presence of prolonged high-dose steroid therapy causes a drug-induced hyperglycaemic state and, consequently, DM and DKA, creating an environment conducive for fungal growth [93].

In patients planned for or having received stem-cell transplantation, the phase of rebuilding immunity can be a perilous one. The complex immune milieu in these patients that results from the cocktail of immunosuppressive therapies, antimicrobial prophylaxis, and graft-versus-host disease (GVHD) multiplies the risk of acquiring invasive fungal infections [88]. This risk does not appear to be ameliorated even with antifungal prophylaxis with some azole and echinocandin medications. Breakthrough mucor infections are not uncommon [94,95]. In the last few years, mortality due to breakthrough mucormycosis has reduced with the advent of prophylaxis regimens with posaconazole. Treatment of mucormycosis in the cohort with liposomal amphotericin B and isavuconazole has also led to improvements in morbidity and mortality statistics [6,44,48].

Patients receiving chemotherapeutic drugs such as folate antagonists, antipurines, and steroids tend to develop oral ulcerations in the oral cavity and gastrointestinal tract. This discontinuity in the mucosa can pave the way for disseminated fungal infection, beginning as mucocutaneous or gastrointestinal disease [96].

5.3. COVID-19 and Cancer

Patients with cancer and concomitant COVID-19 infection have a higher fatality rate (7.6%) as compared to those without cancer (3.8%), with risk doubling [72,97]. There is no clear evidence of any associative relation linking the virus to any modulatory oncogenic pathobiology. However, there are a few theories postulating reasons for this doubling in mortality and high rate of complications [98,99].

Cancer treatments significantly increased risk of COVID-19 infection in the form of radiation therapy, chemotherapy, stem cell transplantation, and chimeric-antigen-receptor T-cell (CAR-T) therapy. All of these therapies cause varying degrees of immunosuppression in the body [100]. These treatments may sometimes cause a suppression of prodrome symptoms such as fever, leading to a delay in diagnosis [101].

Cellular senescence, via its oncogenic mutations, is a proven risk factor for many cancers [102]. Increasing age is a determinant of various other comorbidities, such as DM and hypertension that worsen the outlook of COVID-19 patients [73,102–104]. Inflammaging in a setting of cancer leads to a progressive weakening of the body's defence mechanisms and healing [102,105]. This immune decline predisposes them to COVID-19 and also paves the way for poorer outcomes [104]. Recent immunophenotyping studies also show the virus promotes age-induced immune and genetic dysregulation, and this accounts for the vulnerability of the geriatric population [106]. This is supported by the fact that paediatric oncology patients show a lower infection rate as compared to their adult contacts and caregivers (2.5% as compared to 14.7%) [107,108].

The SARS-CoV-2 virus enters host cells by using its spike protein (S), which binds to the ACE-2 receptors on human cells [109,110]. These receptors are highly populated on type 2 alveolar cells, leading to pulmonary inflammation. They are also present on other tissues, such as the endothelium of the blood vessels, smooth muscles of the gastrointestinal tract, heart, kidneys, and liver [111,112]. Interestingly, the expression of the ACE-2 receptors varies in patients with cancer, i.e., some cancers show an upregulation of the receptor on their cell surfaces. This suggests that oncology patients are more likely to be infected with the virus and also have a poorer prognosis [113,114].

Both cancer and COVID-19 share a few metabolic risk factors. Type 2 DM, obesity, and metabolic syndrome are amongst the widely known hazards implicated, notably, in liver, pancreatic and endometrial cancers [115,116]. The immunologic response in cancer depends on the T helper cells and cytotoxic T cells. This mechanism is impaired by DM [117].

Lymphopenia is noted in a subset of patients with advanced cancers such as pancreatic, breast, lymphomas, and melanomas [118,119]. The lymphocytes also control the progression and therapeutic responses in cancer [120]. Additionally, the immune response of the body, adaptive and innate systems alters the natural history of the malignancy. A cytokine storm is a systemic inflammatory response syndrome that can be triggered in cancer patients by a plethora of factors, such as CAR-T cell therapy, monoclonal antibodies, infections, and some chemotherapeutic drugs [57,121,122]. The virus itself also causes an immune dysregulation, as part of its core pathogenesis, resulting in hypercytokinaemia [3,58]. This double-edged immune response may influence the pathogenesis of both diseases. The therapeutic approaches used in oncology patients to combat this syndrome can be considered as options in COVID-19-induced cytokine storm [123,124].

Coagulopathy is another common link between cancer and COVID-19 infection. Thrombosis and bleeding-related complications cause significant morbidity and mortality in severe COVID-19 infection [125,126]. They are also significantly accountable for cancer-related deaths, due to the direct thrombolytic properties of tumour cells, hypercoagulable state resulting from immune dysregulation, and release of pro-coagulant mediators such as cysteine proteases [127]. A combination of both entities can compound these procoagulant effects [98,99,128,129].

6. Conclusions

Despite there being a higher risk of morbidity, complications, and mortality, there is still a paucity of data emphasising the need for stricter guidelines, prophylactic therapies, and management strategies in cancer patients with COVID and systemic fungal infections such as mucormycosis. Studies into the interactions of the virus with cancer cells and the virus with chemotherapeutic drugs are also required to better understand the pathogenesis of infection and the interplay of the human body with cancer and virus cells. The presence of invasive fungal infection in this setting only complicates the treatment due to it being a superadded infection in a doubly immunocompromised host.

For physicians to effectively control and manage the triple threat posed due to a poor general condition from the malignancy, a viral infection and an invasive systemic fungal infection, a thorough study into the interplay of physical factors, chemical and microbiological factors is the need of the hour. Moreover, not only do COVID-19 and mucormycosis pose a larger risk in cancer patients, the COVID-19 and mucormycosis status of a patient with an underlying malignancy may also have bearing on the cancer chemotherapy/immunotherapy or antibody therapy. Figure 1 illustrates the interwoven relationship between COVID-19, mucormycosis and cancer.

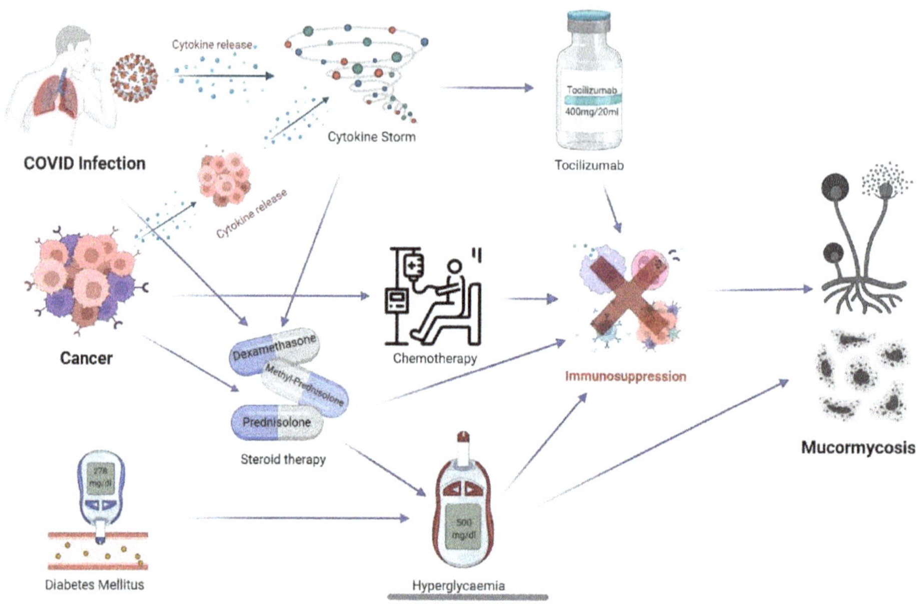

Figure 1. An illustration summarising the interwoven relationship between COVID-19, mucormycosis and cancer.

It is therefore increasingly important to understand the interactions of the three entities to effectively manage them all together to reduce the morbidity presented. We hope this will invite further research to build up an evidence base for this triple threat—COVID-19, mucormycosis and cancer.

Author Contributions: Conceptualization, I.M. and A.G.; methodology, D.G.; software, M.M.; validation, S.B. (Saisha Bhandari), A.D. and E.S.; formal analysis, E.S.; investigation, S.B. (Saisha Bhandari); resources, E.S.; data curation, S.B. (Saisha Bhandari); writing—original draft preparation, I.M. and A.G.; writing—review and editing, S.B. (Stergios Boussios); visualization, D.G.; supervision, S.B. (Stergios Boussios); project administration, M.M.; funding acquisition, E.S. All authors have read and agreed to the published version of the manuscript.

Funding: This research received no external funding.

Institutional Review Board Statement: Not applicable.

Informed Consent Statement: Not applicable.

Conflicts of Interest: The authors declare no conflict of interest.

References

1. Revythis, A.; Shah, S.; Enyioma, S.; Ghose, A.; Patel, M.; Karathanasi, A.; Sanchez, E.; Boussios, S. The Experience of a Single NHS England Trust on the Impact of the COVID-19 Pandemic on Junior and Middle-Grade Doctors: What Is Next? *Int. J. Environ. Res. Public. Health* **2021**, *18*, 10413. [CrossRef] [PubMed]
2. WHO Coronavirus (COVID-19) Dashboard [Internet]. 2022. Available online: https://covid19.who.int (accessed on 4 July 2022).
3. Lansbury, L.; Lim, B.; Baskaran, V.; Lim, W.S. Co-infections in people with COVID-19: A systematic review and meta-analysis. *J. Infect.* **2020**, *81*, 266–275. [CrossRef] [PubMed]
4. Huang, C.; Wang, Y.; Li, X.; Ren, L.; Zhao, J.; Hu, Y.; Zhang, L.; Fan, G.; Xu, J.; Gu, X.; et al. Clinical features of patients infected with 2019 novel coronavirus in Wuhan, China. *Lancet* **2020**, *395*, 497–506. [CrossRef]
5. Zhang, G.; Hu, C.; Luo, L.; Fang, F.; Chen, Y.; Li, J.; Peng, Z.; Pan, H. Clinical features, and short-term outcomes of 221 patients with COVID-19 in Wuhan, China. *J. Clin. Virol.* **2020**, *127*, 104364. [CrossRef]
6. Song, G.; Liang, G.; Liu, W. Fungal Co-infections Associated with Global COVID-19 Pandemic: A Clinical and Diagnostic Perspective from China. *Mycopathologia* **2020**, *185*, 599–606. [CrossRef]
7. Prakash, H.; Ghosh, A.K.; Rudramurthy, S.M.; Singh, P.; Xess, I.; Savio, J.; Pamidimukkala, U.; Jillwin, J.; Varma, S.; Das, A.; et al. A prospective multicenter study on mucormycosis in India: Epidemiology, diagnosis, and treatment. *Med. Mycol.* **2019**, *57*, 395–402. [CrossRef]
8. Prakash, H.; Chakrabarti, A. Global Epidemiology of Mucormycosis. *J. Fungi.* **2019**, *5*, 26. [CrossRef]
9. Patel, A.; Kaur, H.; Xess, I.; Michael, J.S.; Savio, J.; Rudramurthy, S.; Singh, R.; Shastri, P.; Umabala, P.; Sardana, R.; et al. A multicentre observational study on the epidemiology, risk factors, management, and outcomes of mucormycosis in India. *Clin. Microbiol. Infect.* **2020**, *26*, 944.e9–944.e15. [CrossRef]
10. Skiada, A.; Pavleas, I.; Drogari-Apiranthitou, M. Epidemiology and Diagnosis of Mucormycosis: An Update. *J. Fungi.* **2020**, *6*, 265. [CrossRef]
11. Pagano, L.; Dragonetti, G.; De Carolis, E.; Veltri, G.; Del Principe, M.I.; Busca, A. Developments in identifying and managing mucormycosis in hematologic cancer patients. *Expert. Rev. Hematol.* **2020**, *13*, 895–905. [CrossRef]
12. Whiting, D.R.; Guariguata, L.; Weil, C.; Shaw, J. IDF Diabetes Atlas: Global estimates of the prevalence of diabetes for 2011 and 2030. *Diabetes Res. Clin. Pract.* **2011**, *94*, 311–321. [CrossRef] [PubMed]
13. Goldman, N.; Fink, D.; Cai, J.; Lee, Y.N.; Davies, Z. High prevalence of COVID-19-associated diabetic ketoacidosis in UK secondary care. *Diabetes Res. Clin. Pract.* **2020**, *166*, 108291. [CrossRef] [PubMed]
14. John, T.M.; Jacob, C.N.; Kontoyiannis, D.P. When Uncontrolled Diabetes Mellitus and Severe COVID-19 Converge: The Perfect Storm for Mucormycosis. *J. Fungi* **2021**, *7*, 298. [CrossRef] [PubMed]
15. Singh, A.K.; Singh, R.; Joshi, S.R.; Misra, A. Mucormycosis in COVID-19: A systematic review of cases reported worldwide and in India. *Diabetes Metab. Syndr.* **2021**, *15*, 102146. [CrossRef]
16. COVID-19-Associated Mucormycosis: Triple Threat of the Pandemic. Available online: https://asm.org/Articles/2021/July/COVID-19-Associated-Mucormycosis-Triple-Threat-of (accessed on 12 June 2022).
17. Fillmore, N.R.; La, J.; Szalat, R.E.; Tuck, D.P.; Nguyen, V.; Yildirim, C.; Do, N.V.; Brophy, M.T.; Munshi, N.C. Prevalence and Outcome of COVID-19 Infection in Cancer Patients: A National Veterans Affairs Study. *J. Natl. Cancer. Inst.* **2021**, *113*, 691–698. [CrossRef]
18. Lee, L.Y.W.; Cazier, J.B.; Starkey, T.; Briggs, S.E.W.; Arnold, R.; Bisht, V.; Booth, S.; Campton, N.A.; Cheng, V.W.T.; Collins, G.; et al. UK Coronavirus Cancer Monitoring Project Team. COVID-19 prevalence and mortality in patients with cancer and the effect of primary tumour subtype and patient demographics: A prospective cohort study. *Lancet Oncol.* **2020**, *21*, 1309–1316. [CrossRef]
19. Hoenigl, M.; Seidel, D.; Carvalho, A.; Rudramurthy, S.M.; Arastehfar, A.; Gangneux, J.P.; Nasir, N.; Bonifaz, A.; Araiza, J.; Klimko, N.; et al. The emergence of COVID-19 associated mucormycosis: A review of cases from 18 countries. *Lancet Microbe* **2022**. *Epub Ahead of Print*. [CrossRef]
20. Jeong, W.; Keighley, C.; Wolfe, R.; Lee, W.L.; Slavin, M.A.; Kong, D.C.M.; Chen, S.C. The epidemiology and clinical manifestations of mucormycosis: A systematic review and meta-analysis of case reports. *Clin. Microbiol. Infect.* **2019**, *25*, 26–34. [CrossRef]
21. Cornely, O.A.; Alastruey-Izquierdo, A.; Arenz, D.; Chen, S.C.A.; Dannaoui, E.; Hochhegger, B.; Hoenigl, M.; Jensen, H.E.; Lagrou, K.; Lewis, R.E.; et al. Global guideline for the diagnosis and management of mucormycosis: An initiative of the European Confederation of Medical Mycology in cooperation with the Mycoses Study Group Education and Research Consortium. *Lancet Infect. Dis.* **2019**, *19*, e405–e421. [CrossRef]
22. Pakdel, F.; Ahmadikia, K.; Saleh, M.; Tabari, A.; Jafari, R.; Mehrparvar, G.; Rezaie, Y.; Rajaeih, S.; Alijani, N.; Barac, A.; et al. Mucormycosis in patients with COVID-19: A cross-sectional descriptive multicentre study from Iran. *Mycoses* **2021**, *64*, 1238–1252. [CrossRef]
23. Pal, R.; Singh, B.; Bhadada, S.K.; Banerjee, M.; Bhogal, R.S.; Hage, N.; Kumar, A. COVID-19-associated mucormycosis: An updated systematic review of literature. *Mycoses* **2021**, *64*, 1452–1459. [CrossRef] [PubMed]
24. Brunet, K.; Rammaert, B. Murcomycosis treatment: Recommendations, latest advances, and perspectives. *J. Mycol. Med.* **2020**, *30*, 101007. [CrossRef] [PubMed]
25. All India Institute of Medical Sciences, New Delhi, Mucormycosis in COVID-19. Available online: https://covid.aiims.edu/mucormycosis-in-covid-19/ (accessed on 12 June 2022).

26. Garg, D.; Muthu, V.; Sehgal, I.S.; Ramachandran, R.; Kaur, H.; Bhalla, A.; Puri, G.D.; Chakrabarti, A.; Agarwal, R. Coronavirus Disease (Covid-19) Associated Mucormycosis (CAM): Case Report and Systematic Review of Literature. *Mycopathologia* **2021**, *186*, 289–298. [CrossRef] [PubMed]
27. Moorthy, A.; Gaikwad, R.; Krishna, S.; Hegde, R.; Tripathi, K.K.; Kale, P.G.; Rao, P.S.; Haldipur, D.; Bonanthaya, K. SARS-CoV-2, Uncontrolled Diabetes and Corticosteroids-An Unholy Trinity in Invasive Fungal Infections of the Maxillofacial Region? A Retrospective, Multi-centric Analysis. *J. Maxillofac. Oral Surg.* **2021**, *20*, 418–425. [CrossRef]
28. Groppo, E.R.; El-Sayed, I.H.; Aiken, A.H.; Glastonbury, C.M. Computed tomography, and magnetic resonance imaging characteristics of acute invasive fungal sinusitis. *Arch. Otolaryngol. Head Neck Surg.* **2011**, *137*, 1005–1010. [CrossRef]
29. Herrera, D.A.; Dublin, A.B.; Ormsby, E.L.; Aminpour, S.; Howell, L.P. Imaging findings of rhinocerebral mucormycosis. *Skull Base* **2009**, *19*, 117–125. [CrossRef]
30. Nagesh, C.P. The "black fungus" through a gray lens: Imaging COVID-19-associated mucormycosis. *Indian J. Ophthalmol.* **2021**, *69*, 1648–1649. [CrossRef]
31. Pasero, D.; Sanna, S.; Liperi, C.; Piredda, D.; Branca, G.P.; Casadio, L.; Simeo, R.; Buselli, A.; Rizzo, D.; Bussu, F.; et al. A challenging complication following SARS-CoV-2 infection: A case of pulmonary mucormycosis. *Infection* **2021**, *49*, 1055–1060. [CrossRef]
32. Guarner, J.; Brandt, M.E. Histopathologic Diagnosis of Fungal Infections in the 21st Century. *Clin. Microbiol. Rev.* **2011**, *24*, 247–280. [CrossRef]
33. Mishra, Y.; Prashar, M.; Sharma, D.; Kumar, V.P.; Tilak, T.V.S.V.G.K. Diabetes, COVID 19 and mucormycosis: Clinical spectrum and outcome in a tertiary care medical center in Western India. *Diabetes Metab. Syndr. Clin. Res. Rev.* **2021**, *15*, 102196. [CrossRef]
34. Skiada, A.; Lass-Floerl, C.; Klimko, N.; Ibrahim, A.; Roilides, E.; Petrikkos, G. Challenges in the diagnosis and treatment of mucormycosis. *Med. Mycol.* **2018**, *56*, S93–S101. [CrossRef] [PubMed]
35. Dadwal, S.S.; Kontoyiannis, D.P. Recent advances in the molecular diagnosis of mucormycosis. *Expert Rev. Mol. Diagn.* **2018**, *18*, 845–854. [CrossRef]
36. Honavar, S.G.; Sen, M.; Lahane, S.; Lahane, T.P.; Parekh, R. Mucor in a Viral Land: A Tale of Two Pathogens. *Indian J. Ophthalmol.* **2021**, *69*, 244–252. [CrossRef]
37. Patel, A.; Agarwal, R.; Rudramurthy, S.M.; Shevkani, M.; Xess, I.; Sharma, R.; Savio, J.; Sethuraman, N.; Madan, S.; Shastri, P.; et al. Multicenter Epidemiologic Study of Coronavirus Disease–Associated Mucormycosis, India. *Emerg. Infect. Dis.* **2021**, *27*, 2349–2359. [CrossRef] [PubMed]
38. Burnham-Marusich, A.R.; Hubbard, B.; Kvam, A.J.; Gates-Hollingsworth, M.; Green, H.R.; Soukup, E.; Limper, A.H.; Kozel, T.R. Conservation of Mannan Synthesis in Fungi of the Zygomycota and Ascomycota Reveals a Broad Diagnostic Target. *mSphere* **2018**, *3*, e00094-18. [CrossRef] [PubMed]
39. Kung, V.L.; Chernock, R.; Burnham, C.-A.D. Diagnostic accuracy of fungal identification in histopathology and cytopathology specimens. *Eur. J. Clin. Microbiol.* **2018**, *37*, 157–165. [CrossRef] [PubMed]
40. Drogari-Apiranthitou, M.; Panayiotides, I.; Galani, I.; Konstantoudakis, S.; Arvanitidis, G.; Spathis, A.; Gouloumi, A.-R.; Tsakiraki, Z.; Tsiodras, S.; Petrikkos, G. Diagnostic value of a semi-nested PCR for the diagnosis of mucormycosis and aspergillosis from paraffin-embedded tissue: A single center experience. *Pathol. Res. Pract.* **2016**, *212*, 393–397. [CrossRef]
41. Zhang, X.; Tan, Y.; Ling, Y.; Lu, G.; Liu, F.; Yi, Z.; Jia, X.; Wu, M.; Shi, B.; Xu, S.; et al. Viral and host factors related to the clinical outcome of COVID-19. *Nature* **2020**, *583*, 437–440. [CrossRef]
42. Mekonnen, Z.K.; Ashraf, D.C.; Jankowski, T.; Grob, S.R.; Vagefi, M.R.; Kersten, R.C.; Simko, J.P.; Winn, B.J. Acute Invasive Rhino-Orbital Mucormycosis in a Patient With COVID-19-Associated Acute Respiratory Distress Syndrome. *Ophthalmic Plast. Reconstr. Surg.* **2021**, *37*, e40–e80. [CrossRef]
43. Kyvernitakis, A.; Torres, H.; Jiang, Y.; Chamilos, G.; Lewis, R.; Kontoyiannis, D. Initial use of combination treatment does not impact survival of 106 patients with haematologic malignancies and mucormycosis: A propensity score analysis. *Clin. Microbiol. Infect.* **2016**, *22*, 811.e1–811.e8. [CrossRef]
44. Rudramurthy, S.M.; Hoenigl, M.; Meis, J.F.; Cornely, O.A.; Muthu, V.; Gangneux, J.P.; Perfect, J.; Chakrabarti, A.; Isham, E.A. ECMM/ISHAM recommendations for clinical management of COVID-19 associated mucormycosis in low- and middle-income countries. *Mycoses* **2021**, *64*, 1028–1037. [CrossRef] [PubMed]
45. Marty, F.M.; Ostrosky-Zeichner, L.; Cornely, O.A.; Mullane, K.M.; Perfect, J.R.; Thompson, G.R., 3rd; Alangaden, G.J.; Brown, J.M.; Fredricks, D.N.; Heinz, W.J.; et al. VITAL and FungiScope Mucormycosis Investigators. Isavuconazole treatment for mucormycosis: A single-arm open-label trial and case-control analysis. *Lancet Infect. Dis.* **2016**, *16*, 828–837. [CrossRef]
46. Jenks, J.D.; Salzer, H.J.; Prattes, J.; Krause, R.; Buchheidt, D.; Hoenigl, M. Spotlight on isavuconazole in the treatment of invasive aspergillosis and mucormycosis: Design, development, and place in therapy. *Drug Des. Devel. Ther.* **2018**, *12*, 1033–1044. [CrossRef] [PubMed]
47. Indian Council of Medical Research. Evidence Based Advisory in the time of COVID-19 (Screening, Diagnosis & Management of Mucormycosis). Available online: https://www.icmr.gov.in/pdf/covid/techdoc/Mucormycosis_ADVISORY_FROM_ICMR_In_COVID19_time.pdf (accessed on 12 June 2022).
48. Saraiya, H.A. Successful Management of Cutaneous Mucormycosis by Delaying Debridement. *Ann. Plast. Surg.* **2012**, *69*, 301–306. [CrossRef] [PubMed]

49. Jeong, W.; Keighley, C.; Wolfe, R.; Lee, W.L.; Slavin, M.A.; Chen, S.C.-A.; Kong, D.C. Contemporary management and clinical outcomes of mucormycosis: A systematic review and meta-analysis of case reports. *Int. J. Antimicrob. Agents* **2019**, *53*, 589–597. [CrossRef]
50. Prakash, H.; Chakrabarti, A. Epidemiology of Mucormycosis in India. *Microorganisms* **2021**, *9*, 523. [CrossRef]
51. COVID-19 Treatment Guidelines Panel. Coronavirus Disease 2019 (COVID-19) Treatment Guidelines. National Institutes of Health. Available online: https://www.covid19treatmentguidelines.nih.gov/ (accessed on 3 July 2022).
52. Segala, F.; Bavaro, D.; Di Gennaro, F.; Salvati, F.; Marotta, C.; Saracino, A.; Murri, R.; Fantoni, M. Impact of SARS-CoV-2 Epidemic on Antimicrobial Resistance: A Literature Review. *Viruses* **2021**, *13*, 2110. [CrossRef]
53. World Health Organisation. Mucormycosis. Available online: https://www.who.int/india/emergencies/coronavirus-disease-(covid-19)/mucormycosis (accessed on 12 June 2022).
54. Priya, P.; Ganesan, V.; Rajendran, T.; Geni, V.G. Mucormycosis in a Tertiary Care Center in South India: A 4-Year Experience. *Indian J. Crit. Care Med.* **2020**, *24*, 168–171. [CrossRef]
55. Swain, S.K. COVID-19 associated mucormycosis in head and neck region of paediatric patients: A life-threatening disease in current pandemic. *Int. J. Contemp. Pediatrics.* **2021**, *8*, 1322–1327. [CrossRef]
56. Kimmig, L.M.; Wu, D.; Gold, M.; Pettit, N.N.; Pitrak, D.; Mueller, J.; Husain, A.N.; Mutlu, E.A.; Mutlu, G.M. IL-6 Inhibition in Critically Ill COVID-19 Patients Is Associated With Increased Secondary Infections. *Front. Med.* **2020**, *7*, 583897. [CrossRef]
57. Kumar, G.; Adams, A.; Hererra, M.; Rojas, E.R.; Singh, V.; Sakhuja, A.; Meersman, M.; Dalton, D.; Kethireddy, S.; Nanchal, R.; et al. Predictors and outcomes of healthcare-associated infections in COVID-19 patients. *Int. J. Infect. Dis.* **2021**, *104*, 287–292. [CrossRef] [PubMed]
58. Vaninov, N. In the eye of the COVID-19 cytokine storm. *Nat. Rev. Immunol.* **2020**, *20*, 277. [CrossRef] [PubMed]
59. Ruan, Q.; Yang, K.; Wang, W.; Jiang, L.; Song, J. Clinical predictors of mortality due to COVID-19 based on an analysis of data of 150 patients from Wuhan, China. *Intensive Care Med.* **2020**, *46*, 846–848. [CrossRef] [PubMed]
60. Guo, Y.R.; Cao, Q.D.; Hong, Z.S.; Tan, Y.Y.; Chen, S.D.; Jin, H.J.; Tan, K.S.; Wang, D.Y.; Yan, Y. The origin, transmission, and clinical therapies on coronavirus disease 2019 (COVID-19) outbreak—An update on the status. *Mil. Med. Res.* **2020**, *7*, 11. [CrossRef] [PubMed]
61. Dolhnikoff, M.; Duarte-Neto, A.N.; de Almeida Monteiro, R.A.; Da Silva, L.F.F.; De Oliveira, E.P.; Saldiva, P.H.N.; Mauad, T.; Negri, E.M. Pathological evidence of pulmonary thrombotic phenomena in severe COVID-19. *J. Thromb. Haemost.* **2020**, *18*, 1517–1519. [CrossRef]
62. Varga, Z.; Flammer, A.J.; Steiger, P.; Haberecker, M.; Andermatt, R.; Zinkernagel, A.S.; Mehra, M.R.; Schuepbach, R.A.; Ruschitzka, F.; Moch, H. Endothelial cell infection and endotheliitis in COVID-19. *Lancet* **2020**, *395*, 1417–1418. [CrossRef]
63. Favalli, E.G.; Ingegnoli, F.; De Lucia, O.; Cincinelli, G.; Cimaz, R.; Caporali, R. COVID-19 infection and rheumatoid arthritis: Faraway, so close! *Autoimmun. Rev.* **2020**, *19*, 102523. [CrossRef]
64. Schulert, G.S.; Cron, R.Q. The genetics of macrophage activation syndrome. *Genes Immun.* **2020**, *21*, 169–181. [CrossRef]
65. Panigrahy, D.; Gilligan, M.M.; Huang, S.; Gartung, A.; Cortés-Puch, I.; Sime, P.J.; Phipps, R.P.; Serhan, C.N.; Hammock, B.D. Inflammation resolution: A dual-pronged approach to averting cytokine storms in COVID-19? *Cancer Metastasis Rev.* **2020**, *39*, 337–340. [CrossRef]
66. Bhaskar, S.; Sinha, A.; Banach, M.; Mittoo, S.; Weissert, R.; Kass, J.S.; Rajagopal, S.; Pai, A.R.; Kutty, S. Cytokine Storm in COVID-19-Immunopathological Mechanisms, Clinical Considerations, and Therapeutic Approaches: The REPROGRAM Consortium Position Paper. *Front. Immunol.* **2020**, *11*, 1648. [CrossRef]
67. Guo, T.; Fan, Y.; Chen, M.; Wu, X.; Zhang, L.; He, T.; Wang, H.; Wan, J.; Wang, X.; Lu, Z. Cardiovascular Implications of Fatal Outcomes of Patients With Coronavirus Disease 2019 (COVID-19). *JAMA Cardiol.* **2020**, *5*, 811–818. [CrossRef] [PubMed]
68. Rabb, H. Kidney diseases in the time of COVID-19: Major challenges to patient care. *J. Clin. Investig.* **2020**, *130*, 2749–2751. [CrossRef] [PubMed]
69. Banach, M.; Penson, P.E.; Fras, Z.; Vrablik, M.; Pella, D.; Reiner, Ž.; Nabavi, S.M.; Sahebkar, A.; Kayikcioglu, M.; Dac-cord, M. Brief recommendations on the management of adult patients with familial hypercholesterolemia during the COVID-19 pandemic. *Pharmacol. Res.* **2020**, *158*, 104891. [CrossRef] [PubMed]
70. Reiner, Ž.; Hatamipour, M.; Banach, M.; Pirro, M.; Al-Rasadi, K.; Jamialahmadi, T.; Radenkovic, D.; Montecucco, F.; Sahebkar, A. Statins and the COVID-19 main protease: In silico evidence on direct interaction. *Arch. Med. Sci.* **2020**, *16*, 490–496. [CrossRef] [PubMed]
71. Katsiki, N.; Banach, M.; Mikhailidis, D.P. Lipid-lowering therapy and renin-angiotensin-aldosterone system inhibitors in the era of the COVID-19 pandemic. *Arch. Med. Sci.* **2020**, *16*, 485–489. [CrossRef] [PubMed]
72. RECOVERY Collaborative Group; Horby, P.; Lim, W.S.; Emberson, J.R.; Mafham, M.; Bell, J.L.; Linsell, L.; Staplin, N.; Brightling, C.; Ustianowski, A.; et al. Dexamethasone in Hospitalized Patients with Covid-19. *N. Engl. J. Med.* **2021**, *384*, 693–704. [CrossRef]
73. Guan, W.J.; Ni, Z.Y.; Hu, Y.; Liang, W.H.; Qu, C.Q.; He, J.X.; Liu, L.; Shan, H.; Lei, C.L.; Hui, D.S.C.; et al. China medical treatment expert group for COVID-19 2020. Clinical Characteristics of coronavirus disease in China. *N. Engl. J. Med.* **2020**, *382*, 1708–1720. [CrossRef]
74. Müller, J.A.; Groß, R.; Conzelmann, C.; Krüger, J.; Merle, U.; Steinhart, J.; Weil, T.; Koepke, L.; Bozzo, C.P.; Read, C.; et al. SARS-CoV-2 infects and replicates in cells of the human endocrine and exocrine pancreas. *Nat. Metab.* **2021**, *3*, 149–165. [CrossRef]

75. Pal, R.; Bhadada, S.K. COVID-19 and diabetes mellitus: An unholy interaction of two pandemics. *Diabetes Metab. Syndr.* **2020**, *14*, 513–517. [CrossRef]
76. Yonas, E.; Alwi, I.; Pranata, R.; Huang, I.; Lim, M.A.; Yamin, M.; Nasution, S.A.; Setiati, S.; Virani, S.S. Elevated interleukin levels are associated with higher severity and mortality in COVID 19—A systematic review, meta-analysis, and meta-regression. *Diabetes Metab. Syndr.* **2020**, *14*, 2219–2230. [CrossRef]
77. Jose, A.; Singh, S.; Roychoudhury, A.; Kholakiya, Y.; Arya, S.; Roychoudhury, S. Current Understanding in the Pathophysiology of SARS-CoV-2-Associated Rhino-Orbito-Cerebral Mucormycosis: A Comprehensive Review. *J. Maxillofac. Oral Surg.* **2021**, *20*, 373–380. [CrossRef] [PubMed]
78. Perricone, C.; Bartoloni, E.; Bursi, R.; Cafaro, G.; Guidelli, G.M.; Shoenfeld, Y.; Gerli, R. COVID-19 as part of the hyperferritinemic syndromes: The role of iron depletion therapy. *Immunol. Res.* **2020**, *68*, 213–224. [CrossRef] [PubMed]
79. Habib, H.M.; Ibrahim, S.; Zaim, A.; Ibrahim, W.H. The role of iron in the pathogenesis of COVID-19 and possible treatment with lactoferrin and other iron chelators. *Biomed. Pharmacother.* **2021**, *136*, 111228. [CrossRef]
80. Morena, V.; Milazzo, L.; Oreni, L.; Bestetti, G.; Fossali, T.; Bassoli, C.; Torre, A.; Cossu, M.V.; Minari, C.; Ballone, E.; et al. Off-label use of tocilizumab for the treatment of SARS-CoV-2 pneumonia in Milan, Italy. *Eur. J. Intern. Med.* **2020**, *76*, 36–42. [CrossRef]
81. Rossotti, R.; Travi, G.; Ughi, N.; Corradin, M.; Baiguera, C.; Fumagalli, R.; Bottiroli, M.; Mondino, M.; Merli, M.; Bellone, A.; et al. Safety and efficacy of anti-il6-receptor tocilizumab use in severe and critical patients affected by coronavirus disease 2019: A comparative analysis. *J. Infect.* **2020**, *81*, e11–e17. [CrossRef] [PubMed]
82. Pagano, L.; Offidani, M.; Fianchi, L.; Nosari, A.; Candoni, A.; Picardi, M.; Corvatta, L.; D'Antonio, D.; Girmenia, C.; Martino, P.; et al. Mucormycosis in hematologic patients. *Haematologica* **2004**, *89*, 207–214. [PubMed]
83. Meyer, R.D.; Rosen, P.; Armstrong, N. Phycomycosis Complicating Leukemia and Lymphoma. *Ann. Intern. Med.* **1972**, *77*, 871. [CrossRef]
84. Nosari, A.; Oreste, P.; Montillo, M.; Carrafiello, G.; Draisci, M.; Muti, G.; Molteni, A.; Morra, E. Mucormycosis in hematologic malignancies: An emerging fungal infection. *Haematologica* **2000**, *85*, 1068–1071.
85. Kara, I.O.; Tasova, Y.; Uguz, A.; Sahin, B. Mucormycosis-associated fungal infections in patients with haematologic malignancies. *Int. J. Clin. Pract.* **2009**, *63*, 134–139. [CrossRef]
86. Ghuman, H.; Voelz, K. Innate and Adaptive Immunity to Mucorales. *J. Fungi* **2017**, *3*, 48. [CrossRef]
87. Chamilos, G.; Luna, M.; Lewis, R.E.; Bodey, G.P.; Chemaly, R.; Tarrand, J.J.; Safdar, A.; Raad, I.I.; Kontoyiannis, D.P. Invasive fungal infections in patients with hematologic malignancies in a tertiary care cancer center: An autopsy study over a 15-year period (1989–2003). *Haematologica* **2006**, *91*, 986–989.
88. Berenguer, J.; Allende, M.C.; Lee, J.W.; Garrett, K.; Lyman, C.; Ali, N.M.; Bacher, J.; Pizzo, P.A.; Walsh, T.J. Pathogenesis of pulmonary aspergillosis. Granulocytopenia versus cyclosporine and methylprednisolone-induced immunosuppression. *Am. J. Respir. Crit. Care Med.* **1995**, *152*, 1079–1086. [CrossRef]
89. Ben-Ami, R.; Luna, M.; Lewis, R.E.; Walsh, T.J.; Kontoyiannis, D.P. A clinicopathological study of pulmonary mucormycosis in cancer patients: Extensive angioinvasion but limited inflammatory response. *J. Infect.* **2009**, *59*, 134–138. [CrossRef] [PubMed]
90. Funada, H.; Matsuda, T. Pulmonary Mucormycosis in a Hematology Ward. *Intern. Med.* **1996**, *35*, 540–544. [CrossRef] [PubMed]
91. Leppert, W.; Buss, T. The Role of Corticosteroids in the Treatment of Pain in Cancer Patients. *Curr. Pain Headache Rep.* **2012**, *16*, 307–313. [CrossRef] [PubMed]
92. Paulsen, Ø.; Aass, N.; Kaasa, S.; Dale, O. Do corticosteroids provide analgesic effects in cancer patients? A systematic literature review. *J. Pain Symptom Manag.* **2013**, *46*, 96–105. [CrossRef]
93. Lossignol, D. A little help from steroids in oncology. *J. Transl. Int. Med.* **2016**, *4*, 52–54. [CrossRef]
94. Muggeo, P.; Calore, E.; Decembrino, N.; Frenos, S.; De Leonardis, F.; Colombini, A.; Petruzziello, F.; Perruccio, K.; Berger, M.; Burnelli, R.; et al. Invasive mucormycosis in children with cancer: A retrospective study from the Infection Working Group of Italian Pediatric Hematology Oncology Association. *Mycoses* **2019**, *62*, 165–170. [CrossRef]
95. Imhof, A.; Balajee, S.A.; Fredricks, D.; Englund, J.A.; Marr, K.A. Breakthrough Fungal Infections in Stem Cell Transplant Recipients Receiving Voriconazole. *Clin. Infect. Dis.* **2004**, *39*, 743–746. [CrossRef]
96. Lionakis, M.S.; Lewis, R.E.; Kontoyiannis, D.P. Breakthrough Invasive Mold Infections in the Hematology Patient: Current Concepts and Future Directions. *Clin. Infect. Dis.* **2018**, *67*, 1621–1630. [CrossRef]
97. Hutter, R.V.P. Phycomycetous infection (mucormycosis) in cancer patients: A complication of therapy. *Cancer* **1959**, *12*, 330–350. [CrossRef]
98. Hanna, T.P.; Evans, G.A.; Booth, C.M. Cancer, COVID-19 and the precautionary principle: Prioritizing treatment during a global pandemic. *Nat. Rev. Clin. Oncol.* **2020**, *17*, 268–270. [CrossRef] [PubMed]
99. Jyotsana, N.; King, M.R. The Impact of COVID-19 on Cancer Risk and Treatment. *Cell Mol. Bioeng.* **2020**, *13*, 285–291. [CrossRef] [PubMed]
100. Jyotsana, N. COVID-19 and Cancer: Biological Interconnection and Treatment. In *Biotechnology to Combat COVID-19*; Agrawal, M., Biswas, S., Eds.; IntechOpen: London, UK, 2021.
101. Liang, W.-H.; Guan, W.-J.; Li, C.; Li, Y.-M.; Liang, H.-R.; Zhao, Y.; Liu, X.-Q.; Sang, L.; Chen, R.-C.; Tang, C.-L.; et al. Clinical characteristics and outcomes of hospitalised patients with COVID-19 treated in Hubei (epicentre) and outside Hubei (non-epicentre): A nationwide analysis of China. *Eur. Respir. J.* **2020**, *55*, 2000562. [CrossRef] [PubMed]

102. Leonetti, A.; Facchinetti, F.; Zielli, T.; Brianti, E.; Tiseo, M. COVID-19 in lung cancer patients receiving ALK/ROS1 inhibitors. *Eur. J. Cancer* **2020**, *132*, 122–124. [CrossRef]
103. White, M.C.; Holman, D.M.; Boehm, J.E.; Peipins, L.A.; Grossman, M.; Henley, S.J. Age, and cancer risk: A potentially modifiable relationship. *Am. J. Prev. Med.* **2014**, *46*, S7–S15. [CrossRef]
104. Kamboj, M.; Sepkowitz, K.A. Nosocomial infections in patients with cancer. *Lancet Oncol.* **2009**, *10*, 589–597. [CrossRef]
105. Wu, C.; Chen, X.; Cai, Y.; Xia, J.; Zhou, X.; Xu, S.; Huang, H.; Zhang, L.; Zhou, X.; Du, C.; et al. Risk Factors Associated With Acute Respiratory Distress Syndrome and Death in Patients With Coronavirus Disease 2019 Pneumonia in Wuhan, China. *JAMA Intern. Med.* **2020**, *180*, 934–943. [CrossRef]
106. Pawelec, G. Age, and immunity: What is "immunosenescence"? *Exp. Gerontol.* **2018**, *105*, 4–9. [CrossRef]
107. Zheng, Y.; Liu, X.; Le, W.; Xie, L.; Li, H.; Wen, W.; Wang, S.; Ma, S.; Huang, Z.; Ye, J.; et al. A human circulating immune cell landscape in aging and COVID-19. *Protein Cell* **2020**, *11*, 740–770. [CrossRef]
108. Choi, S.-H.; Kim, H.W.; Kang, J.-M.; Kim, D.H.; Cho, E.Y. Epidemiology and clinical features of coronavirus disease 2019 in children. *Clin. Exp. Pediatr.* **2020**, *63*, 125–132. [CrossRef] [PubMed]
109. Boulad, F.; Kamboj, M.; Bouvier, N.; Mauguen, A.; Kung, A.L. COVID-19 in Children With Cancer in New York City. *JAMA Oncol.* **2020**, *6*, 1459. [CrossRef] [PubMed]
110. Xu, H.; Zhong, L.; Deng, J.; Peng, J.; Dan, H.; Zeng, X.; Li, T.; Chen, Q. High expression of ACE2 receptor of 2019-nCoV on the epithelial cells of oral mucosa. *Int. J. Oral Sci.* **2020**, *12*, 8. [CrossRef] [PubMed]
111. Wan, Y.; Shang, J.; Graham, R.; Baric, R.S.; Li, F. Receptor Recognition by the Novel Coronavirus from Wuhan: An Analysis Based on Decade-Long Structural Studies of SARS Coronavirus. *J. Virol.* **2020**, *94*, e00127-20. [CrossRef] [PubMed]
112. Hamming, I.; Timens, W.; Bulthuis, M.L.C.; Lely, A.T.; Navis, G.J.; van Goor, H. Tissue distribution of ACE2 protein, the functional receptor for SARS coronavirus. A first step in understanding SARS pathogenesis. *J. Pathol.* **2004**, *203*, 631–637. [CrossRef]
113. Ziegler, C.G.K.; Allon, S.J.; Nyquist, S.K.; Mbano, I.M.; Miao, V.N.; Tzouanas, C.N.; Cao, Y.; Yousif, A.S.; Bals, J.; Hauser, B.M.; et al. HCA Lung Biological Network. SARS-CoV-2 Receptor ACE2 Is an Interferon-Stimulated Gene in Human Airway Epithelial Cells and Is Detected in Specific Cell Subsets across Tissues. *Cell* **2020**, *181*, 1016–1035.e19. [CrossRef]
114. Dai, Y.J.; Hu, F.; Li, H.; Huang, H.Y.; Wang, D.W.; Liang, Y. A profiling analysis on the receptor ACE2 expression reveals the potential risk of different type of cancers vulnerable to SARS-CoV-2 infection. *Ann. Transl. Med.* **2020**, *8*, 481. [CrossRef]
115. Liang, W.; Guan, W.; Chen, R.; Wang, W.; Li, J.; Xu, K.; Li, C.; Ai, Q.; Lu, W.; Liang, H.; et al. Cancer patients in SARS-CoV-2 infection: A nationwide analysis in China. *Lancet Oncol.* **2020**, *21*, 335–337. [CrossRef]
116. Oberaigner, W.; Ebenbichler, C.; Oberaigner, K.; Juchum, M.; Schönherr, H.R.; Lechleitner, M. Increased cancer incidence risk in type 2 diabetes mellitus: Results from a cohort study in Tyrol/Austria. *BMC Public Health* **2014**, *14*, 1058. [CrossRef]
117. Muniyappa, R.; Gubbi, S. COVID-19 pandemic, coronaviruses, and diabetes mellitus. *Am. J. Physiol. Endocrinol. Metab.* **2020**, *318*, E736–E741. [CrossRef]
118. Hodgson, K.; Morris, J.; Bridson, T.; Govan, B.; Rush, C.; Ketheesan, N. Immunological mechanisms contributing to the double burden of diabetes and intracellular bacterial infections. *Immunology* **2015**, *144*, 171–185. [CrossRef] [PubMed]
119. Péron, J.; Cropet, C.; Tredan, O.; Bachelot, T.; Ray-Coquard, I.; Clapisson, G.; Chabaud, S.; Philip, I.; Borg, C.; Cassier, P.; et al. CD4 lymphopenia to identify end-of-life metastatic cancer patients. *Eur. J. Cancer* **2013**, *49*, 1080–1089. [CrossRef] [PubMed]
120. Trédan, O.; Manuel, M.; Clapisson, G.; Bachelot, T.; Chabaud, S.; Bardin-Dit-Courageot, C.; Rigal, C.; Biota, C.; Bajard, A.; Pasqual, N.; et al. Patients with metastatic breast cancer leading to CD4+ T cell lymphopaenia have poor outcome. *Eur. J. Cancer* **2013**, *49*, 1673–1682. [CrossRef] [PubMed]
121. Galluzzi, L.; Buqué, A.; Kepp, O.; Zitvogel, L.; Kroemer, G. Immunogenic cell death in cancer and infectious disease. *Nat. Rev. Immunol.* **2017**, *17*, 97–111. [CrossRef]
122. Chatenoud, L.; Ferran, C.; Legendre, C.; Thouard, I.; Merite, S.; Reuter, A.; Gevaert, Y.; Kreis, H.; Franchimont, P.; Bach, J.F. In vivo cell activation following OKT3 administration. Systemic cytokine release and modulation by corticosteroids. *Transplantation* **1990**, *49*, 697–702. [CrossRef]
123. Shimabukuro-Vornhagen, A.; Gödel, P.; Subklewe, M.; Stemmler, H.J.; Schlößer, H.A.; Schlaak, M.; Kochanek, M.; Böll, B.; von Bergwelt-Baildon, M.S. Cytokine release syndrome. *J. Immunother. Cancer* **2018**, *6*, 56. [CrossRef]
124. Mehta, P.; McAuley, D.F.; Brown, M.; Sanchez, E.; Tattersall, R.S.; Manson, J.J.; on behalf of theHLH Across Speciality Collaboration, UK. COVID-19: Consider cytokine storm syndromes and immunosuppression. *Lancet* **2020**, *395*, 1033–1034. [CrossRef]
125. Wen, W.; Su, W.; Tang, H.; Le, W.; Zhang, X.; Zheng, Y.; Liu, X.; Xie, L.; Li, J.; Ye, J.; et al. Immune cell profiling of COVID-19 patients in the recovery stage by single-cell sequencing. *Cell Discov.* **2020**, *6*, 31. [CrossRef]
126. Levi, M.; Thachil, J.; Iba, T.; Levy, J.H. Coagulation abnormalities and thrombosis in patients with COVID-19. *Lancet Haematol.* **2020**, *7*, e438–e440. [CrossRef]
127. Tang, N.; Li, D.; Wang, X.; Sun, Z. Abnormal Coagulation parameters are associated with poor prognosis in patients with novel coronavirus pneumonia. *J. Thromb. Haemost.* **2020**, *18*, 844–847. [CrossRef]
128. Falanga, A.; Marchetti, M.; Vignoli, A. Coagulation and cancer: Biological and clinical aspects. *J. Thromb. Haemost.* **2012**, *11*, 223. [CrossRef] [PubMed]
129. Derosa, L.; Melenotte, C.; Griscelli, F.; Gachot, B.; Marabelle, A.; Kroemer, G.; Zitvogel, L. The immuno-oncological challenge of COVID-19. *Nat. Cancer* **2020**, *1*, 946–964. [CrossRef] [PubMed]

Journal of *Personalized Medicine*

MDPI

Article

Vitamin D Levels in Patients with Overlap Syndrome, Is It Associated with Disease Severity?

Kostas Archontogeorgis [1], Athanasios Voulgaris [2], Evangelia Nena [3], Athanasios Zissimopoulos [4], Izolde Bouloukaki [5], Sophia E. Schiza [5] and Paschalis Steiropoulos [1,2,*]

1 MSc Program in Sleep Medicine, Medical School, Democritus University of Thrace, 68100 Alexandroupolis, Greece
2 Department of Pneumonology, Medical School, Democritus University of Thrace, 68100 Alexandroupolis, Greece
3 Laboratory of Social Medicine, Medical School, Democritus University of Thrace, 68100 Alexandroupolis, Greece
4 Laboratory of Nuclear Medicine, Medical School, Democritus University of Thrace, 68100 Alexandroupolis, Greece
5 Sleep Disorders Unit, Department of Respiratory Medicine, Medical School, University of Crete, 71110 Heraklion, Greece
* Correspondence: pstirop@med.duth.gr; Tel.: +30-551352096

Abstract: Background: The coexistence of chronic obstructive pulmonary disease (COPD) and obstructive sleep apnea (OSA) has been defined as overlap syndrome (OVS). Recently, a link between OSA, COPD and Vitamin D (Vit D) serum concentration was reported, however, evidence regarding Vit D status in patients with OVS is scarce. The aim of the present study was to evaluate Vit D serum levels and to explore the association of those levels with anthropometric, pulmonary function and sleep parameters in patients with OVS. Methods: Vit D serum levels were measured in patients diagnosed with OVS, as confirmed by overnight polysomnography and pulmonary function testing. Results: A total of 90 patients (79 males and 11 females) were included in the analysis. The patients were divided into three groups matched for age, gender, and BMI: the control group that included 30 patients (27 males and 3 females), the OSA group that included 30 patients (26 males and 4 females), and the OVS group that included 30 patients (26 males and 4 females). Patients with OVS exhibited decreased serum 25(OH)D levels compared with OSA patients and controls (14.5 vs. 18.6 vs. 21.6 ng/mL, $p < 0.001$). In the OVS group, multiple linear regression analysis identified AHI and FEV_1, as predictors of serum 25(OH)D levels ($p = 0.041$ and $p = 0.038$, respectively). Conclusions: Lower Vit D levels have been observed in patients with OVS compared with OSA patients and non-apneic controls, indicating an increased risk of hypovitaminosis D in this population which might be associated with disease severity.

Keywords: chronic obstructive pulmonary disease; obstructive sleep apnea; overlap syndrome; vitamin D

Citation: Archontogeorgis, K.; Voulgaris, A.; Nena, E.; Zissimopoulos, A.; Bouloukaki, I.; Schiza, S.E.; Steiropoulos, P. Vitamin D Levels in Patients with Overlap Syndrome, Is It Associated with Disease Severity? *J. Pers. Med.* **2022**, *12*, 1693. https://doi.org/10.3390/jpm12101693

Academic Editor: Anne-Marie Caminade

Received: 16 September 2022
Accepted: 6 October 2022
Published: 11 October 2022

1. Introduction

Chronic obstructive pulmonary disease (COPD) and obstructive sleep apnea (OSA) are both highly prevalent pulmonary diseases [1,2]. COPD affects approximately 11.7% of the global adult population, and OSA is estimated to affect 10–17% men and 3–9% women [1,2]. The coexistence of the two conditions has been defined as "overlap syndrome (OVS)", a distinct clinical syndrome, which may be different to the simple aggregate of OSA and COPD [3]. Epidemiological studies indicate an OVS prevalence of approximately 1% among adult males [4]. Both COPD and OSA share common risk factors such as tobacco smoking, and are associated with systemic inflammation and oxidative stress [4]. Additionally, patients with OVS are known to exhibit a more profound hypoxia during sleep compared with patients with COPD or OSA alone, further contributing to inflammatory activation and the potential development and progression of cardiovascular disease and other comorbidities [5,6]. OSA and COPD have also been associated with an increased risk of endocrinal and metabolic conditions, namely insulin resistance, depression of the somatotropic axis, hypogonadism, imbalance of the adrenal axis, bone loss and hypothyroidism [7–9].

Recently, a link between OSA, COPD and Vitamin D serum concentration was reported in the literature [10–12]. Vitamin D (Vit D) is a fat-soluble vitamin that can be found in some foods and is synthesized in the skin after exposure to sunlight [13]. The classical function of Vit D is the regulation of bone metabolism; however, accumulating evidence suggests that it also possesses anti-inflammatory and immune-modulating properties [13]. Serum 25-hydroxyvitamin D (25(OH)D) concentration is recommended as the best indicator of Vit D status [13]. Vit D insufficiency has exponentially increased over the years, with its prevalence estimated between 13% and 40.4% in European countries, depending on the definition of Vit D insufficiency used [14]. Decreased serum 25(OH)D has been associated with a variety of pulmonary diseases, including respiratory infections, asthma and cancer [13]. Recently, OSA has been associated with hypovitaminosis D, and adequate use of CPAP is likely to augment Vit D levels [10,15]. Similarly, decreased serum 25(OH)D levels have been described in COPD patients, while a protective role against exacerbations has been attributed to Vit D supplementation [11,16]. However, evidence regarding Vit D status in patients with OVS is scarce. Therefore, the aim of the present study is to evaluate Vit D serum levels and to explore the association of those levels with anthropometric, pulmonary function and sleep parameters in patients with OVS.

2. Methods

2.1. Study Population

The present study protocol was approved by the Alexandroupolis University General Hospital Ethics Committee and all performances were carried out in accordance with the Helsinki Declaration of Human Rights [17]. Participants who enrolled between January of 2019 and December of 2020 from the sleep laboratory of our institution were referred for evaluation of suspected sleep-disordered breathing. Informed consent was provided prior to each participant's enrollment.

Patients diagnosed with both OSA and COPD, as confirmed by overnight polysomnography and pulmonary function testing were included in the study. Exclusion criteria were as follows: refusal to participate, previously diagnosed OSA under continuous positive airway pressure (CPAP) therapy or evidence of exclusively central apneas documented in polysomnography, Vit D supplementation, conditions known to affect calcium, phosphorus and Vit D metabolism and absorption, serious heart diseases, chronic kidney or hepatic disease, cancer, inflammatory diseases, severe cognitive or psychiatric disorders, or any respiratory disorder other than COPD.

2.2. Study Variables

Information on previous medical history, current medication use (with an emphasis on Vit D supplementation) and tobacco smoking were obtained from all participants. Anthropometrical data including age, sex, neck, waist and hip circumference were also recorded, and a waist-to-hip ratio (WHR) was calculated. Body mass index (BMI) was calculated using the following formula: BMI = Weight (in kilograms)/Height (in meters)^2. Subjective sleepiness was assessed using the Greek version of the Epworth Sleepiness Scale (ESS), which assesses the tendency to fall asleep during eight typical daytime situations [18]. A score >10 was considered indicative of excessive daytime sleepiness [18].

2.3. Polysomnography

All participants underwent an overnight, in-laboratory supervised polysomnography from 22:00 to 06:00 and the data were recorded on a computer system (Alice® 4, Philips

Respironics, Murrysville, PA, USA). In each patient, electroencephalogram, electrooculogram, electromyogram (submental and bilateral tibial), and electrocardiogram were recorded. Airflow was detected using combined oronasal thermal sensors. Additionally, continuous pulse oximetry was performed, and impedance belts were used to assess the thoracic and abdomen respiratory effort. Respiratory events (apneas and hypopneas) and electroencephalogram recordings were manually scored according to international guidelines [19]. Apnea was defined as a cessation of airflow or $\geq$90% of airflow reduction for at least 10 s [19]. Obstructive apneas were defined based on the presence of thoracic efforts. Hypopnea was defined as $\geq$50% decrement in airflow for at least 10 s in combination with oxyhemoglobin desaturation of at least 4% [19]. The apnea-hypopnea index (AHI) was calculated as the average number of apneas and hypopneas per hour of recorded sleep time [19]. A diagnosis of OSA was made when the AHI in the recorded study was >5 events/h of sleep accompanied by related symptoms (sleepiness, snoring, witnessed apneas) [20]. Subjects with AHI <5/h of sleep were considered as controls.

2.4. Pulmonary Function Testing

Pulmonary function testing (Screenmate, Erich Jaeger GmbH & Co., Hochberg, Germany) and arterial blood gasses analysis were performed the day before PSG. Forced expiratory volume in the first second (FEV_1) and forced vital capacity (FVC) were measured and the FEV_1/FVC ratio was calculated. The diagnosis of COPD was based on the presence of persistent airflow limitation, as defined by a post-bronchodilator FEV_1/FVC ratio of less than 70%, associated with symptoms such as dyspnea, cough and/or sputum production and a compatible history of exposure to risk factors [21].

2.5. Blood Samples and Analysis

All participants provided a fasting blood sample upon awakening the morning after polysomnography. Blood was then centrifuged (3000 rpm for 10 min) and the serum obtained was conserved frozen at −80 °C for future analysis. Serum 25(OH)D was measured using a commercial radioimmunoassay kit (DiaSorin, Stillwater, MN, USA). Biochemical examinations regarding renal, hepatic function, as well as lipidemic profile were also performed.

2.6. Statistical Analysis

All analyses were performed using the IBM Statistical Package for Social Sciences version 17.0 (SPSS Inc., Chicago, IL, USA, 2008). Study participants were divided into three groups matched for age, gender and BMI, as controls, OSA and OVS. All variables were presented as median (25th–75th percentile) for continuous variables and proportions for categorical variables. Comparison of baseline characteristics among groups was performed by a chi-square test for categorical variables, or a one-way ANOVA for continuous variables. Post -hoc analysis was performed using the Tukey's post-hoc test when homogeneity of variances was assumed, and the Games-Howell's post-hoc test when the assumption of homogeneity of variances was violated. Multiple linear regression analysis was used to test the association of serum 25(OH)D levels with sleep, anthropometric and respiratory function parameters.

3. Results

A total of 90 patients (79 males and 11 females) were included in the analysis. In general, participants were middle-aged (median age 57 years) and obese (median BMI 36.2 kg/m^2) with a median value of serum 25(OH)D concentration of 18.1 (13.4–24.4) ng/mL. The sample was divided into three groups matched for age, gender and BMI: the control group that included 30 subjects (27 males and 3 females), the OSA group that included 30 patients (26 males and 4 females) and the OVS group that included 30 patients (26 males and 4 females). Neck and waist circumference differed between the groups (p = 0.005 and p = 0.023, respectively), and patients with OVS were more likely to be current or ex-smokers. The baseline characteristics of the three groups are shown in Table 1.

Table 1. Comparison of anthropometric characteristics between control, obstructive sleep apnea (OSA) and overlap syndrome (OVS) groups.

	Control Group (n = 30)	Patients with OSA (n = 30)	Patients with OVS (n = 30)	p
Gender (males/females)	27/3	26/4	26/4	0.902
Age (years)	56 (48.8–64.3)	56 (52.8–65)	60 (54.8–67.3)	0.386
Neck circumference (cm)	41 (38–45.5)	46 (43.3–48.8) *	45 (44–50) *	0.005
Waist circumference (cm)	119 (104–131.5)	122 (119–129)	130 (119–135.5) *	0.023
Hip circumference (cm)	119 (107.5–125.5)	117 (112.3–122.8)	119 (110–123.5)	0.554
WHR	1.01 (0.94–1.05)	1.03 (1–1.08)	1.08 (1.04–1.11) *	0.040
BMI (kg/m^2)	33.6 (29.9–40.2)	36.9 (34.5–41.6)	36.5 (32–42)	0.105
Tobacco smoking				
Non-smokers	36.7%	30%	0% **, #	0.001
Ex-smokers	36.7%	13.3% *	40% #	0.049
Current smokers	26.7%	56.7% *	60% *	0.017

Abbreviations: BMI, body mass index; WHR, waist to hip ratio; *: $p < 0.05$ compared with control group; **: $p < 0.001$ compared with control group; #: $p < 0.05$ compared with OSA group.

Regarding sleep parameters, AHI was similar between OSA and OVS patients (40.2 vs. 37.1 events/h of sleep, $p > 0.05$). Patients with OVS exhibited the worst oxygenation during sleep compared with OSA patients, as assessed by the average oxygen saturation during sleep (90.8% vs. 92%, $p < 0.05$) and the percentage of time with oxygen saturation during sleep < 90% (27.3% vs. 19.7%, $p < 0.05$). No difference was noted between the groups in terms of daytime sleepiness, as expressed by the Epworth sleepiness scale score ($p = 0.106$). Comparisons of sleep parameters between the groups are presented in Table 2.

Table 2. Comparison of sleep characteristics between control, obstructive sleep apnea syndrome (OSA) and overlap syndrome (OVS) groups.

	Control Group (n = 30)	Patients with OSA (n = 30)	Patients with OVS (n = 30)	p
TST (min)	326.3 (280.8–358.5)	310.1 (260.8–346.3)	320.5 (278.3–360.8)	0.843
N1 (%)	10 (5.6–17.2)	11.4 (6.6–23.4)	11 (4.8–18.2)	0.191
N2 (%)	62.2 (56.9–71)	66.7 (56.2–77.3)	66.4 (60.7–71.4)	0.676
N3 (%)	11.7 (4.4–20.5)	7 (2.1–15.8)	9.9 (6.6–19.1)	0.391
REM (%)	13.5 (4.9–15.1)	6.2 (1.2–13.7) *	4.1 (0.3–12.3) *	0.011
Sleep efficiency (%)	84.6 (75–91.9)	86.4 (74.2–91.9)	83.2 (72.4–92.2)	0.871
Arousal index	14.4 (9.1–18.2	36.5 (21.5–47.4) **	26.1 (10.3–44.9) *	<0.001
AHI (events/h)	2.9 (1.5–4)	40.2 (22.1–70) **	37.1 (18–62.6) **	<0.001
Aver SaO_2 (%)	94.3 (93.8–95.3)	92(89–93) **	90.8 (84.9–92.1) **,#	<0.001
Min SaO_2 (%)	88 (84–89)	74.5 (71.8–83.8) **	74.5 (63.5–81.3) **,#	<0.001
T < 90% (%)	0.1 (0–0.4)	19.7 (3.8–39.4) *	27.3 (7.7–83.1) **,#	<0.001
ESS score	8 (4.8–12.3)	11.5 (7.8–15.3)	9 (6–13.8)	0.106

Abbreviations: AHI, apnea hypopnea index; Aver SaO_2, average oxyhemoglobin saturation; ESS, Epworth sleepiness scale; Min SaO_2, minimum oxyhemoglobin saturation; N1, sleep stage 1; N2, sleep stage 2; N3, sleep stage 3; REM, rapid eye movement; TST, total sleep time; T < 90%, time with oxyhemoglobin saturation <90%; *: $p < 0.05$ compared with control group; **: $p < 0.001$ compared with control group; #: $p < 0.05$ compared with OSA group.

In the OVS group, the median COPD assessment test (CAT) score was 9 (5–14.3). According to the GOLD classification, the OVS group included 7 patients with mild, 16 patients with moderate, 6 patients with severe and 1 patient with very severe airflow limitation [21]. Patients with OVS had a lower FEV_1 compared with the OSA patients and control patients (64.7 vs. 89.9 vs. 105.5%, respectively, $p < 0.001$) and presented increased $PaCO_2$ during wake in respect to both the OSA and control patients (46.5 vs. 41 vs. 40.5 mmHg, respectively, $p < 0.001$). Patients with OVS also exhibited decreased serum 25(OH)D levels compared with the OSA patients and control patients (14.5 vs. 18.6 vs.

21.6 ng/mL, $p < 0.001$) (Figure 1). Comparisons of measurements and laboratory analyses between the groups are presented in Table 3.

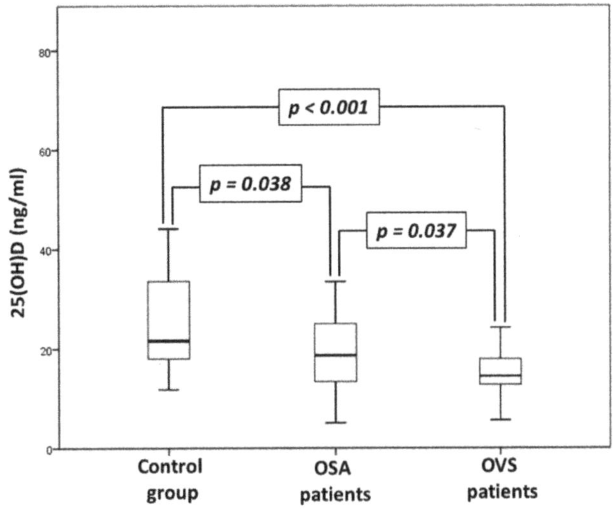

Figure 1. Comparison of 25(OH)D serum levels between control, obstructive sleep apnea syndrome (OSA) and overlap syndrome (OVS) groups.

Table 3. Comparison of measurements and laboratory analyses between control, obstructive sleep apnea syndrome (OSA) and overlap syndrome (OVS) groups.

	Control Group (n = 30)	Patients with OSA (n = 30)	Patients with OVS (n = 30)	p
FEV_1 (% predicted)	105.5 (91.3–113.8)	89.9 (74.6–100.3)	64.7 (48.8–76) **, ##	<0.001
FVC (% predicted)	102.5 (87.5–111.4)	85.6 (71–97.3) *	77.9 (62.4–89.1) **	<0.001
FEV_1/FVC (%)	83 (80.4–87.8)	84.6 (79.1–90.3)	67.1 (62.5–69.2) **, ##	<0.001
pO_2 (mmHg)	79 (73.5–84.3)	75 (66.8–79.5) *	67 (60.8–76) **	<0.001
pCO_2 (mmHg)	40.5 (37–43)	41 (37–44.5)	46.5 (42–52) **, #	<0.001
Glucose (mg/dL)	95 (81–112.5)	112.5 (96,5–151.3)	112.5 (95.3–129.3)	0.080
Creatinine (mg/dL)	0.8 (0.75–1)	0.9 (0.8–1.1)	0.9 (0.8–1.1)	0.707
SGOT (U/L)	20 (17.5–22.3)	22 (17.3–27)	20 (17–24)	0.296
SGPT (U/L)	19.5 (16.8–28)	24 (18–30.5)	21.5 (16.8–26.3)	0.624
Cholesterol(mg/dL)	207 (182–258)	210.5 (162.5–242.8)	185.5 (149–214.5)	0.154
Triglycerides (mg/dL)	158 (117.5–214.3)	158 (120.5–190)	156.5 (117–244.8)	0.166
LDL-C (mg/dL)	128.2 (101.9–159.4)	124.2 (95.4–156.1)	96.1 (77.3–126.2) *	0.030
HDL-C (mg/dL)	46 (41.5–56)	44.5 (35.8–52.5)	50 (43–56.3)	0.411
25(OH)D (ng/mL)	21.6 (17.8–33.6)	18.6 (13.2–25.2) *	14.5 (12.3–17.9) **, #	<0.001

Abbreviations: FEV_1, forced expiratory volume in first second; FVC, forced vital capacity; HDL-C, high density lipoprotein cholesterol; LDL-C, low density lipoprotein cholesterol; pCO_2, carbon dioxide partial pressure; pO_2, oxygen partial pressure; SGOT, serum glutamate-oxaloacetate transaminase; SGPT, serum glutamate-pyruvate transaminase; 25(OH)D, 25-hydroxyvitamin D; *: $p < 0.05$ compared with control group; **: $p < 0.001$ compared with control group; #: $p < 0.05$ compared with OSA group; ##: $p < 0.001$ compared with OSA group.

In order to explore the factors predicting serum 25(OH)D levels, a regression model was created with 25(OH)D serum levels used as a dependent variable. Anthropometric (age, gender, BMI), pulmonary function parameters (FEV_1, FVC and CAT score), and sleep parameters (AHI, average and minimum saturation during sleep, time spent with SaO_2 < 90%, arousal index and ESS score) were used as independent variables. Multiple linear regression identified that serum 25(OH)D levels could be predicted by AHI ($\beta = -0.758$, $p = 0.041$, 95% CI: −0.282–−0.006) and FEV_1 ($\beta = 0.698$, $p = 0.038$, 95% CI: 0.014–0.410).

4. Discussion

The OVS is characterized by the co-existence of COPD and OSA, both conditions that seem to have an adverse effect on calcium homeostasis. In this study, we found that 25(OH)D serum levels were decreased in patients with OVS, compared with both OSA patients and non-apneic controls. Additionally, AHI and FEV_1 were independently associated with 25(OH)D serum levels in OVS patients.

The definition of Vit D status in OSA patients remained controversial for a long time, with some studies reporting decreased Vit D levels in those patients while others concluded otherwise [22–25]. A meta-analysis conducted by Neighbors et al., [26] including 14 studies with 1513 controls and 3424 OSA patients, reported decreased Vit D serum levels in the latter, with Vit D insufficiency being incrementally exacerbated with increasing severity of OSA (mean differences were −2.7% for mild OSA, −10.1% for moderate OSA and −17.4% for severe OSA). Similarly, results from a more recent meta-analysis conducted by Li et al. [10] showed decreased Vit D levels only in patients with moderate ($p = 0.002$) and severe ($p < 0.0001$) OSA compared with non-apneic controls. OSA treatment with CPAP seems to possess a beneficial effect on Vit D serum levels. Liguori et al. found increased levels of 25(OH)D after 7 nights of CPAP therapy in male OSA patients, and after 1 year of treatment mainly in obese OSA patients [27,28]. Similar conclusions were reached in a randomized sham-controlled trial, where 24 weeks of CPAP treatment improved Vit D serum levels ($p = 0.045$) in patients with severe OSA [29]. In the present study, Vit D serum levels were strongly associated with AHI. These results are in line with those from previous studies. In a study that included 139 OSA patients and 30 controls, OSA patients had lower 25(OH)D levels compared with controls (17.8 ± 7.8 vs. 23.9 ± 12.4 ng/mL respectively, $p = 0.019$) and there was a significant inverse association between AHI and 25(OH)D levels ($r = -0.187$, $p = 0.045$) [22]. Similarly, in another study that included 75 OSA patients and 31 controls, serum 25(OH)D levels decreased with OSA severity ($p = 0.003$) and were negatively associated with AHI ($r = -0.40$, $p = 0.0001$) [30]. Overall, current evidence suggests that hypovitaminosis D is frequent among OSA patients. Multiple factors such as BMI, gender and disease severity may play a regulatory role in this relationship.

Previous studies have shown that, compared with controls, patients with COPD frequently presented with decreased Vit D levels, with a prevalence that ranged between 33% and 77% according to disease severity [31–33]. In a recent study that included 1609 COPD patients, Vit D deficiency was present in 21% and was associated with a 4.11% decrease in predicted FEV_1 at enrollment (95% CI: −6.90% to −1.34% predicted FEV_1, $p = 0.004$), a 1.27% predicted greater rate of FEV_1 decline after 1 year (95% CI: −2.32% to −0.22% predicted/year, $p = 0.02$), and increased odds of any COPD exacerbation in the prior year (OR: 1.32, 95% CI: 1.00–1.74, $p = 0.049$) [11]. Studies evaluating the effect of Vit D supplementation in lung function produced contradictory results. In a meta-analysis that included 8 studies and 687 COPD patients, Vit D treatment resulted in no significant improvements in FEV_1 ($p = 0.144$), FVC ($p = 0.299$), and FEV_1/FVC ($p = 0.995$) in COPD patients [34]. On the other hand, another meta-analysis that included 25 studies involving 2670 COPD patients concluded that Vit D supplementation was significantly associated with FEV_1 ($p < 0.01$), FEV_1/FVC ($p < 0.01$), exacerbations ($p < 0.01$), sputum volume ($p < 0.01$), 6-min walk distance ($p = 0.02$) and CAT score ($p < 0.01$) [35].

Vit D insufficiency in OVS is multifactorial and represents the result of the interaction of individual conditions present in COPD and OSA. In particular, altered cutaneous synthesis due to aging, toxic smoke effects and reduced sun exposure because of disability to perform outdoor activities, sequestration in the adipose tissue, increased Vit D catabolism caused by glucocorticoid therapy (together with impaired liver and renal activation and reduced gastrointestinal absorption) are all factors known to reduce Vit D serum levels in COPD patients [36]. Reduced outdoor activities due to exertional dyspnea may also negatively affect Vit D cutaneous synthesis. In addition, low 25(OH)D levels in COPD correlate with genetic variants of the Vit D-binding gene [32]. Conversely, OSA may represent a risk factor for Vit D insufficiency. Lack of outdoor activity due to obesity and excessive daytime sleepiness might result in reduced Vit D synthesis because of insufficient sun exposure [37]. Moreover, Vit D is stored in fat tissue, thus reducing the release of Vit D into the circulation and decreasing its bioavailability [37]. Finally, Vit D receptor gene variations affect both 25(OH)D serum levels and disease susceptibility in OSA [38].

A previous study has shown that OVS patients had a similar Vit D status compared to the controls, OSA, or COPD patients and Vit D deficiency as associated with overall mortality among groups [39]. Diagnosis of Vit D insufficiency and correction of Vit D status has several clinical implications in patients with both OSA and COPD. Notably, hypovitaminosis D is highly prevalent in COPD patients and has been associated with worse lung function, increased symptoms and rate of exacerbations, and worse prognosis [11,35]. Moreover, decreased Vit D serum levels were related to COPD severity, and acute exacerbations may be prevented with adequate Vit D supplementation [16,40]. Finally, there is evidence to suggest that a normal Vit D status could improve prognosis in patients with COPD exacerbations caused by respiratory tract infections [41]. Serum 25(OH)D levels were decreased in OSA patients compared with controls, and have been associated with disease severity [10]. On the other hand, with increasing OSA severity serum 25(OH)D levels further decreased, this suggests that serum 25(OH)D might be a risk factor for OSA [10]. Finally, markers of inflammation and HOMA-IR were significantly decreased after Vit D supplementation in a small group of patients with mild OSA [42].

The relationship between an adequate Vit D status and cardiovascular risk is controversial, with data suggesting a negative association between Vit D levels and the risk of cardiovascular disease. Whereas, other studies failed to demonstrate an association between Vit D serum levels and a reduced risk for adverse cardiovascular events [43,44]. In a study conducted in Southern Italy that included 451 participants, the Vit D level did not correlate with cardiovascular risk, as assessed using the Framingham cardiovascular risk charts [45]. On the contrary, the parathyroid hormone was in direct correlation ($p < 0.001$) with cardiovascular risk, and increased parathyroid hormone levels identified a population with a higher risk for cardiovascular events ($p < 0.001$) [45]. Results from this study suggest increased parathyroid hormone levels might be a better predictor of cardiovascular risk in patients with hypovitaminosis D.

Nonetheless, the present study has limitations. First, the study was cross-sectional in design, and consequently, it is not possible to infer any causal relationships. Second, the sample of OVS patients included was rather small and was not representative of the female population. This is largely due to the low prevalence of OVS in the general population and to the disparity in gender-specific estimates of COPD and OSA prevalence [4,46,47]. Thus, larger scale studies are necessary to better assess Vit D status in OVS patients. Third, 25(OH)D was not deseasonalized and the parathyroid hormone was not measured. However, all participants were Caucasian, living in a single region of northern Greece, a relatively small geographical area, with no variations in latitude and with similar conditions regarding cloudiness and air pollution. Moreover, participants had no variations regarding skin pigmentation and in their way of dressing, had similar dietary habits, similar sun exposure and sun-bathing behavior. Furthermore, OVS patients were middle-aged, so caution is warranted in order to extrapolate this data to elderly OVS patients. Finally, only one patient with very a severe air-flow limitation was included in the OVS group. However, it has been shown that Vit D levels decrease with increasing air-flow limitation severity [11]. An analysis of sub-populations based on disease severity in the OSA group was not performed due to the limited sample size.

5. Conclusions

In conclusion, lower Vit D levels have been observed in patients with OVS compared with OSA patients and non-apneic controls, indicating an increased risk of hypovitaminosis D in this population. This is expected since both the components of OVS, COPD and OSA are also associated with decreased Vit D levels. Given the protective role of Vit D against FEV_1 decline and COPD exacerbations, as well as the association of low Vit D levels with cardiovascular and metabolic diseases, screening of OVS patients for Vit D might be beneficial in the management of this population of patients.

Author Contributions: Conceptualization, P.S.; writing—original draft preparation, K.A. and A.V.; writing—review and editing, E.N., A.Z., I.B., S.E.S. and P.S. All authors have read and agreed to the published version of the manuscript.

Funding: This research received no external funding.

Institutional Review Board Statement: The study was conducted in accordance with the Declaration of Helsinki, and approved by the Ethics Committee of Alexandroupolis University General Hospital (approval code 1/27-01-2017, date of approval 27 January 2017).

Informed Consent Statement: Informed consent was obtained from all subjects involved in the study.

Data Availability Statement: Not applicable.

Conflicts of Interest: The authors declare no conflict of interest.

References

1. Adeloye, D.; Chua, S.; Lee, C.; Basquill, C.; Papana, A.; Theodoratou, E.; Nair, H.; Gasevic, D.; Sridhar, D.; Campbell, H.; et al. Global and regional estimates of COPD prevalence: Systematic review and meta-analysis. *J. Glob. Health* **2015**, *5*, 020415. [CrossRef] [PubMed]
2. Peppard, P.E.; Young, T.; Barnet, J.H.; Palta, M.; Hagen, E.W.; Hla, K.M. Increased prevalence of sleep-disordered breathing in adults. *Am. J. Epidemiol.* **2013**, *177*, 1006–1014. [CrossRef] [PubMed]
3. Owens, R.L.; Malhotra, A. Sleep-disordered breathing and COPD: The overlap syndrome. *Respir. Care* **2010**, *55*, 1333–1344, discussion 1344–1336. [PubMed]
4. McNicholas, W.T. Chronic obstructive pulmonary disease and obstructive sleep apnea: Overlaps in pathophysiology, systemic inflammation, and cardiovascular disease. *Am. J. Respir. Crit. Care Med.* **2009**, *180*, 692–700. [CrossRef] [PubMed]
5. Weitzenblum, E.; Chaouat, A.; Kessler, R.; Canuet, M. Overlap syndrome: Obstructive sleep apnea in patients with chronic obstructive pulmonary disease. *Proc. Am. Thorac. Soc.* **2008**, *5*, 237–241. [CrossRef] [PubMed]
6. Voulgaris, A.; Archontogeorgis, K.; Papanas, N.; Pilitsi, E.; Nena, E.; Xanthoudaki, M.; Mikhailidis, D.P.; Froudarakis, M.E.; Steiropoulos, P. Increased risk for cardiovascular disease in patients with obstructive sleep apnoea syndrome-chronic obstructive pulmonary disease (overlap syndrome). *Clin. Respir. J.* **2019**, *13*, 708–715. [CrossRef]
7. Ruchala., M.; Brominska, B.; Cyranska-Chyrek, E.; Kuznar-Kaminska, B.; Kostrzewska, M.; Batura-Gabryel, H. Obstructive sleep apnea and hormones—A novel insight. *Arch. Med. Sci.* **2017**, *13*, 875–884. [CrossRef]
8. Archontogeorgis, K.; Papanas, N.; Nena, E.; Tzouvelekis, A.; Tsigalou, C.; Voulgaris, A.; Xanthoudaki, M.; Mouemin, T.; Froudarakis, M.; Steiropoulos, P. Insulin Sensitivity and Insulin Resistance in Non-Diabetic Middle-Aged Patients with Obstructive Sleep Apnoea Syndrome. *Open Cardiovasc. Med. J.* **2017**, *11*, 159–168. [CrossRef]
9. Laghi, F.; Adiguzel, N.; Tobin, M.J. Endocrinological derangements in COPD. *Eur. Respir. J.* **2009**, *34*, 975–996. [CrossRef]
10. Li, X.; He, J.; Yun, J. The association between serum vitamin D and obstructive sleep apnea: An updated meta-analysis. *Respir. Res.* **2020**, *21*, 294. [CrossRef]
11. Burkes, R.M.; Ceppe, A.S.; Doerschuk, C.M.; Couper, D.; Hoffman, E.A.; Comellas, A.P.; Barr, R.G.; Krishnan, J.A.; Cooper, C.; Labaki, W.W.; et al. Associations Among 25-Hydroxyvitamin D Levels, Lung Function, and Exacerbation Outcomes in COPD: An Analysis of the SPIROMICS Cohort. *Chest* **2020**, *157*, 856–865. [CrossRef]
12. Bouloukaki, I.; Tsiligianni, I.; Mermigkis, C.; Bonsignore, M.R.; Markakis, M.; Pataka, A.; Steiropoulos, P.; Ermidou, C.; Alexaki, I.; Tzanakis, N.; et al. Vitamin D deficiency in patients evaluated for obstructive sleep apnea: Is it associated with disease severity? *Sleep Breath* **2021**, *25*, 1109–1117. [CrossRef]
13. Umar, M.; Sastry, K.S.; Chouchane, A.I. Role of Vitamin D Beyond the Skeletal Function: A Review of the Molecular and Clinical Studies. *Int. J. Mol. Sci.* **2018**, *19*, 1618. [CrossRef]
14. Cashman, K.D.; Dowling, K.G.; Skrabakova, Z.; Gonzalez-Gross, M.; Valtuena, J.; de Henauw, S.; Moreno, L.; Damsgaard, C.T.; Michaelsen, K.F.; Molgaard, C.; et al. Vitamin D deficiency in Europe: Pandemic? *Am. J. Clin. Nutr.* **2016**, *103*, 1033–1044. [CrossRef]
15. Siachpazidou, D.I.; Stavrou, V.; Zouridis, S.; Gogou, E.; Economou, N.T.; Pastaka, C.; Hatzoglou, C.; Gourgoulianis, K.I. 25-hydroxyvitamin D levels in patients with obstructive sleep apnea and continuous positive airway pressure treatment: A brief review. *Sleep Sci.* **2020**, *13*, 78–83.
16. Martineau, A.R.; James, W.Y.; Hooper, R.L.; Barnes, N.C.; Jolliffe, D.A.; Greiller, C.L.; Islam, K.; McLaughlin, D.; Bhowmik, A.; Timms, P.M.; et al. Vitamin D3 supplementation in patients with chronic obstructive pulmonary disease (ViDiCO): A multicentre, double-blind, randomised controlled trial. *Lancet Respir. Med.* **2015**, *3*, 120–130. [CrossRef]
17. World Medical Association. World Medical Association Declaration of Helsinki: Ethical principles for medical research involving human subjects. *JAMA* **2013**, *310*, 2191–2194. [CrossRef]

18. Tsara, V.; Serasli, E.; Amfilochiou, A.; Constantinidis, T.; Christaki, P. Greek version of the Epworth Sleepiness Scale. *Sleep Breath* **2004**, *8*, 91–95. [CrossRef]
19. Berry, R.B.; Brooks, R.; Gamaldo, C.E.; Gamaldo, C.E.; Harding, S.M.; Lloyd, R.M.; Marcus, C.L.; Vaughn, B.V.; American Academy of Sleep Medicine. *The AASM Manual for the Scoring of Sleep and Associated Events: Rules, Terminology and Technical Specifications*; Version 2.3; American Academy of Sleep Medicine: Darien, IL, USA, 2016.
20. McNicholas, W.T.; Strohl, K.P.; White, D.P.; Levy, P.; Schmidt, W.; Wheatley, J.R.; Redline, S.; Carley, D.; Buysse, J.D.; Young, T.; et al. Sleep-related breathing disorders in adults: Recommendations for syndrome definition and measurement techniques in clinical research. The Report of an American Academy of Sleep Medicine Task Force. *Sleep* **1999**, *22*, 667–689.
21. Mirza, S.; Clay, R.D.; Koslow, M.A.; Scanlon, P.D. COPD Guidelines: A Review of the 2018 GOLD Report. *Mayo Clin. Proc.* **2018**, *93*, 1488–1502. [CrossRef]
22. Archontogeorgis, K.; Nena, E.; Papanas, N.; Zissimopoulos, A.; Voulgaris, A.; Xanthoudaki, M.; Manolopoulos, V.; Froudarakis, M.; Steiropoulos, P. Vitamin D Levels in Middle-Aged Patients with Obstructive Sleep Apnoea Syndrome. *Curr. Vasc. Pharmacol.* **2018**, *16*, 289–297. [CrossRef]
23. Salepci, B.; Caglayan, B.; Nahid, P.; Parmaksiz, E.T.; Kiral, N.; Fidan, A.; Comert, S.S.; Dogan, C.; Gungor, G.A. Vitamin D Deficiency in Patients Referred for Evaluation of Obstructive Sleep Apnea. *J. Clin. Sleep Med.* **2017**, *13*, 607–612. [CrossRef]
24. Siachpazidou, D.I.; Kotsiou, O.S.; Stavrou, V.; Pastaka, C.; Gogou, E.; Kechagia, M.; Varsamas, C.; Economou, N.T.; Zouridis, S.; Patrikioy, E.; et al. Serum vitamin D levels in patients with obstructive sleep apnea syndrome and level changes after continuous positive airway pressure therapy. *Sleep Breath* **2021**, *25*, 657–668. [CrossRef]
25. Ma, D.; Zheng, X.; Dong, L.; Zheng, C.; Chen, Y.; Chen, Z.; Lin, M.; Li, X.; Li, Z.; Liu, C. The Relationship of Serum 25-Hydroxyvitamin-D Level with Severity of Obstructive Sleep Apnea in Patients with Type 2 Diabetes Mellitus. *Diabetes Metab Syndr Obes* **2020**, *13*, 1391–1398. [CrossRef] [PubMed]
26. Neighbors, C.L.P.; Noller, M.W.; Song, S.A.; Zaghi, S.; Neighbors, J.; Feldman, D.; Kushida, C.A.; Camacho, M. Vitamin D and obstructive sleep apnea: A systematic review and meta-analysis. *Sleep Med.* **2018**, *43*, 100–108. [CrossRef] [PubMed]
27. Liguori, C.; Romigi, A.; Izzi, F.; Mercuri, N.B.; Cordella, A.; Tarquini, E.; Giambrone, M.P.; Marciani, M.G.; Placidi, F. Continuous Positive Airway Pressure Treatment Increases Serum Vitamin D Levels in Male Patients with Obstructive Sleep Apnea. *J. Clin. Sleep Med.* **2015**, *11*, 603–607. [CrossRef] [PubMed]
28. Liguori, C.; Izzi, F.; Mercuri, N.B.; Romigi, A.; Cordella, A.; Tarantino, U.; Placidi, F. Vitamin D status of male OSAS patients improved after long-term CPAP treatment mainly in obese subjects. *Sleep Med.* **2017**, *29*, 81–85. [CrossRef] [PubMed]
29. Theorell-Haglow, J.; Hoyos, C.M.; Phillips, C.L.; Yee, B.J.; Herrmann, M.; Brennan-Speranza, T.C.; Grunstein, R.R.; Liu, P.Y. Changes of vitamin D levels and bone turnover markers after CPAP therapy: A randomized sham-controlled trial. *J. Sleep Res.* **2018**, *27*, e12606. [CrossRef]
30. Kerley, C.P.; Hutchinson, K.; Bolger, K.; McGowan, A.; Faul, J.; Cormican, L. Serum Vitamin D Is Significantly Inversely Associated with Disease Severity in Caucasian Adults with Obstructive Sleep Apnea Syndrome. *Sleep* **2016**, *39*, 293–300. [CrossRef]
31. Forli, L.; Halse, J.; Haug, E.; Bjortuft, O.; Vatn, M.; Kofstad, J.; Boe, J. Vitamin D deficiency, bone mineral density and weight in patients with advanced pulmonary disease. *J. Intern. Med.* **2004**, *256*, 56–62. [CrossRef]
32. Janssens, W.; Bouillon, R.; Claes, B.; Carremans, C.; Lehouck, A.; Buysschaert, I.; Coolen, J.; Mathieu, C.; Decramer, M.; Lambrechts, D. Vitamin D deficiency is highly prevalent in COPD and correlates with variants in the vitamin D-binding gene. *Thorax* **2010**, *65*, 215–220. [CrossRef]
33. Persson, L.J.; Aanerud, M.; Hiemstra, P.S.; Hardie, J.A.; Bakke, P.S.; Eagan, T.M. Chronic obstructive pulmonary disease is associated with low levels of vitamin D. *PLoS ONE* **2012**, *7*, e38934. [CrossRef]
34. Chen, F.Y.; Xiao, M.; Ling, B.; Liu, L.; Chen, L. Vitamin D does not improve lung function decline in COPD: A meta-analysis. *Eur. Rev. Med. Pharmacol. Sci.* **2019**, *23*, 8637–8644.
35. Li, X.; He, J.; Yu, M.; Sun, J. The efficacy of vitamin D therapy for patients with COPD: A meta-analysis of randomized controlled trials. *Ann. Palliat. Med.* **2020**, *9*, 286–297. [CrossRef]
36. Janssens, W.; Mathieu, C.; Boonen, S.; Decramer, M. Vitamin D deficiency and chronic obstructive pulmonary disease: A vicious circle. *Vitam. Horm.* **2011**, *86*, 379–399.
37. Archontogeorgis, K.; Nena, E.; Papanas, N.; Steiropoulos, P. The role of vitamin D in obstructive sleep apnoea syndrome. *Breathe* **2018**, *14*, 206–215. [CrossRef]
38. Ragia, G.; Archontogeorgis, K.; Simmaco, M.; Gentile, G.; Borro, M.; Zissimopoulos, A.; Froudarakis, M.; Manolopoulos, V.G.; Steiropoulos, P. Genetics of Obstructive Sleep Apnea: Vitamin D Receptor Gene Variation Affects Both Vitamin D Serum Concentration and Disease Susceptibility. *OMICS* **2019**, *23*, 45–53. [CrossRef]
39. Du, W.; Liu, J.; Zhou, J.; Ye, D.; OuYang, Y.; Deng, Q. Obstructive sleep apnea, COPD, the overlap syndrome, and mortality: Results from the 2005-2008 National Health and Nutrition Examination Survey. *Int. J. Chron. Obstruct. Pulmon. Dis.* **2018**, *13*, 665–674. [CrossRef]
40. Zhu, M.; Wang, T.; Wang, C.; Ji, Y. The association between vitamin D and COPD risk, severity, and exacerbation: An updated systematic review and meta-analysis. *Int. J. Chron. Obstruct. Pulmon. Dis.* **2016**, *11*, 2597–2607. [CrossRef]
41. Ginde, A.A.; Mansbach, J.M.; Camargo, C.A., Jr. Vitamin D, respiratory infections, and asthma. *Curr. Allergy Asthma Rep.* **2009**, *9*, 81–87. [CrossRef]

42. Ayyildiz, F.; Yildiran, H.; Afandiyeva, N.; Gulbahar, O.; Kokturk, O. The effects of vitamin D supplemantation on prognosis in patients with mild obstructive sleep apnea syndrome. *Turk. J. Med. Sci.* **2021**, *51*, 2524–2533. [CrossRef]
43. Barbarawi, M.; Kheiri, B.; Zayed, Y.; Barbarawi, O.; Dhillon, H.; Swaid, B.; Yelangi, A.; Sundus, S.; Bachuwa, G.; Alkotob, M.L.; et al. Vitamin D Supplementation and Cardiovascular Disease Risks in More Than 83,000 Individuals in 21 Randomized Clinical Trials: A Meta-analysis. *JAMA Cardiol.* **2019**, *4*, 765–776. [CrossRef]
44. Zhou, Y.; Jiang, M.; Sun, J.Y.; Cheng, C.; Shen, H.; Sun, W.; Kong, X.Q. The Association Between Vitamin D Levels and the 10-Year Risk of Atherosclerotic Cardiovascular Disease: A Population-Based Study. *J. Cardiovasc. Nurs.* **2022**. [CrossRef]
45. Pascale, A.V.; Finelli, R.; Giannotti, R.; Visco, V.; Fabbricatore, D.; Matula, I.; Mazzeo, P.; Ragosa, N.; Massari, A.; Izzo, R.; et al. Vitamin D, parathyroid hormone and cardiovascular risk: The good, the bad and the ugly. *J. Cardiovasc. Med.* **2018**, *19*, 62–66. [CrossRef]
46. Ntritsos, G.; Franek, J.; Belbasis, L.; Christou, M.A.; Markozannes, G.; Altman, P.; Fogel, R.; Sayre, T.; Ntzani, E.E.; Evangelou, E. Gender-specific estimates of COPD prevalence: A systematic review and meta-analysis. *Int. J. Chron. Obstruct. Pulmon. Dis.* **2018**, *13*, 1507–1514. [CrossRef]
47. Wimms, A.; Woehrle, H.; Ketheeswaran, S.; Ramanan, D.; Armitstead, J. Obstructive Sleep Apnea in Women: Specific Issues and Interventions. *BioMed Res. Int.* **2016**, *2016*, 1764837. [CrossRef]

Journal of
Personalized Medicine

Article

Diagnostic Efficacy of Ultra-Short Term HRV Analysis in Obstructive Sleep Apnea

Seung-Su Ha [1] and Dong-Kyu Kim [1,2,*]

[1] Department of Otorhinolaryngology-Head and Neck Surgery, Chuncheon Sacred Heart Hospital, Hallym University College of Medicine, Chuncheon 24253, Korea
[2] Institute of New Frontier Research, Division of Big Data and Artificial Intelligence, Chuncheon Sacred Heart Hospital, Hallym University College of Medicine, Chuncheon 24253, Korea
* Correspondence: doctordk@naver.com; Tel.: +82-33-240-5180

Abstract: Heart rate variability (HRV) is the standard method for assessing autonomic nervous system (ANS) activity and is considered a surrogate marker for sympathetic overactivity in obstructive sleep apnea (OSA). Although HRV features are usually obtained from the short-term segment method, it is impossible to evaluate rapid dynamic changes in ANS activity. Herein, we propose the ultra-short-term analysis to detect the balance of ANS activity in patients with OSA. In 1021 OSA patients, 10 min HRV target datasets were extracted from polysomnographic data and analyzed by shifting the 2 min (ultra-short-term) and 5 min (short-term) segments. We detected frequency-domain parameters, including total power (Ln TP), very low frequency (Ln VLF), low frequency (Ln LF), and high frequency (Ln HF). We found that overall HRV feature alterations indicated sympathetic overactivity dependent on OSA severity, and that this was more pronounced in the ultra-short-term methodology. The apnea-hypopnea index, oxygen desaturation index, and Epworth sleepiness scale correlated with increased sympathetic activity and decreased parasympathetic activity, regardless of the methodology. The Bland-Altman plot analyses also showed a higher agreement of HRV features between the two methodologies. This study suggests that ultra-short-term HRV analysis may be a useful method for detecting alterations in ANS function in OSA patients.

Keywords: heart rate variability; autonomic nervous system; sleep; obstructive sleep apnea

Citation: Ha, S.-S.; Kim, D.-K. Diagnostic Efficacy of Ultra-Short Term HRV Analysis in Obstructive Sleep Apnea. *J. Pers. Med.* **2022**, *12*, 1494. https://doi.org/10.3390/jpm12091494

Academic Editor: Anne-Marie Caminade

Received: 24 August 2022
Accepted: 9 September 2022
Published: 13 September 2022

1. Introduction

Obstructive sleep apnea (OSA) is the most common sleep-related breathing disorder and is characterized by total or partial upper airway collapse. Patients with OSA usually experience frequent hypoxic events and sleep fragmentation, resulting in repeated exposure to hypoxemia and hypercapnia. Additionally, this phenomenon could induce sympathetic overactivity in patients [1–3]. These alterations in autonomic nervous system (ANS) activity also contribute to the development of cardiovascular diseases in patients with OSA [4–6]. Heart rate variability (HRV) is a noninvasive method for the assessment of changes in ANS activity and represents the balance between the parasympathetic nervous system (PNS) and sympathetic nervous system (SNS) [7]. Most studies regarding HRV have shown a higher low-frequency power (LF) and a higher ratio of LF to high-frequency power (HF) in patients with OSA than in healthy subjects [8–10]. These findings support the idea that the balance shift in ANS activity moves toward sympathetic overactivity in these patients. In fact, evidence suggests that HRV can assess the shift in ANS activity and could be a useful marker of OSA severity [11–14].

To date, linear and non-linear methods of HRV analysis have been used. There are several algorithms of the non-linear method, such as Hurst exponent and Higuchi fractal dimension. However, HRV analysis is usually performed as a function of the time or temporal frequencies from linear analysis methods: peaks below 0.04 Hz (very low frequency, VLF), between 0.04 and 0.15 Hz (low frequency, LF), and between 0.15 and 0.4 (high frequency, HF). Previously, the majority of these studies used a short-term window, which investigates the degree of fluctuations in the time and frequency domains, to evaluate the HRV analysis. Short-term segment HRV is typically calculated over 5 min, and the magnitude of the frequency components obtained by spectral analysis is useful for assessing ANS activity [15]. However, HRV analysis based on short-term segments has inevitable limitations in clinical applications because its spectral estimates cannot reflect rapid ANS fluctuations. For this reason, several recent studies have suggested that ultra-short-term segments (shorter than 5 min) may be effective alternatives to short-term HRV analysis for mental stress detection [16–18]. Herein, we investigated the frequency spectral analysis of HRV features using the ultra-short-term methodology and evaluated whether the methodology could be more effective as a diagnostic marker for the alteration of ANS activity in patients with OSA. To the best of our knowledge, no studies have investigated ultra-short HRV as a valid surrogate in patients with OSA.

2. Materials and Methods

We consecutively enrolled patients diagnosed with OSA at the Sleep Center of Hallym Medical University Chuncheon Sacred Hospital and the informed consent was obtained from all subjects involved in the study. We excluded patients who had previously been diagnosed with a psychological disorder, a history of any medical condition that could influence sympathovagal activity, and serious comorbidities (e.g., cancer, severe depression or insomnia, severe cardiac or respiratory failure, and severe renal or hepatic insufficiency). We performed overnight full polysomnography (PSG) and obtained the Epworth Sleepiness Scale (ESS) scores for all enrolled patients to assess daytime sleepiness. In this study, sleep stage and respiratory events were scored according to the American Academy of Sleep Medicine [19]. The following sleep parameters were assessed: total sleep time (TST), sleep efficiency (TST/time in bed $\times$ 100), sleep latency, oxygen desaturation index (ODI), and apnea-hypopnea index (AHI). ODI was defined as the number of episodes of oxygen desaturation per hour of sleep, with oxygen desaturation defined as a decrease in blood oxygen saturation (SpO_2) to lower than 3% below baseline. The AHI was defined as the sum of the number of apneas and hypopneas per hour of sleep. In this study, we defined apnea as when the peak signal excursions, measured using an oronasal thermal sensor, dropped by $\geq$90% of baseline before the event, lasting for more than 10 s. Conversely, hypopnea was scored when the peak signal excursions using nasal pressure dropped by $\geq$30% of baseline before the event for at least 10 s, followed by a $\geq$ 3% decrease in oxygen desaturation or accompanied by arousal. Based on AHI, we classified the severity of OSA as follows: normal (AHI < 5), mild OSA ($5 \leq$ AHI < 15), moderate OSA ($15 \leq$ AHI < 30), and severe OSA (AHI $\geq$ 30).

HRV is characterized by the variability in beat-to-beat intervals of the heart and is typically measured using RR intervals extracted from an ECG signal. Thus, in this study, we extracted the electrocardiography data from PSG recordings visually inspected for accuracy and quality. Specifically, we eliminated ectopic beats and artifacts and selected only normal-to-normal beats for the HRV analysis. Generally, time-domain HRV parameters showed variability in beat-to-beat intervals, which influenced both sympathetic and parasympathetic activities [20]. Thus, these parameters are not typically used to discriminate the balance of ANS activity (sympathetic and parasympathetic functions). For this reason, we obtained the frequency-domain HRV parameters. We used four spectral power analyses by shifting predetermined segments (ultra-short-term, 2 min and short-term, 5 min) forward by 2 s during the entire 10 min HRV target dataset, which were collected from each patient: total power (Ln TP), very low frequency (Ln VLF), low frequency (Ln LF), and high frequency (Ln HF) in frequency-domain measures.

Numerical variables are expressed as mean ± standard deviation, and Student's t-test was used to compare HRV parameters after the normality test. ANOVA with post hoc analysis was performed to compare the specific values according to the OSA severity group. Pearson's coefficient test was used to identify the relationships between HRV and sleep parameters according to OSA severity. To compare the agreement of the HRV parameters between the ultra-short-term and short-term segments, we used Bland-Altman plots. All statistical analyses were conducted using R version 3.5.0 (R Foundation for Statistical Computing, Vienna, Austria). Statistical significance was set at $p < 0.05$.

3. Results

In total, 1021 patients with OSA were enrolled in this study. The patients were divided into three groups: mild OSA (n = 238), moderate OSA (n = 331), and severe OSA (n = 452). The demographics and PSG parameters are presented in Table 1. Age and ESS were significantly higher in patients with severe OSA than in the other groups ($p < 0.001$), whereas there were no significant group differences in TST, sleep latency, and sleep efficiency.

Table 1. Baseline characteristics and polysomnographic parameters.

	Mild	Moderate	Severe
Number	238 (23.3%)	331 (32.4%)	452 (44.3%)
Age	53.87 ± 7.55	56.14 ± 7.91	57.06 ± 8.53
Sex			
Male	206 (86.6%)	237 (71.6%)	403 (89.2%)
Female	32 (13.4%)	94 (28.4%)	49 (10.8%)
Body mass index (kg/m^2)	24.2 ± 3.8	26.4 ± 4.8	27.2 ± 3.1
Hypertension	168	216	285
ESS	5.06 ± 1.34	5.56 ± 2.24	11.11 ± 4.04
Total sleep Time (min)	363.93 ± 56.94	367.12 ± 56.43	375.67 ± 55.23
Sleep latency (min)	44.70 ± 9.64	45.94 ± 8.19	45.76 ± 7.72
Sleep efficiency (%)	79.74 ± 6.63	79.59 ± 6.42	79.30 ± 6.53
AHI	9.34 ± 2.37	20.50 ± 3.86	60.69 ± 17.38
ODI	7.87 ± 2.23	9.43 ± 3.96	33.62 ± 17.21

Epworth sleepiness scale, ESS; apnea-hypopnea index, (AHI); oxygen desaturation index, (ODI).

A 10 min HRV target segment was divided into consecutive individual ultra-short-term and short-term segments every 2 s, and all frequency-domain HRV parameters were averaged according to OSA severity (Tables 2–4). In mild and moderate OSA, Ln TP and Ln HF were significantly higher in the ultra-short term than in the short term (Tables 2 and 3), whereas the values of Ln TP, Ln VLF, and Ln LF based on the ultra-short term were significantly higher in those based on the short term (Table 4). However, there was no significant difference in the value of Ln LF/Ln HF between the ultra-short-term and short-term segments, regardless of OSA severity. We also confirmed that patients with OSA had overall alterations in HRV measures, indicating sympathetic overactivity, and this tendency was more pronounced in those with severe OSA. Additionally, these findings were more clearly detected in the ultra-short-term analysis than in the short-term analysis.

Table 2. Heart rate variability parameters between ultra-short and short term in mild obstructive sleep apnea.

	Ultra-Short Term	Short Term	p Value
Ln TP	7.47 ± 0.34	7.29 ± 0.47	<0.001
Ln VLF	6.97 ± 0.57	6.93 ± 0.61	0.504
Ln LF	7.24 ± 0.91	7.20 ± 1.21	0.733
Ln HF	7.10 ± 0.29	6.97 ± 0.61	0.002
Ln LF/Ln HF	1.02 ± 0.15	1.04 ± 0.20	0.221

Total power, Ln TP; very low frequency, Ln VLF; low frequency, Ln LF; high frequency, Ln HF.

Table 3. Heart rate variability parameters between ultra-short and short term in moderate obstructive sleep apnea.

	Ultra-Short Term	Short Term	p Value
Ln TP	7.43 ± 0.35	7.30 ± 0.46	<0.001
Ln VLF	7.19 ± 0.58	7.13 ± 0.53	0.152
Ln LF	7.56 ± 1.08	7.41 ± 1.00	0.071
Ln HF	5.18 ± 1.24	4.71 ± 1.45	<0.001
Ln LF/Ln HF	1.60 ± 0.64	1.78 ± 0.75	0.001

Total power, Ln TP; very low frequency, Ln VLF; low frequency, Ln LF; high frequency, Ln HF.

Table 4. Heart rate variability parameters between ultra-short and short term in severe obstructive sleep apnea.

	Ultra-Short Term	Short Term	p Value
Ln TP	7.45 ± 0.34	7.22 ± 0.50	<0.001
Ln VLF	7.19 ± 0.52	7.02 ± 0.58	<0.001
Ln LF	8.36 ± 0.97	7.95 ± 0.57	<0.001
Ln HF	1.80 ± 0.92	1.82 ± 1.06	0.849
Ln LF/Ln HF	6.93 ± 7.17	6.33 ± 4.14	0.125

Total power, Ln TP; very low frequency, Ln VLF; low frequency, Ln LF; high frequency, Ln HF.

The correlation between sleep parameters and HRV parameters is shown in Figures 1–3. We found that, in both HRV analyses (ultra-short or short-term segments), AHI was positively correlated with Ln LF and Ln LF/Ln HF, whereas there was a negative correlation between AHI and Ln HF. This trend was also detected in the ODI and ESS, although the Pearson's correlation coefficients slightly decreased from AHI to the ODI and ESS. This means that AHI, ODI, and ESS were well correlated with increased sympathetic activity and decreased parasympathetic activity, a tendency that was more prominent in the ultra-short-term analysis.

Moreover, we used Bland-Altman plots to analyze the consistency of the HRV parameters. The Bland-Altman plots for HRV analyses based on ultra-short and short-term segments are shown in Figure 4. We found that the major HRV parameters, such as Ln LF, Ln HF, and Ln LF/Ln HF, were mostly located within the upper and lower 2.0 SD lines. This means that the agreement of HRV parameters between the ultra-short and short-term segments was substantial.

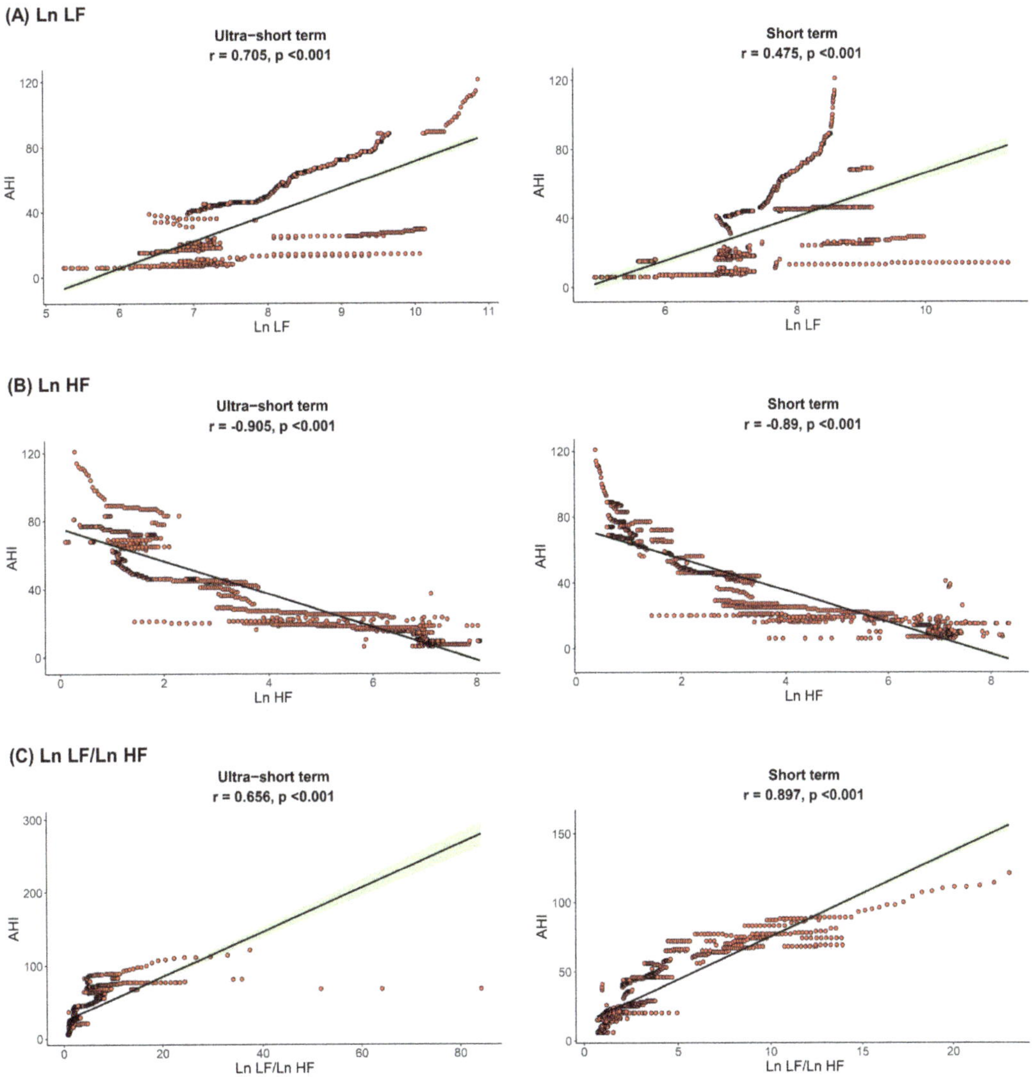

Figure 1. Correlation between AHI (**A**) Ln LF, (**B**) Ln HF, (**C**) Ln LF/Ln HF. Apnea-hypopnea index, AHI; total power, Ln TP; very low frequency, Ln VLF; low frequency, Ln LF; high frequency, Ln HF.

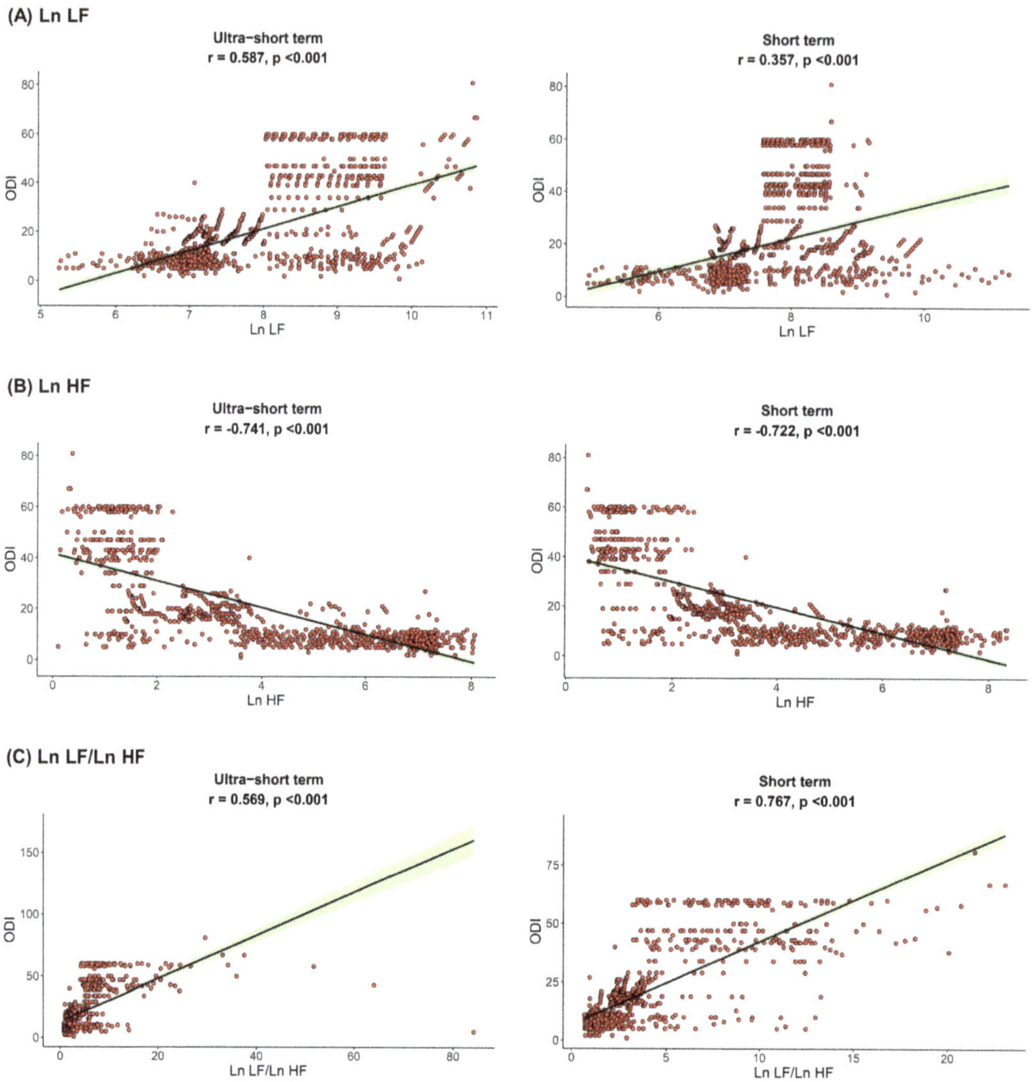

Figure 2. Correlation between ODI and (**A**) Ln LF, (**B**) Ln HF, (**C**) Ln LF/Ln HF. Oxygen desaturation index, ODI; total power, Ln TP; very low frequency, Ln VLF; low frequency, Ln LF; high frequency, Ln HF.

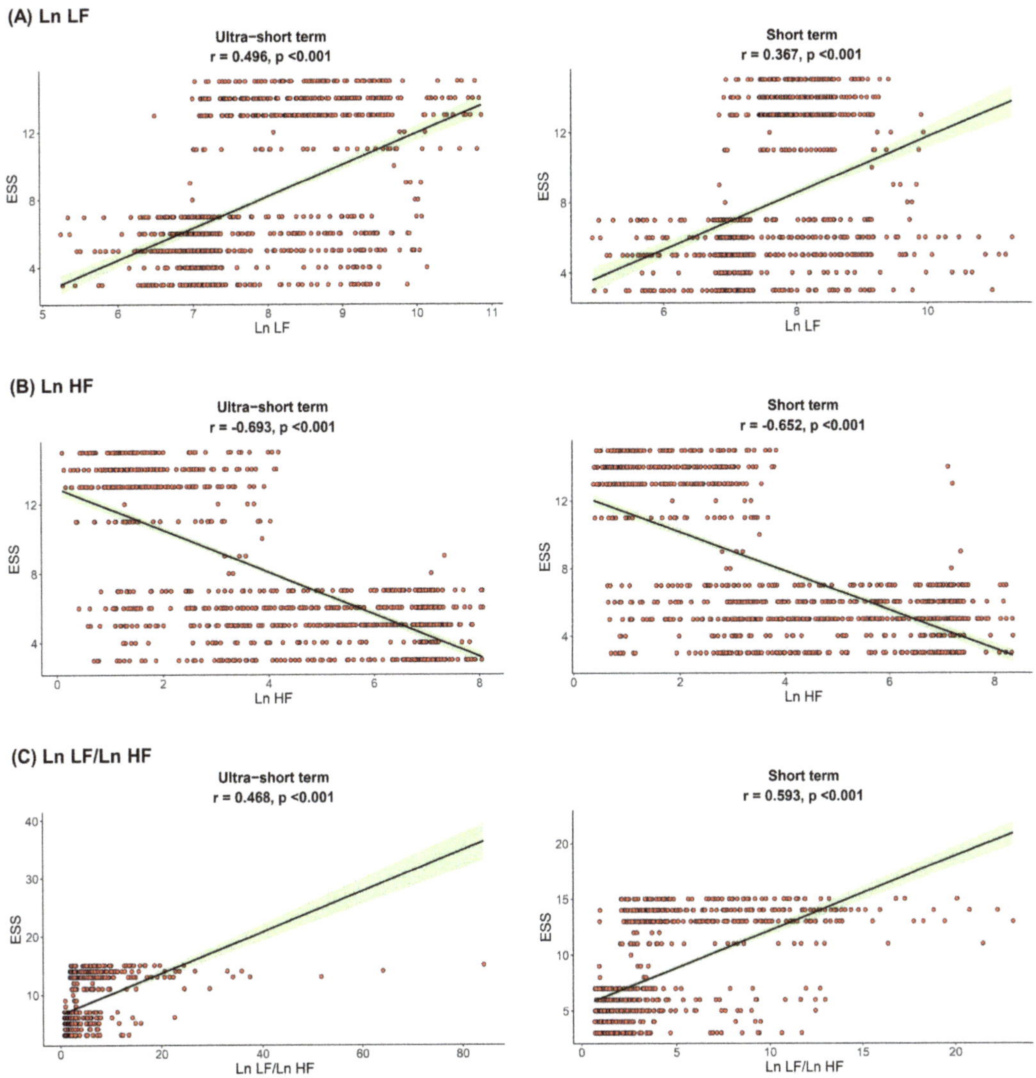

Figure 3. Correlation between ESS and (**A**) Ln LF, (**B**) Ln HF, (**C**) Ln LF/Ln H. Epworth sleepiness scale ESS, total power; Ln TP, very low frequency; Ln VLF, low frequency; Ln LF, high frequency; Ln HF, low frequency.

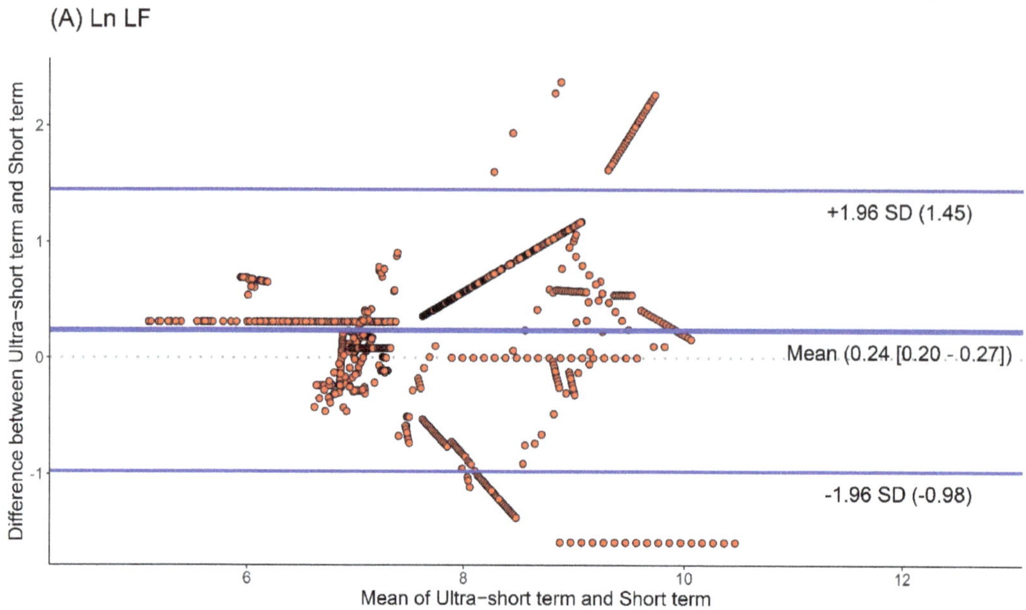

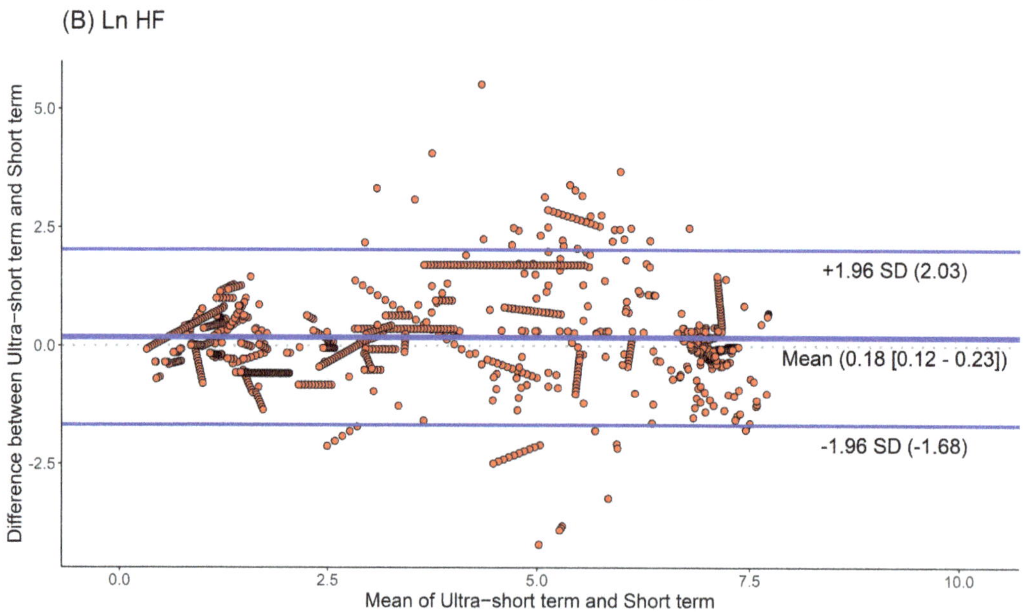

Figure 4. *Cont.*

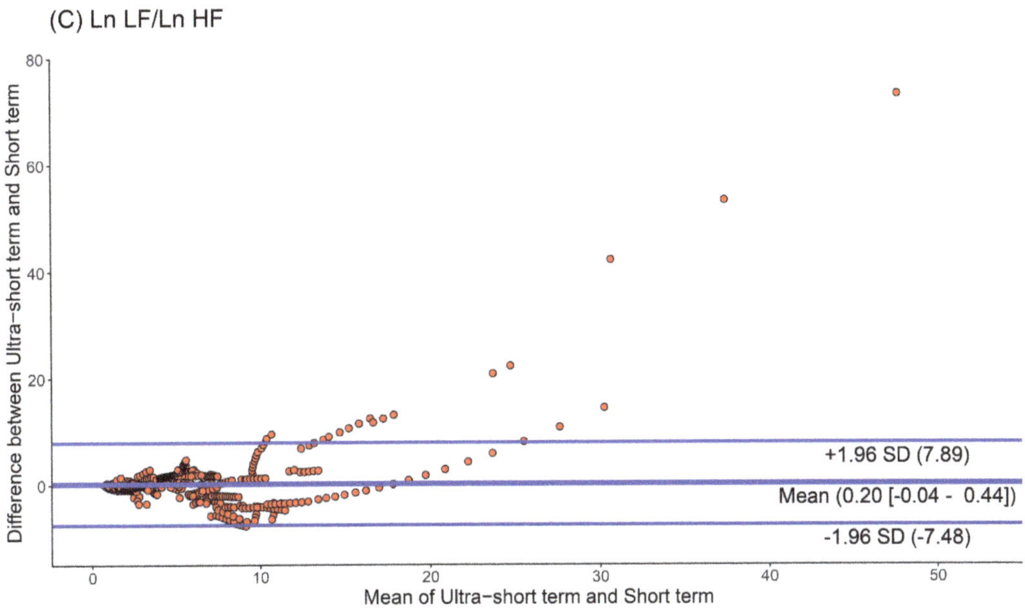

Figure 4. Agreement rate of heart rate variability parameters between ultra-short- and short-term methodology using Bland-Altman plots: (**A**) Ln LF, (**B**) Ln HF, (**C**) Ln LF/Ln HF (standard deviation, SD; low frequency, Ln LF; high frequency, Ln HF).

4. Discussion

HRV is widely used as a marker related to the ANS and is classically divided into two categories based on the length of the data recording. One is short-term HRV, typically calculated over 5 min, and the magnitude of its frequency components obtained by spectral analysis is used for the assessment of ANS function [21–24]. Another is long-term HRV computed over a nominal 24 h and its time-frequency domain is used mainly for mortality risk prediction [25–28]. However, the demand for rapid detection of ANS alterations has increased in several medical areas with the development of wearable devices [29,30]. Thus, ultra-short-term HRV analysis has been actively applied in various medical fields [16,17,31–33]. In the present study, to overcome the shortcomings of short-term methodology in HRV analysis, we investigated the effectiveness of ultra-short-term methodology in the detection of ANS alterations in patients with OSA. We found that, in both ultra-short- and short-term methodologies, all HRV parameters showed a similar tendency, which indicated sympathetic overactivity along with increased OSA severity. Additionally, we detected a significant correlation between HRV parameters and AHI, ODI, and ESS scores. The HRV feature also showed a higher agreement between the ultra-short- and short-term methodologies.

The frequency analysis of HRV typically uses three parameters: Ln VLF, Ln LF, and Ln HF. Additionally, Ln TP can be measured by the sum of the total spectral power, and it is thought to represent the global autonomic function. The Ln LF/HF ratio, as a major marker for sympathovagal balance, can also be calculated during HRV analysis. Among HRV studies of OSA, there is widespread evidence demonstrating that the alteration of ANS balance toward sympathetic predominance is well reflected in the Ln LF/HF ratio, Ln LF [8,10,34–36]. In addition, a shift in ANS function towards parasympathetic predominance was observed with increases in Ln HF [34–36]. Thus, the shift toward sympathetic predominance demonstrated by HRV analysis of patients with OSA could suggest a higher cardiovascular risk during waking status.

It is well known that HRV features obtained from short-term segments show smoother spectral profiles than those obtained from ultra-short-term segments. Thus, short-term methodology may be easy to evaluate in an entirely autonomic state during sleep. However, short-term HRV analysis has inherent limitations in real-world settings. First, short-term HRV-based studies are highly susceptible to noise. Second their failure to detect dynamic ANS activity, especially the rapid changes in Ln LF and Ln HF. However, ultra-short-term HRV analysis could assess the immediate changes in RR intervals on electrocardiograms before and after the period of respiratory events. For these reasons, to investigate sophisticated sleep studies, short-term HRV analysis is not recommended. Although ultra-short-term HRV analyses have not yet been adopted as a standard tool in OSA, a recent study showed that higher ultra-short-term HRV was strongly related to longer respiratory event durations [32]. In this study, we detected substantial agreement in HRV parameters between ultra-short and short-term methodologies. Interestingly, this alteration in HRV value was more prominent with the ultra-short-term method, regardless of OSA severity. Consistent with our findings, ultra-short-term HRV parameters were generally higher with longer respiratory events, regardless of the respiratory event type [37]. Therefore, our findings may be used as a diagnostic marker for the alteration of ANS activity in patients with OSA.

However, our study had several limitations. First, only time-domain HRV parameters were assessed for ANS function, because frequency-domain HRV parameters require a longer RR interval segment for the analysis. Second, this study did not include data on HRV parameters in control subjects. However, to include the control group, we should match the age, sex, BMI, and comorbidities; thus, it is very difficult in the real-world study. Third, in this study, we obtained HRV features from only the linear method, not using the non-linear method. Thus, our further study should include the comparison of HRV features between the linear and the non-linear methodologies. Fourth, several comorbidities could influence the inflammatory status of OSA patients and these conditions could be confounding factors in the HRV analysis [38,39]. Finally, our data need to be used to further evaluate the effect of OSA treatment on HRV features.

5. Conclusions

In conclusion, ultra-short-term HRV features are valid surrogates of short-term HRV features for investigating ANS function in OSA patients. Therefore, this study provides valuable insights into the balance of ANS activity in OSA patients.

Author Contributions: Conceptualization, D.-K.K.; methodology, S.-S.H.; software, S.-S.H.; validation, D.-K.K.; formal analysis, S.-S.H.; resources, S.-S.H.; data curation, D.-K.K.; writing—original draft preparation, D.-K.K.; writing—review and editing, D.-K.K.; visualization, S.-S.H.; supervision, D.-K.K.; project administration, S.-S.H.; and funding acquisition, D.-K.K. All authors have read and agreed to the published version of the manuscript.

Funding: This research was funded by the Bio and Medical Technology Development Program of the National Research Foundation, the Korean Government (MSIT, NRF-2021R1C1C1005746), and a grant from the Hallym University Research Fund.

Institutional Review Board Statement: This study adhered to the tenets of the Declaration of Helsinki and was approved by the institutional review board of Hallym Medical University Chuncheon Sacred Hospital (IRB No. 2020-03-022).

Informed Consent Statement: Informed consent was obtained from all subjects involved in the study.

Data Availability Statement: The authors confirm that data supporting the findings of this study are available within the article.

Acknowledgments: The authors would like to thank all the members of the New Frontier Research Team for their assistance.

Conflicts of Interest: The authors declare no conflict of interest.

References

1. Somers, V.K.; Dyken, M.E.; Clary, M.P.; Abboud, F.M. Sympathetic neural mechanisms in obstructive sleep apnea. *J. Clin. Investig.* **1995**, *96*, 1897–1904. [CrossRef] [PubMed]
2. Partinen, M.; Jamieson, A.; Guilleminault, C. Long-term outcome for obstructive sleep apnea syndrome patients. *Mortal. Chest* **1988**, *94*, 1200–1204. [CrossRef] [PubMed]
3. Smith, M.L.; Muenter, N.K. Effects of hypoxia on sympathetic neural control in humans. *Respir. Physiol.* **2000**, *121*, 163–171. [CrossRef]
4. Garvey, J.F.; Taylor, C.T.; McNicholas, W.T. Cardiovascular disease in obstructive sleep apnoea syndrome: The role of intermittent hypoxia and inflammation. *Eur. Respir. J.* **2009**, *33*, 1195–1205. [CrossRef] [PubMed]
5. Bradley, T.D.; Floras, J.S. Obstructive sleep apnoea and its cardiovascular consequences. *Lancet* **2009**, *373*, 82–93. [CrossRef]
6. Kasai, T.; Floras, J.S.; Bradley, T.D. Sleep apnea and cardiovascular disease: A bidirectional relationship. *Circulation* **2012**, *126*, 1495–1510. [CrossRef]
7. Heart rate variability: Standards of measurement, physiological interpretation and clinical use. Task Force of the European Society of Cardiology and the North American Society of pacing and electrophysiology. *Circulation* **1996**, *93*, 1043–1065. [CrossRef]
8. Kim, Y.S.; Kim, S.Y.; Park, D.Y.; Wu, H.W.; Hwang, G.S.; Kim, H.J. Clinical implication of heart rate variability in obstructive sleep apnea syndrome patients. *J. Craniofac. Surg.* **2015**, *26*, 1592–1595. [CrossRef]
9. Gula, L.J.; Krahn, A.D.; Skanes, A.; Ferguson, K.A.; George, C.; Yee, R.; Klein, G.J. Heart rate variability in obstructive sleep apnea: A prospective study and frequency domain analysis. *Ann. Noninvasive Electrocardiol.* **2003**, *8*, 144–149. [CrossRef]
10. Aydin, M.; Altin, R.; Ozeren, A.; Kart, L.; Bilge, M.; Unalacak, M. Cardiac autonomic activity in obstructive sleep apnea: Time-dependent and spectral analysis of heart rate variability using 24-h Holter electrocardiograms. *Tex. Heart Inst. J.* **2004**, *31*, 132–136.
11. Qin, H.; Keenan, B.T.; Mazzotti, D.R.; Vaquerizo-Villar, F.; Kraemer, J.F.; Wessel, N.; Tufik, S.; Bittencourt, L.; Cistulli, P.A.; de Chazal, P.; et al. Heart rate variability during wakefulness as a marker of obstructive sleep apnea severity. *Sleep* **2021**, *44*. [CrossRef] [PubMed]
12. Noda, A.; Hayano, J.; Ito, N.; Miyata, S.; Yasuma, F.; Yasuda, Y. Very low frequency component of heart rate variability as a marker for therapeutic efficacy in patients with obstructive sleep apnea: Preliminary study [Preliminary study]. *J. Res. Med. Sci.* **2019**, *24*, 84. [CrossRef]
13. Ravelo-García, A.G.; Saavedra-Santana, P.; Juliá-Serdá, G.; Navarro-Mesa, J.L.; Navarro-Esteva, J.; Álvarez-López, X.; Gapelyuk, A.; Penzel, T.; Wessel, N. Symbolic dynamics marker of heart rate variability combined with clinical variables enhance obstructive sleep apnea screening. *Chaos* **2014**, *24*, 024404. [CrossRef] [PubMed]
14. Hilton, M.F.; Bates, R.A.; Godfrey, K.R.; Chappell, M.J.; Cayton, R.M. Evaluation of frequency and time-frequency spectral analysis of heart rate variability as a diagnostic marker of the sleep apnoea syndrome. *Med. Biol. Eng. Comput.* **1999**, *37*, 760–769. [CrossRef] [PubMed]
15. Hayano, J.; Yuda, E. Pitfalls of assessment of autonomic function by heart rate variability. *J. Physiol. Anthropol.* **2019**, *38*, 3. [CrossRef]
16. Castaldo, R.; Montesinos, L.; Melillo, P.; James, C.; Pecchia, L. Ultra-short term HRV features as surrogates of short term HRV: A case study on mental stress detection in real life. *BMC Med. Inform. Decis. Mak.* **2019**, *19*, 12. [CrossRef]
17. Lee, S.; Hwang, H.B.; Park, S.; Kim, S.; Ha, J.H.; Jang, Y.; Hwang, S.; Park, H.K.; Lee, J.; Kim, I.Y. Mental stress assessment using ultra short term HRV analysis based on non-linear method. *Biosensors* **2022**, *12*, 465. [CrossRef]
18. Castaldo, R.; Xu, W.; Melillo, P.; Pecchia, L.; Santamaria, L.; James, C. Detection of mental stress due to oral academic examination via ultra-short-term HRV analysis. *Annu. Int. Conf. IEEE Eng. Med. Biol. Soc.* **2016**, *2016*, 3805–3808. [CrossRef]
19. Berry, R.B.; Budhiraja, R.; Gottlieb, D.J.; Gozal, D.; Iber, C.; Kapur, V.K.; Marcus, C.L.; Mehra, R.; Parthasarathy, S.; Quan, S.F.; et al. Rules for scoring respiratory events in sleep: Update of the 2007 AASM Manual for the Scoring of Sleep and Associated Events. Deliberations of the sleep Apnea Definitions Task Force of the American Academy of Sleep Medicine. *J. Clin. Sleep Med.* **2012**, *8*, 597–619. [CrossRef]
20. Rajendra Acharya, U.; Paul Joseph, K.; Kannathal, N.; Lim, C.M.; Suri, J.S. Heart rate variability: A review. *Med. Biol. Eng. Comput.* **2006**, *44*, 1031–1051. [CrossRef]
21. Altuncu, M.E.; Baspinar, O.; Keskin, M. The use of short-term analysis of heart rate variability to assess autonomic function in obese children and its relationship with metabolic syndrome. *Cardiol. J.* **2012**, *19*, 501–506. [CrossRef] [PubMed]
22. Lutfi, M.F. Autonomic modulations in patients with bronchial asthma based on short-term heart rate variability. *Lung India* **2012**, *29*, 254–258. [CrossRef] [PubMed]
23. Na, K.S.; Lee, N.Y.; Park, S.H.; Park, C.K. Autonomic dysfunction in normal tension glaucoma: The short-term heart rate variability analysis. *J. Glaucoma* **2010**, *19*, 377–381. [CrossRef] [PubMed]
24. He, B.; Li, W.; Zhang, X.; Wu, Y.; Liu, J.; Brewer, L.M.; Yu, L. The analysis of how apnea influences the autonomic nervous system using short-term heart rate variability indices. *J. Healthc. Eng.* **2020**, *2020*, 6503715. [CrossRef] [PubMed]
25. Fei, L.; Copie, X.; Malik, M.; Camm, A.J. Short- and long-term assessment of heart rate variability for risk stratification after acute myocardial infarction. *Am. J. Cardiol.* **1996**, *77*, 681–684. [CrossRef]
26. Fei, L.; Statters, D.J.; Anderson, M.H.; Malik, M.; Camm, A.J. Relationship between short- and long-term measurements of heart rate variability in patients at risk of sudden cardiac death. *Pacing Clin. Electrophysiol.* **1994**, *17*, 2194–2200. [CrossRef]

27. Melillo, P.; De Luca, N.; Bracale, M.; Pecchia, L. Classification tree for risk assessment in patients suffering from congestive heart failure via long-term heart rate variability. *IEEE J. Biomed. Health Inform.* **2013**, *17*, 727–733. [CrossRef]
28. Shahbazi, F.; Asl, B.M. Generalized discriminant analysis for congestive heart failure risk assessment based on long-term heart rate variability. *Comput. Methods Programs Biomed.* **2015**, *122*, 191–198. [CrossRef]
29. Kumari, P.; Mathew, L.; Syal, P. Increasing trend of wearables and multimodal interface for human activity monitoring: A review. *Biosens. Bioelectron.* **2017**, *90*, 298–307. [CrossRef]
30. Finžgar, M.; Podržaj, P. Feasibility of assessing ultra-short-term pulse rate variability from video recordings. *PeerJ* **2020**, *8*, e8342. [CrossRef]
31. Moya-Ramon, M.; Mateo-March, M.; Peña-González, I.; Zabala, M.; Javaloyes, A. Validity and reliability of different smartphones applications to measure HRV during short and ultra-short measurements in elite athletes. *Comput. Methods Programs Biomed.* **2022**, *217*, 106696. [CrossRef] [PubMed]
32. Hietakoste, S.; Korkalainen, H.; Kainulainen, S.; Sillanmäki, S.; Nikkonen, S.; Myllymaa, S.; Duce, B.; Töyräs, J.; Leppänen, T. Longer apneas and hypopneas are associated with greater ultra-short-term HRV in obstructive sleep apnea. *Sci. Rep. Sci. Rep.* **2020**, *10*, 21556. [CrossRef] [PubMed]
33. Wu, L.; Shi, P.; Yu, H.; Liu, Y. An optimization study of the ultra-short period for HRV analysis at rest and post-exercise. *J. Electrocardiol.* **2020**, *63*, 57–63. [CrossRef] [PubMed]
34. Aytemir, K.; Deniz, A.; Yavuz, B.; Ugur Demir, A.; Sahiner, L.; Ciftci, O.; Tokgozoglu, L.; Can, I.; Sahin, A.; Oto, A. Increased myocardial vulnerability and autonomic nervous system imbalance in obstructive sleep apnea syndrome. *Respir. Med.* **2007**, *101*, 1277–1282. [CrossRef] [PubMed]
35. Xie, J.; Yu, W.; Wan, Z.; Han, F.; Wang, Q.; Chen, R. Correlation analysis between obstructive sleep apnea syndrome (OSAS) and heart rate variability. *Iran J. Public Health.* **2017**, *46*, 1502–1511.
36. Busek, P.; Vanková, J.; Opavský, J.; Salinger, J.; Nevsímalová, S. Spectral analysis of the heart rate variability in sleep. *Physiol. Res.* **2005**, *54*, 369–376. [CrossRef]
37. Chouchou, F.; Pichot, V.; Barthélémy, J.C.; Bastuji, H.; Roche, F. Cardiac sympathetic modulation in response to apneas/hypopneas through heart rate variability analysis. *PLoS ONE* **2014**, *9*, e86434. [CrossRef]
38. Kim, D.K.; Lee, B.C.; Park, K.J.; Son, G.M. Effect of Obstructive Sleep Apnea on Immunity in Cases of Chronic Rhinosinusitis With Nasal Polyps. *Clin. Exp. Otorhinolaryngol.* **2021**, *14*, 390–398. [CrossRef]
39. Liu, X.; Ma, Y.; Ouyang, R.; Zeng, Z.; Zhan, Z.; Lu, H.; Cui, Y.; Dai, Z.; Luo, L.; He, C.; et al. The relationship between inflammation and neurocognitive dysfunction in obstructive sleep apnea syndrome. *J. Neuroinflammation* **2020**, *17*, 229. [CrossRef]

Journal of *Personalized Medicine*

MDPI

Article

Risk Factors of 30-Day All-Cause Mortality in Patients with Carbapenem-Resistant *Klebsiella pneumoniae* Bloodstream Infection

Keh-Sen Liu [1], Yao-Shen Tong [2], Ming-Tsung Lee [3,4], Hung-Yu Lin [3] and Min-Chi Lu [5,6,*]

[1] Division of Infectious Diseases, Department of Internal Medicine, Show Chwan Memorial Hospital, Changhua 500, Taiwan; liumilka2@gmail.com
[2] Department of Medical Laboratory, Show Chwan Memorial Hospital, Changhua 500, Taiwan; yaushen51@gmail.com
[3] Research Assistant Center, Show Chwan Memorial Hospital, Changhua 500, Taiwan; lee6717kimo@yahoo.com.tw (M.-T.L.); linhungyu700218@gmail.com (H.-Y.L.)
[4] Department of Nursing, Hungkuang University, Taichung 433, Taiwan
[5] Division of Infectious Diseases, Department of Internal Medicine, China Medical University Hospital, Taichung 404, Taiwan
[6] Department of Microbiology and Immunology, School of Medicine, China Medical University, Taichung 404, Taiwan
* Correspondence: luminchi@gmail.com

Citation: Liu, K.-S.; Tong, Y.-S.; Lee, M.-T.; Lin, H.-Y.; Lu, M.-C. Risk Factors of 30-Day All-Cause Mortality in Patients with Carbapenem-Resistant *Klebsiella pneumoniae* Bloodstream Infection. *J. Pers. Med.* **2021**, *11*, 616. https://doi.org/10.3390/jpm11070616

Academic Editor: Anne-Marie Caminade

Received: 12 May 2021
Accepted: 27 June 2021
Published: 29 June 2021

Abstract: An optimal antimicrobial regimen for the treatment of patients with carbapenem-resistant *Klebsiella pneumoniae* (CRKP) bloodstream infection (BSI) is currently unavailable. This study aimed to identify the appropriate antibiotics and the risk factors of all-cause mortality for CRKP BSI patients. This retrospective cohort study included the hospitalized patients with CRKP BSI. Primary outcome was 30-day all-cause mortality. Cox regression analysis was used to evaluate the risk factors of 30-day mortality. A total of 89 patients were included with a 30-day mortality of 52.1%. A total of 52 (58.4%) patients were treated with appropriate antimicrobial regimens and 58 (65.2%) isolates carried $bla_{\text{KPC-2}}$ genes. Microbiologic eradication within 7 days (adjusted hazard ratio [HR] = 0.09, $p < 0.001$), platelet count (per $1 \times 10^4/\text{mm}^3$, adjusted HR = 0.95, $p = 0.002$), and Pitt bacteremia scores (adjusted HR = 1.40, $p < 0.001$) were independently associated with 30-day all-cause mortality. No effective antimicrobial regimens were identified. In conclusion, risk factors of 30-day mortality in patients with CRKP BSI included microbiologic eradication > 7 days, lower platelet count, and a higher Pitt bacteremia score. These findings render a new insight into the clinical landscape of CRKP BSI.

Keywords: bacteremia; carbapenemase; carbapenem-resistant *Enterobacteriaceae*; *Klebsiella pneumoniae*; mortality; risk factors

1. Introduction

The rapidly increasing prevalence of antibiotic resistance in *Enterobacteriaceae* is currently a major threat to public health worldwide [1]. Recently, the European Survey of Carbapenemase-Producing *Enterobacteriaceae* (EuSCAPE) Working Group investigated 2703 clinical isolates of carbapeneme-resistant *Enterobacteriaceae* (CRE) submitted from 455 sentinel hospitals in 36 countries and reported that 15% of these were *Escherichia coli* and 85% were *Klebsiella pneumoniae*. Among these Klebsiella isolates, 37% were carbapenemase-producers [2]. Four gene families encoding carbapenemase-production have been identified: *Klebsiella pneumoniae* carbapenemase (KPC), New Delhi metallo-β-lactamase, oxacillinase 48-like, and Verona integron-encoded metallo-β-lactamase [2]. Carbapenem-resistant *Klebsiella pneumoniae* (CRKP) has attracted particular attention since it was first identified as one of the multidrug resistant bacteria strains [3]. Previous studies reported that the

KPC-producing *Klebsiella pneumoniae* was independently associated with higher mortality [4,5]. Mortality rate of CRKP bloodstream infection (BSI) is reported to range from 42% to 84% [6]. Mechanistically, the acquisition of genes encoding carbapenemase-production is the predominant mechanism of carbapenem-resistance through inactivation or degradation of carbapenems [7].

Although the innovation of new drugs combating resistant bacteria may create a glimmer of hope, the evolution of drug-resistant genes soon overwhelms progression. For instance, drug resistance-combating ceftazidime-avibactam was first approved under the Generating Antibiotic Incentives Now (GAIN) Act, whereas its clinical use was soon followed by reports of resistance caused by CRKP [8–11]. Due to the rapid progression and high mortality rate, the role of CRKP can be better understood by an early examination. Nonetheless, the clinical value of CRKP and its related risk factors are yet to be clarified.

Many studies found that CRKP BSI patients treated with combined antibiotics had better survival outcomes than those treated with monotherapy [12,13]. However, potentially effective regimens are varied across studies and those that are available are very limited [14–17]. Since there is no consensus on optimal antimicrobial regimens for CRKP infection, it is crucial to understand the possible risk factors for unfavourable outcomes and potentially optimal antibiotics when treating CRKP BSI patients. In this retrospective cohort study, we attempted to identify the appropriate antimicrobial regimens and to investigate the risk factors of all-cause mortality for CRKP bloodstream infection (BSI) patients.

2. Materials and Methods

2.1. Study Design and Patients

This retrospective cohort study was ethically approved by the Institutional Review Board of Show Chwan Memorial Hospital, which is a regional hospital offering an estimated 750 beds in Taiwan (approval number: SCMH_IRB No. 1061202). Hospitalized patients diagnosed with CRKP BSI (number of patients) between June 2014 and August 2017 in Show Chwan Memorial Hospital in Taiwan were included in the study. CRKP BSI was defined by at least one blood-culture positive for a CRKP strain. The patients with polymicrobial BSI and those aged < 20 years were excluded.

2.2. Data Collection

All patient data were collected through a retrospective review of medical records. Variables included demographics, medical history, Charlson Comorbidity Index [18], admission history, initial presentations (quick sepsis related organ failure assessment [qSOFA] score [19], systemic inflammatory response syndrome [SIRS] criteria [20], Pitt bacteremia score [21], vital signs, and laboratory data), length of hospitalization, antibiotic regimens, microbiologic results, and drug susceptibilities. Sepsis was defined by qSOFA $\geq$ 2 [22] and shock was as mean arterial pressure $\leq$ 65 mmHg [23]. Only antimicrobial agents used for more than 48 h would be included for analysis. Antimicrobial regimens were classified as empiric or definitive. Empiric regimen included the prescription of antibiotics before the blood culture result was available. Definitive regimen referred to prescription of antibiotics based on drug susceptibility results. An appropriate regimen was defined as including one or more in vitro active drugs against the CRKP isolates. Primary outcome was 30-day all-cause mortality, measured from the day on which the first blood culture revealing CRKP was taken. Microbiologic eradication was defined as negative blood cultures for CRKP during follow-up.

2.3. Microbiology and Antimicrobial Susceptibilities

The CRKP isolates were collected from the first positive blood culture of each patient. Bacterial identification and antimicrobial susceptibility tests were performed using Phoenix Automated Microbiology System (Becton, Dickinson and Company, USA) and the interpretative criteria of the Clinical and Laboratory Standards Institute (CLSI) guidelines was applied. CRKP was defined as an isolate with a minimum inhibitory concentration (MIC) of $\geq$2 μg/mL for ertapenem, $\geq$4 μg/mL for meropenem, or $\geq$4 μg/mL for imipenem. Due to no available CLSI breakpoints for *Enterobacteriaceae*, MIC breakpoints by the European Committee on Antimicrobial Susceptibility Testing were used for the interpretation of colistin susceptibility (susceptible: $\leq$2 mg/L, resistant: >2 mg/L) [24] and tigecycline susceptibility was determined by MIC breakpoints by the USA Food and Drug Administration (susceptible: MIC $\leq$ 2 μg/mL, intermediate: 4 μg/mL, resistant: MIC $\geq$ 8 μg/mL) [25]. Detection for bla_{KPC} gene and $bla_{OXA\text{-}48}$ gene were performed by polymerase chain reaction with specific primers, as previously described [26]. Pulse-field gel electrophoresis (PFGE) was used to detect the relatedness of the CRKP strains. The profiles of XbaI macro-restricted fragments of each strain were determined by a standardized PulseNet PFGE protocol [27]. The BioNumerics version 6.6 (Applied Maths, Belgium) was used to analyze the PFGE profiles. The relatedness was based on PFGE profiles using the dice coefficients and the unweighted pair group method with arithmetic mean algorithm. The optimization value and position tolerance were set at 1.5% and 0.75%, respectively.

2.4. Statistical Analysis

Continuous variables were presented as median with interquartile range (IQR) and categorical variables were presented as a percentage. The comparison between the survival group and mortality group was performed using the Mann–Whitney U test and Fisher's exact test for continuous and categorical variables, respectively. Mortality rate was estimated using the Kaplan–Meier method and compared using the log-rank test. A univariate and multivariate Cox proportional hazard model was used to evaluate the risk factors of 30-day mortality. The variables with $p < 0.1$ in the univariate model were manually selected into the multivariate model in a backward stepwise manner. The results of the Cox proportional hazard model were presented as hazard ratio (HR) with 95% confidence interval (CI). A p-value of <0.05 was considered statistically significant. IBM SPSS Statistics for Windows, version 24.0 (IBM Corp., Armonk, NY, USA) was used for the statistical analyses.

3. Results

3.1. Patient Characteristics and Antimicrobial Treatment

As shown in Table 1, a total of 89 patients (59.6% male, median age 75.6 years) with CRKP BSI were included: 43 (48.3%) patients in the 30-day survival group and 46 (51.7%) in the 30-day mortality group. The 7-day, 14-day, and 30-day all-cause mortality of all patients were 32.6%, 43.9%, and 51.7%, respectively. Within 48 h of admission, 23 out of 89 (25.8%) had a positive sign of blood culture testing for CRKP, indicating that it was community acquired. The 66 out of 89 (74.1%) patients who presented a positive sign of blood culture testing for CRKP after 48 h of admission were regarded as nosocomial infections. Survival group patients had a significantly lower Charlson comorbidity index, Pitt bacteremia score, and albumin levels. The proportion of sepsis and septic shock was lower in the survival group, while platelet counts were higher. Pre-existing cardiovascular disease and hospitalizations during the prior year was more uncommon among survivors (Table 1).

The results of antimicrobial susceptibility revealed that 52 (58.4%) were treated with the appropriate antimicrobial regimens, including 9 (10.1%) as empiric therapy and additionally 50 (56.2%) as definite regimens. Colistin (31.5%) and carbapenem (41.6%) were the most commonly prescribed antibiotics. Patients in the survival group had a significant proportion of microbiological eradication within 7 days, as compared to the mortality group (72.1% vs. 13.0%, $p < 0.001$) (Table 2). However, the proportion of appropriate regimen, length of hospitalization, and the pattern of antibiotics utilization did not significantly differ between the two groups (Table 2).

Table 1. Demographics and baseline characteristics in patients with carbapenem-resistant *Klebsiella pneumoniae* bloodstream infection according to 30-day mortality.

Variables	All (n = 89)	Survival Group (n = 43) (48.3%)	Mortality Group (n = 46) (51.7%)	p Value
Demographics				
Male; n (%)	53 (59.6%)	26 (60.5%)	27 (58.7%)	1.000
Age; y/o (IQR)	75.6 (63.8–83.7)	74.5 (62.3–83.4)	80.5 (65.8–85.2)	0.153
BMI; kg/m^2 (IQR)	21.3 (18.3–25.9)	20.8 (18.2–24.8)	21.4 (19.2–30.0)	0.223
Comorbidities				
Charlson comorbidity index; score (IQR)	8 (6–10)	7 (6–9)	8 (7–11)	0.008
Diabetes mellitus; n (%)	53 (59.6%)	24 (55.8%)	29 (63.0%)	0.523
Cardiovascular disease; n (%)	57 (64.0%)	34 (79.1%)	23 (50.0%)	0.008
Chronic obstructive pulmonary disease; n (%)	21 (23.6%)	9 (21.4%)	12 (26.1%)	0.628
Chronic liver disease; n (%)	15 (16.9%)	4 (9.5%)	11 (23.9%)	0.092
Chronic kidney disease; n (%)	15 (16.9%)	9 (21.4%)	6 (13.0%)	0.397
Malignancy; n (%)	18 (20.2%)	7 (16.7%)	11 (23.9%)	0.439
Steroid use $\geq$ 3 months; n (%)	15 (16.9%)	4 (9.5%)	11 (23.9%)	0.092
Immunocompromised condition; n (%)	8 (9.0%)	3 (7.1%)	5 (10.9%)	0.716
Events in the prior year				
Hospitalization; events (%)	65 (73.0%)	36 (85.7%)	29 (64.4%)	0.028
Admitted to intensive care units; events (%)	30 (33.7%)	16 (38.1%)	14 (30.4%)	0.504
Nursing home residence; events (%)	31 (34.8%)	14 (32.6%)	17 (37.0%)	0.824
CRKP colonization; events (%)	14 (15.7%)	7 (16.7%)	7 (15.2%)	1.000
Surgery; events (%)	24 (27.0%)	12 (27.9%)	12 (26.1%)	1.000
Initial presentation				
Pitt bacteremia score; score (IQR)	4 (2–6)	3 (1–4)	6 (4–8)	< 0.001
SIRS; score (IQR)	3 (2–4)	3 (2–3)	3 (2–4)	0.273
SIRS $\geq$ 2; n (%)	76 (85.4%)	35 (81.4%)	41 (89.1%)	0.375
qSOFA; score (IQR)	2 (1–3)	1 (1–2)	2 (2–3)	< 0.001
Sepsis; n (%)	54 (60.7%)	18 (41.9%)	36 (78.3%)	0.001
Septic shock; n (%)	27 (30.3%)	7 (16.3%)	20 (43.5%)	0.006
Body temperature $\geq$38 °C; n (%)	54 (60.7%)	27 (62.8%)	27 (58.7%)	0.828
White blood cell count (10^3/mm^3); count (IQR)	13.2 (8.7–17.8)	13.7 (10.1–17.8)	10.7 (6.4–18.0)	0.249
Hemoglobin (g/dL); value (IQR)	9.8 (9.0–10.6)	9.8 (9.1–11)	9.7 (8.6–10.5)	0.352
Platelet count (10^4/mm^3); count (IQR)	12.4 (5.2–23.8)	17.5 (10.8–26.3)	8.15 (3.8–18.2)	0.002
Creatinine (mg/dL); value (IQR)	1.3 (0.8–2.8)	1 (0.7–2.8)	1.4 (0.9–2.8)	0.088
Albumin (g/dL); value (IQR)	2.6 (2.2–2.9)	2.9 (2.5–3.0)	2.4 (2.1–2.7)	< 0.001
C-reactive protein (mg/dL); value (IQR)	11.6 (4.9–19.4)	10.8 (3.6–19.3)	12.7 (8.0–19.6)	0.256
Microbiology				
Presence of KPC gene; n (%)	58 (65.2%)	22 (51.2%)	36 (78.3%)	0.008
Clonal relatedness of CRKP strain				0.006
Cluster I; n (%)	11 (12.4%)	8 (18.6%)	3 (6.5%)	
Cluster II; n (%)	62 (69.7%)	23 (53.5%)	39 (84.8%)	
Others; n (%)	16 (18.0%)	12 (27.9%)	4 (8.7%)	

Abbreviations: BMI, body mass index; CRKP, carbapenem-resistant *Klebsiella pneumoniae*; IQR, interquartile range; KPC, *Klebsiella pneumoniae* carbapenemase; qSOFA, quick sepsis related organ failure assessment; SIRS, systemic inflammatory response syndrome.

3.2. Microbiology Revealed Redcued Proportion of KPC Genes and KPC Cluster II in the Survival Group

Among the 89 CRKP strains, 58 (65.2%) isolates carried KPC genes, and all were bla_{KPC-2} genes. According to the results of PFGE, two distinct clusters could be identified from the CRKP strains (Figure 1): 11 (12.4%) strains in the cluster I, 62 (69.7%) in cluster II, and others (18.0%). All the strains carrying KPC genes were in cluster II. None of the cluster I and II strains had bla_{OXA-48} genes. Notably, we noted that the survival group had a reduced proportion of KPC genes (51.2% vs. 78.3%, p = 0.008) and the cluster II strains (53.5% vs. 84.8%, p = 0.006) as compared to the mortality group (Table 1).

Table 2. Treatment in patients with carbapenem-resistant *Klebsiella pneumoniae* bloodstream infection according to 30-day mortality.

Variables	All (n = 89)	Survival Group (n = 43)	Mortality Group (n = 46)	p Value
Appropriate antimicrobial regimen; n (%)	52 (58.4%)	28 (65.1%)	24 (52.2%)	0.283
Colistin included; n (%)	28 (31.5%)	10 (23.3%)	18 (39.1%)	0.006
Amikacin included; n (%)	13 (14.6%)	9 (20.9%)	4 (8.7%)	0.336
Carbapenem included; n (%)	37 (41.6%)	17 (39.5%)	20 (43.5%)	0.124
Tigecycline included; n (%)	3 (3.4%)	0 (0%)	3 (6.5%)	0.092
Monotherapy; n (%)	12 (13.5%)	11 (25.6%)	1 (2.2%)	0.003
Appropriate empiric regimen; n (%)	9 (10.1%)	3 (7.0%)	6 (13.0%)	0.487
Appropriate definitive regimen; n (%)	50 (56.2%)	28 (65.1%)	22 (47.8%)	0.135
Microbiologic eradication within 7 days; n (%)	37 (41.6%)	31 (72.1%)	6 (13.0%)	<0.001
Length of hospitalization; days (IQR)	21.5 (13.0–34.0)	21.5 (14.0–40.0)	21.5 (13.0–33.0)	0.483

Abbreviations: CRKP, carbapenem-resistant *Klebsiella pneumoniae*; IQR, interquartile range; KPC, *Klebsiella pneumoniae* carbapenemase.

3.3. Comparison of Antimicrobial Susceptibility

All of the 89 isolates were resistant to at least one of the three carbapenems. For the 89 isolates, colistin exerted the highest susceptibility rate (96.6%), followed by amikacin (88.8%), tigecycline (82.0%), meropenem (22.5%), levofloxacin (11.2%), ertapenem (7.9%), and imipenem (5.6%) (Figure 2A). The pattern of drug susceptibility in the survival group differed from that of the mortality group (Figure 2A). The pattern difference was observed between CRKP strains with and without KPC genes (Figure 2B). The survival group had a significantly higher susceptibility rate of imipenem (11.6% vs. 0%, $p = 0.023$) and meropenem (34.9% vs. 10.9%, $p = 0.01$) than the mortality group (Figure 2A). Except for colistin, tigecycline, and amikacin, the susceptibility rate of all drugs were significantly lower in the CRKP strains with KPC genes than in the strains without KPC genes.

3.4. Risk Factors of 30-Day All-Cause Mortality

For patients with CRKP BSI, multivariate analysis revealed that microbiologic eradication within 7 days (adjusted HR = 0.09, $p < 0.001$), platelet count (per $1 \times 10^4/\text{mm}^3$, adjusted HR = 0.95, $p = 0.002$), and Pitt bacteremia score (adjusted HR = 1.40, $p < 0.001$) were independently associated with 30-day all-cause mortality. The univariate Kaplan-Meier method revealed that CRKP patients with KPC genes had significantly higher 30-day mortality than those without KPC genes (62.6% vs. 32.4%, $p = 0.022$) (Figure 3). However, the predictive role of the presence of KPC genes was not confirmed by the multivariate model. None of the antimicrobial regimens was significantly associated with 30-day mortality (Table 3).

Table 3. Multivariate analysis for risk factors of 30-day mortality in patients with carbapenem-resistant *Klebsiella pneumoniae* bloodstream infection.

Variables	Adjusted Hazard Ratio	95% Confidence Interval	p
Microbiologic eradication within 7 days (yes vs. no)	0.09	0.03–0.26	<0.001
Platelet count (per $1 \times 10^4/\text{mm}^3$)	0.95	0.92–0.98	0.002
Pitt bacteremia score (per 1 unit)	1.40	1.21–1.61	<0.001

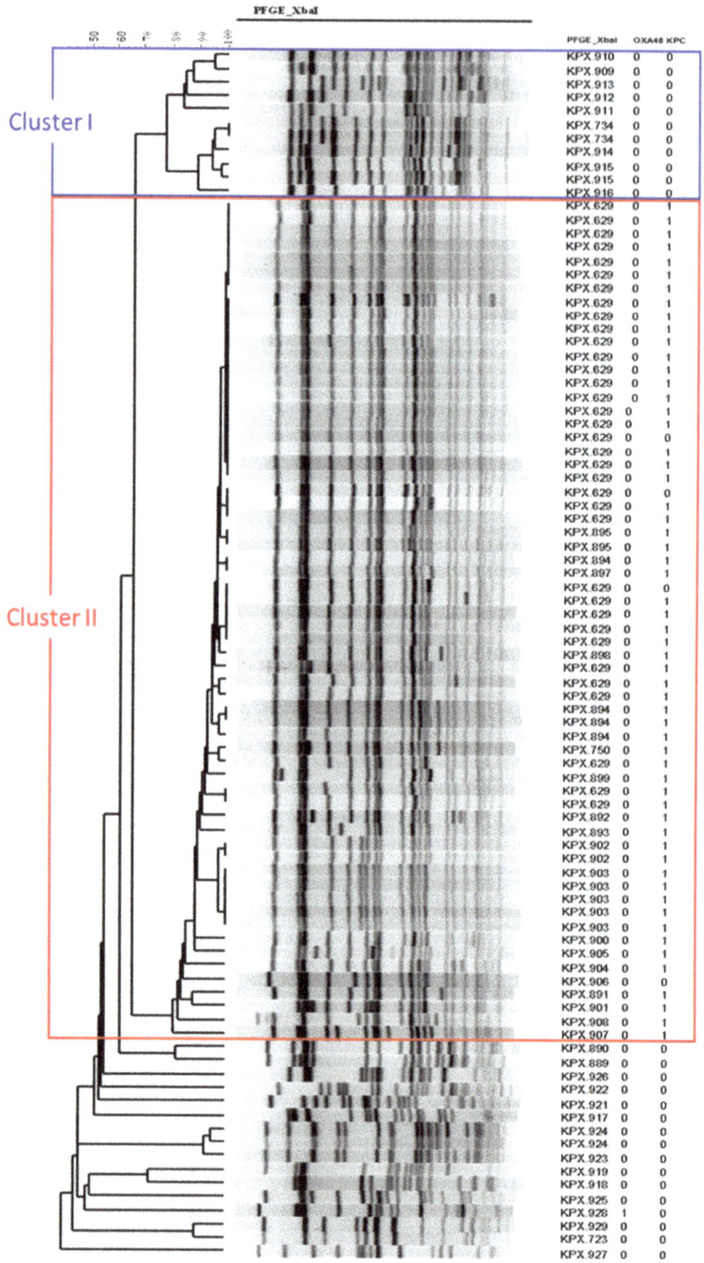

Figure 1. Clonal relatedness of the carbapenemase-producing *Klebsiella pneumoniae* strains. Pulse-field gel electrophoresis (PFGE) was used to determine the profiles of XbaI macro-restricted fragments of each strain were determined by a standardized PulseNet PFGE protocol.

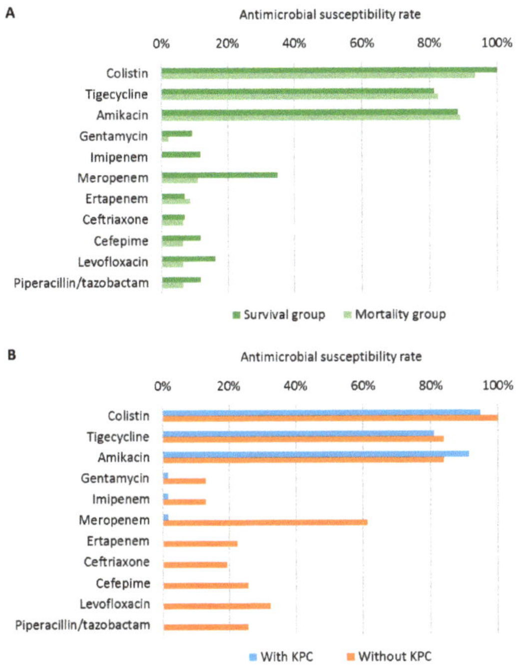

Figure 2. Antimicrobial susceptibility of carbapenemase-producing *Klebsiella pneumonia* isolates according to 30-day mortality (**A**) and the presence of *Klebsiella pneumonia* carbapenemase (KPC) genes (**B**). KPC, *Klebsiella pneumoniae* carbapenemase.

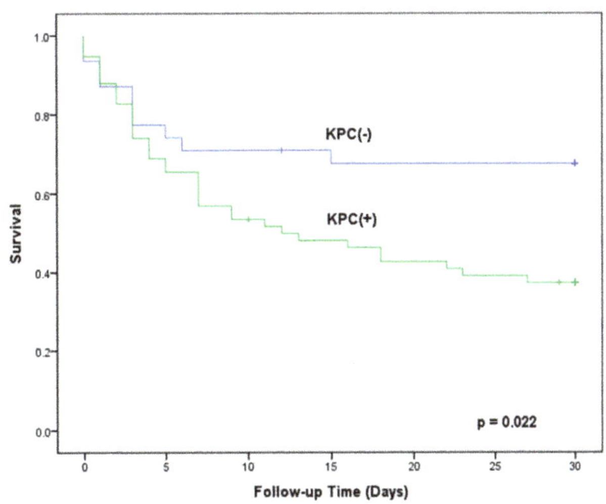

Figure 3. Kaplan-Meier curve showing the 30-day in-hospital survival for bacteremic patients with carbapenem-resistant *Klebsiella pneumoniae* (CRKP) carrying KPC gene as compared to CRKP not carrying KPC gene.

4. Discussion

CRKP BSI is a clinical challenge as no effective antimicrobial regimens are currently available. This study observed a 30-day all-cause mortality of 52.1% in patients with CRKP BSI. The lack of microbiologic eradication within 7 days, a lower platelet count, and a higher Pitt bacteremia score were independently associated with higher 30-day mortality. Notably, the predictive role of KPC genes and appropriate antibiotic regimens were not identified in this study.

All the KPC genes detected in this study were $bla_{KPC\text{-}2}$, which was consistent with the previous nationwide surveillance in Taiwan that $bla_{KPC\text{-}2}$ accounted for the majority of KPC genes [28]. However, in the present study, the prevalence of $bla_{KPC\text{-}2}$ gene among CRKP strains was 65.2%, much higher than the previous prevalence of 36.2% in 2011–2015 [29]. The results suggested the KPC-2-producing CRKP strains disseminate rapidly at an alarming rate in Taiwan. In addition, this study reveals that non-KPC-producing strains present high susceptibility to imipenem, meropenem, ertapenem, ceftriaxone, cefepime, levofloxacin, and piperacillin/tazobactam, as compared to that of KPC-producing strains. Although the results might suggest the role of KPC in the treatment of CRKP BSI, multivariate analysis did not find an association of KPC with mortality. Wang et al. reported similar results that KPC-gene positive and negative strains had different MICs of antibiotics, but mortality did not differ between these two CRKP strains [30]. Some studies suggested CRKP infection as an important risk factor of hospital mortality [4,5]; however, the role of KPC deserves further investigations for patients with CRKP BSI.

This study revealed that a higher Pitt bacteremia score was independently associated with 30-day mortality in patients with CRKP BSI. The result was comparable with the study of Shen et al. and Xiao et al. [31,32] Gomez-Simmonds et al. and Lee et al. also reported that the patients with Pitt bacteremia score > 4 had significantly higher mortality [15,17] Additionally, our study revealed an association of lower platelet count with higher mortality, which might also reflect the severity of illness [33]. Some studies used different tools to evaluate the severity of illness at the presence of CRKP BSI, for example, APACHE II score; a higher APACHE II score was reported to be independently associated with 30-day mortality [13,34]. All these results supported the predictive value of the severity of illness for 30-day morality in patients with CRKP BSI.

Both our results and that of Nguyen et al. revealed that the patients with microbiologic eradication within 7 days had a significantly better survival rate [35]. Falcone et al. further evaluated the effect of time to appropriate antibiotic therapy on 30-day mortality and showed that time to appropriate antibiotic therapy was an independent predictor of 30-day mortality in patients with CRKP BSI. Falcone et al. suggested that appropriate antibiotic therapy was preferably initiated within the first 24 h after collection of the blood culture [36]. From our study and the previous research, the importance of timely control of the disease is evident.

Many studies have attempted to identify the best combination of antibiotics for the treatment of CRKP BSI, but the results are mixed and there is no consistent conclusion [14–17]. Our results revealed that the most susceptible drug was colistin, followed by tigecycline and amikacin [13,15,31,35,37]. Medeiros et al. observed that the combination of colistin and amikacin could provide survival benefits for the patients with CRKP BSI [38]. However, both colistin [39] and amikacin [40] are nephrotoxic agents. Most patients with CRKP BSI are older, have multiple comorbid diseases, and frequently present with septic shock [13,35,41]. The use of colistin or amikacin, or in combination, can further aggravate nephrotoxicity [39,40]. This combination is reasonable on the basis of only antimicrobial susceptibility without consideration of patients' condition and drug adverse effects. Clinicians are usually reluctant to use the combination because of the high risk of renal injury, which may lead to dialysis and increase subsequent morbidity and mortality. Although tigecycline causes minimal organ toxicity, the use of tigecycline for Gram-negative bacteremia remains controversial. The serum concentrations provided by standard doses of tigecycline are below the MICs of most Gram-negative pathogens [42]. Therefore, the most common combinations are colistin/tigecycline, aminoglycoside/tigecycline, colistin/carbapenem, aminoglycoside/carbapenem, and a combination of three drugs (e.g., colistin/tigecycline/carbapenem) [12,13]. However, it lacks consensus on the best combination for the treatment of CRKP BSI. Ceftazidime-avibactam is a new combination of third generation cephalosporin and non-β-lactam β-lactamase inhibitor, and has promising in vitro activity against many Gram-negative pathogens, including KPC-producing *Enterobacteriaceae* [43]. It is hoped that the new drugs will not be misused or overused to avoid the development of antibacterial resistance [44].

The major caveat of this study is the observational design, in which potential reporting bias and selection could not be avoided. Secondly, all patients were diagnosed and treated in a single institute, which limited the external validity of the results. Thirdly, this study focused on the patients with CRKP BSI instead of other types of infection, such as urinary tract infection, pneumonia, or intra-abdominal infection. Fourthly, the definition of appropriate regimens in this study was the inclusion of one or more efficient antibiotics, instead of two or more, and the time to appropriate regimens was not evaluated. This might explain the results that the use of appropriate regimen was not significantly associated with 30-day mortality.

5. Conclusions

This retrospective cohort study observed that the risk factors of 30-day all-cause mortality in patients with CRKP BSI included microbiologic eradication > 7 days, a lower platelet count, and a higher Pitt bacteremia score. These findings render new insights into the clinical landscape of CRKP BSI.

Author Contributions: Conceptualization, K.-S.L. and M.-C.L.; methodology, Y.-S.T.; software, M.-T.L.; validation, H.-Y.L.; formal analysis, Y.-S.T.; investigation, K.-S.L.; resources, K.-S.L.; data curation, Y.-S.T.; writing—original draft preparation, K.-S.L.; writing—review and editing, M.-C.L.; visualization, H.-Y.L.; supervision, M.-C.L.; project administration, K.-S.L.; funding acquisition, K.-S.L. All authors have read and agreed to the published version of the manuscript.

Funding: This research received no external funding.

Institutional Review Board Statement: The study was conducted according to the guidelines of the Declaration of Helsinki and approved by the Institutional Review Board of Show Chwan Memorial Hospital, Taiwan (protocol code SCMH_IRB No. 1061202; date of approval: 2 December 2017).

Informed Consent Statement: Patient consent was waived due to the retrospective study design.

Data Availability Statement: The data used to support the findings of this study are available from the corresponding author upon request.

Acknowledgments: The authors would like to thank the technical support from Hsing-Ju Wu, director of the Research Assistant Center, Show Chwan Memorial Hospital, Taiwan.

Conflicts of Interest: The authors declare no conflict of interest.

References

1. World Health Organization. *Antimicrobial Resistance: Global Report on Surveillance*; World Health Organization: Geneva, Switzerland, 2014.
2. Grundmann, H.; Glasner, C.; Albiger, B.; Aanensen, D.M.; Tomlinson, C.T.; Andrasević, A.T.; Cantón, R.; Carmeli, Y.; Friedrich, A.W.; Giske, C.G.; et al. Occurrence of carbapenemase-producing *Klebsiella pneumoniae* and *Escherichia coli* in the European survey of carbapenemase-producing *Enterobacteriaceae* (EuSCAPE): A prospective, multinational study. *Lancet Infect. Dis.* **2017**, *17*, 153–163. [CrossRef]
3. Yigit, H.; Queenan, A.M.; Anderson, G.J.; Domenech-Sanchez, A.; Biddle, J.W.; Steward, C.D.; Alberti, S.; Bush, K.; Tenover, F.C. Novel carbapenem-hydrolyzing beta-lactamase, KPC-1, from a carbapenem-resistant strain of *Klebsiella pneumoniae. Antimicrob. Agents Chemother.* **2001**, *45*, 1151–1161. [CrossRef]
4. Gasink, L.B.; Edelstein, P.; Lautenbach, E.; Synnestvedt, M.; Fishman, N.O. Risk Factors and Clinical Impact of *Klebsiella pneumoniae* Carbapenemase-Producing *K. pneumoniae. Infect. Control. Hosp. Epidemiol.* **2009**, *30*, 1180–1185. [CrossRef]

5. Mouloudi, E.; Protonotariou, E.; Zagorianou, A.; Iosifidis, E.; Karapanagiotou, A.; Giasnetsova, T.; Tsioka, A.; Roilides, E.; Sofianou, D.; Gritsi-Gerogianni, N. Bloodstream Infections Caused by Metallo-β-Lactamase/*Klebsiella pneumoniae* Carbapenemase–Producing *K. pneumoniae* among Intensive Care Unit Patients in Greece: Risk Factors for Infection and Impact of Type of Resistance on Outcomes. *Infect. Control. Hosp. Epidemiol.* **2010**, *31*, 1250–1256. [CrossRef]
6. Xu, L.; Sun, X.; Ma, X. Systematic review and meta-analysis of mortality of patients infected with carbapenem-resistant *Klebsiella pneumoniae. Ann. Clin. Microbiol. Antimicrob.* **2017**, *16*, 1–12. [CrossRef]
7. Nguyen, M.; Joshi, S.G. Carbapenem resistance in *Acinetobacter baumannii*, and their importance in hospital-acquired infections: A scientific review. *J. Appl. Microbiol.* **2021**. [CrossRef] [PubMed]
8. Gaibani, P.; Campoli, C.; Lewis, R.E.; Volpe, S.L.; Scaltriti, E.; Giannella, M.; Pongolini, S.; Berlingeri, A.; Cristini, F.; Bartoletti, M.; et al. In vivo evolution of resistant subpopulations of KPC-producing *Klebsiella pneumoniae* during ceftazidime/avibactam treatment. *J. Antimicrob. Chemother.* **2018**, *73*, 1525–1529. [CrossRef] [PubMed]
9. Humphries, R.M.; Hemarajata, P. Resistance to Ceftazidime-Avibactam in *Klebsiella pneumoniae* Due to Porin Mutations and the Increased Expression of KPC-3. *Antimicrob. Agents Chemother.* **2017**, *61*, 6. [CrossRef] [PubMed]
10. Nelson, K.; Hemarajata, P.; Sun, D.; Rubio-Aparicio, D.; Tsivkovski, R.; Yang, S.; Sebra, R.; Kasarskis, A.; Nguyen, H.; Hanson, B.M.; et al. Resistance to Ceftazidime-Avibactam Is Due to Transposition of KPC in a Porin-Deficient Strain of *Klebsiella pneumoniae* with Increased Efflux Activity. *Antimicrob. Agents Chemother.* **2017**, *61*, e00989-17. [CrossRef] [PubMed]
11. Tumbarello, M.; Trecarichi, E.M.; Corona, A.; DE Rosa, F.G.; Bassetti, M.; Mussini, C.; Menichetti, F.; Viscoli, C.; Campoli, C.; Venditti, M.; et al. Efficacy of Ceftazidime-Avibactam Salvage Therapy in Patients with Infections Caused by *Klebsiella pneumoniae* Carbapenemase–producing *K. pneumoniae. Clin. Infect. Dis.* **2019**, *68*, 355–364. [CrossRef]
12. Gutiérrez-Gutiérrez, B.; Salamanca, E.; de Cueto, M.; Hsueh, P.-R.; Viale, P.; Paño-Pardo, J.R.; Venditti, M.; Tumbarello, M.; Daikos, G.; Cantón, R.; et al. Effect of appropriate combination therapy on mortality of patients with bloodstream infections due to carbapenemase-producing *Enterobacteriaceae* (INCREMENT): A retrospective cohort study. *Lancet Infect. Dis.* **2017**, *17*, 726–734. [CrossRef]
13. Tumbarello, M.; Viale, P.; Viscoli, C.; Trecarichi, E.M.; Tumietto, F.; Marchese, A.; Spanu, T.; Ambretti, S.; Ginocchio, F.; Cristini, F.; et al. Predictors of Mortality in Bloodstream Infections Caused by *Klebsiella pneumoniae* Carbapenemase-Producing *K. pneumoniae*: Importance of Combination Therapy. *Clin. Infect. Dis.* **2012**, *55*, 943–950. [CrossRef]
14. Geng, T.T.; Xu, X.; Huang, M. High-dose tigecycline for the treatment of nosocomial carbapenem-resistant *Klebsiella pneumoniae* bloodstream infections: A retrospective cohort study. *Medicine* **2018**, *97*, e9961. [CrossRef]
15. Gomez-Simmonds, A.; Nelson, B.; Eiras, D.P.; Loo, A.; Jenkins, S.G.; Whittier, S.; Calfee, D.P.; Satlin, M.J.; Kubin, C.J.; Furuya, E.Y. Combination Regimens for Treatment of Carbapenem-Resistant *Klebsiella pneumoniae* Bloodstream Infections. *Antimicrob. Agents Chemother.* **2016**, *60*, 3601–3607. [CrossRef]
16. Gonzalez-Padilla, M.; Torre-Cisneros, J.; Rivera-Espinar, F.; Pontes-Moreno, A.; López-Cerero, L.; Pascual, A.; Natera, C.; Rodríguez, M.; Salcedo, I.; Rodríguez-López, F.; et al. Gentamicin therapy for sepsis due to carbapenem-resistant and colistin-resistant *Klebsiella pneumoniae. J. Antimicrob. Chemother.* **2014**, *70*, 905–913. [CrossRef]
17. Lee, N.-Y.; Tsai, C.-S.T.; Syue, L.-S.; Chen, P.-L.; Li, C.-W.; Li, M.-C.; Ko, W.-C. Treatment Outcome of Bacteremia Due to Non–Carbapenemase-producing Carbapenem-Resistant *Klebsiella pneumoniae* Bacteremia: Role of Carbapenem Combination Therapy. *Clin. Ther.* **2020**, *42*, e33–e44. [CrossRef]
18. Charlson, M.E.; Pompei, P.; Ales, K.L.; MacKenzie, C. A new method of classifying prognostic comorbidity in longitudinal studies: Development and validation. *J. Chronic Dis.* **1987**, *40*, 373–383. [CrossRef]
19. Seymour, C.W.; Liu, V.X.; Iwashyna, T.J.; Brunkhorst, F.M.; Rea, T.D.; Scherag, A.; Rubenfeld, G.; Kahn, J.M.; Shankar-Hari, M.; Singer, M.; et al. Assessment of Clinical Criteria for Sepsis: For the Third International Consensus Definitions for Sepsis and Septic Shock (Sepsis-3). *JAMA* **2016**, *315*, 762–774. [CrossRef] [PubMed]
20. Bone, R.C.; Balk, R.A.; Cerra, F.B.; Dellinger, R.P.; Fein, A.M.; Knaus, W.A.; Schein, R.M.; Sibbald, W.J. Definitions for sepsis and organ failure and guidelines for the use of innovative therapies in sepsis. *Chest* **1992**, *101*, 1644–1655. [CrossRef] [PubMed]
21. Rhee, J.Y.; Kwon, K.T.; Ki, H.K.; Shin, S.Y.; Jung, D.S.; Chung, D.R.; Ha, B.C.; Peck, K.R.; Song, J.H. Scoring systems for prediction of mortality in patients with intensive care unit-acquired sepsis: A comparison of the Pitt bacteremia score and the Acute Physiology and Chronic Health Evaluation II scoring systems. *Shock* **2009**, *31*, 146–150. [CrossRef] [PubMed]
22. Singer, M.; Deutschman, C.S.; Seymour, C.W.; Shankar-Hari, M.; Annane, D.; Bauer, M.; Bellomo, R.; Bernard, G.R.; Chiche, J.-D.; Coopersmith, C.M.; et al. The Third International Consensus Definitions for Sepsis and Septic Shock (Sepsis-3). *JAMA* **2016**, *315*, 801–810. [CrossRef] [PubMed]
23. Rhodes, A.A.; Evans, L.E.; Alhazzani, W.; Levy, M.M.; Antonelli, M.; Ferrer, R.; Kumar, A.; Sevransky, J.E.; Sprung, C.L.; Nunnally, M.E.; et al. Surviving Sepsis Campaign: International Guidelines for Management of Sepsis and Septic Shock: 2016. *Crit. Care Med.* **2017**, *45*, 486–552. [CrossRef]
24. The European Committee on Antimicrobial Susceptibility Testing. Breakpoint Tables for Interpretation of MICs and Zone Diameters, Version 11.0. 2021. Available online: https://www.eucast.org/clinical_breakpoints/ (accessed on 10 April 2021).
25. Wyeth Pharmaceuticals Inc. *TYGACIL® (Tigecycline)*; Wyeth Pharmaceuticals Inc.: Madison, NJ, USA, 2005.
26. Poirel, L.; Walsh, T.; Cuvillier, V.; Nordmann, P. Multiplex PCR for detection of acquired carbapenemase genes. *Diagn. Microbiol. Infect. Dis.* **2011**, *70*, 119–123. [CrossRef] [PubMed]

27. Kuo, H.-C.; Lauderdale, T.-L.; Lo, D.-Y.; Chen, C.-L.; Chen, P.-C.; Liang, S.-Y.; Kuo, J.-C.; Liao, Y.-S.; Liao, C.-H.; Tsao, C.-S.; et al. An Association of Genotypes and Antimicrobial Resistance Patterns among Salmonella Isolates from Pigs and Humans in Taiwan. *PLoS ONE* **2014**, *9*, e95772. [CrossRef] [PubMed]
28. Chiu, S.K.; Wu, T.L.; Chuang, Y.C.; Lin, J.C.; Fung, C.P.; Lu, P.L.; Wang, J.T.; Wang, L.S.; Siu, L.K.; Yeh, K.M. National surveillance study on carbapenem non-susceptible *Klebsiella pneumoniae* in Taiwan: The emergence and rapid dissemination of KPC-2 carbapenemase. *PLoS ONE* **2013**, *8*, e69428.
29. Lu, M.-C.; Tang, H.-L.; Chiou, C.-S.; Wang, Y.-C.; Chiang, M.-K.; Lai, Y.-C. Clonal dissemination of carbapenemase-producing *Klebsiella pneumoniae*: Two distinct sub-lineages of Sequence Type 11 carrying blaKPC-2 and blaOXA-48. *Int. J. Antimicrob. Agents* **2018**, *52*, 658–662. [CrossRef]
30. Wang, Z.; Qin, R.-R.; Huang, L.; Sun, L.-Y. Risk Factors for Carbapenem-resistant *Klebsiella pneumoniae* Infection and Mortality of *Klebsiella pneumoniae* Infection. *Chin. Med. J.* **2018**, *131*, 56–62. [CrossRef]
31. Shen, L.; Lian, C.; Zhu, B.; Yao, Y.; Yang, Q.; Zhou, J.; Zhou, H. Bloodstream Infections due to Carbapenem-Resistant *Klebsiella pneumoniae*: A Single-Center Retrospective Study on Risk Factors and Therapy Options. *Microb. Drug Resist.* **2021**, *27*, 227–233. [CrossRef]
32. Xiao, T.; Zhu, Y.; Zhang, S.; Wang, Y.; Shen, P.; Zhou, Y.; Yu, X.; Xiao, Y. A Retrospective Analysis of Risk Factors and Outcomes of Carbapenem-Resistant *Klebsiella pneumoniae* Bacteremia in Nontransplant Patients. *J. Infect. Dis.* **2020**, *221* (Suppl. S2), S174–S183. [CrossRef] [PubMed]
33. Zhang, S.; Cui, Y.-L.; Diao, M.-Y.; Chen, D.-C.; Lin, Z.-F. Use of Platelet Indices for Determining Illness Severity and Predicting Prognosis in Critically Ill Patients. *Chin. Med. J.* **2015**, *128*, 2012–2018. [CrossRef]
34. Chuang, C.; Su, C.-F.; Lin, J.-C.; Lu, P.-L.; Huang, C.-T.; Wang, J.-T.; Chuang, Y.-C.; Siu, L.K.; Fung, C.-P.; Lin, Y.-T. Does Antimicrobial Therapy Affect Mortality of Patients with Carbapenem-Resistant *Klebsiella pneumoniae* Bacteriuria? A Nationwide Multicenter Study in Taiwan. *Microorganisms* **2020**, *8*, 2035. [CrossRef]
35. Nguyen, M.; Eschenauer, G.A.; Bryan, M.; O'Neil, K.; Furuya, E.Y.; Della-Latta, P.; Kubin, C.J. Carbapenem-resistant *Klebsiella pneumoniae* bacteremia: Factors correlated with clinical and microbiologic outcomes. *Diagn. Microbiol. Infect. Dis.* **2010**, *67*, 180–184. [CrossRef]
36. Falcone, M.; Bassetti, M.; Tiseo, G.; Giordano, C.; Nencini, E.; Russo, A.; Graziano, E.; Tagliaferri, E.; Leonildi, A.; Barnini, S.; et al. Time to appropriate antibiotic therapy is a predictor of outcome in patients with bloodstream infection caused by KPC-producing *Klebsiella pneumoniae*. *Crit. Care* **2020**, *24*, 1–12. [CrossRef]
37. Patel, G.; Huprikar, S.; Factor, S.H.; Jenkins, S.G.; Calfee, D.P. Outcomes of Carbapenem-Resistant *Klebsiella pneumoniae* Infection and the Impact of Antimicrobial and Adjunctive Therapies. *Infect. Control. Hosp. Epidemiol.* **2008**, *29*, 1099–1106. [CrossRef]
38. Medeiros, G.S.; Rigatto, M.H.; Falci, D.R.; Zavascki, A.P. Combination therapy with polymyxin B for carbapenemase-producing *Klebsiella pneumoniae* bloodstream infection. *Int. J. Antimicrob. Agents* **2019**, *53*, 152–157. [CrossRef] [PubMed]
39. Javan, A.O.; Shokouhi, S.; Sahraei, Z. A review on colistin nephrotoxicity. *Eur. J. Clin. Pharmacol.* **2015**, *71*, 801–810. [CrossRef]
40. Oliveira, J.F.P.; Silva, C.A.; Barbieri, C.D.; Oliveira, G.M.; Zanetta, D.M.T.; Burdmann, E.A. Prevalence and Risk Factors for Aminoglycoside Nephrotoxicity in Intensive Care Units. *Antimicrob. Agents Chemother.* **2009**, *53*, 2887–2891. [CrossRef] [PubMed]
41. Vardakas, K.Z.; Matthaiou, D.K.; Falagas, M.E.; Antypa, E.; Koteli, A.; Antoniadou, E. Characteristics, risk factors and outcomes of carbapenem-resistant *Klebsiella pneumoniae* infections in the intensive care unit. *J. Infect.* **2015**, *70*, 592–599. [CrossRef] [PubMed]
42. Stein, G.E.; Babinchak, T. Tigecycline: An update. *Diagn. Microbiol. Infect. Dis.* **2013**, *75*, 331–336. [CrossRef]
43. Shirley, M. Ceftazidime-Avibactam: A Review in the Treatment of Serious Gram-Negative Bacterial Infections. *Drugs* **2018**, *78*, 675–692. [CrossRef]
44. Holmes, A.H.; Moore, L.S.P.; Sundsfjord, A.; Steinbakk, M.; Regmi, S.; Karkey, A.; Guerin, P.J.; Piddock, L.J.V. Understanding the mechanisms and drivers of antimicrobial resistance. *Lancet* **2016**, *387*, 176–187. [CrossRef]

Journal of
Personalized Medicine

Article

Thiopurine S-Methyltransferase Polymorphisms Predict Hepatotoxicity in Azathioprine-Treated Patients with Autoimmune Diseases

Heh-Shiang Sheu [1], Yi-Ming Chen [1,2,3,4,5,6], Yi-Ju Liao [6], Chia-Yi Wei [2], Jun-Peng Chen [2], Hsueh-Ju Lin [2], Wei-Ting Hung [1,4,7,*], Wen-Nan Huang [1,3,4,8] and Yi-Hsing Chen [1,3]

1 Division of Allergy, Immunology and Rheumatology, Department of Internal Medicine, Taichung Veterans General Hospital, Taichung 40705, Taiwan
2 Department of Medical Research, Taichung Veterans General Hospital, Taichung 40705, Taiwan
3 School of Medicine, National Yang-Ming Chiao Tung University, Taipei 30010, Taiwan
4 Department of Post-Baccalaureate Medicine, College of Medicine, National Chung Hsing University, Taichung 40227, Taiwan
5 Rong Hsing Research Center for Translational Medicine & Ph.D. Program in Translational Medicine, National Chung Hsing University, Taichung 40227, Taiwan
6 Department of Pharmacy, Taichung Veterans General Hospital, Taichung 40705, Taiwan
7 Department of Medical Education, Taichung Veterans General Hospital, Taichung 40705, Taiwan
8 College of Business and Management, Ling Tung University, Taichung 408284, Taiwan
* Correspondence: wtinghung@gmail.com; Tel.: +886-4-2359-2525 (ext. 4304)

Abstract: Thiopurine methyltransferase (TPMT) is the rate-limiting enzyme in Azathioprine (AZA) metabolization. Although studies have discussed the association between the TPMT polymorphisms and myelosuppression, the data about the relationship between TPMT genotypes and hepatoxicity in Asian patients remain limited. This study investigated the correlation between TPMT polymorphisms and AZA-related hepatotoxicity. This study enrolled the patients who had prior exposure to AZA from the Taichung Veterans General Hospital (TCVGH)-Taiwan Precision Medicine Initiative (TPMI) cohort. Genetic variants were determined using a single nucleotide polymorphism (SNP) array. Participants were accordingly categorized into normal metabolizer (NM) and non-normal metabolizer (non-NM) groups. From the TCVGH-TPMI cohort, we included 50 TPMT non-NM patients, including 1 poor metabolizer (PM), 49 intermediate metabolizers (IMs), and 1000 NM patients. The non-NM genotype was associated with hepatotoxicity compared with the NM genotype (hazard ratio (HR): 3.85, 95% confidence interval (CI): 1.83–8.10). In the non-NM group, the 3-year cumulative incidence of hepatotoxicity was higher than that in the NM group at 8.5% in the first year and 18.6% in the second and third years ($p < 0.001$). A TPMT non-NM genotype was associated with the occurrence of hepatotoxicity following AZA therapy. Preemptive testing helps individualize AZA therapy by minimizing the risk of hepatotoxicity.

Keywords: TPMT genotype; TPMT poor metabolizers; TPMT intermediate metabolizers; AZA; hepatotoxicity; cumulative incidence of hepatotoxicity; pharmacogenomics; autoimmune disease; preemptive genotyping; TPMI

Citation: Sheu, H.-S.; Chen, Y.-M.; Liao, Y.-J.; Wei, C.-Y.; Chen, J.-P.; Lin, H.-J.; Hung, W.-T.; Huang, W.-N.; Chen, Y.-H. Thiopurine S-Methyltransferase Polymorphisms Predict Hepatotoxicity in Azathioprine-Treated Patients with Autoimmune Diseases. *J. Pers. Med.* **2022**, *12*, 1399. https://doi.org/10.3390/jpm12091399

Academic Editor: Anne-Marie Caminade

Received: 15 July 2022
Accepted: 26 August 2022
Published: 28 August 2022

1. Introduction

Azathioprine (AZA) is a valuable steroid-sparing immunosuppressant that is broadly used for systemic lupus erythematosus (SLE) [1–3], severe rheumatoid arthritis (RA), inflammatory bowel disease (IBD) [4–7], autoimmune hepatitis (AIH) [8], dermatomyositis (DM)/polymyositis, pemphigus [9], and post-transplant rejection. Azathioprine is a prodrug. Initially, AZA is metabolized to 6-mercaptopurine (6-MP) by glutathione S-transferase (GST). Three competitive enzymatic pathways metabolize 6-MP. First, xanthine oxidase (XO) catalyzes 6-MP to thiouric acid, an inactive metabolite. Second, thiopurine S-methyltransferase (TPMT), methylates 6-MP into 6-methylmercaptopurine (6-MMP).

Third, hypoxanthine guanine phosphoribosyltransferase (HPRT) converts 6-MP into 6-thioinosine monophosphate (6-TIMP). Inosine monophosphate dehydrogenase (IMPDH) dehydrogenizes 6-TIMP into 6-thioxanthosine monophosphate (6-TXMP). Subsequently, guanosine monophosphate synthetase (GMPS) metabolizes 6-TXMP to 6-thioguanine nucleotides (6TGNs). 6TGNs, the primary active metabolites, integrate into DNA and RNA molecules to generate cytotoxic and therapeutic effects [10,11]. Patients with intermediate TPMT activity show 50% more 6-TGNs than those with normal or high TPMT activity. Low or absent TPMT activity can lead to further accumulation of 6-TGNs. Deficient enzyme activity may increase AZA-related side effects. AZA-related side effects include myelosuppression, which occurs in 3–17% of the patients [12,13], hepatotoxicity in up to 10% of patients [14], gastrointestinal adverse reactions (nausea, vomiting, and diarrhea), and hair loss.

Genetic polymorphisms of the TMPT gene have been studied extensively [15]. In the Caucasian population, approximately 4–11% of individuals have intermediate TPMT activity, while approximately 1 in 300 (0.3%) have very low or absent TPMT activity [16,17]. In the Asian population, the frequency of TPMT gene mutations is approximately 1.5–3%. TPMT polymorphisms were significantly associated with AZA-induced bone marrow toxicity [18–20]. Two meta-analyses by Liu et al. [19,20] reported the incidence of the TPMT polymorphisms was not associated with AZA-induced hepatotoxicity in patients with autoimmune diseases and IBD. However, the populations in these studies discussing AZA-induced hepatotoxicity were mainly Caucasian. The enrolled Asian population was limited. We may need a larger Asian population to evaluate the association between the incidence of AZA-induced hepatotoxicity and TPMT polymorphisms.

This retrospective case-control study aimed to evaluate whether there is a relationship between TPMT polymorphisms and the incidence of AZA-related hepatotoxicity.

2. Material and Methods

2.1. Study Design

This retrospective case-control study was performed in a single medical center in Taiwan with the approval of the Ethics Committee of Taichung Veterans General Hospital (SF19153A). Each participant offered written informed consent before study participation. This study was conducted according to the Declaration of Helsinki.

2.2. Study Population

The study participants were from the Taiwan Precision Medicine Initiative (TPMI), which collected information and specimens from Taiwanese volunteers from 15 hospitals throughout the nation. The tertiary referral medical center, Taichung Veterans General Hospital (TCVGH), contributed to the majority of the TPMI cohort. A total of 43,035 patients who were >20 years old were enrolled in the TPMI study from June 2019 to August 2021. A total of 2128 patients with present or previous exposure to AZA were selected. Patients who had not yet started therapy at the time of laboratory analysis were excluded. We excluded patients without a white blood cell (WBC) count, those without alanine transaminase (ALT) data, and those with a cancer diagnosis before taking AZA.

2.3. Data Collection

The date of AZA prescription was defined as index day. Clinical data after the index day, including age, gender, azathioprine dosage, first biochemistry profile, use of other immunosuppressive agents, comorbidities, and AZA prescription department were collected from the electric health record.

Comorbidities diagnosed once during hospitalization or at least twice in the outpatient department and six months before and after AZA prescription were included. These included hepatitis B (International classification of diseases (ICD) 10/9: B16, B18.0–B18.1, B19.10–B19.11/070.2–070.3), SLE (M32/710), Sjogren's syndrome (M35/710.2), DM(M33.0–33.1, M33.9, M36.0/710.3), polymyositis (M33.2/710.4), pemphigus (L10.0–10.9/694.4), RA (M05.7–M05.9,

M06.0, M06.2–M06.3, M06.8–M06.9, M08/714), systemic sclerosis (SSc) (M34/710.1), mixed connective tissue disease (M35.1/710.8), vasculitis (M30.0, M31.0, M31.3-M31.7/446.0, 446.2, 446.4–446.5, 446.7), pemphigoid (L12.1–12.9, K05.11/694.5–694.6), Behcet's disease (M35.2/136.1), and IBD (K50, K51, K52.9/555.0–555.9, 556.0–556.9, 558.9).

Following the AZA therapy index day, the first onset time of leukopenia and hepatotoxicity was analyzed. Leukopenia was defined as a WBC count $\leq$ 4000/uL. Hepatotoxicity was defined as an ALT level $\geq$ 150 U/L. The AZA prescription division was also analyzed and included the divisions of immunology and rheumatology (IMRH), nephrology (NEPH), neurology (NEUR), dermatology (DERM), chest medicine (CM), cardiology (CV), gastroenterology and hepatology (GI), hematology (HEMA), metabolism, endocrinology and nutrition (META), gastrointestinal surgery (GS), Thoracic Surgery (CS), colorectal surgery (CRS), otolaryngology (ENT), ophthalmology (OPH), and pediatrics (PENP).

2.4. TMPT Genotyping

Genotyping was performed using the Taiwan Biobank (TWB, established in 2012) version 2 array [21], a single nucleotide polymorphism (SNP) array designed for the Taiwanese population by choosing optimized SNPs for imputation from the whole-genome sequencing (WGS) of the Taiwan Biobank participants. The TWBv2 array straightly genotypes more than 100,000 functional variants. The SNPs rs1142345 representing TPMT alleles *3A or *3C were analyzed. Alleles *3A and *3C were the common inactivating alleles in East Asia, accounting for 98% of all no-function alleles [22]. Genotyping tests including these variants are likely to be informative for the TPMT phenotype. A patient carrying one allele with the rs1142345 variant was reported as *1/*3 and was defined as intermediate metabolizer (IM). The patient carrying two alleles with the rs1142345 variant was reported as *3/*3 and was defined as a poor metabolizer (PM). According to genotyping, the patients were classified into TPMT normal metabolizers (NMs) and non-normal metabolizers (non-NMs), which included IMs and PMs.

2.5. Statistical Analysis

Data are presented as either numbers (percentages) or as the mean $\pm$ standard deviation (SD) or as the median (interquartile range). Comparisons were performed using the Mann–Whitney U test. Categorical data are presented as numbers and percentages and were compared using the chi-square test. Cox regression analysis was used to analyze which factors are related to hepatotoxicity with adjustment for covariates (TPMT genotype, age, gender, AZA dosage, other medication (methotrexate (MTX)), and comorbidity of hepatitis B). Kaplan–Meier curves were used to determine differences in cumulative incidence of hepatotoxicity between the TPMT non-NM and NM groups. A p-value $\leq$ 0.05 was considered statistically significant. All data were analyzed using the Statistical Package for the Social Sciences (SPSS, IBM Corp., Armonk, NY, USA) version 22.0.

3. Results

3.1. Study Patient Flow

Of the 43,035 patients enrolled in the TPMI, 2128 patients with AZA exposure were selected (Figure 1). A total of 628 patients without a WBC count, without ALT data, or with a cancer diagnosis before taking AZA were excluded. The TPMT non-NM and NM groups were matched by age and gender. There were 50 TPMT non-NM patients, including 49 IMs and 1 PM, and 1000 TPMT NM patients among the AZA users.

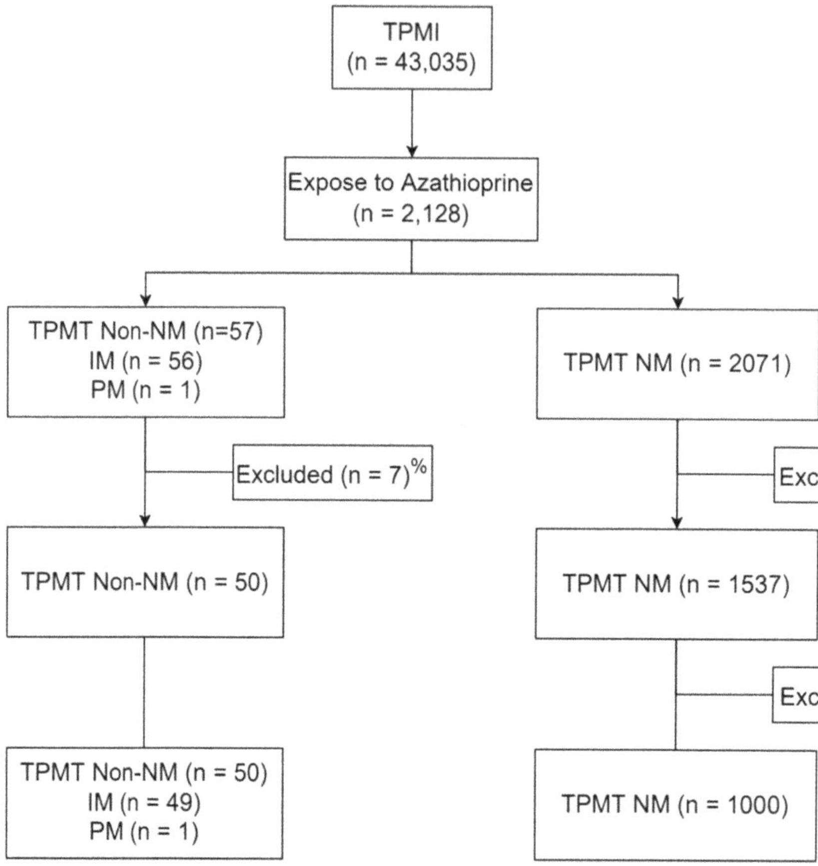

Figure 1. Patient enrollment flowchart. % 1. Exclude patients without WBC count and ALT data; 2. Exclude patients with a diagnosis of cancer before taking AZA. & Match by age and sex at 1:20. TPMI, Taiwan Precision Medicine Initiative; TPMT, thiopurine S-methyltransferase; NM, normal metabolizer; IM, intermediate metabolizer; PM, poor metabolizer; WBC, white blood cell; ALT, alanine transaminase; AZA, azathioprine.

3.2. Demographics of the Selected Patients

The patient characteristics are summarized in Table 1. Patients were classified into TPMT non-normal metabolizer (non-NM) (n = 50) and normal metabolizer (NM) (n = 1000) groups. Age, gender, AZA dosage, post-AZA treatment first mean WBC count and ALT, and AZA prescription division were similar among the TPMT non-NM and NM groups. In terms of comorbidities, more patients had hepatitis B in the TPMT non-NM group than in the NM group.

Table 1. Clinical and demographic characteristics of patients in the TPMP non-NM and NM groups.

Variable	TPMT Non-NM (n = 50)	TPMT NM (n = 1000)	p-Value
Age	51.5 ± 11.7	51.4 ± 11.6	0.95
Gender			
Female	43 (86.0)	858 (85.8)	0.97
Male	7 (14.0)	142 (14.2)	
AZA dose (mg)	36.4 ± 17.9	39.6 ± 23.47	0.35

Table 1. *Cont.*

Variable	TPMT Non-NM (n = 50)	TPMT NM (n = 1000)	p-Value
Biochemistry			
WBC count (U/L)	3920 (3600–5400)	3900 (3500–5660)	0.73
ALT (U/L)	13.5 (11–27)	13 (10–21)	0.16
Medication			
MTX	6 (12.0)	125 (12.5)	0.92
Cyclophosphamide	5 (10.0)	137 (13.7)	0.46
Comorbidity			
Hepatitis B	6 (12.0)	36 (3.6)	0.01
SLE	30 (60.0)	647 (64.7)	0.50
SS	24 (48.0)	456 (45.6)	0.74
DM and polymyositis	7 (14.0)	129 (12.9)	0.82
Pemphigus	3 (6.0)	25 (2.5)	0.14
Other autoimmune diseases ^	12 (24.0)	244 (24.4)	0.95
Division			
IMRH	42 (84)	864 (86.4)	
NEPH	5 (10)	59 (5.9)	
NEUR	2 (4)	24 (2.4)	0.52
DERM	0 (0)	27 (2.7)	
Others &	1 (2)	26 (2.6)	

Data are expressed as n (%) or the mean ± SD or the median (interquartile range). ^ Other autoimmune diseases: RA, MCTD, vasculitis, SSc, Behcet's disease, IBD. & Other divisions: CM, CV, GI, HEMA, META, GS, CS, CRS, ENT, OPH, PNEP. TPMT, thiopurine S-methyltransferase; NM, normal metabolizer; AZA, azathioprine; WBC, white blood cell; ALT, alanine transaminase; MTX, methotrexate; SLE, systemic lupus erythematosus; SS, Sjogren's syndrome; DM, dermatomyositis; RA, rheumatoid arthritis; MCTD, mixed connective tissue disease; SSc, systemic sclerosis; IBD, inflammatory bowel disease; IMRH, immunology and rheumatology; NEPH, nephrology; NEUR, neurology; DERM, dermatology; CM, chest medicine; CV, cardiology; GI, gastroenterology and hepatology; HEMA, hematology; META, metabolism, endocrinology, and nutrition; GS, gastrointestinal surgery; CS, thoracic surgery; CRS, colorectal surgery; ENT, otolaryngology; OPH, ophthalmology; PENP, pediatrics.

3.3. TPMT Phenotypes and Risks of Hepatotoxicity and Leukopenia

The average dosage of AZA was not different between the TPMT non-NM and NM groups. However, more than half of the patients took less than 25 mg daily. A hepatotoxicity event was defined by an ALT level of more than 150 U/L, which showed no difference between the TPMT non-NM and NM groups. However, the onset time of hepatotoxicity was shorter in the TPMT non-NM group compared with the TPMT NM group (Table 2). Cox proportional hazards regression analysis was applied to evaluate the independent risk factors for hepatotoxicity following AZA treatment. Among 1050 AZA users, male gender (hazard ratio (HR): 1.77, 95% confidence interval (CI): 1.07–2.91), non-NM genotype (HR: 3.85, 95% CI: 1.83–8.10), and MTX use (HR: 1.62, 95% CI: 1.03–2.57) were associated with hepatotoxicity compared with the NM genotype. Age, AZA dosage, and hepatitis B carrier status were not associated with hepatotoxicity (Table 3). The cumulative incidence of hepatoxicity was significantly higher in the first three years in the TPMP non-NM group than in the NM group ($p < 0.001$). In the TPMP non-NM group, the 1-year cumulative incidence rate was 8.5%, the 2-year cumulative incidence rate was 18.6%, and the 3-year cumulative incidence rate was 18.6% (Figure 2).

Table 2. Outcome post-AZA treatment.

Variable	TPMT non-NM (n = 50)	TPMT NM (n = 1000)	p-Value
	AZA dose (mg)		
≤25	27 (54.0)	563 (56.3)	
25–50	21 (42.0)	333 (33.3)	0.21
>50	2 (4.0)	104 (10.4)	

Table 2. *Cont.*

Variable	TPMT non-NM (n = 50)	TPMT NM (n = 1000)	p-Value
	Outcome postAZA exposure		
Leukopenia ^ cases	28 (59.5)	611 (63.3)	0.60
Hepatitis & cases	10 (20.0)	150 (15.0)	0.33
Lowest WBC (U/L)	4375 (3210–5420)	4200 (3150–5390)	0.52
Highest ALT (U/L)	38 (22–84)	39 (26–73)	0.68
Onset of leukopenia ^ (days)	1359.3 ± 1709.7	1597.3 ± 1606.6	0.44
Onset of hepatitis & (days)	676.3 ± 837.9	2395.8 ± 1911.0	<0.0001

Data are expressed as n (%) or the mean ± SD or the median (interquartile range). ^ WBC count ≤ 4000 (U/L). & ALT level ≥ 150 (U/L). TPMT, thiopurine S-methyltransferase; NM, normal metabolizer; AZA, azathioprine; WBC, white blood cell; ALT, alanine transaminase.

Table 3. Cox regression analysis of risk factors for hepatotoxicity following AZA treatment.

	Hepatotoxicity ^ HR	95% CI	p-Value
Age	0.99	0.98–1.01	0.60
Gender (reference female)	1.77	1.07–2.91	0.03
TPMT non-NM genotype	3.85	1.83–8.10	0.0004
AZA dose (mg)	1.00	1.00–1.01	0.31
MTX	1.62	1.03–2.57	0.04
Hepatitis B carrier	1.18	0.40–3.48	0.76

^ Development of an ALT level ≥ 150 U/L after AZA treatment. ALT, alanine transaminase; TPMT, thiopurine S-methyltransferase; NM, normal metabolizer; AZA, azathioprine; MTX, methotrexate; HR: hazard ratio; CI: confidence interval.

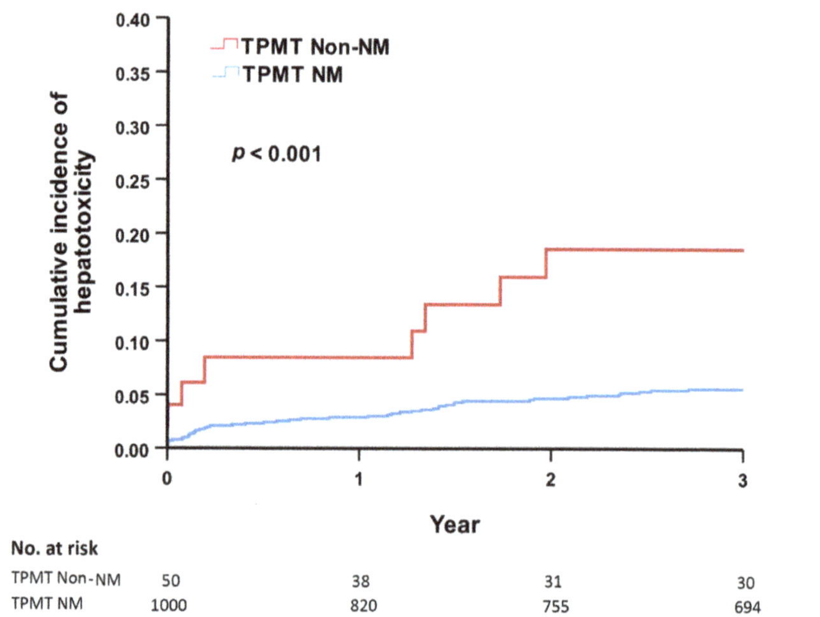

Figure 2. Three-year cumulative incidence of hepatotoxicity between the TMPT non-NM and NM groups. TPMT, thiopurine S-methyltransferase; NM, normal metabolizer.

4. Discussion

This was a hospital-wide retrospective case-control study. The patients were selected from the TPMI cohort in which the TPMT gene status was tested. We enrolled all AZA users. Among 1050 AZA users, the non-NM genotype, male gender, and MTX usage were independent risk factors for hepatotoxicity. The cumulative incidence of hepatoxicity was significantly higher in the first three years in the TPMT non-NM group than in the NM group. Our result shows that pre-emptive TPMT genotyping before starting AZA therapy may minimize the risk of hepatotoxicity.

Two meta-analyses by Liu et al. [19,20] reported that the incidence of TPMT polymorphisms was not associated with AZA-induced hepatotoxicity in patients with autoimmune diseases and IBD. Unlike the previous studies, our study revealed an association between TPMT polymorphisms and thiopurine-induced hepatotoxicity. First, our study design was different from those of the previous studies. We classified patients by TPMT genotyping before evaluating hepatotoxicity events. However, the earlier studies classified patients with hepatotoxicity events before assessing the TPMT genotype. Our study showed no difference in the first hepatotoxicity events between the TPMT non-NM and the NM groups. However, as the onset time of hepatotoxicity was considered and eliminating the interaction effect of variables by Cox regression analysis, the TPMT non-NM group was more prominent. Our study design was closer to the prospective concept. Second, our study population was different. Our patients had various autoimmune diseases, but the previous studies mainly analyzed patients with IBD. The subjects in the earlier studies discussing AZA-induced hepatotoxicity were primarily Caucasians. The Asian population was limited, but our study population was all Asians and was larger. As genetic testing is becoming less expensive and more accessible, pre-emptive TPMT testing is suggested in Asians to prevent post-AZA hepatotoxicity.

Previous studies revealed a significant association between TPMT polymorphisms and bone marrow toxicity. However, in our study, this was not noted. The previous studies primarily discussed Caucasians with IBD. The median dose of AZA in the previous studies was more than 1–2 mg/kg/day [23–28]. In contrast, our study enrolled patients with various autoimmune diseases, particularly SLE and SS. Our average AZA dosage was 36.4 mg in the TPMT non-NM group and 39.6 mg in the TPMT NM group. Approximately 90% of patients took less than 50 mg per day. We followed the AZA treatment strategy of starting with a lower dose and increasing the dose at a slow pace. Such a strategy seemed to keep the patients safe from myelosuppression even in those with the TPMT non-NM genotype. However, our study results show that taking AZA for a longer duration still has risks of hepatotoxicity.

In addition to the TPMT genotype, methotrexate usage and male gender were independent risk factors for hepatotoxicity. Clinical studies have widely reported the hepatotoxicity of MTX [29]. According to the FAERS database, females were less likely to develop liver injury than males [30]. A cross-sectional community study in Taiwan revealed that hepatitis with an elevated ALT was more common in men than in women [31]. According to our research, male patients with the TPMP non-NM genotype using MTX would have the highest risk of hepatotoxicity. These patients may take benefit from pre-emptive genotyping. This finding provides insights into the potential utilization of pharmacogenomics in individualized medical care.

There are several limitations to this study. First, the study design was retrospective. Missing data were inevitable. After three years of observation, the cumulative incidence of hepatotoxicity related to the TMPT non-NM genotype was noted. However, it is impossible to define a causal relationship between AZA and hepatotoxicity. The result was still significant. Second, we may have underestimated the number of cases of hepatotoxicity. There are three types of AZA-related hepatotoxicity: hepatocellular, cholestatic, and mixed. However, we only enrolled patients with hepatocellular-type hepatotoxicity. According to research by Siramolpiwat et al., the AZA-related hepatotoxicity type was predominantly a mixed type [32]. Evaluating only the ALT level may not miss patients with mixed-type hepatotoxicity. Thus, our hepatotoxicity event number was still reliable. Third, we only discussed the TMPT genotype. In addition to the TPMT genotype, our study did not discuss many other genetic polymorphisms, such as nucleoside diphosphate-liked moiety X motif 15 (NUDT15). NUDT15 was reported to be associated with more frequent adverse events than TPMT polymorphisms in the Asian population [33]. Further studies are needed to verify our results.

5. Conclusions

This hospital-wide retrospective case-control study supported the hypothesis that the TPMT non-NM genotype is associated with hepatotoxicity following AZA therapy. Pre-emptive testing for TPMT polymorphisms could help to individualize AZA therapy by minimizing the risk of hepatotoxicity, especially in male patients taking MTX.

Author Contributions: H.-S.S. was involved in the conceptualization of this study, methodology, original draft preparation, and reviewing and editing of the manuscript. Y.-M.C. was involved in the conceptualization of this study, methodology, data generation, curation, resource acquisition, original draft preparation, and reviewing and editing of the manuscript. Y.-J.L. was involved in the methodology, data analysis, and review and editing of the manuscript editing. C.-Y.W. was involved in the methodology, data analysis, and reviewing and editing of the manuscript. J.-P.C. was involved in data curation, statistical analysis, and reviewing, and editing of the manuscript. H.-J.L. was involved in the data analysis. W.-T.H. was involved in the conceptualization of this study, methodology, data generation, curation, resource acquisition, original draft preparation, and reviewing and editing of the manuscript. W.-N.H. was involved in interpreting the results, resource acquisition, and reviewing and editing of the manuscript. Y.-H.C. participated in the study design, methodology, data interpretation, resource acquisition, and reviewing and editing of the manuscript. All authors have read and agreed to the published version of the manuscript.

Funding: This study was funded by Academia Sinica 40-05-GMM and AS-GC-110-MD02.

Institutional Review Board Statement: The study was conducted in accordance with the Declaration of Helsinki and approved by the Ethics Committee of Taichung Veterans General Hospital (SF19153A).

Informed Consent Statement: Informed consent was obtained from all subjects involved in the study. Written informed consent was obtained from the patients to publish this paper.

Data Availability Statement: Not applicable.

Acknowledgments: We thank all the participants and investigators of the Taiwan Precision Medicine Initiative.

Conflicts of Interest: The authors declare that they have no conflict of interest to declare.

References

1. Pego-Reigosa, J.M.; Cobo-Ibáñez, T.; Calvo-Alén, J.; Loza-Santamaría, E.; Rahman, A.; Muñoz-Fernández, S.; Rúa-Figueroa, Í. Efficacy and Safety of Nonbiologic Immunosuppressants in the Treatment of Nonrenal Systemic Lupus Erythematosus: A Systematic Review. *Arthritis Care Res.* **2013**, *65*, 1775–1785. [CrossRef] [PubMed]
2. Bertsias, G.; Ioannidis, J.P.A.; Boletis, J.; Bombardieri, S.; Cervera, R.; Dostal, C.; Font, J.; Gilboe, I.M.; Houssiau, F.; Huizinga, T.; et al. EULAR recommendations for the management of systemic lupus erythematosus. Report of a Task Force of the EULAR Standing Committee for International Clinical Studies Including Therapeutics. *Ann. Rheum. Dis.* **2007**, *67*, 195–205. [CrossRef] [PubMed]
3. Østensen, M.; Khamashta, M.; Lockshin, M.; Parke, A.; Brucato, A.; Carp, H.; Doria, A.; Rai, R.; Meroni, P.; Cetin, I.; et al. Anti-inflammatory and immunosuppressive drugs and reproduction. *Arthritis Res. Ther.* **2006**, *8*, 209. [CrossRef]
4. Ran, Z.; Wu, K.; Matsuoka, K.; Jeen, Y.T.; Wei, S.C.; Ahuja, V.; Chen, M.; Hu, P.J.; Andoh, A.; Kim, H.J.; et al. Asian Organization for Crohn's and Colitis and Asia Pacific Association of Gastroenterology practice recommendations for medical management and monitoring of inflammatory bowel disease in Asia. *J. Gastroenterol. Hepatol.* **2020**, *36*, 637–645. [CrossRef]
5. Lichtenstein, G.R.; Loftus, E.V.; Isaacs, K.L.; Regueiro, M.D.; Gerson, L.B.; Sands, B.E. ACG Clinical Guideline: Management of Crohn's Disease in Adults. *Am. J. Gastroenterol.* **2018**, *113*, 481–517. [CrossRef]
6. Gomollón, F.; Dignass, A.; Annese, V.; Tilg, H.; Van Assche, G.; Lindsay, J.O.; Peyrin-Biroulet, L.; Cullen, G.J.; Daperno, M.; Kucharzik, T.; et al. 3rd European Evidence-based Consensus on the Diagnosis and Management of Crohn's Disease 2016: Part 1: Diagnosis and Medical Management. *J. Crohn's Colitis* **2017**, *11*, 3–25. [CrossRef]

7. Harbord, M.; Eliakim, R.; Bettenworth, D.; Karmiris, K.; Katsanos, K.; Kopylov, U.; Kucharzik, T.; Molnár, T.; Raine, T.; Sebastian, S.; et al. Third European Evidence-based Consensus on Diagnosis and Management of Ulcerative Colitis. Part 2: Current Management. *J. Crohn's Colitis* **2017**, *11*, 769–784. [CrossRef]
8. Mack, C.L.; Adams, D.; Assis, D.N.; Kerkar, N.; Manns, M.P.; Mayo, M.J.; Vierling, J.M.; Alsawas, M.; Murad, M.H.; Czaja, A.J. Diagnosis and Management of Autoimmune Hepatitis in Adults and Children: 2019 Practice Guidance and Guidelines From the American Association for the Study of Liver Diseases. *Hepatology* **2019**, *72*, 671–722. [CrossRef]
9. Joly, P.; Horvath, B.; Patsatsi, A.; Uzun, S.; Bech, R.; Beissert, S.; Bergman, R.; Bernard, P.; Borradori, L.; Caproni, M.; et al. Updated S2K guidelines on the management of pemphigus vulgaris and foliaceus initiated by the european academy of dermatology and venereology (EADV). *J. Eur. Acad. Dermatol. Venereol.* **2020**, *34*, 1900–1913. [CrossRef]
10. Chouchana, L.; Narjoz, C.; Beaune, P.; Loriot, M.A.; Roblin, X. Review article: The benefits of pharmacogenetics for improving thiopurine therapy in inflammatory bowel disease. *Aliment. Pharmacol. Ther.* **2011**, *35*, 15–36. [CrossRef]
11. Lennard, L. The clinical pharmacology of 6-mercaptopurine. *Eur. J. Clin. Pharm.* **1992**, *43*, 329–339. [CrossRef] [PubMed]
12. Ngo, S.; Sauvetre, G.; Vittecoq, O.; Lévesque, H.; Marie, I. Azathioprine-associated severe myelosuppression: Indication of routine determination of thiopurine S-methyltransferase variant? *Rev. Med. Int.* **2011**, *32*, 373–376. [CrossRef] [PubMed]
13. Boonsrirat, U.; Angsuthum, S.; Vannaprasaht, S.; Kongpunvijit, J.; Hirankarn, N.; Tassaneeyakul, W.; Avihingsanon, Y. Azathioprine-induced fatal myelosuppression in systemic lupus erythematosus patient carrying TPMT*3C polymorphism. *Lupus* **2008**, *17*, 132–134. [CrossRef] [PubMed]
14. Bastida, G.; Nos, P.; Aguas, M.; Beltrán, B.; Rubín, A.; Dasí, F.; Ponce, J. Incidence, risk factors and clinical course of thiopurine-induced liver injury in patients with inflammatory bowel disease. *Aliment. Pharm.* **2005**, *22*, 775–782. [CrossRef] [PubMed]
15. Appell, M.L.; Berg, J.; Duley, J.; Evans, W.E.; Kennedy, M.A.; Lennard, L.; Marinaki, T.; McLeod, H.L.; Relling, M.V.; Schaeffeler, E.; et al. Nomenclature for alleles of the thiopurine methyltransferase gene. *Pharm. Genom.* **2013**, *23*, 242–248. [CrossRef]
16. Engen, R.M.; Marsh, S.; Van Booven, D.J.; McLeod, H.L. Ethnic differences in pharmacogenetically relevant genes. *Curr. Drug Targets* **2006**, *7*, 1641–1648. [CrossRef]
17. Corominas, H.; Domènech, M.; del Río, E.; Gich, I.; Domingo, P.; Baiget, M. Frequency of thiopurine S-methyltransferase alleles in different ethnic groups living in Spain. *Med. Clin.* **2006**, *126*, 410–412. [CrossRef]
18. Dong, X.W.; Zheng, Q.; Zhu, M.M.; Tong, J.L.; Ran, Z.H. Thiopurine S-methyltransferase polymorphisms and thiopurine toxicity in treatment of inflammatory bowel disease. *World J. Gastroenterol.* **2010**, *16*, 3187–3195. [CrossRef]
19. Liu, Y.-P.; Wu, H.-Y.; Yang, X.; Xu, H.-Q.; Li, Y.-C.; Shi, D.-C.; Huang, J.-F.; Huang, Q.; Fu, W.-L. Association between Thiopurine S-methyltransferase Polymorphisms and Thiopurine-Induced Adverse Drug Reactions in Patients with Inflammatory Bowel Disease: A Meta-Analysis. *PLoS ONE* **2015**, *10*, e0121745. [CrossRef]
20. Liu, Y.-P.; Xu, H.-Q.; Li, M.; Yang, X.; Yu, S.; Fu, W.-L.; Huang, Q. Association between Thiopurine S-Methyltransferase Polymorphisms and Azathioprine-Induced Adverse Drug Reactions in Patients with Autoimmune Diseases: A Meta-Analysis. *PLoS ONE* **2015**, *10*, e0144234. [CrossRef]
21. Wei, C.-Y.; Yang, J.-H.; Yeh, E.-C.; Tsai, M.-F.; Kao, H.-J.; Lo, C.-Z.; Chang, L.-P.; Lin, W.-J.; Hsieh, F.-J.; Belsare, S.; et al. Genetic profiles of 103,106 individuals in the Taiwan Biobank provide insights into the health and history of Han Chinese. *NPJ Genom. Med.* **2021**, *6*, 1–10. [CrossRef] [PubMed]
22. Relling, M.V.; Schwab, M.; Whirl-Carrillo, M.; Suarez-Kurtz, G.; Pui, C.H.; Stein, C.M.; Moyer, A.M.; Evans, W.E.; Klein, T.E.; Antillon-Klussmann, F.G.; et al. Clinical Pharmacogenetics Implementation Consortium Guideline for Thiopurine Dosing Based on TPMT and NUDT15 Genotypes: 2018 Update. *Clin. Pharm.* **2018**, *105*, 1095–1105. [CrossRef]
23. Ansari, A.; Arenas, M.; Greenfield, S.M.; Morris, D.; Lindsay, J.; Gilshenan, K.; Smith, M.; Lewis, C.; Marinaki, A.; Duley, J.; et al. Prospective evaluation of the pharmacogenetics of azathioprine in the treatment of inflammatory bowel disease. *Aliment. Pharmacol. Ther.* **2008**, *28*, 973–983. [CrossRef] [PubMed]
24. Palmieri, O.; Latiano, A.; Bossa, F.; Vecchi, M.; D'Incà, R.; Guagnozzi, D.; Tonelli, F.; Cucchiara, S.; Valvano, M.R.; Latiano, T.; et al. Sequential evaluation of thiopurine methyltransferase, inosine triphosphate pyrophosphatase, and HPRT1 genes polymorphisms to explain thiopurines' toxicity and efficacy. *Aliment. Pharmacol. Ther.* **2007**, *26*, 737–745. [CrossRef] [PubMed]
25. Schwab, M.; Schäffeler, E.; Marx, C.; Fischer, C.; Lang, T.; Behrens, C.; Gregor, M.; Eichelbaum, M.; Zanger, U.M.; Kaskas, B.A. Azathioprine therapy and adverse drug reactions in patients with inflammatory bowel disease: Impact of thiopurine S-methyltransferase polymorphism. *Pharmacogenetics* **2002**, *12*, 429–436. [CrossRef]
26. Winter, J.W.; Gaffney, D.; Shapiro, D.; Spooner, R.J.; Marinaki, A.M.; Sanderson, J.D.; Mills, P.R. Assessment of thiopurine methyltransferase enzyme activity is superior to genotype in predicting myelosuppression following azathioprine therapy in patients with inflammatory bowel disease. *Aliment. Pharmacol. Ther.* **2007**, *25*, 1069–1077. [CrossRef]
27. Wroblova, K.; Kolorz, M.; Batovsky, M.; Zboril, V.; Suchankova, J.; Bartos, M.; Ulicny, B.; Pav, I.; Bartosova, L. Gene Polymorphisms Involved in Manifestation of Leucopenia, Digestive Intolerance, and Pancreatitis in Azathioprine-Treated Patients. *Dig. Dis. Sci.* **2012**, *57*, 2394–2401. [CrossRef]
28. Zabala-Fernández, W.; Barreiro-de Acosta, M.; Echarri, A.; Carpio, D.; Lorenzo, A.; Castro, J.; Martínez-Ares, D.; Pereira, S.; Martin-Granizo, I.; Corton, M.; et al. A pharmacogenetics study of TPMT and ITPA genes detects a relationship with side effects and clinical response in patients with inflammatory bowel disease receiving Azathioprine. *J. Gastrointestin Liver Dis.* **2011**, *20*, 247–253.
29. Ezhilarasan, D. Hepatotoxic potentials of methotrexate: Understanding the possible toxicological molecular mechanisms. *Toxicology* **2021**, *458*, 152840. [CrossRef]

30. Rana, P.; Aleo, M.D.; Wen, X.; Kogut, S. Hepatotoxicity reports in the FDA adverse event reporting system database: A comparison of drugs that cause injury via mitochondrial or other mechanisms. *Acta. Pharm. Sin. B* **2021**, *11*, 3857–3868. [CrossRef]
31. Chen, C.H.; Huang, M.H.; Yang, J.C.; Nien, C.K.; Yang, C.C.; Yeh, Y.H.; Yueh, S.K. Prevalence and etiology of elevated serum alanine aminotransferase level in an adult population in Taiwan. *J. Gastroenterol. Hepatol.* **2007**, *22*, 1482–1489. [CrossRef] [PubMed]
32. Siramolpiwat, S.; Sakonlaya, D. Clinical and histologic features of Azathioprine-induced hepatotoxicity. *Scand. J. Gastroenterol.* **2017**, *52*, 876–880. [CrossRef] [PubMed]
33. Jena, A.; Jha, D.K.; Kumar, M.P.; Kasudhan, K.S.; Kumar, A.; Sarwal, D.; Mishra, S.; Singh, A.K.; Bhatia, P.; Patil, A.; et al. Prevalence of polymorphisms in thiopurine metabolism and association with adverse outcomes: A South Asian region-specific systematic review and meta-analysis. *Expert Rev. Clin. Pharm.* **2021**, *14*, 491–501. [CrossRef] [PubMed]

Journal of
*Personalized
Medicine*

Article

Short- and Long-Term Effects of a Prebiotic Intervention with Polyphenols Extracted from European Black Elderberry—Sustained Expansion of *Akkermansia* spp.

Simon Reider [1,2], Christina Watschinger [1,2], Julia Längle [1], Ulrike Pachmann [3], Nicole Przysiecki [1,4], Alexandra Pfister [1,4], Andreas Zollner [1,4], Herbert Tilg [4], Stephan Plattner [5] and Alexander R. Moschen [1,2,*]

1 Christian Doppler Laboratory for Mucosal Immunology, Faculty of Medicine, Johannes Kepler University, 4020 Linz, Austria
2 Department of Internal Medicine 2, Faculty of Medicine, Johannes Kepler University, 4020 Linz, Austria
3 VASCage (Research Centre on Vascular Ageing and Stroke) GmbH, 6020 Innsbruck, Austria
4 Division of Internal Medicine I (Gastroenterology, Hepatology, Endocrinology, and Metabolism), Department of Medicine, Medical University Innsbruck, 6020 Innsbruck, Austria
5 IPRONA AG/SPA, 39011 Lana, Italy
* Correspondence: alexander.moschen@jku.at

Abstract: (1) Background: The intestinal microbiome has emerged as a central factor in human physiology and its alteration has been associated with disease. Therefore, great hopes are placed in microbiota-modulating strategies. Among various approaches, prebiotics, substrates with selective metabolization conferring a health benefit to the host, are promising candidates. Herein, we studied the prebiotic properties of a purified extract from European black elderberries, with a high and standardized content of polyphenols and anthocyanins. (2) Methods: The ELDERGUT trial represents a 9-week longitudinal intervention study divided into 3 distinct phases, namely a baseline, an intervention and a washout period, three weeks each. The intervention consisted of capsules containing 300 mg elderberry extract taken twice a day. Patient-reported outcomes and biosamples were collected weekly. Microbiome composition was assessed using 16S amplicon metagenomics. (3) Results: The supplementation was well tolerated. Microbiome trajectories were highly individualized with a profound shift in diversity indices immediately upon initiation and after termination of the compound. This was accompanied by corresponding changes in species abundance over time. Of particular interest, the relative abundance of *Akkermansia* spp. continued to increase in a subset of participants even beyond the supplementation period. Associations with participant metadata were detected.

Keywords: black elderberry; polyphenols; prebiotic; microbiota

Citation: Reider, S.; Watschinger, C.; Längle, J.; Pachmann, U.; Przysiecki, N.; Pfister, A.; Zollner, A.; Tilg, H.; Plattner, S.; Moschen, A.R. Short- and Long-Term Effects of a Prebiotic Intervention with Polyphenols Extracted from European Black Elderberry—Sustained Expansion of *Akkermansia* spp. *J. Pers. Med.* **2022**, *12*, 1479. https://doi.org/10.3390/jpm12091479

Academic Editor: Anne-Marie Caminade

Received: 15 August 2022
Accepted: 6 September 2022
Published: 9 September 2022

Publisher's Note: MDPI stays neutral with regard to jurisdictional claims in published maps and institutional affiliations.

1. Introduction

Although to date no distinct pattern of a healthy microbiota has emerged, certain alterations of microbiota composition and function are described as markers of disease-associated states. Collectively, these changes are summarized as dysbiosis, although a clear definition of the term is lacking and it has grown to be regarded increasingly unhelpful by experts in the field [1]. Nevertheless, loss of microbial diversity, a reduction in the number of beneficial microbes and an expansion of pathobionts are regarded as hallmarks of dysbiosis [2,3].

Because dysbiotic states of the microbiota are observed in and have been associated with numerous disease conditions [4], the field of microbiota modulation is receiving utmost attention. One potential means to influence this important determinant of health and disease are dietary interventions. The influence of long-term nutritional patterns on the microbiota has been studied and described extensively [5,6]. However, evidence suggests that drastic and sustained modifications of the diet are necessary to achieve long-term effects [7]. In a more focused approach, prebiotic supplementation could address this gap. The International Association for Probiotics and Prebiotics (ISAPP) defines prebiotics as nutritional compounds or substances which undergo selective metabolization by specific members of the intestinal microbiota, ultimately conferring a health benefit on the host [8]. While traditionally, non-digestible carbohydrates such as inulin, fructooligosaccharides and galactooligosaccharides have been considered prebiotics, the ISAPP has pointed out that some plant polyphenols might also fulfill these criteria and therefore might be considered prebiotics [8]. However, this is currently not reflected in the legal and regulatory status of these compounds within the European Union.

Polyphenols are a heterogenous class of secondary plant metabolites present in fruit and vegetables. Phenolic compounds possess one or more aromatic rings with two or more hydroxyl groups. They can be present in free forms or conjugated with sugars, acid and other biomolecules [9–11].

Plant polyphenols are not readily taken up in the small intestine and reach the colon in considerable amounts, where they can serve as substrate for the colonic bacterial microbiota [9]. As potential health effects of polyphenols seem to depend on the resulting metabolites from degradation by the colonic microbiota [12], the individual microbiome configuration seems to be a key determinant of potential effects of polyphenol-rich black elderberry extracts. The bioavailability of polyphenols is variable and depends on the nutritional setting in which the substance is presented and an extensive secondary metabolism, largely by the intestinal microbiome [13–19]. Thus, microbiome composition might influence polyphenol effects in given individuals through diverse modification [20,21]. Bacterial secondary metabolization of dietary polyphenols improves bioavailability and hydrophilicity of these compounds and enables detection in the urine [20]. One systematic review identified data hinting at potential antibacterial and anti-inflammatory effects, although underlying mechanisms remained elusive and the overall level of evidence was classified as weak [22]. Anthocyanins of black elderberry extract are taken up and incorporated into endothelial cell membranes and cytosol, hinting at potential antioxidative benefits [23]. On the other hand, microbiota-modifying properties of polyphenols have been reported and include an expansion of Bifidobacteria [24] and Lactobacilli [12,25] and alterations of the Bacteroides–Firmicutes ratio, possibly reflecting an adaption to a higher glycan load [26]. Besides direct prebiotic effects on the microbiota, an environmental mechanism has also been proposed for polyphenol-associated microbiota modulation. Certain polyphenols have been shown to exert anti-microbial activity in vitro [27]. Therefore, broad antimicrobial activity of certain polyphenol compounds could serve to open up niches in the gastrointestinal tract and allow for expansion of other taxa. To include this dualistic effect, the term "duplibiotics" has been coined [28,29].

Food-derived anthocyanins exhibit a low bioavailability in their native form [15] but are partly metabolized and taken up in the intestine and can be recovered from the urine [13–19,30,31].

Effectively, the polyphenol–microbiota axis exerts influence on host physiology. Certain polyphenol metabolites of microbial origin serve to strengthen barrier function and boost secretion of antimicrobial peptides [32] and effects on secondary bile acid metabolism have been described [33]. In a mouse model, food rich in polyphenols attenuated detrimental metabolic and inflammatory responses [34]. Short-chain fatty acids are also produced from polyphenols by the intestinal microbiota [35] and may serve as regulators of colonic mucosal homeostasis [36].

European black elderberries (Sambucus nigra) and their extracts have been used traditionally in parts of Europe for centuries as a remedy for common colds and influenza-like symptoms. Recently, an alleviating effect of an extract of black elderberries on symptoms and duration of influenza infections has been shown [37–39]. As research has mostly focused on the effects of black elderberry on upper respiratory tract infections, little is known on its microbiota-directed effects. Black elderberries are rich in polyphenols [27] and also contain various immunomodulating polysaccharides such as pectin [40,41]. The main group of polyphenols in black elderberry fruits are anthocyanins (cyanidin 3- sambubioside-5-glucoside, cyanidin 3,5-diglucoside, cyanidin 3-sambubioside, cyanidin 3-glucoside). Additional polyphenols present in black elderberry are chlorogenic acids, rutin, and isoquercitrin [42].

This study aimed to provide a comprehensive in depth characterization of the microbiota-modulating effects of a 3-week intervention with black elderberry extract, standardized to anthocyanin and polyphenol levels.

2. Materials and Methods

2.1. Trial Design and Study Cohort

The ELDERGUT trial was a 9-week longitudinal intervention study with 3 distinct study phases of 3-week duration each. A baseline phase serving to obtain robust initial measurements was followed by a 3-week intervention period during which a daily dose of 600 mg of highly purified black elderberry extract (300 mg twice daily) was consumed by the participants. Participants logged digestive symptoms, bowel movements and possible adverse effects in a study diary provided by the investigational team. Participants could be either male or female and had to be 18 to 50 years old. These individuals had to report a history of vaginal delivery and being breast-fed for at least 3 months after birth. Exclusion criteria were recent antibiotic treatment within 3 months of the beginning of the study, a history of gastrointestinal disease or other acute or chronic medical conditions with the potential to affect the intestinal microbiome, a history of major abdominal surgery (appendectomy excluded), baseline consumption of dietary supplements or probiotics, a strictly vegan diet, relevant pathologic abnormalities in baseline clinical biochemistry tests and a known allergy to black elderberry.

The ELDERGUT trial (Figure 1A,B) included n = 30 healthy individuals (15 females, 15 males) that volunteered to participate in this study. All study subjects completed the 9-week study period and provided biological samples and clinical information at weekly study visits as indicated in Figure 1A. No study drop-outs occurred. All patients underwent a thorough clinical and biochemical evaluation at the baseline. Fecal calprotectin—a marker of intestinal inflammation—was within the normal range in all study subjects. Notably, two study participants happened to have a positive screening test for IgA antibodies against tissue transglutaminase and the diagnosis of celiac disease was confirmed histologically in both participants. These were excluded from the analysis as celiac disease was a predefined exclusion criterion. During conduction of the trial, 2 male participants contracted SARS-CoV2 during week 5 of the study period. The disease course was mild and the infection did not result in exclusion from the study or patients not providing samples.

The mean age of the trial population was 23.7 years (±3.05 years, range 19–31 years) and the mean body mass index was 22.3 kg/m^2 (±2.49, range 17.7–27.3). Baseline questionnaire-based nutritional assessment revealed a well-balanced nutritional pattern with differences between female and male participants. Male participants reported less intake of dietary fiber ($p < 0.01$, chi-square test), and a considerable but not significantly larger proportion of daily intake of animal fat (Figure 1C).

2.2. Intervention

Capsules containing 300 mg European black elderberry extract (ElderCraft®) were provided by IPRONA AG/SPA (Lana, Italy). ElderCraft® is a full spectrum water extract standardized to 14% anthocyanins (primarily cyanidin 3-sambubioside-5-glucoside, cyanidin 3,5-diglucoside, cyanidin 3-sambubioside, cyanidin 3-glucosid) and 18% polyphenols (primarily chlorogenic acids, rutin, isoquercitrin).

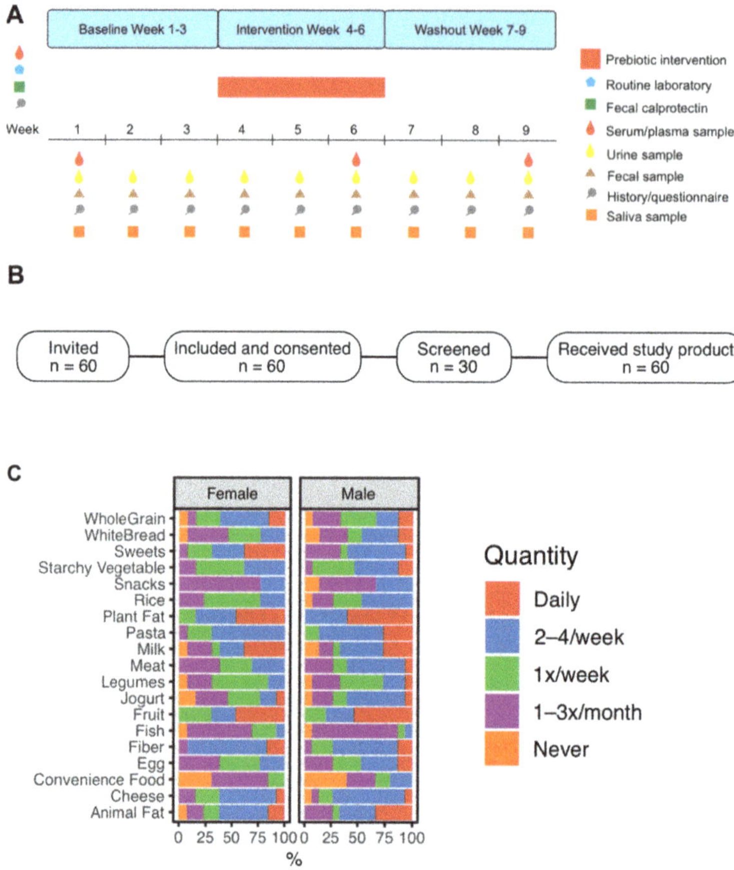

Figure 1. ELDERGUT trial design and baseline nutrition data: (**A**) Timeline of interventions and sampling during the 9-week study period, depicting the baseline, the intervention and the washout phase of the study. During the study, biological samples including urine, faeces, and saliva were collected once a week, blood samples were drawn once per study phase (i.e., in weeks 1, 6 and 9). Questionnaires on well-being and gastrointestinal symptoms were completed by the participants during weekly visits. (**B**) Overview of the inclusion and screening strategy. (**C**) Data from questionnaires at baseline, indicating nutritional patterns within the cohort as well as gender-associated differences in food intake.

2.3. Biosample Acquisition and Biobanking

During the trial period, weekly biospecimen sampling (feces, urine, saliva) was performed. Additionally, blood samples were drawn once per study phase (at weeks 1, 6 and 9). Fecal samples were collected simultaneously into tubes pre-filled with nucleic acid stabilizing solution (INVITEK, Berlin, Germany) and standard fecal collection tubes (Sarstedt, Nümbrecht, Germany). Stabilized fecal samples were stored at −20 °C, unstabilized samples were stored in aliquots at −80 °C. Urine samples were centrifuged and the supernatant stored at −80 °C. Saliva samples were directly stored at −80 °C. Blood samples were collected into EDTA- and Serum-Tubes (Sarstedt, Nümbrecht, Germany), centrifuged and the supernatants were stored at −80 °C.

2.4. Fecal DNA Extraction

Fecal DNA was extracted from stabilized samples, according to the instructions provided by the manufacturer using a spin-column based kit (Qiagen Fast DNA Stool Kit, Qiagen, Germany). Briefly, homogenized suspensions were heated to 95 °C for 10 min, centrifuged and treated with Proteinase K. Then, samples were incubated with the buffer provided in the kit and pure ethanol. After vortexing, the preparations were transferred onto spin columns and centrifuged, followed by multiple washing steps on the column using the washing buffer from the isolation kit. Finally, DNA was eluated in 100 µL of a TE-based buffer. Nucleic acid concentration was measured using a Nanodrop 1000 spectrophotometer (Eppendorf, Hamburg, Germany).

2.5. Metagenomic Analysis

Fecal DNA isolates were analyzed using 16S amplicon based metagenomics in cooperation with a commercial provider. DNA concentration and purity was monitored on 1% agarose gels and DNA was amplified using a PCR with primers targeting the V3-V4 variable regions of the bacterial 16S rRNA gene (primer pair 314F-806R) All PCR reactions were carried out with Phusion® High-Fidelity PCR Master Mix (New England Biolabs). PCR products were mixed at equal ratios and purified using Qiagen Gel extraction kit (Qiagen, Germany). Libraries were prepared using the NEBNext Ultra DNA Library Prep Kit for Illumina and quantification was done using Qubit and qPCR. Sequencing was performed on the Illumina Miseq platform with PE250 chemistry.

Paired end reads were assigned to samples using specific barcodes added in the PCR amplification step. Barcode sequences were removed and reads were merged using FLASH (v.1.2.7; Baltimore, MD, USA) [43]. Resulting raw tags were quality filtered using QIIME (v.1.7, Boulder, CO, USA [44]). Chimera detection and removal was performed using UCHIME (Boulder, CO, USA) against its Gold reference database [45,46]. OTU clustering was performed using UPARSE version 7 [47] using effective tags at a clustering level of 97%. Representative sequences were annotated for every OTU using mothur [48] and the SILVA database [49]. Further downstream analysis including alpha- and beta-diversity and analysis of differential abundance is detailed in the statistics section.

2.6. Statistics

Data from study diaries were summarized using descriptive statistics and significance was tested using ANOVA with the post hoc Tukey test and the Kruskal–Wallis/Wilcoxon test for continuous numeric variables depending on the underlying distribution. The frequency of symptoms and data from the nutrition questionnaire was assessed using chi-square statistics. Amplicon-based metagenomic data were imported into R v. 4.1.2 (R Core Team, Vienna, Austria) [50] and RStudio using phyloseq [51] and then analyzed according to published workflows. Permutational analysis of variance (PERMANOVA) was performed using Vegan (v.2.5-7, Helsinki, Finland) [52]. Trends in differential abundance of bacterial species over the duration of the study period were explored using a linear mixed model implemented in Maaslin2 [53]. For these approaches, appropriate variance and occurrence filters were implemented (supplementary material and analysis scripts are provided). The tidyverse [54], rstatix [55] and ggpubr [56] packages were used for data cleaning and statistical procedures.

3. Results

3.1. Patient-Reported Outcomes during the Trial Period

Participants completed symptom-based questionnaires on a weekly basis, assessing typical gastrointestinal symptoms as well as measures of general health and well-being. Additionally, information on the number of bowel movements and their consistency (Bristol Stool Scale, BSS value) was recorded.

Overall perception of health and quality of life of the study participants was consistently within the upper third of the 7-point visual analogue scale used for assessment.

No dynamic was observed along baseline, intervention and washout phase of the trial. Likewise, no differences between female and male participants were observed within these parameters (Figure 2A).

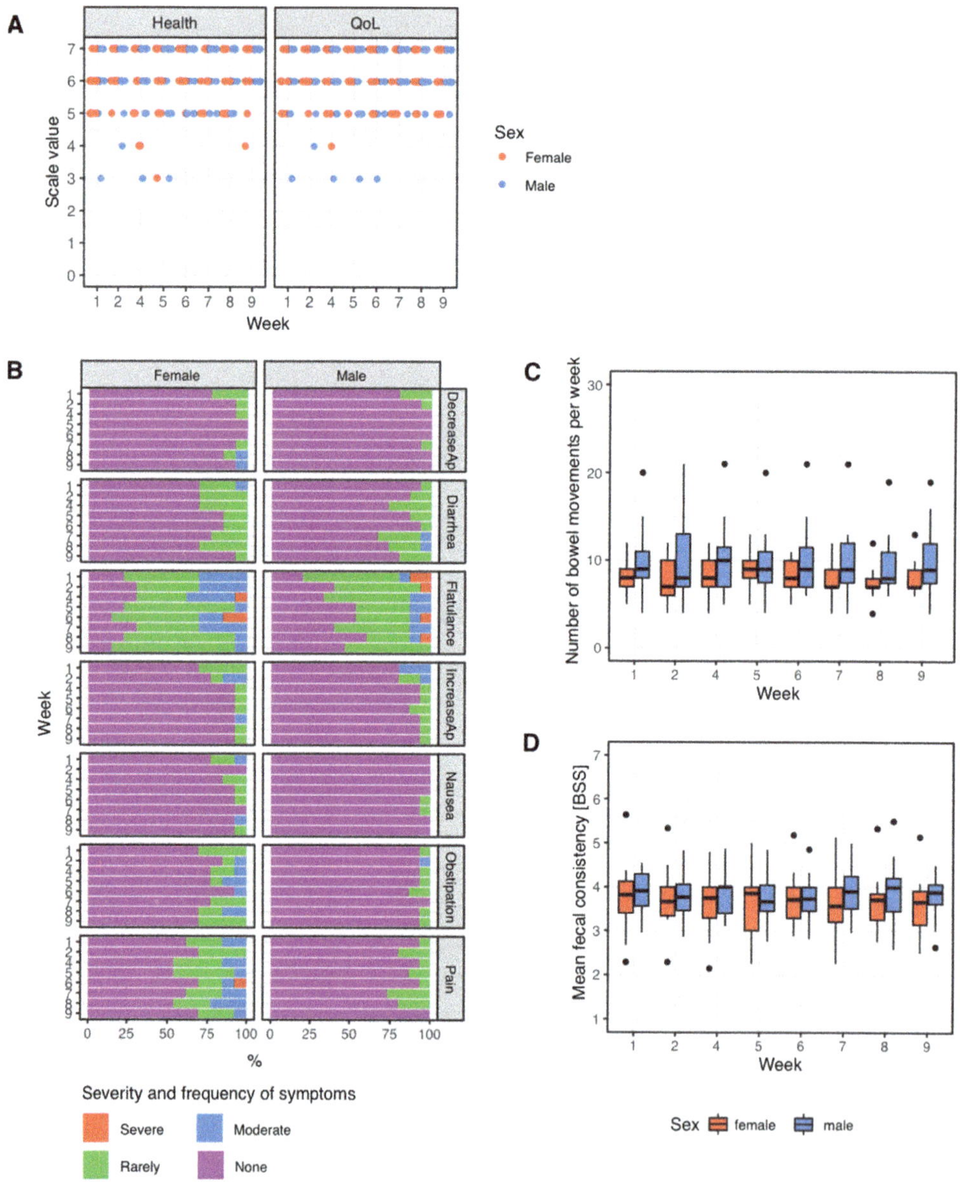

Figure 2. Patient-reported outcomes. (**A**) Weekly scores for health and quality of life (QoL), (**B**) results from weekly symptom-based questionnaires including common gastrointestinal symptoms. (**C**) Number of bowel movements per week in female and male participants. (**D**) Mean fecal consistency, i.e., the Bristol Stool Scale (BSS) value per week. Data were compared by the Kruskal–Wallis test, no significant differences were observed.

Study subjects were asked for the presence of gastrointestinal symptoms including increase or decrease in appetite, the occurrence of flatulence and diarrhea, nausea, constipation and abdominal discomfort. Of these, only flatulence was a regular complaint, but no clear association with the intervention emerged. Abdominal pain was slightly more severe and more frequent in female participants, although in most instances this complaint coincided with specific events of the menstrual cycle. Overall, no symptom-based signal associated with the intervention was observed (Figure 2B).

The number of bowel openings was recorded and calculated as mean number of bowel movements per week. No changes were observed on a week-to-week basis, particularly not within the intervention period (Table 1, Figure 2C).

Table 1. Mean number of bowel movements and mean BSS values. No values for week 3 are available.

	Mean Number of Bowel Movements/Week		Mean BSS Value/Week	
Week	**Mean**	**std. Deviation**	**Mean**	**std. Deviation**
1	9.1	3.2	3.8	0.6
2	9.1	4	3.7	0.6
4	9.1	3.4	3.7	0.6
5	9.4	3	3.7	0.6
6	9.1	3.3	3.7	0.6
7	8.9	3.5	3.8	0.7
8	8.7	2.9	3.8	0.7
9	9	3.3	3.7	0.6

The Bristol Stool Scale (BSS) is a widely-used and validated tool [57,58] to assess the consistency of a bowel movement. Higher values on the BSS indicate more liquid stool, low values represent very hard consistency. Similar to bowel movements, the mean weekly BSS value (mean of the BSS value of all bowel movements of a single study participant during every week of the trial) was calculated and happened to remain in the middle part of the scale and unchanged during the whole trial period (Table 1, Figure 2D).

3.2. Alterations in 16S Amplicon-Based Metagenomic Profiles at the Beginning and after the End of the Prebiotic Intervention

A 16S amplicon based metagenomic approach was applied to investigate temporal alterations in microbiota composition associated with an intervention with ElderCraft®. Weekly fecal samples from 28 participants were analyzed (252 samples overall) and the resulting profiles were compared on a weekly basis to the baseline configuration, i.e., median abundances from weeks 1 to 3.

3.3. Changes in Intra-Sample Diversity

In the first week of the intervention period an immediate and strong increase in measures of α-diversity (i.e., intra-sample diversity) was observed (Figure 3A). The number of observed species increased from 623.4 ± 58.0 (mean and standard deviation) during baseline to 925.3 ± 348.2 at the end of week 4. Similarly, the Shannon index increased from 4.1 ± 0.27 to 4.45 ± 0.49 ($p < 0.001$ for both comparisons; pairwise Wilcoxon test). The most prevalent OTUs were mapped to the genus *Bacteroides* (phylum *Bacteroidetes*) and *Faecalibacterium* (phylum *Firmicutes*). Additionally, other genera from the phylum *Firmicutes* were highly abundant both at baseline and throughout the study period (i.e., *Agathobacter, Ruminococcus, Roseburia*). This was also observed for the genenera *Bifidobacterium* (phylum *Actinobacteria*) and *Akkermansia* (phylum *Verrucomicrobia*).

In the later weeks of the intervention period, namely weeks 5 and 6, measures of α-diversity decreased but remained significantly higher in terms of the number of observed species (818.5 ± 356.6, 629.8 ± 43.0). The Shannon index was significantly different to baseline only were not significantly different from baseline apart from week 4 (Figure 3A). After stopping the prebiotic intervention in week 7, again an increase in the number of observed species was evident (897.9 ± 228.9, $p < 0.001$, pairwise Wilcoxon test). There was no significant difference of Shannon indices.

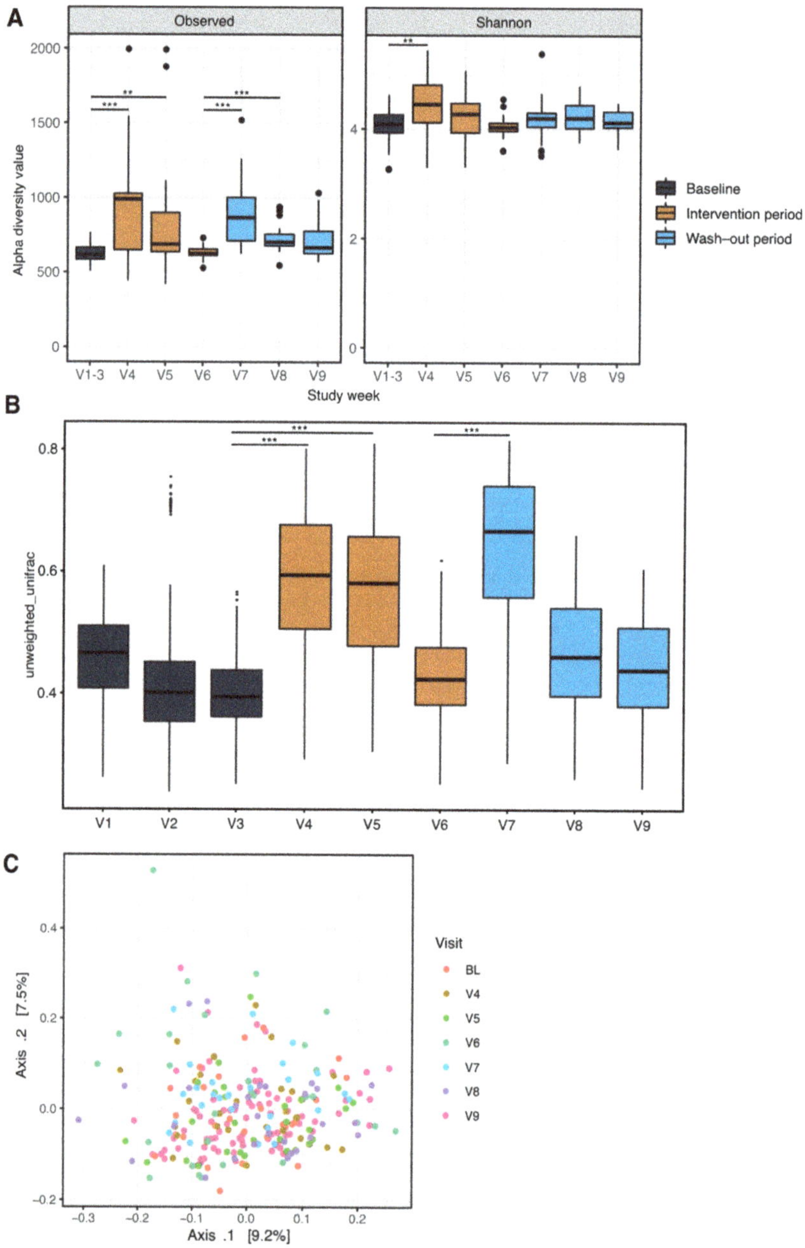

Figure 3. Microbial ecology measures of diversity. (**A**) α-diversity—describing intra-group diversity, i.e., the number of observed species—and Shannon indices. (**B**,**C**) β-diversity as unweighted Unifrac values by study week (**B**) and a Principal Coordinate Analysis of unweighted Unifrac distances (**C**). Statistical comparisons were made using pairwise Wilcoxon tests between the median of the baseline values (weeks 1–3) and the subsequent weeks. p values ** < 0.01, *** < 0.001.

3.4. Compositional Differences over Time

Differences of overall microbial composition between the different study weeks were investigated using the unweighted unifrac index of α-diversity as a measure of difference between groups of samples (Figure 3B,C). Both paired data and principle coordinate analysis (PCoA) revealed changes during, most pronounced in weeks 4 and 5, and in the first week after the prebiotic intervention. These differences met the prespecified significance level. Additionally, a significant effect of sex on microbiome composition was evident ($p < 0.01$ for both timepoint and sex, R-squared 0.07 for timepoint and 0.02 for sex, permutational analysis of variance).

3.5. Alterations on a Species Level Driving Overall Changes in Bacterial Ecology

Next, we employed linear modelling—taking the longitudinal nature of our data into account—to identify differences in the abundance at the species level responsible for driving the observed microbial ecologic dynamics.

Overall, on the genus level 36 genera and on the species level 71 differently abundant microbial taxa were detected when compared to the baseline abundances (q value < 0.05, Maaslin2 linear random mixed-effects model, Table S1). The taxa identified could be attributed to all major microbial phyla of the human fecal microbiota. Confirming findings from diversity analyses (Figure 3), most species exhibited significantly different abundance in week 4 and week 7 compared to baseline abundances, i.e., at the beginning of the intervention and washout periods, respectively. Most of the detected species showed transient increases such as *Butyricicoccus* spp., Fusobacterium mortiferum, certain species of Ruminococci or decreases including different Roseburia species and Bifidobacterium adolescentis in response to the supply or withdrawal of black elderberry extract. However, some taxa including Akkermannsia were found to have a sustainable expansion throughout the washout period. The most relevant genera and species identified are outlines in Figure 4A,B and a summary statistic for relative abundances as well as information on significance levels derived from Maaslin2 are shown in Table 1.

3.6. Factors Determining Response to the Intervention

Considerable interest is directed towards inter-individual differences related to differential effects of prebiotic interventions. We therefore associated baseline demographic, behavioral and nutritional covariates with increases in *Akkermannsia* spp. of at least 3-fold during the trial period (i.e., 10 out of 28 participants). Daily intake of foods rich in plant fat was associated with a relevant increase in the abundance of *Akkermansia* spp. in response to the intervention ($p = 0.03$, Fisher's exact test). However, no associations of baseline α- or β-diversity with such an increase were detected. Likewise, we detected no baseline microbial signature predicting a relevant increase in *Akkermansia* abundance.

An increase in the number of observed species or in Shannon index values at week 4 and week 7 of the study compared to baseline was considered a second marker of individual response to the prebiotic. For the number of observed species, this was defined as an increase by at least 50% while for Shannon indices the 3rd quartile value (1.1687 for week 4 and 1.0568) was applied. Thus, based on observed species count, 17 and 8 participants were classified as α-diversity-responders at week 4 and week 7, corresponding to 60.7% and 28.6% of participants. Regarding Shannon index cutoffs, seven individuals were classified as α-diversity-responders at both timepoints (25%). Univariate correlation with baseline nutritional patterns revealed a significant association of α-diversity-response at week 4 with reduced consumption of dairy products such as cheese, eggs and milk (for Shannon index values) and rice (based on observed species counts). Increases in α-diversity at week 7 were associated with less baseline consumption of whole grains (Shannon) and pasta (observed species count). However, there was no association of β-diversity response at weeks 4 and/or 7 with baseline microbiome composition (i.e., Bray–Curtis dissimilarity indices comparing β-diversity responders at baseline) and specific bacterial taxa on the genus and species level (linear mixed effect model Maaslin2; data not shown).

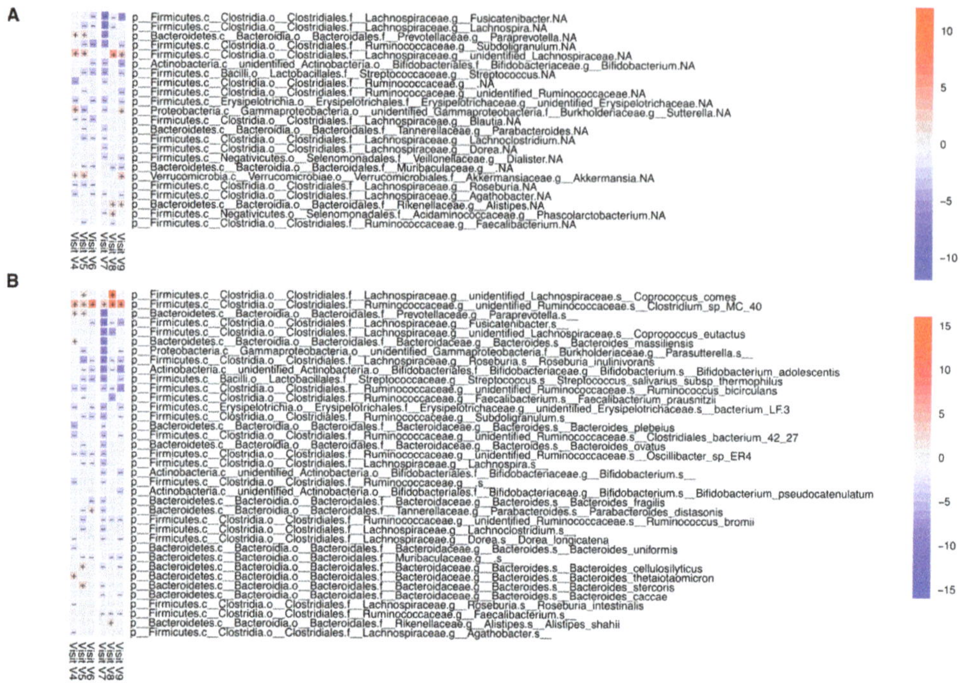

Figure 4. Analysis for differential abundance of bacterial taxa at the genus (A) and species (B) levels. Blue and red squares depict coefficients of significantly differential abundance for every week of the trial, grey squares represent non-significance (Maaslin2 linear mixed model, q-value < 0.25, p value < 0.05; detailed results provided in Table S1).

4. Discussion

Prebiotic interventions are an attractive and promising platform towards modification of the intestinal microbiome. Their comparatively broad effect on different, albeit rather specific bacterial strains, holds the potential to provide symptom relief yet a potentially mechanistic approach in microbiome-mediated diseases [8,59]. To better understand the interactions of a polyphenol-containing prebiotic based on black a elderberry extract, the ELDERGUT trial has been designed and performed. The design of this 9-week longitudinal trial with a sequence of baseline, intervention and washout periods allowed for assessment of intra-individual dynamics and mitigates the considerable baseline variation that is a common characteristic of studies of the intestinal microbiome [60,61].

Dietary polyphenols have multifaceted and often diverging effects on members of the intestinal microbiota. Extensive metabolization by the microbial enzymatic machinery results in active secondary metabolites [12,15,35]. The aim of the ELDERGUT trial was to investigate the prebiotic properties of a phenol-rich elderberry extract. This highly purified preparation enables a polyphenol content of 18% a dose corresponding to a daily consumption of 30 g raw fruit.

In our population the intake of this concentration of black elderberry extract did not result in any adverse effects. No changes in overall perception of health and quality of life as an array of digestion-associated symptoms was observed. This confirms the good tolerability of the substance which is a prerequisite for further investigation as a prebiotic. Notably, no effects on fecal consistency and the number of bowel openings were observed. These are a typical symptoms induced by many other prebiotics through osmotic and bulk-forming effects [8,62,63].

Although no effects on patient-reported outcomes were detected, the introduction and withdrawal of the study medication at week 4 and week 7 of the trial resulted in changes of measures of microbial α- and β-diversity. As a measure of within-sample variability, indices of α-diversity (observed species count, Shannon index) spiked in these moments of perturbation. Between sample variability, i.e., β-diversity, was quantified using Bray–Curtis dissimilarity. Primary coordinate analysis and PERMANOVA statistical analysis revealed significant, although comparably small, associations with sex and week of the trial, the latter corresponding to the status of supplementation. Overall, 92 bacterial taxa were identified at the species level to be differentially abundant in fecal samples when comparing the weeks during the intervention and/or washout period with the median abundance at baseline. As sudden changes in diversity measures indicate perturbation of microbial community structures this points toward a relevant microbiome-shaping capacity of this prebiotic intervention. Interestingly however, these changes in diversity were only observed immediately after the start and end of the intervention period, respectively. This finding could reflect a quick stabilizing response within the microbial communities. In the context of well-known properties of the intestinal microbiome [7], this observation is likely to reflect the resilience and stability of a healthy microbiome configuration in this very uniform study population of student volunteers. This also underscores that the stability of microbiome compositions might result in obstacles to all variations of microbiome-targeted therapies, underlining the relevance of individual baseline composition and microbiome properties for prebiotic effects.

As the study population in the ELDERGUT trial included only healthy young volunteers, applicability of these findings in the general population and especially in certain disease groups such as patients suffering from inflammatory bowel disease or irritable bowel syndrome might be limited. Overall quality of life and subjective perception of health were already high at baseline in this cohort. Therefore, there is a possibility of a beneficial effect of black elderberry extract on these measures that could not be detected in this trial. Furthermore, minor adverse events which could be potentially connected to the intervention are difficult to validate in the absence of a true placebo control group and without adequate statistical power to address this issue. Another possible limitation is, that the study intervention consisted of a natural, although highly purified mixture of different polyphenols and cannot be considered a chemically well-defined pure substance. Therefore, attribution of effects to single compounds is not possible.

Statistical methods to assess differential abundance in microbiome studies are notoriously non-robust, with different methods resulting in often differing and even contradicting conclusions. These discrepancies can depend on choices made during processing of the data and intrinsic properties of the dataset. Furthermore, some methods which have been used in the field have been reported to suffer from poor specificity and a broadly accepted 'state of the art' workflow is lacking [64,65]. Nevertheless, recent studies have made the effort to systematically compare the different methods that are currently in use [64]. On the basis of these published findings, we chose a method that has been scoring consistently in these artificial settings and also conceptionally addresses the nature of this specific dataset [53]. Thus, we are confident, that the taxa that are identified in the present work represent true hits with biologic significance. In line with earlier reports [66], we detected a significant increase in the abundance of *Akkermansia* spp. associated with the study intervention. This bacterial taxon is known for its beneficial effects on inflammation and metabolism [67,68]. It can influence and strengthen the intestinal barrier despite its mucolytic properties. Notably, *Akkermansia* has already been reported to bloom under certain polyphenol-rich diets, although no polyphenol degrading enzymatic capabilities of this genus could be delineated. It is assumed, that this effect could be due to indirect interactions between dietary polyphenols and *Akkermansia* [25,28,35], mediated by other members of the microbiota. Whether this specific effect can be considered truly prebiotic by the most recent ISAPP definitions is controversial [28]. The bacterial genus Suterella from the phylum Proteobacteria was significantly increased immediately in the first week of supplementation, although it returned to values non-significantly different from the baseline during the intervention period. This taxon has been beneficially associated with glucose metabolism and metabolic changes after Roux-Y-gastric bypass [69]. As members of this genus possess epithelium-adhering and mild pro-inflammatory properties without causing barrier disruption, it has been postulated, that this bacterial group might exert immunomodulatory effects [70]. On the other hand, in studies on fecal microbiota transplantation for Ulcerative Colitis, prevalence of this genus in the fecal transplant has been associated with a lack of remission after the procedure [71]. This detrimental effect on intestinal inflammation was linked to its ability to degrade host IgA [72].

Within the genus Bacteroides (Phylum Bacteroidetes) a more nuanced effect was detected with granular differential effects at the species level. Bacteroides cellulosolyticus increased early during the intervention. This carbohydrate-degrading species [73] has been associated with plant based nutrition [74], better outcomes of anti PD-L1 immuno-oncologic therapy [75] and disease course in COVID-19 [76]. Furthermore, Bacteroides thetaiotaomicron [77] has been implicated in the context of inflammatory bowel diseases [78,79] immune system maturation and obesity [80] and is well adapted to its existence as a gut commensal [81]. The genus Bacteroides seems to emerge as a key target genus of polyphenol supplementation in the setting of this study, probably reflecting the extensive enzymatic repertoire encoded by this bacterial taxon.

Observing the high inter-individual variability of changes in abundance during the intervention, we hypothesized that there might be underlying differences between participants at baseline that could be used to predict probability of taxonomic changes during supplementation. We identified an increase in *Akkermansia* spp. as most relevant to host health and tested for associations of baseline demographic (i.e., sex) and nutritional variables with a 3x increase in *Akkermansia* by week 4. This analysis revealed a significant association of an increase in *Akkermansia* with daily consumption of plant fats. Of note, baseline microbiome diversity indices and taxonomic composition were not significantly associated with this increase in *Akkermansia* abundance. Likewise, we detected associations of changes in α-diversity upon supplementation of black elderberry extract with baseline nutritional features but not with a distinct microbial signature at the baseline. Nevertheless, further analyses could provide better mechanistical insights in the interaction of dietary polyphenol intake, microbial abundance, enzymatic functions within the intestinal microbiota and resulting effects on the host.

5. Conclusions

The ELDERGUT trial provided convincing evidence for the prebiotic properties of a polyphenol-rich black elderberry extract in healthy individuals. While intake of this extract was associated with specific and individualized taxonomic changes within the fecal microbiota, no convincing associations with baseline factors emerged in this analysis.

Supplementary Materials: The following supporting information can be downloaded at: https://www.mdpi.com/article/10.3390/jpm12091479/s1.

Author Contributions: Conceptualization, S.R. and S.P.; data curation, S.R.; formal analysis, S.R.; funding acquisition, A.R.M.; investigation, S.R., C.W., J.L., N.P., A.P. and A.Z.; methodology, S.R., A.P. and A.R.M.; project administration, S.R., J.L., U.P. and A.R.M.; resources, H.T., S.P. and A.R.M.; supervision, A.R.M.; visualization, S.R.; writing—original draft, S.R. and S.P.; Writing—review and editing, H.T. and A.R.M. All authors have read and agreed to the published version of the manuscript.

Funding: This project was planned and conducted within the framework of the VASCage COMET Centre as a non-funded project. Funding was provided by IPRONA AG/SPA. Additionally, this work was supported by the Christian Doppler Research Association (to A.R.M) and we gratefully acknowledge the support by the Austrian Federal Ministry of Science, Research, and Economy and the National Foundation for Research, Technology, and Development.

Institutional Review Board Statement: This study was conducted according to the guidelines of the Declaration of Helsinki, and approved by the Institutional Ethics Committee of the Medical University Innsbruck (protocol code 1020/2020, 20 February 2020).

Informed Consent Statement: Informed consent was obtained from all subjects involved in this study.

Data Availability Statement: Sequence data are publicly available at EMBL-EBI and scripts detailing the reported statistical analysis and output have been deposited at github (https://github.com/reider-si/ELDERGUT).

Acknowledgments: Black elderberry extract was supplied at no additional charge by IPRONA AG/SPA who also provided financial support regarding the analyses in this study. We gratefully acknowledge the support of the VASCAGE trial platform and would like to thank Ulrike Pachmann for invaluable support regarding regulatory approval and project administration. Open Access Funding by the University of Linz.

Conflicts of Interest: S.P. is employed at IPRONA AG/SPA. The other authors declare no conflict of interest.

References

1. Brüssow, H. Problems with the concept of gut microbiota dysbiosis. *Microb. Biotechnol.* **2020**, *13*, 423–434. [CrossRef]
2. Eckburg, P.B.; Bik, E.M.; Bernstein, C.N.; Purdom, E.; Dethlefsen, L.; Sargent, M.; Gill, S.R.; Nelson, K.E.; Relman, D.A. Diversity of the human intestinal microbial flora. *Science* **2005**, *308*, 1635–1638. [CrossRef] [PubMed]
3. Human Microbiome Project Consortium. Structure, function and diversity of the healthy human microbiome. *Nature* **2012**, *486*, 207–214. [CrossRef]
4. Levy, M.; Kolodziejczyk, A.A.; Thaiss, C.A.; Elinav, E. Dysbiosis and the immune system. *Nat. Rev. Immunol.* **2017**, *17*, 219–232. [CrossRef] [PubMed]
5. Claesson, M.J.; Jeffery, I.B.; Conde, S.; Power, S.E.; O'Connor, E.M.; Cusack, S.; Harris, H.M.; Coakley, M.; Lakshminarayanan, B.; O'Sullivan, O.; et al. Gut microbiota composition correlates with diet and health in the elderly. *Nature* **2012**, *488*, 178–184. [CrossRef] [PubMed]
6. Zimmer, J.; Lange, B.; Frick, J.S.; Sauer, H.; Zimmermann, K.; Schwiertz, A.; Rusch, K.; Klosterhalfen, S.; Enck, P. A vegan or vegetarian diet substantially alters the human colonic faecal microbiota. *Eur. J. Clin. Nutr.* **2012**, *66*, 53–60. [CrossRef]
7. Lozupone, C.A.; Stombaugh, J.I.; Gordon, J.I.; Jansson, J.K.; Knight, R. Diversity, stability and resilience of the human gut microbiota. *Nature* **2012**, *489*, 220–230. [CrossRef]
8. Gibson, G.R.; Hutkins, R.; Sanders, M.E.; Prescott, S.L.; Reimer, R.A.; Salminen, S.J.; Scott, K.; Stanton, C.; Swanson, K.S.; Cani, P.D.; et al. Expert consensus document: The International Scientific Association for Probiotics and Prebiotics (ISAPP) consensus statement on the definition and scope of prebiotics. *Nat. Rev. Gastroenterol. Hepatol.* **2017**, *14*, 491–502. [CrossRef]
9. Plamada, D.; Vodnar, D.C. Polyphenols-Gut Microbiota Interrelationship: A Transition to a New Generation of Prebiotics. *Nutrients* **2022**, *14*, 137.
10. Nazzaro, F.; Fratianni, F.; De Feo, V.; Battistelli, A.; Da Cruz, A.G.; Coppola, R. Chapter Two-Polyphenols, the new frontiers of prebiotics. In *Advances in Food and Nutrition Research*; da Cruz, A.G., Prudencio, E.S., Esmerino, E.A., da Silva, M.C., Eds.; Academic Press: Cambridge, MA, USA, 2020; Volume 94, pp. 35–89.
11. Thilakarathna, W.P.D.W.; Langille, M.G.I.; Rupasinghe, H.P.V. Polyphenol-based prebiotics and synbiotics: Potential for cancer chemoprevention. *Curr. Opin. Food Sci.* **2018**, *20*, 51–57. [CrossRef]
12. Dueñas, M.; Muñoz-González, I.; Cueva, C.; Jiménez-Girón, A.; Sánchez-Patán, F.; Santos-Buelga, C.; Moreno-Arribas, M.V.; Bartolomé, B. A survey of modulation of gut microbiota by dietary polyphenols. *Biomed. Res. Int.* **2015**, *2015*, 850902. [CrossRef]
13. Netzel, M.; Strass, G.; Herbst, M.; Dietrich, H.; Bitsch, R.; Bitsch, I.; Frank, T. The excretion and biological antioxidant activity of elderberry antioxidants in healthy humans. *Food Res. Int.* **2005**, *38*, 905–910. [CrossRef]
14. Bitsch, I.; Janßen, M.; Netzel, M.; Straß, G.; Frank, T. Bioavailability of anthocyanidin-3-glycosides following consumption of elderberry extract and blackcurrant juice. *Int. J. Clin. Pharmacol. Ther.* **2004**, *42*, 293–300. [CrossRef] [PubMed]
15. Frank, T.; Janßen, M.; Netzel, G.; Christian, B.; Bitsch, I.; Netzel, M. Absorption and excretion of elderberry (*Sambucus nigra* L.) anthocyanins in healthy humans. *Methods Find. Exp. Clin. Pharmacol.* **2007**, *29*, 525–533. [CrossRef]
16. Frank, T.; Sonntag, S.; Strass, G.; Bitsch, I.; Bitsch, R.; Netzel, M. Urinary pharmacokinetics of cyanidin glycosides in healthy young men following consumption of elderberry juice. *Int. J. Clin. Pharmacol. Res.* **2005**, *25*, 47–56. [PubMed]
17. Milbury, P.E.; Cao, G.; Prior, R.L.; Blumberg, J. Bioavailablility of elderberry anthocyanins. *Mech. Ageing Dev.* **2002**, *123*, 997–1006. [CrossRef]
18. Czank, C.; Cassidy, A.; Zhang, Q.; Morrison, D.J.; Preston, T.; Kroon, P.A.; Botting, N.P.; Kay, C.D. Human metabolism and elimination of the anthocyanin, cyanidin-3-glucoside: A 13C-tracer study. *Am. J. Clin. Nutr.* **2013**, *97*, 995–1003. [CrossRef]
19. Mülleder, U.; Murkovic, M.; Pfannhauser, W. Urinary excretion of cyanidin glycosides. *J. Biochem. Biophys. Methods* **2002**, *53*, 61–66. [CrossRef]
20. Marín, L.; Miguélez, E.M.; Villar, C.J.; Lombó, F. Bioavailability of dietary polyphenols and gut microbiota metabolism: Antimicrobial properties. *Biomed. Res. Int.* **2015**, *2015*, 905215. [CrossRef]

21. Aura, A.-M.; Mattila, I.; Seppänen-Laakso, T.; Miettinen, J.; Oksman-Caldentey, K.-M.; Orešič, M. Microbial metabolism of catechin stereoisomers by human faecal microbiota: Comparison of targeted analysis and a non-targeted metabolomics method. *Phytochem. Lett.* **2008**, *1*, 18–22. [CrossRef]
22. Vlachojannis, J.E.; Cameron, M.; Chrubasik, S. A systematic review on the sambuci fructus effect and efficacy profiles. *Phytother. Res. PTR* **2010**, *24*, 1–8. [CrossRef] [PubMed]
23. Youdim, K.A.; Martin, A.; Joseph, J.A. Incorporation of the elderberry anthocyanins by endothelial cells increases protection against oxidative stress. *Free. Radic. Biol. Med.* **2000**, *29*, 51–60. [CrossRef]
24. Gwiazdowska, D.; Juś, K.; Jasnowska-Małecka, J.; Kluczyńska, K. The impact of polyphenols on Bifidobacterium growth. *Acta Biochim. Pol.* **2015**, *62*, 895–901. [CrossRef] [PubMed]
25. Roopchand, D.E.; Carmody, R.N.; Kuhn, P.; Moskal, K.; Rojas-Silva, P.; Turnbaugh, P.J.; Raskin, I. Dietary polyphenols promote growth of the gut bacterium akkermansia muciniphila and attenuate high-fat diet-induced metabolic syndrome. *Front. Immunol.* **2015**, *64*, 2847–2858. [CrossRef] [PubMed]
26. Rastmanesh, R. High polyphenol, low probiotic diet for weight loss because of intestinal microbiota interaction. *Chem. Biol. Interact.* **2011**, *189*, 1–8. [CrossRef]
27. Przybylska-Balcerek, A.; Szablewski, T.; Szwajkowska-Michałek, L.; Świerk, D.; Cegielska-Radziejewska, R.; Krejpcio, Z.; Suchowilska, E.; Tomczyk, Ł.; Stuper-Szablewska, K. Sambucus nigra extracts–natural antioxidants and antimicrobial compounds. *Molecules* **2021**, *26*, 2910. [CrossRef]
28. Rodríguez-Daza, M.C.; Pulido-Mateos, E.C.; Lupien-Meilleur, J.; Guyonnet, D.; Desjardins, Y.; Roy, D. Polyphenol-Mediated Gut Microbiota Modulation: Toward Prebiotics and Further. *Front. Nutr.* **2021**, *8*, 689456. [CrossRef]
29. Tihăuan, B.-M.; Axinie, M.; Marinaș, I.-C.; Avram, I.; Nicoară, A.-C.; Grădișteanu-Pîrcălăbioru, G.; Dolete, G.; Ivanof, A.-M.; Onisei, T.; Cășărică, A.; et al. Evaluation of the Putative Duplicity Effect of Novel Nutraceuticals Using Physico-Chemical and Biological in Vitro Models. *Foods* **2022**, *11*, 1636.
30. de Ferrars, R.M.; Cassidy, A.; Curtis, P.; Kay, C.D. Phenolic metabolites of anthocyanins following a dietary intervention study in post-menopausal women. *Mol. Nutr. Food Res.* **2014**, *58*, 490–502. [CrossRef]
31. Hidalgo, M.; Oruna-Concha, M.J.; Kolida, S.; Walton, G.E.; Kallithraka, S.; Spencer, J.P.E.; Gibson, G.R.; De Pascual-Teresa, S. Metabolism of anthocyanins by human gut microflora and their influence on gut bacterial growth. *J. Agric. Food Chem.* **2012**, *60*, 3882–3890. [CrossRef]
32. Wan, M.L.Y.; Ling, K.H.; Wang, M.F.; El-Nezami, H. Green tea polyphenol epigallocatechin-3-gallate improves epithelial barrier function by inducing the production of antimicrobial peptide pBD-1 and pBD-2 in monolayers of porcine intestinal epithelial IPEC-J2 cells. *Mol. Nutr. Food Res.* **2016**, *60*, 1048–1058. [CrossRef] [PubMed]
33. Chambers, K.F.; Day, P.E.; Aboufarrag, H.T.; Kroon, P.A. Polyphenol effects on cholesterol metabolism via bile acid biosynthesis, CYP7A1: A review. *Nutrients* **2019**, *11*, 2588. [CrossRef] [PubMed]
34. Masumoto, S.; Terao, A.; Yamamoto, Y.; Mukai, T.; Miura, T.; Shoji, T. Non-absorbable apple procyanidins prevent obesity associated with gut microbial and metabolomic changes. *Sci. Rep.* **2016**, *6*, 31208. [CrossRef] [PubMed]
35. Alves-Santos, A.M.; Sugizaki, C.S.A.; Lima, G.C.; Naves, M.M.V. Prebiotic effect of dietary polyphenols: A systematic review. *J. Funct. Foods* **2020**, *74*, 104169. [CrossRef]
36. Koh, A.; De Vadder, F.; Kovatcheva-Datchary, P.; Backhed, F. From Dietary Fiber to Host Physiology: Short-Chain Fatty Acids as Key Bacterial Metabolites. *Cell* **2016**, *165*, 1332–1345. [CrossRef]
37. Tiralongo, E.; Wee, S.S.; Lea, R.A. Elderberry Supplementation Reduces Cold Duration and Symptoms in Air-Travellers: A Randomized, Double-Blind Placebo-Controlled Clinical Trial. *Nutrients* **2016**, *8*, 182. [CrossRef]
38. Zakay-Rones, Z.; Thom, E.; Wollan, T.; Wadstein, J. Randomized study of the efficacy and safety of oral elderberry extract in the treatment of influenza A and B virus infections. *J. Int. Med. Res.* **2004**, *32*, 132–140. [CrossRef]
39. Zakay-Rones, Z.; Varsano, N.; Zlotnik, M.; Manor, O.; Regev, L.; Schlesinger, M.; Mumcuoglu, M. Inhibition of Several Strains of Influenza Virus in Vitro and Reduction of Symptoms by an Elderberry Extract (*Sambucus nigra* L.) during an Outbreak of Influenza B Panama. *J. Altern. Complement. Med.* **1995**, *1*, 361–369. [CrossRef]
40. Ho, G.T.T.; Ahmed, A.; Zou, Y.F.; Aslaksen, T.; Wangensteen, H.; Barsett, H. Structure-activity relationship of immunomodulating pectins from elderberries. *Carbohydr. Polym.* **2015**, *125*, 241–248. [CrossRef]
41. Ho, G.T.T.; Wangensteen, H.; Barsett, H. Elderberry and elderflower extracts, phenolic compounds, and metabolites and their effect on complement, RAW 264.7 macrophages and dendritic cells. *Int. J. Mol. Sci.* **2017**, *18*, 584. [CrossRef]
42. Lee, J.; Finn, C.E. Anthocyanins and other polyphenolics in American elderberry (*Sambucus canadensis*) and European elderberry (*S. nigra*) cultivars. *J. Sci. Food Agric.* **2007**, *87*, 2665–2675. [CrossRef] [PubMed]
43. Altschul, S.F.; Gish, W.; Miller, W.; Myers, E.W.; Lipman, D.J. Basic local alignment search tool. *J. Mol. Biol.* **1990**, *215*, 403–410. [CrossRef]
44. Caporaso, J.G.; Kuczynski, J.; Stombaugh, J.; Bittinger, K.; Bushman, F.D.; Costello, E.K.; Fierer, N.; Peña, A.G.; Goodrich, J.K.; Gordon, J.I.; et al. QIIME allows analysis of high-throughput community sequencing data. *Nat. Methods* **2010**, *7*, 335–336. [CrossRef] [PubMed]
45. Edgar, R.C.; Haas, B.J.; Clemente, J.C.; Quince, C.; Knight, R. UCHIME improves sensitivity and speed of chimera detection. *Bioinformatics* **2011**, *27*, 2194–2200. [CrossRef] [PubMed]

46. Haas, B.J.; Gevers, D.; Earl, A.M.; Feldgarden, M.; Ward, D.V.; Giannoukos, G.; Ciulla, D.; Tabbaa, D.; Highlander, S.K.; Sodergren, E.; et al. Chimeric 16S rRNA sequence formation and detection in Sanger and 454-pyrosequenced PCR amplicons. *Genome Res.* **2011**, *21*, 494–504. [CrossRef]
47. Edgar, R.C. UPARSE: Highly accurate OTU sequences from microbial amplicon reads. *Nat. Methods* **2013**, *10*, 996–998. [CrossRef] [PubMed]
48. Schloss, P.D.; Westcott, S.L.; Ryabin, T.; Hall, J.R.; Hartmann, M.; Hollister, E.B.; Lesniewski, R.A.; Oakley, B.B.; Parks, D.H.; Robinson, C.J.; et al. Introducing mothur: Open-source, platform-independent, community-supported software for describing and comparing microbial communities. *Appl. Environ. Microbiol.* **2009**, *75*, 7537–7541. [CrossRef]
49. Quast, C.; Pruesse, E.; Yilmaz, P.; Gerken, J.; Schweer, T.; Yarza, P.; Peplies, J.; Glockner, F.O. The SILVA ribosomal RNA gene database project: Improved data processing and web-based tools. *Nucleic Acids Res.* **2013**, *41*, D590–D596. [CrossRef]
50. R Core Team. *R: A Language and Environment for Statistical Computing*; R Core Team: Vienna, Austria, 2020.
51. McMurdie, P.J.; Holmes, S. phyloseq: An R package for reproducible interactive analysis and graphics of microbiome census data. *PLoS ONE* **2013**, *8*, e61217. [CrossRef]
52. Oksanen, J.; Blanchet, F.G.; Friendly, M.; Kindt, R.; Legendre, P.; McGlinn, D.; Minchin, P.R.; O'Hara, R.B.; Simpson, G.L.; Solymos, P.; et al. Vegan: Community Ecology Package; R Package Version 2.4-6. 2018. Available online: https://cran.r-project.org/web/packages/vegan/vegan.pdf (accessed on 1 September 2022).
53. Mallick, H.; Rahnavard, A.; McIver, L.J.; Ma, S.; Zhang, Y.; Nguyen, L.H.; Tickle, T.L.; Weingart, G.; Ren, B.; Schwager, E.H.; et al. Multivariable association discovery in population-scale meta-omics studies. *PLoS Comput. Biol.* **2021**, *17*, e1009442. [CrossRef]
54. Wickham, H.; Averick, M.; Bryan, J.; Chang, W.; McGowan, L.; François, R.; Grolemund, G.; Hayes, A.; Henry, L.; Hester, J.; et al. Welcome to the Tidyverse. *J. Open Source Softw.* **2019**, *4*, 1686. [CrossRef]
55. Kassambara, A. Rstatix: Pipe-Friendly Framework for Basic Statistical Tests; 0.6.0. 2020. Available online: https://rpkgs.datanovia.com/rstatix/ (accessed on 1 September 2022).
56. Kassambara, A. Ggpubr: 'Ggplot2' Based Publication Ready Plots; 0.4.0. 2020. Available online: https://rpkgs.datanovia.com/ggpubr/ (accessed on 1 September 2022).
57. Lewis, S.J.; Heaton, K.W. Stool form scale as a useful guide to intestinal transit time. *Scand. J. Gastroenterol.* **1997**, *32*, 920–924. [CrossRef] [PubMed]
58. Blake, M.R.; Raker, J.M.; Whelan, K. Validity and reliability of the Bristol Stool Form Scale in healthy adults and patients with diarrhoea-predominant irritable bowel syndrome. *Aliment. Pharmacol. Ther.* **2016**, *44*, 693–703. [CrossRef] [PubMed]
59. Caruso, R.; Lo, B.C.; Núñez, G. Host-microbiota interactions in inflammatory bowel disease. *Nat. Rev. Immunol.* **2020**, *20*, 411–426. [CrossRef] [PubMed]
60. Schirmer, M.; Franzosa, E.A.; Lloyd-Price, J.; McIver, L.J.; Schwager, R.; Poon, T.W.; Ananthakrishnan, A.N.; Andrews, E.; Barron, G.; Lake, K.; et al. Dynamics of metatranscription in the inflammatory bowel disease gut microbiome. *Nat. Microbiol.* **2018**, *3*, 337–346. [CrossRef]
61. Mehta, R.S.; Abu-Ali, G.S.; Drew, D.A.; Lloyd-Price, J.; Subramanian, A.; Lochhead, P.; Joshi, A.D.; Ivey, K.L.; Khalili, H.; Brown, G.T.; et al. Stability of the human faecal microbiome in a cohort of adult men. *Nat. Microbiol.* **2018**, *3*, 347–355. [CrossRef]
62. Marcobal, A.; Barboza, M.; Froehlich, J.W.; Block, D.E.; German, J.B.; Lebrilla, C.B.; Mills, D.A. Consumption of human milk oligosaccharides by gut-related microbes. *J. Agric. Food Chem.* **2010**, *58*, 5334–5340. [CrossRef]
63. Yasukawa, Z.; Inoue, R.; Ozeki, M.; Okubo, T.; Takagi, T.; Honda, A.; Naito, Y. Effect of Repeated Consumption of Partially Hydrolyzed Guar Gum on Fecal Characteristics and Gut Microbiota: A Randomized, Double-Blind, Placebo-Controlled, and Parallel-Group Clinical Trial. *Nutrients* **2019**, *11*, 2170. [CrossRef] [PubMed]
64. Nearing, J.T.; Douglas, G.M.; Hayes, M.G.; MacDonald, J.; Desai, D.K.; Allward, N.; Jones, C.M.A.; Wright, R.J.; Dhanani, A.S.; Comeau, A.M.; et al. Microbiome differential abundance methods produce different results across 38 datasets. *Nat. Commun.* **2022**, *13*, 342. [CrossRef]
65. Khomich, M.; Måge, I.; Rud, I.; Berget, I. Analysing microbiome intervention design studies: Comparison of alternative multivariate statistical methods. *PLoS ONE* **2021**, *16*, e0259973. [CrossRef]
66. Anhê, F.F.; Roy, D.; Pilon, G.; Dudonné, S.; Matamoros, S.; Varin, T.V.; Garofalo, C.; Moine, Q.; Desjardins, Y.; Levy, E.; et al. A polyphenol-rich cranberry extract protects from diet-induced obesity, insulin resistance and intestinal inflammation in association with increased Akkermansia spp. population in the gut microbiota of mice. *Gut* **2015**, *64*, 872–883. [CrossRef] [PubMed]
67. Cani, P.D.; Depommier, C.; Derrien, M.; Everard, A.; de Vos, W.M. Akkermansia muciniphila: Paradigm for next-generation beneficial microorganisms. *Nat. Rev. Gastroenterol. Hepatol.* **2022**. *epub ahead of print*. [CrossRef] [PubMed]
68. Grander, C.; Adolph, T.E.; Wieser, V.; Lowe, P.; Wrzosek, L.; Gyongyosi, B.; Ward, D.V.; Grabherr, F.; Gerner, R.R.; Pfister, A.; et al. Recovery of ethanol-induced Akkermansia muciniphila depletion ameliorates alcoholic liver disease. *Gut* **2018**, *67*, 891–901. [CrossRef] [PubMed]
69. Wang, C.; Zhang, H.; Liu, H.; Zhang, H.; Bao, Y.; Di, J.; Hu, C. The genus Sutterella is a potential contributor to glucose metabolism improvement after Roux-en-Y gastric bypass surgery in T2D. *Diabetes Res. Clin. Pract.* **2020**, *162*, 108116. [CrossRef] [PubMed]
70. Hiippala, K.; Kainulainen, V.; Kalliomäki, M.; Arkkila, P.; Satokari, R. Mucosal Prevalence and Interactions with the Epithelium Indicate Commensalism of Sutterella spp. *Front. Microbiol.* **2016**, *7*, 1706. [CrossRef]
71. Paramsothy, S.; Nielsen, S.; Kamm, M.A.; Deshpande, N.P.; Faith, J.J.; Clemente, J.C.; Paramsothy, R.; Walsh, A.J.; van den Bogaerde, J.; Samuel, D.; et al. Specific Bacteria and Metabolites Associated with Response to Fecal Microbiota Transplantation in Patients with Ulcerative Colitis. *Gastroenterology* **2019**, *156*, 1440–1454. [CrossRef]

72. Moon, C.; Baldridge, M.T.; Wallace, M.A.; Burnham, C.-A.D.; Virgin, H.W.; Stappenbeck, T.S. Vertically transmitted faecal IgA levels determine extra-chromosomal phenotypic variation. *Nature* **2015**, *521*, 90–93. [CrossRef]
73. Robert, C.; Chassard, C.; Lawson, P.A.; Bernalier-Donadille, A. Bacteroides cellulosilyticus sp. nov., a cellulolytic bacterium from the human gut microbial community. *Int. J. Syst. Evol. Microbiol.* **2007**, *57*, 1516–1520. [CrossRef]
74. McNulty, N.P.; Wu, M.; Erickson, A.R.; Pan, C.; Erickson, B.K.; Martens, E.C.; Pudlo, N.A.; Muegge, B.D.; Henrissat, B.; Hettich, R.L.; et al. Effects of diet on resource utilization by a model human gut microbiota containing Bacteroides cellulosilyticus WH2, a symbiont with an extensive glycobiome. *PLoS Biol.* **2013**, *11*, e1001637. [CrossRef]
75. Dees, K.J.; Koo, H.; Humphreys, J.F.; Hakim, J.A.; Crossman, D.K.; Crowley, M.R.; Nabors, L.B.; Benveniste, E.N.; Morrow, C.D.; McFarland, B.C. Human gut microbial communities dictate efficacy of anti-PD-1 therapy in a humanized microbiome mouse model of glioma. *Neurooncol. Adv.* **2021**, *3*, vdab023. [CrossRef]
76. Xu, X.; Zhang, W.; Guo, M.; Xiao, C.; Fu, Z.; Yu, S.; Jiang, L.; Wang, S.; Ling, Y.; Liu, F.; et al. Integrated analysis of gut microbiome and host immune responses in COVID-19. *Front. Med.* **2022**, *16*, 263–275. [CrossRef] [PubMed]
77. Porter, N.T.; Luis, A.S.; Martens, E.C. Bacteroides thetaiotaomicron. *Trends Microbiol.* **2018**, *26*, 966–967. [CrossRef]
78. Delday, M.; Mulder, I.; Logan, E.T.; Grant, G. Bacteroides thetaiotaomicron Ameliorates Colon Inflammation in Preclinical Models of Crohn's Disease. *Inflamm. Bowel. Dis.* **2019**, *25*, 85–96. [CrossRef]
79. Li, K.; Hao, Z.; Du, J.; Gao, Y.; Yang, S.; Zhou, Y. Bacteroides thetaiotaomicron relieves colon inflammation by activating aryl hydrocarbon receptor and modulating CD4(+)T cell homeostasis. *Int. Immunopharmacol.* **2021**, *90*, 107183. [CrossRef] [PubMed]
80. Liu, R.; Hong, J.; Xu, X.; Feng, Q.; Zhang, D.; Gu, Y.; Shi, J.; Zhao, S.; Liu, W.; Wang, X.; et al. Gut microbiome and serum metabolome alterations in obesity and after weight-loss intervention. *Nat. Med.* **2017**, *23*, 859–868. [CrossRef] [PubMed]
81. Cullen, T.W.; Schofield, W.B.; Barry, N.A.; Putnam, E.E.; Rundell, E.A.; Trent, M.S.; Degnan, P.H.; Booth, C.J.; Yu, H.; Goodman, A.L. Gut microbiota. Antimicrobial peptide resistance mediates resilience of prominent gut commensals during inflammation. *Science* **2015**, *347*, 170–175. [CrossRef]

Journal of *Personalized Medicine*

MDPI

Review

Clinical Application of Ultra-High-Frequency Ultrasound

Anna Russo [1], Alfonso Reginelli [1], Giorgia Viola Lacasella [2], Enrico Grassi [3], Michele Ahmed Antonio Karaboue [4], Tiziana Quarto [5], Gian Maria Busetto [6], Alberto Aliprandi [7], Roberta Grassi [8,9] and Daniela Berritto [4,*]

1 Department of Precision Medicine, University of Campania "Luigi Vanvitelli", 80138 Naples, Italy
2 Department of Anatomical, Histological, Forensic and Orthopedic Sciences, Sapienza University of Rome, 00185 Rome, Italy
3 Department of Orthopedics, University of Florence, 50121 Florence, Italy
4 Department of Clinical and Experimental Medicine, University of Foggia, 71122 Foggia, Italy
5 Department of Law, University of Foggia, 71100 Foggia, Italy
6 Department of Urology and Renal Transplantation, University of Foggia Policlinico Riuniti of Foggia, 71122 Foggia, Italy
7 Department of Radiology, Istituti Clinici Zucchi, 20900 Monza, Italy
8 Department of Precision Oncology, University of Campania "Luigi Vanvitelli", 80138 Naples, Italy
9 Italian Society of Medical and Interventional Radiology (SIRM), SIRM Foundation, 20122 Milano, Italy
* Correspondence: berritto.daniela@gmail.com

Abstract: Musculoskeletal ultrasound involves the study of many superficial targets, especially in the hands, wrists, and feet. Many of these areas are within the first 3 cm of the skin surface and are ideal targets for ultra-high-frequency ultrasound. The high spatial resolution and the superb image quality achievable allow foreseeing a wider use of this novel technique, which has the potential to bring innovation to diagnostic imaging.

Keywords: ultrasound; ultra-high frequency; musculoskeletal; imaging

Citation: Russo, A.; Reginelli, A.; Lacasella, G.V.; Grassi, E.; Karaboue, M.A.A.; Quarto, T.; Busetto, G.M.; Aliprandi, A.; Grassi, R.; Berritto, D. Clinical Application of Ultra-High-Frequency Ultrasound. *J. Pers. Med.* **2022**, *12*, 1733. https://doi.org/10.3390/jpm12101733

Academic Editor: Anne-Marie Caminade

Received: 18 August 2022
Accepted: 27 September 2022
Published: 19 October 2022

1. Introduction

Ultra-high frequency ultrasound (UHFUS) is a method with the advantage of examining very small structures and providing high-resolution images with a dynamic, real-time, and comparative, where possible, evaluation [1]. UHFUS was established in clinical imaging in the 2000s, with its predominant application in dermatology and angiology [2–4]. In 1979, Alexander and Miller first introduced US as a noninvasive technique to measure normal skin thickness, and, in the 1980s and 1990s, high-resolution ultrasonography (HRUS) was used for the noninvasive assessment of skin nodules and cutaneous diseases [5,6]. The notion of UHFUS is often misunderstood due to a lack of consensus regarding the cutoff of frequencies for the classification of "very high frequencies" and "ultra-high frequencies". The conventional ultrasonography (CUS) technique refers to the use of probes with frequencies ranging from 10 to 15 MHz. Bhatta et al. [7–10] considered high-frequency US (HFUS) to involve frequencies >10 MHz, while Polanska et al. [7–10] selected 20 MHz as the cutoff for HFUS. Finally, Shung et al. [7–10] defined HFUS for probes with frequencies >30 MHz. UHFUS, with frequencies as high as 50–70 MHz and capable of resolutions as fine as 30 microns, could permit new diagnostic applications in superficial small parts' examination. A variety of superficial targets within the first 1 cm of the skin surface could be imaged, including for dermatological evaluation, such as the assessment of skin layers, hair follicles, nail units, and vascular, musculoskeletal, intraoral, and small-parts applications [11–14]. Hayaski et al. carried out a description of their experience in the identification and localization of lymphatic vessels in patients with secondary lymphedema. UHFUS provided images with extremely high resolution, demonstrating new characteristics of the lymphatic vessels [15,16]. The introduction of UHFUS in the study of oral lesions is one of the biggest innovations in US imaging of the head and neck region. The available frequencies of 48 and 70 MHz can effectively image the superficial layers of the mucosa as well as lesions of the oral cavity [17–21]. In comparison to CUS, UHFUS has a superior spatial resolution, even with the limitation of a low penetration depth, which is within 1–3 cm. In fact, 48 MHz probes have a penetration depth of 23.5 mm, while 70 MHz probes allow the imaging of the first 10.0 mm below the scanning surface.

The study of vascularization can be accomplished with the combination of color Doppler, which allows the observation of the distribution and size of vessels, and the power Doppler, which is usually more sensitive for detecting slow flow.

More than others, the method depends on the operator, as the probes are extremely sensitive to the examiner's hand movements. The literature on UHFUS is still evolving, but this technology seems to be the answer to several diagnostic limitations related to the need for high-resolution investigation for both normal anatomy and disease processes. In this paper, we present the role and potential clinical applications of UHFUS in musculoskeletal imaging based on our experience with a commercially available ultrasound system (Vevo MD; FUJIFILM VisualSonics, Amsterdam, The Netherlands): an ultra-high-frequency clinical ultrasound system equipped with a 48 to 70 MHz linear-array transducer. The system yields an image resolution up to 30 μm, which is equivalent to half a grain of sand, within the first 3 cm from the body surface.

Musculoskeletal evaluation with UHFUS includes, in both adult and pediatric patients, physiological and pathological disorders of superficial joints, tendons, and nerves, as well as presurgical planning and post-treatment follow-up and a guide for non-surgical interventional procedures [22,23].

2. Tendons

UHFUS encompasses wide applications, especially in the evaluation of hand anatomy and pathological disorders of the upper arm, including the study of pulleys, fascia, retinacula, and other superficial structures of the soft tissues (Figures 1–3). Particularly concerning the flexor tendon pulleys, the added value of UHFUS becomes crucial due to the subtle thickness of these structures, with a potential role in the identification of traumatic injuries [24–26]. UHFUS can identify pulley ruptures, especially with the dynamic assessment of the tendon-to-bone distance in flexion against resistance. In the evaluation prior to hand surgery, the analysis of hand anatomy can become essential for preoperative planning, postoperative management, and follow-up [27]. Tenosynovitis is defined as hypoechoic or anechoic thickened tissue with or without fluid within the tendon sheath with possible signs of Doppler signals, which are seen in two perpendicular planes [14]. Both tendon disease and tenosynovitis are important features of rheumatoid arthritis (RA), and US represents an ideal tool for their investigation. Tenosynovitis is seen as a combination of synovial thickening within the tendon sheath and tendon sheath effusion [28–36].

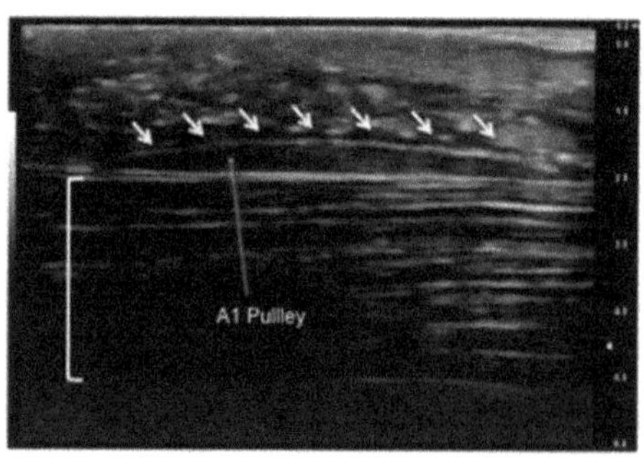

Figure 1. Sagittal view of A1 pulley (white arrows) presenting as a fusiform structure with a hypo-echoic signal contoured by a thin hyperechoic line. The superficial flexor tendon is visible (square parenthesis).

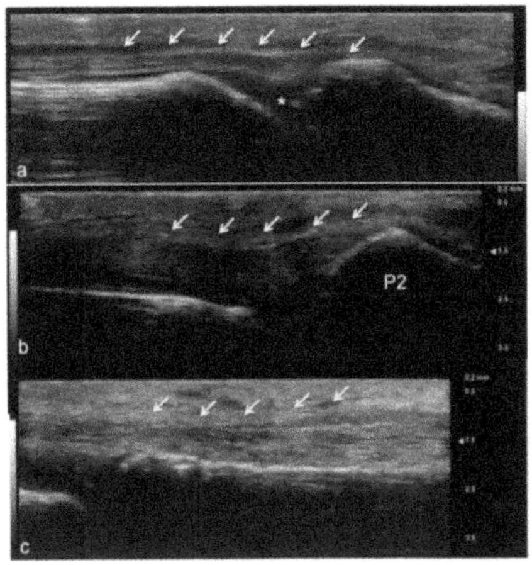

Figure 2. Extensor tendon of a finger. UHFUS gives a detailed and magnified representation of small structures such as the extensor tendon of the finger, even allowing the visualization of partial lesions. In (**a**), the sagittal view of the terminal extensor tendon (white arrows) at the level of the distal interphalangeal joint (white asterisk). In (**b**), the median band of the extensor tendon inserting at the level of the middle phalanx (P2). In (**c**), the thin sagittal band of the extensor tendon at the level of the metacarpal head.

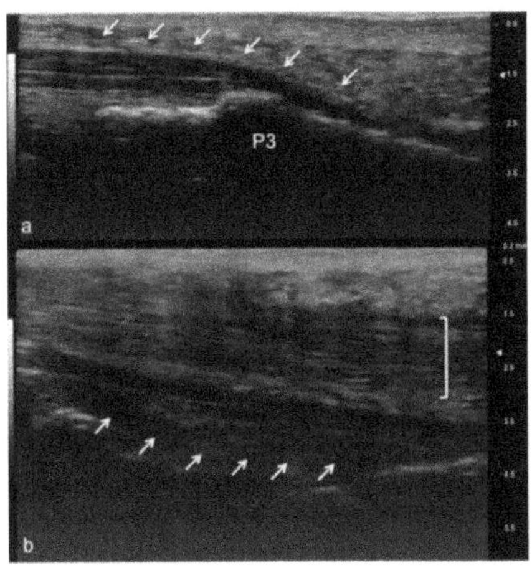

Figure 3. Flexor tendon of a finger. Using UHFUS at the flexor tendons, the same spatial resolution can be achieved as in the imaging of the extensor tendons. In (**a**), the sagittal view of the deep flexor tendon component inserting on the basis of the distal phalanx (P3). In (**b**), the superficial component of the flexor tendon (white arrows) lying near the deep component (white square parenthesis).

3. Small Joints

US is used routinely in rheumatologic and traumatic disorders to evaluate the joints of the wrist, hand, ankle, and foot. The use of UHFUS in the small joints of the hand and foot can demonstrate alterations such as synovial hyperplasia, the presence of calcifications, osteophytes, and an increased vascular signal and can be used to assess the thickness, homogeneity, and sharpness of the articular cartilage of peripheral joints, which is of great help in clinical practice [28–31]. In this field, above all, pediatric patients could benefit from such small transducers with great resolution, and this technology is designed for the smallest of patients with the greatest details and resolution possible (Figures 4 and 5).

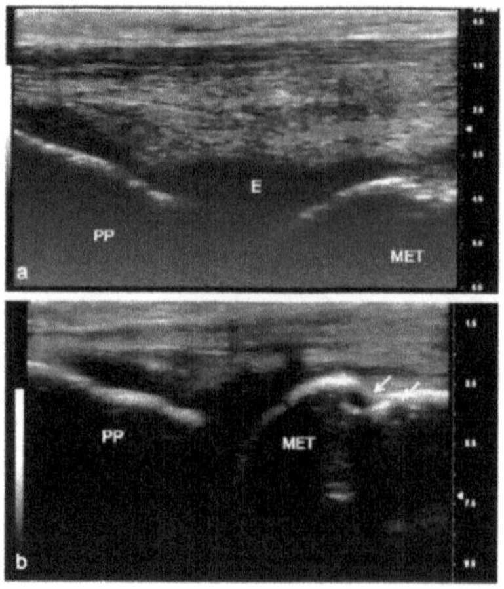

Figure 4. Juvenile idiopathic arthritis (JIA). UHFUS gives clear details of pathological findings in pediatric patients with JIA. In (**a**), articular effusion (E) at the level of the metacarpal–phalangeal joint (MET-PP). In (**b**), osseous erosions at the level of the metacarpal head (white arrows).

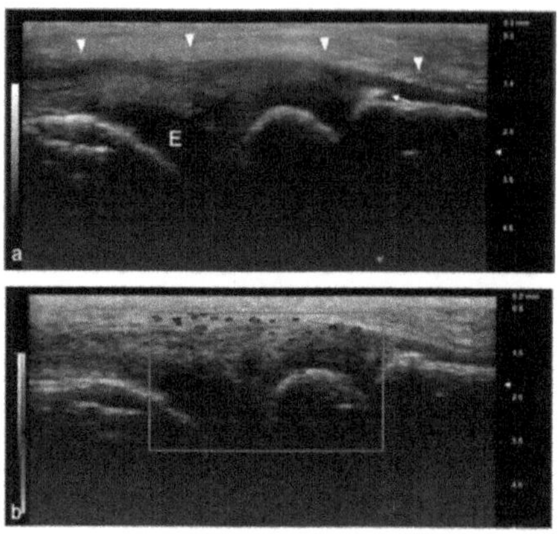

Figure 5. Juvenile idiopathic arthritis. UHFUS gives clear details of pathological findings in pediatric patients with JIA. In (**a**), articular effusion (E) at the level of wrist and enthesophyte (white arrow). Thickened capsule (arrowheads) in (**a**) presenting with an increased Doppler signal (white square) in (**b**).

Over the last 10 years, US research in rheumatology has been standardized for early diagnosis and therapy monitoring [32–35].

The term synovitis is used to indicate the presence of synovial hypertrophy with a power Doppler signal and joint effusion, either proliferative or exudative. Its quantification via grayscale US usually uses a semiquantitative scale with three levels of intensity, indicating mild, moderate, or marked synovial changes [28–36]. According to the OMERACT indications, synovial fluid is defined as abnormal hypoechoic or anechoic intra-articular material that is displaceable and compressible and that does not exhibit a Doppler signal.

Erosions appear on US as focal discontinuities in the bone cortex. US assessment can provide detailed imaging of the hyaline cartilage, identifying small cartilage abnormalities in patients affected by RA, especially when dynamic US is performed in flexion of the finger joints and extension. Effusion is defined in US as abnormal hypoechoic or anechoic intra-articular material that is displaceable and compressible, but it does not exhibit a Doppler signal [37]. In the evaluation of crystalline arthropathies such as gouty arthropathy, US allows the evaluation of intra- and para-articular tophaceous deposits as well as the typical double-contour collateral ligaments. In calcium pyrophosphate deposition disease, US allows the evaluation of crystal deposits as well as hyperechoic dots or an irregular line within the cartilage layer [37–40]. UHFUS with 50 MHz probes is able to accurately identify cartilage thinning, cartilage echogenicity, and subchondral bone.

Collateral Digital Ligaments

UHFUS offers a detailed and magnified representation of collateral ligaments of the hand (Figure 6).

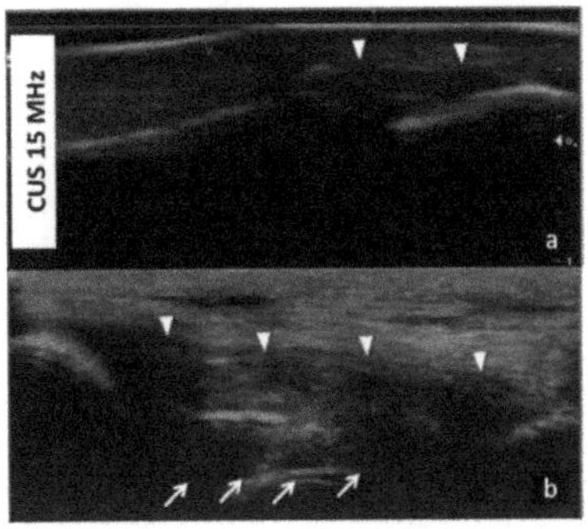

Figure 6. Collateral ligaments of interphalangeal joints: comparison between 50 MHz probes in (**a**) and 15 MHz CUS (arrowheads) in (**b**). In (**a**), UHFUS gives a more detailed and magnified representation of both deep (white arrows) and superficial (white arrowheads) components of the ligament.

This could become useful in demonstrating even partial injuries of these structures. In particular, injury to the ulnar collateral ligament (UCL) complex of the thumb is a common traumatic lesion that requires prompt imaging evaluation for adequate treatment. UHFUS could have a key role in the illustration of both static and dynamic findings related to UCL injuries with even more details than MRI imaging (Figures 7 and 8).

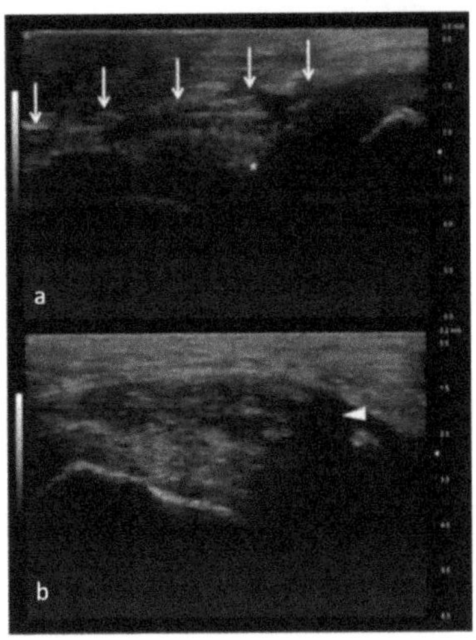

Figure 7. Full-thickness UCL tear. In (**a**), normal appearance of adductor pollicis aponeurosis (white arrows) and below the ulnar collateral ligament (asterisk) of the first metacarpophalangeal joint. In (**b**), the non-visualization of the ulnar collateral ligament and the presence of a mass-like area (arrowhead) proximal to the joint have high accuracy in depicting a displaced full-thickness ligament tear.

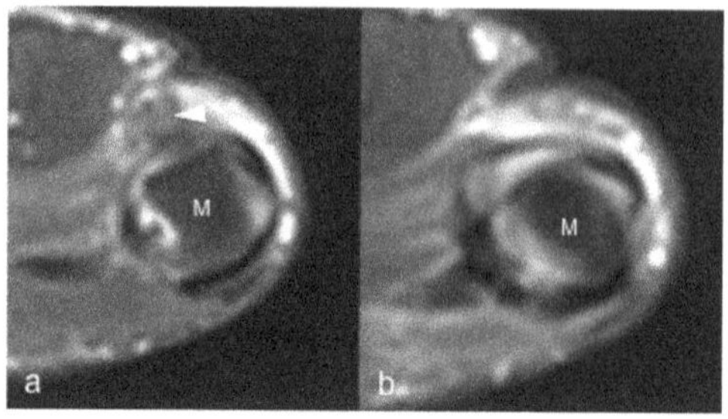

Figure 8. Same patient as in Figure 7. 1.5 T MRI shows a mass-like area (arrowhead) proximal to the joint representing the injured and displaced ligament in (**a**). In (**b**), the contralateral metacarpal phalangeal joint of the thumb. M: metacarpal head.

4. Nerves

Compared to 5–20 MHz probes, UHFUS equipped with 80 MHz probes allows a more detailed evaluation of nerve anatomy in small cutaneous nerve branches, giving a spatial resolution down to 30 microns [6,8,10,41–43] (Figure 9).

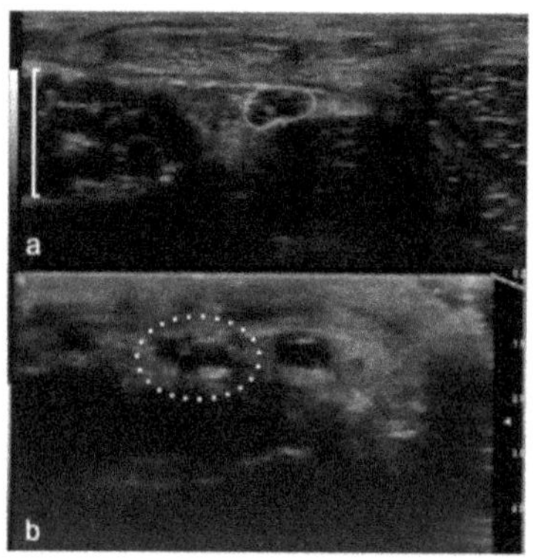

Figure 9. Visualization of small peripheral nerves using UHFUS. Recurrent branch (line) of the median nerve (square parenthesis) at the level of the thenar eminence in (**a**). Digital branch (dot circle) of the median nerve at the level of the finger in (**b**).

UHFUS allows the precise evaluation of the normal anatomical appearance as well as the echo-structural alteration of the honeycomb appearance represented by hypoechoic fascicles on a hyperechoic background. Entrapment neuropathies are generally characterized by nerve thickening, the loss of the fascicular pattern, and decreased echogenicity, mainly due to edema (in the acute phase) and fibrosis (in the chronic phase). It is also possible to see the thinning of the nerve at the site of compression. Conversely, in other disorders, such as Charcot–Marie–Tooth and some nerve neoplasms, we can see the enlargement of selected fascicles, an imaging feature recognizable using UHFUS [44–46]. UHFUS allows the determination of the location, size, and type of lesion, nerve swelling, and inflammation or the identification of compressive structures such as calcifications and scar tissue [47,48]. Nerves that are often involved in traumatic and iatrogenic injuries are the superficial cutaneous branches of the median, ulnar, and radial nerves in the hand and wrist and the digital nerves. UHFUS allows the visualization of the small cutaneous branches of peripheral nerves. The advantages of using US in the evaluation of the anatomy and disorders of the median nerve even after surgery are known in the literature: in cases of post-traumatic neoformations, such as a neuroma, both intra- and extra-operative applications are used for the preservation of the nerve after surgical excision. UHFUS could be useful in the study of pediatric patients and in neuromuscular disorders such as chronic inflammatory demyelinating diseases, thanks to the evaluation of the number of morphologies and the size of the nerve fascicles [49–51]. Benign lesions of the peripheral nerves, such as a lipofibromatous hamartoma and the infiltration of the nerve fascicles, appear as non-homogeneous nerve enlargements, especially in proximal median nerve segments and the brachial plexus [43,52], with variable echogenicity patterns depending on the stage of the disease. UFUS has been used for non-surgical guidance in percutaneous procedures for partial and total wrist denervation. In the treatment of chronic wrist pain due to post-traumatic injuries, degenerative disorders, or arthritis, denervation is the treatment for chronic wrist pain, without impairing motor function, and avoids the need for postoperative immobilization to decrease the risk of stiffness. Under US guidance, the dorsal extensor tendon compartment is visualized, and, using a transverse anatomic projection along the short axis of the tendons, radiofrequency ablation of the posterior and anterior interosseous nerves is performed. The posterior interosseous nerve appears as a thin, hypoechoic, and non-compressible ovoid structure from 1 to 3 mm in diameter (Figure 10) [53–55].

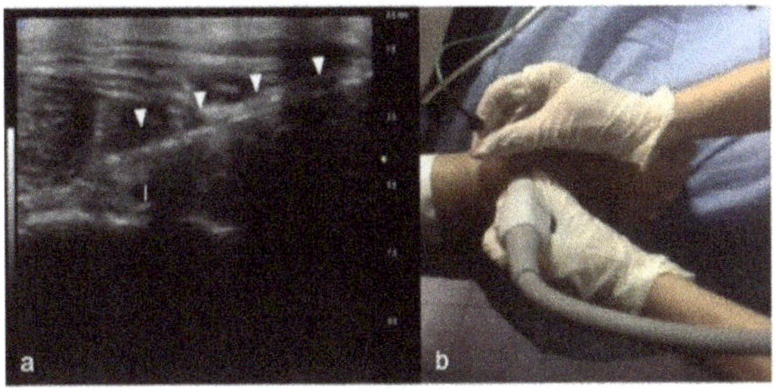

Figure 10. Percutaneous radiofrequency ablation of the posterior interosseous nerve for chronic wrist pain: In (**a**), cannula's insertion (white arrowheads) under real-time ultrasound. Guidance for direct visualization of the posterior interosseous nerve (white caliber). The procedure is performed on an awake patient using a noninvasive approach (**b**).

5. Soft-Tissue Masses

UHFUS could represent an advantageous technique for the identification of small and soft-tissue neoformations such as glomus tumors (Figures 11 and 12).

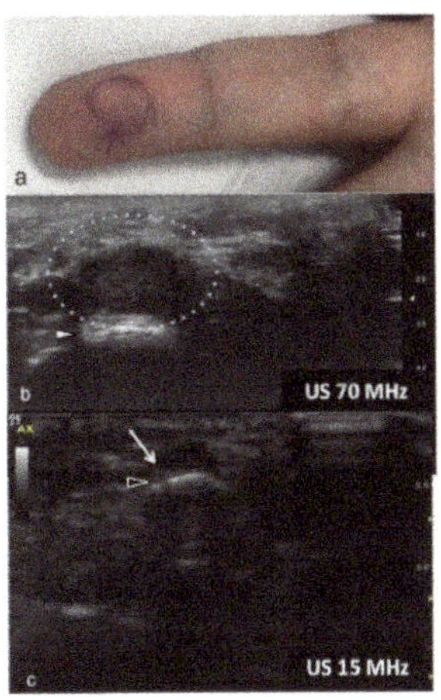

Figure 11. Glomus tumor. The location corresponds to the pulp of finger IV (**a**). The patient had the classical triad of symptoms: paroxysmal pain, pinpoint pain, and cold hypersensitivity, lasting for two years. In (**b**), UHFUS shows a well-delimited nodule in contact with the adjacent phalangeal bone (arrowhead), but no cortical deformity is present. No significant hyperemia on color Doppler was noticed. On CUS (**c**), the nodule was delineated only thanks to the help of the preliminary UHFUS exam.

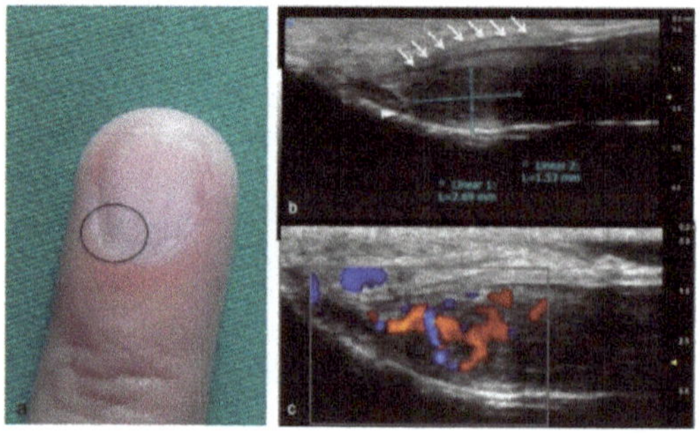

Figure 12. Glomus tumor in the subungual space. No clear alterations were visible during clinical examination at the site of pain (circle in (**a**)). In (**b**), UHFUS effectively demonstrates the presence of a hypo-isoechogenic nodule (calibers) in contact with the adjacent phalangeal bone (arrowhead) under the nail plate (arrows). Note the small deformation of the nail plate. Mild vascularization on color Doppler was present (**c**).

Glomus tumors are rare, benign, vascular neoplasms arising from the glomus body, which is a contractile neuromyoarterial structure found in the reticular dermis. This structure controls blood pressure and temperature by regulating blood flow in the cutaneous vasculature. Hyperplasia in any of these parts can lead to tumor formation, which is extremely painful. Glomus tumors account for 1–5% of soft-tissue tumors of the hand, and 75% of them are subungual in location. Other less commonly involved sites in the hand are the nail matrix, nail bed, and pulp of the finger. The delay in diagnosing these tumors for many years is a significant problem. It is not uncommon that patients are easily misdiagnosed with conditions such as neuropathic complaints, arthritis, or neuralgia and undergo unsuitable treatment. For these reasons, when the clinical examination is equivocal, noninvasive imaging techniques may be needed to aid in the diagnosis and delineate the anatomy preoperatively. Complete surgical excision of the tumor is the only effective treatment. Incomplete excision is considered the main cause of recurrence. US follow-up and/or intraoperative US may be useful for reducing recurrence and ensuring adequate resection, and the UHFUS tool is promising in this field [56].

6. Conclusions

Musculoskeletal evaluation with UHFUS includes, in both adult and pediatric patients, physiological and pathological disorders of superficial joints, tendons, and nerve structures, as well as presurgical planning and post-treatment follow-up and a guide for non-surgical interventional procedures. The high spatial resolution and the superb image quality achievable allow foreseeing the wider use of this novel technique, which has the potential to bring innovation to diagnostic imaging.

Author Contributions: Conceptualization and data collection, D.B. and A.R. (Anna Russo); writing, R.G. and E.G.; reviewing, T.Q. and A.A.; supervision, A.R. (Alfonso Reginelli) and G.M.B.; English editing, G.V.L. and M.A.A.K.; data curation, A.A. All authors have read and agreed to the published version of the manuscript.

Funding: This research received no external funding.

Institutional Review Board Statement: Not applicable.

Informed Consent Statement: Not applicable.

Data Availability Statement: Not applicable.

Conflicts of Interest: The authors declare no conflict of interest.

References

1. Izzetti, R.; Vitali, S.; Aringhieri, G.; Nisi, M.; Oranges, T.; Dini, V.; Ferro, F.; Baldini, C.; Romanelli, M.; Caramella, D.; et al. Ultra-High Frequency Ultrasound, A Promising Diagnostic Technique: Review of the Literature and Single-Center Experience. *Can. Assoc. Radiol. J.* **2020**, *72*, 418–431. [CrossRef] [PubMed]
2. Pierfelice, T.J.; Gaiano, N. Ultrasound-Guided Microinjection into the Mouse Forebrain In Utero at E9.5. *J. Vis. Exp.* **2010**, *45*, e2047. [CrossRef] [PubMed]
3. Foster, F.S. Micro-ultrasound for precli- nical imaging. *Interface Focus* **2011**, *1*, 576–601. [CrossRef] [PubMed]
4. Vogt, M.; Knüttel, A.; Hoffmann, K.; Altmeyer, P.; Ermert, H. Comparison of High Frequency Ultrasound and Optical Coherence Tomography as Modalities for High Resolution and Non Invasive Skin Imaging. Vergleich von hochfrequentem Ultraschall und optischer Kohärenztomographie als Modalitäten für die hochauflösende und nichtinvasive Abbildung der Haut. *Biomed. Eng./Biomed. Tech.* **2003**, *48*, 116–121. [CrossRef]
5. Vogt, M.; Ermert, H. In Vivo Ultrasound Biomicroscopy of Skin: Spectral System Characteristics and Inverse Filtering Optimization. *IEEE Trans. Ultrason. Ferroelectr. Freq. Control* **2007**, *54*, 1551–1559. [CrossRef] [PubMed]
6. Alexander, H.; Miller, D.L. Determining Skin Thickness with Pulsed Ultra Sound. *J. Investig. Dermatol.* **1979**, *72*, 17–19. [CrossRef] [PubMed]
7. Visconti, G.; Bianchi, A.; Hayashi, A.; Salgarello, M. Ultra-high frequency ultrasound preoperative planning of the rerouting method for lymphaticovenular anastomosis in incisions devoid of vein. *Microsurgery* **2020**, *40*, 717–718. [CrossRef]
8. Hwang, M.; Piskunowicz, M.; Darge, K. Advanced Ultrasound Techniques for Pediatric Imaging. *Pediatrics* **2019**, *143*. [CrossRef] [PubMed]
9. Bhatta, A.K. Application of high frequency ultra- sound in dermatology. *Discov. Med.* **2018**, *26*, 237–242. [PubMed]
10. Polańska, A.; Dańczak-Pazdrowska, A.; Jałowska, M.; Żaba, R.; Adamski, Z. Current applications of high-frequency ultrasonography in dermatology. *Adv. Dermatol. Allergol.* **2017**, *34*, 535–542. [CrossRef]
11. Shung, K.K. High Frequency Ultrasonic Imaging. *J. Med. Ultrasound* **2009**, *17*, 25–30. [CrossRef]
12. Berritto, D.; Iacobellis, F.; Rossi, C.; Reginelli, A.; Cappabianca, S.; Grassi, R. Ultra high-frequency ultrasound: New capabilities for nail anatomy exploration. *J. Dermatol.* **2017**, *44*, 43–46. [CrossRef] [PubMed]
13. Wortsman, X.; Carreño, L.; Ms, C.F.-W.; Ms, R.P.; Pizarro, K.; Morales, C.; Calderon, P.; Castro, A. Ultrasound Characteristics of the Hair Follicles and Tracts, Sebaceous Glands, Montgomery Glands, Apocrine Glands, and Arrector Pili Muscles. *J. Ultrasound Med.* **2019**, *38*, 1995–2004. [CrossRef] [PubMed]
14. Reginelli, A.; Belfiore, M.P.; Russo, A.; Turriziani, F.; Moscarella, E.; Troiani, T.; Brancaccio, G.; Ronchi, A.; Giunta, E.F.; Sica, A.; et al. A Preliminary Study for Quantitative Assessment with HFUS (High- Frequency Ultrasound) of Nodular Skin Melanoma Breslow Thickness in Adults Before Surgery: Interdisciplinary Team Experience. *Curr. Radiopharm.* **2020**, *13*, 48–55. [CrossRef] [PubMed]
15. Hayashi, A.; Visconti, G. Intraoperative imaging of lymphatic vessel using ultra high-frequency ultrasound. *J. Plast. Reconstr. Aesthet. Surg.* **2018**, *71*, 778–780. [CrossRef] [PubMed]
16. Hayashi, A.; Giacalone, G.; Yamamoto, T.; Belva, F.; Visconti, G.; Hayashi, N.; Handa, M.; Yoshimatsu, H.; Salgarello, M. Ultra High-frequency Ultrasonographic Imaging with 70 MHz Scanner for Visualization of the Lymphatic Vessels. *Plast. Reconstr. Surg. Glob. Open* **2019**, *7*, e2086. [CrossRef] [PubMed]
17. Bobadilla, F. Pre-surgical high resolution ultrasound of facial basal cell carcinoma: Correlation with histology. *Cancer Imaging* **2008**, *8*, 163–172. [CrossRef] [PubMed]
18. Izzetti, R.; Vitali, S.; Aringhieri, G.; Oranges, T.; Dini, V.; Nisi, M.; Graziani, F.; Gabriele, M.; Caramella, D. Discovering a new anatomy: Exploration of oral mucosa with ultra-high frequency ultrasound. *Dentomaxillofacial Radiol.* **2020**, *49*, 20190318. [CrossRef]
19. Izzetti, R.; Vitali, S.; Aringhieri, G.; Caramella, D.; Nisi, M.; Oranges, T.; Dini, V.; Graziani, F.; Gabriele, M. The efficacy of Ultra-High Frequency Ultrasonography in the diagnosis of intraoral lesions. *Oral Surg. Oral Med. Oral Pathol. Oral Radiol.* **2020**, *129*, 401–410. [CrossRef]
20. Izzetti, R.; Vitali, S.; Oranges, T.; Dini, V.; Romanelli, M.; Caramella, D.; Gabriele, M. Intraoral Ultra-High Frequency Ultrasound study of oral lichen planus: A pictorial review. *Ski. Res. Technol.* **2020**, *26*, 200–204. [CrossRef]
21. Izzetti, R.; Vitali, S.; Gabriele, M.; Caramella, D. Feasibility of a combination of intraoral UHFUS and CBCT in the study of peri-implantitis. *Oral Surg. Oral Med. Oral Pathol. Oral Radiol.* **2019**, *127*, e89–e94. [CrossRef]
22. Chianca, V.; Orlandi, D.; Messina, C.; Albano, D.; Corazza, A.; Rapisarda, S.; Pozzi, G.; Cazzato, R.L.; Mauri, G.; Silvestri, E.; et al. Interventional therapeutic procedures to treat degenerative and inflammatory musculoskeletal conditions: State of the art. *La Radiol. Med.* **2019**, *124*, 1112–1120. [CrossRef] [PubMed]
23. Salerno, S.; Laghi, A.; Cantone, M.-C.; Sartori, P.; Pinto, A.; Frija, G. Overdiagnosis and overimaging: An ethical issue for radiological protection. *La Radiol. Med.* **2019**, *124*, 714–720. [CrossRef] [PubMed]
24. Rossi, F.; Romano, N.; Muda, A.; Martinoli, C.; Tagliafico, A. Wrist and Hand Ultrasound: Reliability of Side-to-Side Comparisons of Very Small (<2-mm) Clinically Relevant Anatomic Structures. *J. Ultrasound Med.* **2018**, *37*, 2785–2795. [CrossRef]
25. Albano, D.; Aringhieri, G.; Messina, C.; De Flaviis, L.; Sconfienza, L.M. High-Frequency and Ultra-High Frequency Ultrasound: Musculoskeletal Imaging up to 70 MHz. *Semin. Musculoskelet. Radiol.* **2020**, *24*, 125–134. [CrossRef] [PubMed]

26. Sconfienza, L.M.; Albano, D.; Allen, G.; Bazzocchi, A.; Bignotti, B.; Chianca, V.; de Castro, F.F.; Drakonaki, E.E.; Gallardo, E.; Gielen, J.; et al. Clinical indications for musculoskeletal ultrasound updated in 2017 by European Society of Musculoskeletal Radiology (ESSR) consensus. *Eur. Radiol.* **2018**, *28*, 5338–5351. [CrossRef] [PubMed]
27. Li, H.; Allen, J.C.; Thumboo, J.; Tan, Y.K. Receiver operating characteristic analysis of ultrasound joint inflammation in relation to structural damage and disease activity in rheumatoid arthritis. *La Radiol. Med.* **2019**, *124*, 1037–1042. [CrossRef]
28. Salaffi, F.; Carotti, M.; Barile, A. Musculoskeletal imaging of the inflammatory and degenerative joints: Current status and perspectives. *La Radiol. Med.* **2019**, *124*, 1067–1070. [CrossRef] [PubMed]
29. Wang, M.-Y.; Wang, X.-B.; Sun, X.-H.; Liu, F.-L.; Huang, S.-C. Diagnostic value of high-frequency ultrasound and magnetic resonance imaging in early rheumatoid arthritis. *Exp. Ther. Med.* **2016**, *12*, 3035–3040. [CrossRef]
30. Filippucci, E.; Cipolletta, E.; Mirza, R.M.; Carotti, M.; Giovagnoni, A.; Salaffi, F.; Tardella, M.; Di Matteo, A.; Di Carlo, M. Ultrasound imaging in rheumatoid arthritis. *La Radiol. Med.* **2019**, *124*, 1087–1100. [CrossRef]
31. Ohrndorf, S.; Backhaus, M. Musculoskeletal ultrasonography in patients with rheumatoid arthritis. *Nat. Rev. Rheumatol.* **2013**, *9*, 433–437. [CrossRef] [PubMed]
32. Hartung, W.; Backhaus, M. Joint sonography in rheumatology. *Z. Rheumatol.* **2013**, *72*, 791–803. [CrossRef] [PubMed]
33. Arend, C.F. Ultrasonography in rheumatoid arthritis: What rheumatologists should know. *Rev. Bras. De Reum.* **2013**, *53*, 88–100. [CrossRef]
34. Gelse, K.; Olk, A.; Eichhorn, S.; Swoboda, B.; Schoene, M.; Raum, K. Quantitative ultrasound biomicroscopy for the analysis of healthy and repair cartilage tissue. *Eur. Cells Mater.* **2010**, *19*, 58–71. [CrossRef]
35. Kang, T.; Emery, P.; Wakefield, R.J. A brief history of ultrasound in rheumatology: Where we are now. *Clin. Exp. Rheumatol.* **2014**, *32*, S7–S11.
36. Girish, G. Imaging appearances in gout. *Arthritis* **2013**, *2013*, 673401. [CrossRef]
37. Filippou, G.; Adinolfi, A.; A Cimmino, M.; A Scirè, C.; Carta, S.; Lorenzini, S.; Santoro, P.; Sconfienza, L.M.; Bertoldi, I.; Picerno, V.; et al. Diagnostic accuracy of ultrasound, conventional radiography and synovial fluid analysis in the diagnosis of calcium pyrophosphate dihydrate crystal deposition disease. *Clin. Exp. Rheumatol.* **2016**, *34*, 254–260.
38. Filippou, G. Ultrasound in the diagnosis of calcium pyrophosphate dihydrate deposition dis- ease. A systematic literature review and a meta-analysis. *Osteoarthr. Cartil.* **2016**, *24*, 973–981. [CrossRef]
39. Zappia, M.; Maggialetti, N.; Natella, R.; Reginelli, A.; Bruno, F.; Di Pietto, F.; Brunese, L. Diagnostic imaging: Pitfalls in rheumatology. *La Radiol. Med.* **2019**, *124*, 1167–1174. [CrossRef]
40. Cannaò, P.M.; Vinci, V.; Caviggioli, F.; Klinger, M.; Orlandi, D.; Sardanelli, F.; Serafini, G.; Sconfienza, L.M. Technical feasibility of real-time elastography to assess the peri-oral region in patients affected by systemic sclerosis. *J. Ultrasound* **2014**, *17*, 265–269. [CrossRef]
41. Stokvis, A.; Van Neck, J.W.; Van Dijke, C.F.; Van Wamel, A.; Coert, J.H. High-resolution ultrasonography of the cutaneous nerve branches in the hand and wrist. *J. Hand Surg.* **2009**, *34*, 766–771. [CrossRef] [PubMed]
42. Cartwright, M.S.; Baute, V.; Caress, J.B.; Walker, F.O. Ultrahigh-frequency ultrasound of fascicles in the median nerve at the wrist. *Muscle Nerve* **2017**, *56*, 819–822. [CrossRef] [PubMed]
43. Forte, A.J.; Boczar, D. Ultra-high- frequency ultrasound to assess nerve fascicles in median nerve traumatic neuroma. *Cureus* **2019**, *11*, e4871. [CrossRef] [PubMed]
44. Yuk, J.I.; Walker, F.O.; Cartwright, M.S. Ultrasonography of peripheral nerves. *Curr. Neurol. Neurosci. Rep.* **2013**, *13*, 328.
45. Tagliafico, A.S. Peripheral nerve imaging: Not only cross-sectional area. *World J. Radiol.* **2016**, *8*, 726–728. [CrossRef]
46. Martinoli, C.; Schenone, A.; Bianchi, S.; Mandich, P.; Caponetto, C.; Abbruzzese, M.; Derchi, L.E. Sonography of the Median Nerve in Charcot-Marie-Tooth Disease. *Am. J. Roentgenol.* **2002**, *178*, 1553–1556. [CrossRef]
47. Regensburger, A.P.; Wagner, A.L.; Hanslik, G.; Schüssler, S.C.; Fahlbusch, F.B.; Woelfle, J.; Jüngert, J.; Trollmann, R.; Knieling, F. Ultra–high-frequency ultrasound in patients with spinal muscular atrophy: A retrospective feasibility study. *Muscle Nerve* **2020**, *61*, E18–E21. [CrossRef] [PubMed]
48. Puma, A.; Azulay, N.; Grecu, N.; Suply, C.; Panicucci, E.; Cambieri, C.; Villa, L.; Raffaelli, C.; Sacconi, S. Comparison of high-frequency and ultrahigh-frequency probes in chronic inflammatory demyelinating polyneuropathy. *J. Neurol.* **2019**, *266*, 2277–2285. [CrossRef] [PubMed]
49. Piccolo, C.L.; Galluzzo, M.; Ianniello, S.; Trinci, M.; Russo, A.; Rossi, E.; Zeccolini, M.; Laporta, A.; Guglielmi, G.; Miele, V. Pediatric musculoskeletal injuries: Role of ultrasound and magnetic resonance imaging. *Musculoskelet. Surg.* **2017**, *101*, 85–102. [CrossRef]
50. Aringhieri, G.; Vitali, S.; Rossi, P.; Caramella, D. The new frontier of imaging: The micron. *Clin. Exp. Rheumatol.* **2018**, *36*, 169.
51. Viviano, S.L.; Chandler, L.K.; Keith, J.D. Ultrahigh Frequency Ultrasound Imaging of the Hand: A New Diagnostic Tool for Hand Surgery. *Hand* **2018**, *13*, 720–725. [CrossRef] [PubMed]
52. Boczar, D.; Forte, A.J.; Serrano, L.P.; Trigg, S.D.; Clendenen, S.R. Use of Ultra-high-frequency Ultrasound on Diagnosis and Management of Lipofibromatous Hamartoma: A Technical Report. *Cureus* **2019**, *11*, e5808. [CrossRef] [PubMed]
53. Smeraglia, F.; Berritto, D.; Basso, M.A.; Mosillo, G.; Grassi, R.; Mariconda, M. Percutaneous Radiofrequency Ablation of the Posterior and Anterior Interosseous Nerves for Chronic Wrist Pain. *Tech. Hand Up. Extrem. Surg.* **2020**. [CrossRef] [PubMed]

54. Sconfienza, L.M.; Adriaensen, M.; Albano, D.; Gómez, M.P.A.; Bazzocchi, A.; Beggs, I.; Bignotti, B.; Chianca, V.; Corazza, A.; Dalili, D.; et al. Clinical indications for image-guided interventional procedures in the musculoskeletal system: A Delphi-based consensus paper from the European Society of Musculoskeletal Radiology (ESSR)—Part II, elbow and wrist. *Eur. Radiol.* **2019**, *30*, 2220–2230. [CrossRef]
55. Arrigoni, F.; Bruno, F.; Zugaro, L.; Splendiani, A.; Di Cesare, E.; Barile, A.; Masciocchi, C. Role of interventional radiology in the management of musculoskeletal soft-tissue lesions. *La Radiol. Med.* **2019**, *124*, 253–258. [CrossRef] [PubMed]
56. Vandhuick, O.; Forlodou, P.; Quintin, I.; Le Nen, D.; Guias, B.; Bressollette, L. Tumeur glomique du DOIGT. *J. Des Mal. Vasc.* **2005**, *30*, 235–236. [CrossRef]

Journal of *Personalized Medicine*

MDPI

Review

Non-Oncological Radiotherapy: A Review of Modern Approaches

Valerio Nardone [1,*,†], Emma D'Ippolito [1,†], Roberta Grassi [1], Angelo Sangiovanni [1], Federico Gagliardi [1], Giuseppina De Marco [1], Vittorio Salvatore Menditti [1], Luca D'Ambrosio [1], Fabrizio Cioce [1], Luca Boldrini [2], Viola Salvestrini [3], Carlo Greco [4], Isacco Desideri [3], Francesca De Felice [5], Ida D'Onofrio [6], Roberto Grassi [1], Alfonso Reginelli [1,‡] and Salvatore Cappabianca [1,‡]

1. Department of Precision Medicine, University of Campania "L. Vanvitelli", 80138 Naples, Italy
2. Radiation Oncology, Fondazione Policlinico Universitario A. Gemelli, IRCCS, Largo Agostino Gemelli, 00168 Rome, Italy
3. Radiation Oncology, Azienda Ospedaliero-Universitaria Careggi, Department of Experimental and Clinical Biomedical Sciences, University of Florence, 50134 Florence, Italy
4. Department of Radiation Oncology, Università Campus Bio-Medico di Roma, Fondazione Policlinico Universitario Campus Bio-Medico, Via Alvaro del Portillo, 00128 Rome, Italy
5. Radiation Oncology, Policlinico Umberto I "Sapienza" University of Rome, Viale Regina Elena 326, 00161 Rome, Italy
6. Radiation Oncology, Ospedale del Mare, ASL Napoli 1 Centro, 80147 Naples, Italy

* Correspondence: valerio.nardone@unicampania.it
† These authors contributed equally to this work.
‡ These authors contributed equally to this work.

Abstract: Despite being usually delivered in oncological patients, radiotherapy can be used as a successful treatment for several non-malignant disorders. Even though this use of radiotherapy has been scarcely investigated since the 1950s, more recent interest has actually shed the light on this approach. Thus, the aim of this narrative review is to analyze the applications of non-oncological radiotherapy in different disorders. Key references were derived from a PubMed query. Hand searching and clinicaltrials.gov were also used. This review contains a narrative report and a critical discussion of non-oncological radiotherapy approaches. In conclusion, non-oncological radiotherapy is a safe and efficacious approach to treat several disorders that needs to be further investigated and used in clinical practice.

Keywords: radiotherapy; non-malignant disorders; non-oncological radiotherapy

Citation: Nardone, V.; D'Ippolito, E.; Grassi, R.; Sangiovanni, A.; Gagliardi, F.; De Marco, G.; Menditti, V.S.; D'Ambrosio, L.; Cioce, F.; Boldrini, L.; et al. Non-Oncological Radiotherapy: A Review of Modern Approaches. *J. Pers. Med.* **2022**, *12*, 1677. https://doi.org/10.3390/jpm12101677

Academic Editor: Anne-Marie Caminade

Received: 22 September 2022
Accepted: 6 October 2022
Published: 9 October 2022

1. Introduction

Radiation therapy (RT) represents one of the cornerstones of cancer management, together with surgery and systemic therapy. It is reported that almost half of all cancer patients will receive RT during their treatment [1].

In the last decades, RT has undergone several advances driven by the increase in knowledge of radiobiology, use of advanced imaging, and treatment delivery approaches [2–5].

Actually, RT can be delivered with great accuracy, to reach an increasing dose to the targets and at the same time sparing the surrounding organs at risk [2,3].

Non-oncological radiotherapy can be used to treat several disorders and accounts for 20% of all treated patients in Germany [6–8].

In many other countries, including Italy, the use of RT for non-oncological diseases is not very common among the RT centers and often unknown among the other specialists.

The knowledge and the promotion of non-oncological RT, thus, could counterbalance the expected loss of patients or RT fractions foregone in the next future due to the use of hypofractionation.

Herein, we will discuss the different non oncological diseases that can be treated by RT. Following a literature search, we will provide a narrative overview of these topics.

2. Search Strategy

A literature search was conducted to retrieve potential eligible studies using PubMed and the clinicaltrials.gov electronic database.

The literature search was performed in September 2022 (from 1980 to August 2022) using the keywords "Radiotherapy AND non oncological OR non malignant OR benign".

Additionally, we manually searched the reference lists of studies and review papers to identify other relevant studies. No limits were applied to publication type. The results are grouped according to the non-oncological disorders and discussed qualitatively.

3. Results

3.1. Heart

Cardiac arrhythmias affected 8.8 million people in Europe and is an independent risk factor for stroke, [9], heart disease [10] and chronic kidney disease [11]. Recently, RT has been investigated as a potential treatment for recurrent ventricular tachycardia (VT) and atrial fibrillation (AF). The technological improvement of RT techniques might arguably open up to new treatments in this setting. We report the main data in the literature regarding RT treatment of VT and AF. A summary of collected evidence is reported in Table 1.

3.1.1. Ventricular Tachycardia

Stereotactic radiotherapy (SBRT) is under investigation as a treatment option for patients with VT who do not respond to antiarrhythmic drugs and/or catheter ablation. The evidence is scarce, but most studies suggest the dose of 25 Gy in a single session.

Cuculich et al. combined techniques of electrocardiographic imaging to map arrhythmogenic scar regions in patients with refractory VT and non-invasive delivery of precise ablation with SBRT to perform noninvasive cardiac radioablation. SBRT dose was a single fraction of 25 Gy. Treatment efficacy was measured using the number of VT episodes recorded by ICDs [12]. Of the nine evaluated patients, only five underwent SBRT, with no acute high-grade toxicity or complications. Sixty percent of the patients showed mild fatigue and all the patients showed a reduction of VT [12]. Similarly, in the Polish SMART-VT Trial, Kurzelowski et al. used Stereotactic Arrhythmia Radioablation (STAR) for two patients with refractory VT, using the same dose (25 Gy in single fx). After 6 months, ICD showed no VT episodes in one patient and a good response in the other patient [13].

Wight et al., similarly, treated with the same dose 14 patients with refractory VT. In this trial, the clinical target volume was delineated according to the individual patient's characteristics, based on electroanatomic mapping. Two patients died after SBRT, whereas one received a heart transplant, and another patient did not respond. Of the other 10 patients, VT was reduced in 59%, ATP was reduced in 39%, and shocks were reduced in 60% [14].

In the UK multicenter experience, seven patients were treated with the same dose and technique similarly to the previous experience. After 6 months, for the five patients analysed VT burden was reduced by 85%, with no high-grade acute toxicity and three deaths due to heart failure. [15].

The STRA-MI-VT was a phase Ib/II trial that evaluated the feasibility of Cyberknife tracking in treatment planning [16].

Table 1. Summary of evidence regarding ventricular tachycardia (VT) and atrial fibrillation (AF) treatment with stereotactic ablative radiotherapy (SABR).

Authors	Year	N pts	Diagnosis	End-Point	Dose tot/fx	Results
Cuculich PS [12]	2017	5	VT	Efficacy and safety of treatment	25 Gy/1 fx	No complications during treatment. Fatigue after treatment (three patients), with no acute heart-failure. Marked reduction in the burden of ventricular tachycardia after treatment.
Kurzelowski R [13]	2022	2	VT	Efficacy and safety of treatment	25 Gy/1 fx	No problem in the first patient. The second one experienced acute side effects with an increase in VT that gradually improved at the end of the follow-up period.
Wight J [14]	2022	14	VT	Efficacy and safety of treatment	25 Gy/1 fx	VT was reduced in 59%, ATP was reduced in 39%, and shocks were reduced in 60%.
Lee J [15]	2021	7	VT	Reduction of VT and safety of treatment	25 Gy/1 fx	VT responded in all patients. After 6 months, VT burden was reduced by 85%. No high grade acute toxicity.
Piccolo C [16]	2022	Phantom study	VT	Feasibility of Cyberknife on cardiac lesions by tracking as a single marker the lead tip of an implantable cardioverter defibrillator.	25 Gy/1 fx	Tracking with a single marker is feasible considering adequate residual planning margins. The volumes could be further reduced by using additional markers.
Bonaparte I [17]	2021	Dosimetric study	VT	STAR is efficacy in terms of BDT and MUs.	25 Gy/1 fx	Several plans were evaluated for dosimetric considerations.
Kovacs B [18]	2021	57	VT/FA	STAR's effectivity and safety for structural VT/VF	25 Gy/1 fx	Significant short-term reduction of sustained VT/VF-burden, but recurrences are common.
Akdag O [19]	2022	Phantom study	VT	First experimental evidence for real-time cardiorespiratory motion-mitigated MRI-guided STAR on the 1.5 T Unity MRlinac aimed at simultaneously compensating cardiac and respiratory motions.	25 Gy/1 fx	Cardiac motion was successfully mitigated using gating, which was demonstrated in the phantom and in-silico experiment.
Kautzner J [20]	2021	3	VT	postmortem immunohistochemical was performed early and late after SBRT	25 Gy/1 fx	Apoptosis and subsequent fibrosis was shown to be not immediate, thus the antiarrhythmic effects may be delayed after SBRT.
Di Monaco A [21]	2022	5	AF	Side effects at 1 month after STAR	25 Gy/1 fx25 Gy/1 fx	No acute treatment-related adverse events (>G1)

Abbreviation: N: number, Pts: patients, RT: radiation therapy, Fx: fractions, VT: Ventricular Tachycardia, AF Atrial Fibrillation, Gy: Gray, STAR: Stereotactic Arrhythmia Radioablation.

Bonaparte et al. [17] also performed a dosimetric analysis for Linac-based STAR for VT, using different treatment planning approaches. The authors concluded that among the different techniques and energies, the 10 MeV Flattening Filter Free (FFF) approach was the faster but not suitable in patient with cardiac implantable electronic devices [17].

A systematic review including 13 studies and 57 patients confirmed efficacy and safety of STAR for refractory VT/ventricular fibrillation (VF). Thirty-one patients (54%) had ischemic cardiomyopathy and fifty patients (88%) had prior catheter ablation (CA) for VT/VF. A single dose of 25 Gy was delivered to a mean PTV of 64.4 cc (range 3.5–238) with a mean safety margin of 3.3 mm (0–5). Electrical storm was shown in 7% after SABR. VT burden was reduced in all patients, but recurrence affected most of the patients (75%), with several adverse events (81%) and no treatment-related deaths. The authors concluded that STAR preliminary experience appears safe and efficacious despite the recurrence rate and deserve to be further investigated [18]. Despite this, there are still few conflicting data on the follow-up of patients and on the influence of respiratory movement on the cardiac dose. In fact, cardiac motion presents an important challenge because the VT isthmus is subject to both respiratory and cardiac motion. So, Akdag et al. provided first experimental evidence for real-time cardiorespiratory motion-mitigated MRI-guided STAR on the 1.5 T Unity MR-linac. A real-time cardiorespiratory motion-mitigated radiotherapy workflow was developed on the Unity MR-linac. A 15-beam intensity-modulated radiation therapy treatment plan (1 × 25 Gy) was created in Monaco v.5.40.01 (Elekta AB, Stockolm, Sweden) for the Quasar MRI^{4D}phantom (ModusQA, Modus Medical Devices, London, Ontario, Canada). Simulations showed that cardiac motion decreased the target's D98% dose between 0.1 and 1.3 Gy, with gating providing effective mitigation. So, real-time MRI-guided cardiorespiratory motion management greatly reduces motion-induced dosimetric uncertainty and warrants further research and development for potential future use in STAR aimed at simultaneously compensating cardiac and respiratory motions [19].

Kautzner et al. have published case series with the first postmortem immunohistochemical analysis of morphologic changes in the myocardium early and late after SBRT. The authors have found apoptosis followed by fibrosis that could explain the timing of the SBRT efficacy on VT [20].

In conclusion, STAR had reasonable VT suppression in patients where conventional treatment had failed.

3.1.2. Atrial Fibrillation

AF affects about 40 million people in the world and increased the risk of stroke and heart failure. Current clinical management include antiarrhythmic therapy and eventually catheter ablation in drug refractory patients. Despite that, a subset of elderly patients are not responsive to systemic therapies and have an high risk of complications following catheter ablation, thus a non-invasive approach as STAR should be investigated.

A systematic review collected available evidence (both preclinical and clinical setting) on the feasibility and efficacy of STAR, including photon RT (XRT) and particle beam therapy (PBT), in the treatment of AF. Twenty-one works (17 for XRT, 3 for PBT, 1 both) published between 2010 and 2021 were included. The main favorable finding consisted in the detection of electrical scar in 4/4 patients undergoing specific evaluation, whereas the minimum dose for efficacy was 25 Gy. No acute complications were observed below this dose and a great heterogeneity was observed among the included studies [22].

Di Monaco et al. performed a phase II trial that reported a preliminary experience of five patients treated for AF. The preliminary results showed no high-grade side effects and a good response in terms of AF control and no further use of antiarrhythmic drugs [21].

Ultimately, STAR represents a safe and effective non-invasive approach in the treatment of drug-refractory arrhythmias, but there is still poor data in the literature regarding long-term efficacy and follow-up. These two aspects, together with the choice of specific patient settings, represent the most important challenge to be investigated with well-defined studies.

3.2. *Soft Tissue Disorders*

Non-malignant, proliferative, soft tissue disorders are a very heterogeneous group of diseases, with a tumor-like phenotype, although they do not have malignant characteristics. These pathologies can reach large dimensions and cause serious organs involved deficits. Abnormal growth of fibroblasts or hypertrophic scar tissue (keloids), or inflammatory factors over-expression, such as bFGF, TGF-β, PDGF, EGF, and CTGF in Dupuytren's disease, can be found at the basis of the onset of these disorders [23,24]. RT may play a role in these diseases control thanks to its anti-inflammatory, anti-proliferative and immunomodulatory effects, as well as has been confirmed by numerous reports, especially when conventional treatments have not achieved sufficient control of symptoms. We report the main data in the literature regarding RT treatment in keloids, Dupuytren's disease, Peyronie's disease, and fibromatosis.

3.2.1. Keloids

Keloids are benign skin disorders based on excessive connective tissue proliferation during the normal scarring process. They typically appear after repeated surgeries at the same site, burns, after trauma, and deep dermis injuries. Pathogenetic mechanisms are not fully known, but the fibroblasts present in keloids have different characteristics compared to the fibroblasts present in normal skin [25]. Keloids are common, occurring in 5% to 15% of wounds and affecting both sexes equally. They mainly affect people 10 to 30 years old and are more commonly seen in those with family history of keloids [26]. The treatment of choice for keloids is surgery, which, however, shows 80% local recurrence; therefore, adjuvant RT can lower the risk of recurrence, as well as being the main treatment in cases of scar inoperability. Adjuvant RT can lower the risk of recurrence, as well as being the main treatment in cases of scar inoperability. Brachytherapy is effective to prevent keloid formation: the first session should start the same day as surgery and recommended doses are 5–6 Gy in three fractions or 5 Gy in four fractions [27,28]. Complete response rate of RT (either with electrons or orthovoltage techniques and brachytherapy) range from 50 to 98% according to the literature data (Table 2).

Table 2. Soft tissue disorders: included study and radiotherapy parameter.

Author	Year	N pts	Diagnosis	End Point	Dose	Results
Jiang [29]	2018	29	Keloids	Control rate	18 Gy/3 fx	Response rate 91.9%
Kim [30]	2015	28	Keloids	Control rate	12–15 Gy/3 fx	Response rate 50%
Shen [31]	2015	568	Keloids	Control rate	18 Gy/3 fx	Response rate 90.41%
Emad [32]	2010	26	Keloids	Control rate	12 Gy/3 fx	Response rate 70.4%
Malaker [33]	2004	64	Keloids	Control rate	37.5 Gy/5 fx	Response rate 97%
Lo [34]	1990	199	Keloids	Control rate	2–20 Gy/1 fx	Response rate 87% for Dose > 9 Gy, 43% for Dose < 9 Gy.
Borok [35]	1988	250	Keloids	Control rate	4–16 Gy/various fx	Response rate 98%
Van de Kar [36]	2007	21	Keloids	Control rate	12 Gy/3–4 fx	Response rate 71.9%
Arneja [37]	2008	25	Keloids	Control rate	HDR BT 5 Gy/3 fx	Response rate 92%
Van Leeuwen [38]	2014	67	Keloids	Control rate	HDR BT 6 Gy/2 fx	Response rate 96.9%
Jiang [39]	2016	32	Keloids	Control rate	HDR BT 6 Gy/3 fx	Response rate 94%
Hafkamp [40]	2017	29	Keloids	Control rate	HDR BT 13 Gy/1 fx	Response rate 75.9%
Kadhum [41]	2017	698	Dupuytren's disease	Control rate	21–42 Gy in 3–14 fx	Good ratio of regressions (6–20% depending on staging), stability (12–81%) and low ratio of progressions (13–65%, depending on staging).

Table 2. *Cont.*

Author	Year	N pts	Diagnosis	End Point	Dose	Results
Seegenschmiedt [7]	2015	1762	Dupuytren's disease	Control rate	15–21 Gy in 5–7 fx, 30 Gy split in 2 series of 5fx with a 3 months interval	Stability of disease in 84% for N stage and 67% for N/I stage
Betz [42]	2010	135	Dupuytren's disease	Control rate	30 Gy split in 2 series of 5 fx separated by a 6- to 8-week interval	Stability of disease in 59%, 10% improved, and 31% progressed. In stage N 87% and in stage N/I 70% remained stable or regressed
Seegenschmiedt [8]	2015	8732	Peyronie's disease	Pain, improvement	10–20 Gy (2–10 fx)	Pain regression in 50–90%, Improvement of penile deviation in 30–70%
Seinen [43]	2015	155 RT alone, 815 Surgery + RT	Fibromatosis	Local control	30–74 Gy	Local control in 78% of the patients treated with surgery and RT versus 85% in patients treated with RT alone

Abbreviation: N: number, Pts: patients, RT: radiation therapy, Fx: fractions, BT: brachytherapy.

3.2.2. Dupuytren's Disease

Dupuytren's disease is a fibrotic hyperplasia of connective tissue structures at the level of the finger band and palm. It is a rare condition, with a prevalence of about 2% and with a higher incidence in males (3:1 ratio). The pathogenesis is probably due to factors such as repeated trauma, alcohol and nicotine abuse, and hereditary factors. The diagnosis is purely clinical and sees around the 4th decade of life the presence of fibrotic nodules at the level of the hand that result in digito-palmar contracture leading to severe functional limitations. Dupuytren's disease can be staged by Tubiana classification, which subdivides pathology according to symptomatology. The gold standard treatments for early stages are medical treatment with intralesional applications of xanthine oxidase/dehydrogenase inhibitor, allopurinol, or cytotoxic agents such as vinblastine or colchicine. Surgery is usually reserved for more advanced stages (Stage 3 Tubiana); the role of RT in the treatment of Dupuytren's disease tends to be rather preventive and prophylactic than curative. Thus, the goal is to avoid future functional impairment and a future need for surgery [6]. The efficacy of RT is higher in early stages, inducing a significant reduction in fibroblast proliferation. RT is usually delivered with electron beam at 6MeV energy, while in other centers orthovoltage can be delivered alternatively. The volume treated includes palpable nodules with a safety margin of at least 10mm and the uninvolved structures are protected with lead-based shielding. Among the RT fractionations present in the literature, the most used, with better results both in terms of efficacy and limited toxicity, are the hypofractionations. The main results of the literature are shown in Table 2. Ledderhose disease is a rare type of plantar fibromatosis histologically related to Dupuytren's disease, and it has been also effectively treated with RT [44].

3.2.3. Peyronie's Disease

Peyronie's disease (PD) is a benign condition leading to plaque formation at the level of the tunica albuginea of the penis, leading to local pain and a change in curvature during erection. It usually affects men between the 4th and 6th decades of life with an incidence of 0.3–3%. The most likely mechanism is penile trauma causing inflammation of the tunica albuginea and eventually scarring with fibrotic plaque formation. The pathology presents with penile pain, plaque formation, deformity during erection, and subsequent erectile dysfunction. Diagnosis is clinical with identification and measurement of the plaque. Medical treatment involves the use of drugs for oral treatment such as vitamin E, tamoxifen, and colchicine and for intralesional injection verapamil and collagenase. If drug therapy fails, surgery remains the best option, especially in the cases of severe curvature or angulation and erectile dysfunction [6]. The use of RT is considered indicated in early-stage disease with soft, noncalcified plaques. RT techniques such as x-rays, photons, or low-energy electrons can be used; to have better dose distribution at the surface level, a bolus is also applied over the plaques. The target volume involves the entire plaque with a safety margin of 1 cm, protecting pubic hair, testes, and penile bulb with shielding. Irradiation can be in antero-posterior or latero-lateral projection, the latter with vertical fixation of the penis. The recommended schedule is single dose of 2–3 Gy for a total dose of 10–20 Gy. With this regard, Seegenschmiedt et al. focused their investigation on non-oncological RT publishing numerouscase series [7].

3.3. Muscle-Skeletal Disorders

Radiation therapy is consolidating over the years its role in the treatment of inflammatory or degenerative skeletal disease. Due to the control of pain, the two most frequent indications for radiotherapy in benign diseases are osteoarthritis and periarthritis. Moreover, RT is a non-invasive approach. Irradiation can be provided by LINAC or orthovoltage. A summary of the collected evidence is reported in Table 3.

Table 3. Muscle-skeletal disorders: included study and radiotherapy parameter.

Author	Year	N pts	Diagnosis	End Point	Dose	Results
Hautmann [45]	2019	124	epicondylitis humeri	pain relief	6 Gy(1 Gy)–3 Gy (0.5 Gy)	complete response 64% at 24 months
Rogers [46]	2020	157	epicondylitis, plantar fasciitis, and finger osteoarthritis	pain relief	4 Gy (0.5 Gy)–8 Gy Orthovoltage	pain relief at rest and during activity and a corresponding objective improvement in handgrip strength in epicondylitis. Pain relief at rest, during activity and improvement in walking time were demonstrated in plantar fasciitis
Hautmann [47]	2020	86	Humeral epicondylitis	pain relief	3 Gy/2.5 Gy (0.5 Gy/fx); 6 Gy (1 Gy/fx)	
Micke [48]	2018	703	Calcaneodynia, Achillodynia, Bursitis trochanterica, Shoulder Syndrome, Gonarthrosis	pain relief	6 Gy (0.5–1 Gy)	At follow up, good response: Calcaneodynia 80.7%, Achillodynia 88.9%, Bursitis trochanterica 46.3%, Shoulder Syndrome 60%; only Gonarthrosis 29.2%
Alvarez [49]	2019	108	OADD	pain relief	6 Gy (1 Gy)–12 Gy	Overall, and with a follow-up of 8 months (range 1–31 months), 91% of patients experienced pain relief. The pain reported according to the VAS scale was 0–3 in 32.6% of the patients, 4–6 in 36.7% and greater or equal to 7 in 20.1% of treated patients.
Mahler [50]	2018	55	knee osteoarthritis	pain relief	6 Gy	At 3 months follow-up: no substantial beneficial effect on symptoms and inflammatory signs of LDRT in patients knee OA, compared with sham treatment
Ott [51]	2015	112	Achillodynia	pain relief	6 Gy/3 Gy	Pain control:Early 84% Middle-term 88% Long-term 95%
Rudat [52]	2021	666	Heel Spur	pain relief	3 Gy (Re-irradiation possible)	Good local control (>75%) and good response to reirradiation
Hautmann [53]	2014	110	Heel Spur (Re-irradiation)	pain relief	3 Gy (Re-irradiation possible)	73.6% of Pain control after 24 months
Niewald [54]	2020	236	Kneel and Hand Osteoarthritis	pain relief	3 Gy/0.3 Gy	Good pain control with no difference between the two schemes

Abbreviation: N: number, Pts: patients, RT: radiation therapy, Fx: fractions, OADD: osteoarticular degenerative disorders, LCH: Langerhans cell histiocytosis, LDRT: low dose radiotherapy.

3.3.1. Osteoarthritis and Osteoarthrosis

Epicondylitis is commonly divided into two main branches: lateral, also commonly known as tennis elbow, and medial, the golfer's elbow. Epicondylitis has a negative impact on patients quality of life with symptoms such as pain, joint mobility restriction, and even sometimes edema, limiting everyday activities and independence [55].

Hautmann et al. [45] irradiated 138 epicondylitis humeri, with linear accelerator, unlike most studies in the literature that used orthovoltage [46]. Patients had a median NRS (pain numeric rating scale) 7. Total dose was 6 Gy or 3 Gy (every other day). Between elbow treatments, 30% repeated RT for a partial or absent response. The authors reported a complete response of pain, NRS 0, with a follow-up at 24 months and demonstrated a comparable results of linear accelerator RT compared with orthovoltage. The same authors published in 2020 the results of reirradiation at a dose of 3 Gy (0.5 Gy/fx) and 6 Gy (1 Gy/fx), reporting a good control of pain at 24 months of 50.9% and a non-difference statically valid between the two dosages [47]. From the Micke et al. study, in addition, the effectiveness of low-dose RT using LINAC can be inferred compared to RT using orthovoltage. Both techniques manage to obtain an excellent control of the pain, especially in the long term. From the series, only the group of patients with gonarthrosis did not have a good response, probably due to the process of irreversible degeneration typical of this lesion [48].

Alvarez et al. [49] have treated 184 degenerative osteoarticular disorders with a dose of 6 Gy (1 Gy/fr every other day). Patients who did not benefit from the first round of RT repeated treatment, for a total dose of 12 Gy (52% of cases). Median follow-up was 8 months. Although 91% of patients reported improved pain, only 32.6% reported VAS pain 0–3.

The effectiveness of RT is to be found in the anti-inflammatory mechanism of low doses (0.5–1.5 Gy/fx), which determine inhibition of the interactions between leukocytes and endothelial cells, a decrease in the production of adhesion molecules to the endothelium, a decrease of mediators of inflammation, and less expression of pro-inflammatory cytokines. Of note, there is no guideline regarding regimen and dose fractionation, but the strategies have been determined empirically and it has been reported that various doses such as 0.5 Gy performed in six fractions had the same effect of a dose of 1 Gy in six fractions. The choice of fractionation is mostly based on in vitro experiments, which have shown that the anti-inflammatory effect of low doses RT was maximum at 48 h after irradiation, and it was lost after 72 h [56].

In the study of the Netherlands, there were reported negative results for RT, this time in the treatment of the knee osteoarthritis [50]. The authors conducted a randomized, double-blind, controlled study that showed that LDRT (6 Gy) does not lead to a substantial reduction of symptoms in patients with knee osteoarthritis. Note that unlike previous experiences, the follow-up in this study is relatively short and does not allow the long-term benefit to be assessed.

More recently, a prospective trial started in 2020 has been started by Niewald et al. including 236 patients (64 knees and 172 hands) all with a diagnosis of osteoarthritis (OA). The aim of this study was to compare two schemes of treatment (3 Gy vs. 0.3 Gy). The study showed no statistically difference in pain control between the two schemes. Further studies need to be performed because there's no evidence about the effect of low doses such as 0.3 Gy in the literature [54].

Given the increasing interest in benign osteoarticular pathologies in recent years, Alvarez and colleagues have outlined a CT-based contouring atlas for non-malignant skeletal and soft tissue disorders [57]. The aim of the authors is to suggest a correct PTV delineation based on simulation CT, for treatment of painful shoulder syndrome, such as periarthritis humero-scapularis, epicondylitis humeri, finger joint osteoarthritis, trochanteric bursitis, gonarthrosis, plantar fasciitis, and Achiles tendinopathy.

3.3.2. Achillodynia

Ott et al. in 2015 evaluated the long-term efficacy of two different RT schemes of 1 Gy/0.5 Gy over 3 weeks, twice per week used for 112 patients with diagnosis of achillodynia. The overall early (right at the end of RT), delayed (6 weeks after RT), and long-term (2 years after RT) response rates for all patients were 84 %, 88 %, and 95 %, respectively. This confirmed the positive role of RT for this kind of treatment [51]. In 2021, a review made by Rudat et al. analysed 666 patients for a total of 864 heels treated between 2009 and 2020. Since 2015, a questionnaire was given to all patients in order to measure the local pain control. For newly treated patients, after 3 months follow-up in case of an unsatisfactory pain control it was offered the possibility of a re-irradiation. Re-irradiation showed an improvement in local pain control of approximately 40%. More than 75% of the patients reported a good pain control, confirming the role of radiotherapy in this field [52]. This confirms the results of another study done by Hautmann in 2014 where 110 heel spur syndrome with lack of pain control has been re-irradiated and results were measured with NRS score system. Re-irradiation confirmed his role because 73.6% of the patients were free from pain 24 months after the treatment [53].

3.3.3. Heterotopic Ossification

The role of RT on Heterotopic Ossification (HO) was supported by weak evidence. In 2014, a systematic review made by Ploumis et al. on a total of 27 studies of elbow HO showed that in most cases RT was stopped due to safety reasons. This review confirmed the lack of high-quality findings in the literature about RT in HO syndromes [58]. Thanks to the review made by Galietta et al. in 2022, it was finally confirmed the positive role of RT in the prevention of hip HO. Despite the numerous schemes available, from a single fractionated to several multiple fractionated schemes, no difference has been reported between the different schemes over the surgery alone [59].

3.4. Neurological Disorders

Radiation therapy has been widely adopted in the treatment of various neurological disorders, for different aims such as pain relief, control of the symptoms, and obliteration of brain arteriovenous malformations. Moreover, RT is a non-invasive approach that can be safely adopted also in elderly patients. A summary of the collected evidence is reported in Table 4.

Table 4. Neurological disorders: included study and radiotherapy parameter.

Authors	Year	N pts	Diagnosis	End Point	Dose	Results
Rauch [60]	2012	11	Epilepsy	Tolerability and seizure frequency.	26.3–58.3 Gy	Treatment led to an improvement in the frequency of seizures in 63%.
Liang [61]	2010	7	Epilepsy	Seizure frequency	12 Gy	Reduction of seizure frequency was 50% in two cases, 30% in one case, and 0% in two cases, and seizure frequency increased more than 100% in two cases.
Bartolomei [62]	2008	15	Epilepsy	Seizure frequency	24 Gy	A total of 60% pts were considered seizure free. All patients who were initially seizure free experienced a relapse of isolated aura (66%) or complex partial seizures (66%) during antiepileptic drug tapering.
Barbaro [63]	2009	28	Epilepsy	Seizure frequency	24 Gy high dose vs. 20 Gy low dose	At the 36-month follow-up evaluation, 67% of patients were free of seizures for the prior 12 months (high dose: 10/13, 76.9%; low dose 10/17, 58.8%)

Table 4. *Cont.*

Authors	Year	N pts	Diagnosis	End Point	Dose	Results
Smith [64]	2011	169	TN	Pain relief	70–85 Gy, 90 Gy	A total of 79.3% experienced significant relief. A total of 19.0% had recurrent pain. Of 87 patients with idiopathic TN without prior procedures, 79 (90.8%) had initial relief. Among 28 patients treated with 70 Gy, 18 patients (64.3%) had significant relief. Of the patients with 90 Gy at the brainstem, 59 (79.0%) had significant relief.
Rashid [65]	2018	55	TN	Pain relief	90 Gy	After 30 months median follow-up, 69% of patients were pain free.
Romanelli [66]	2019	387	TN	Pain relief	60 Gy (80% isodose)	Pain relief rate at 6, 12, 18, 24, 30, and 36 months was, respectively, 92, 87, 87, 82, 78, and 76%.
Lovo [67]	2019	14	TN	Pain relief	140 Gy	A total of 90% pts reported some form of relief. A total of 60% reached the threshold of 50% pain relief, and for 40% the pain never improved.
Kundu [68]	2022	41	TN	Pain relief	90 Gy	There has been a significant improvement in the post-radiation pain score in 72% of patients.
Starke [69]	2016	2236	AVM	Obliteration rate	20.5 Gy (mean margin dose)	Overall obliteration rate was 64.7%.
Ding [70]	2017	232	AVM	Obliteration rates, hemorrhage rate	22.5 Gy	The actuarial obliteration rates at 5 and 10 years were 72% and 87%, respectively. Annual post-SRS hemorrhage rate was 1.0%
Patibandla [71]	2017	233	AVM	Obliteration rates, hemorrhage rate in Grade III-IV AVMs	Mean dose 17.3 Gy	The actuarial obliteration rates at 3, 7, 10, and 12 years were 15%, 34%, 37%, and 42%, respectively. The annual post-SRS hemorrhage rate was 3.0%
Matsuo [72]	2014	51	AVM	Obliteration rate	15 Gy (80% isodose)	The actuarial obliteration rates at 3, 5, 10, and 15 years were 46.9%, 54.0%, 64.4%, and 68.0%, respectively
Matthiesen [73]	2012	211	GO	Symptomatic improvement	20 Gy/10 fx	A total of 84.2% pts reported a symptomatic improvement
Kouloulias [74]	2013	17	GO	Symptomatic improvementand tolerability	20 Gy/10 fx	Stabilization of the disease without recurrence was achieved in 12/17 patients. At the end of radiotherapy, the CAS regressed to 4.82 ± 2.24 ($p < 0.001$, Wilcoxon test). Extraocular motility and pain behind the globe were improved in 14/17 and 16/17 patients, respectively. Five patients developed recurrent signs and symptoms and they underwent surgical decompression
Li Yim [75]	2011	59	GO	duration of symptoms, clinical activity score (CAS)	20 Gy/12 fx (over 2 weeks)	Response (change in CAS) to orbital radiotherapy was statistically significant from 3.17 ± 1.75 standard deviation (SD) to 0.73 ± 0.92 SD ($p < 0.001$)
Kahaly [76]	2000	65	GO	Symptomatic improvement and toxicity	A: 20 Gy/20 fx (over 20 weeks) B: 10 Gy/10 fx (over 2 weeks) C: 20 Gy/10 fx (over 2 weeks)	Response to therapy, defined as a significant amelioration of three objective parameters, was noted in 12 A (67%), 13 B (59%), and 12 C (55%) subjects (C vs. A, $p = 0.007$). Ophthalmic symptoms and signs regressed most in group A
Cardoso [77]	2012	18	GO	Symptomatic improvement and Radiologic response	10 Gy/10 fx (over 10 weeks)	Significant decrease in symptoms such as tearing ($p < 0.001$), diplopia ($p = 0.008$), and conjunctival hyperemia ($p = 0.002$). Magnetic resonance imaging showed decrease in ocular muscle thickness and in the intensity of the T2 sequence signal in the majority of patients

Abbreviation: CAS: clinically activity score, SRS: stereotactic radiosurgery, AVM: arteriovenous malformation, GO: Graves ophthalmopathy, TN: Trigeminal neuralgia.

3.4.1. Epilepsy

A total of 0.5% of the world's population suffers from epilepsy. About 30–40% of the patients do not benefit from pharmacological therapy and are eligible for surgical treatment. In patients where the pharmacological and surgical alternatives have been exhausted, RT is an option. RT, compared to surgery, has the advantage of not being invasive with low risk of neurological damage to the patient. Rauch C. et al. [60] reported first-time long-term outcome (median 10 years) of fractionated stereotactic radiotherapy (FSRT) in 11 patients with drug-resistant epilepsy. The biologically equivalent dose ranged from 26.3 to 58.3 Gy ($\alpha/\beta = 10$). None of the patients developed temporary or permanent neurological deficits. Treatment resulted in improvement of seizure frequency in seven patients: five of them had a decrease in seizure frequency, and two of them were seizure-free at last follow-up.

Liang S et al. [61] reported the long-term outcome of seven patients with temporal lobe epilepsy (TLE) treated with very low-dose LINAC based FSRT, treated with marginal dose of 12 Gy at the 85% isodose line. Reduction of seizure frequency post-FSRT was 50% in two cases, 30% in one case, and 0% in two cases, and seizure frequency increased more than 100% in two cases. No patient was seizure free at the last follow up. Two cases presented transitory complications and two cases showed an obvious drop in IQ, memory decline, and permanent neurologic complications, including partial aphasia and mild hemiplegia in one case, and progressive ataxia and cognition decline in another case. Bartolomei F et al. [62] reported outcome of 15 patients with TLE with median follow-up of 8 years (range 6–10 years) treated with gamma-knife with a marginal dose of 24 Gy. At the last follow-up, 9 of 16 patients (60%) were considered seizure free. A total of 60% of the patients experienced mild headache and were placed on corticosteroid treatment for a short period. All patients who were initially seizure free experienced a relapse of isolated aura (10/15, 66%) or complex partial seizures (10/15, 66%) during antiepileptic drug tapering. Restoration of treatment resulted in good control of seizures. Results are maintained over time with no additional side effects. Long-term results are comparable with conventional surgery.

Barbaro et al. [63] evaluated the effectiveness of radiosurgery, at the level of the amygdala, hippocampus, and parahippocampal gyrus as an alternative to surgery. Thirteen high-dose (24 Gy) and seventeen low-dose (20 Gy) patients were treated. Both groups showed significant reductions in seizures within a year of treatment, without observing major safety issues with high-dose stereotactic radiosurgery (SRS) versus low-dose SRS.

On these bases, radiotherapy has the potential to control the frequency and intensity of seizures in patients with pharmacoresistant epilepsy with mild long-term side effects if administered with proper fractionation, dose prescription, and target volume definition.

3.4.2. Trigeminal Neuralgia

Trigeminal neuralgia (TN) is a neurological disease that cause intense facial pain, usually paroxysmal and excruciating, due to an alteration of the V cranial nerve (predominantly the mandibular and maxillary branches). In patients with medications failure/significant adverse events, radiotherapy (SRT, mainly SRS) may be an important therapeutic option. For SRS Gamma-Knife, linear accelerators and Cyber-Knife can be used. The three techniques showed no differences in terms of pain control, whereas the time to recurrence ranged from 6 to 48 months [78].

Smith et al. [64] performed a retrospective study that evaluated cohort of 169 patients treated with LIANC-SRS. The authors investigated different doses and volumes, concluding that increased dose and volume of brainstem irradiation improve clinical outcomes. Similarly, Rashid et al. [65] evaluated 55 patients with TN and treated with LINAC SRS up to a total dose of 90 Gy with 20% isodose line constraint to brainstem. After 30 months median follow-up, 69% of patients were pain free. Another important study has reported the outcomes of using Cyber-Knife SRS administered to 527 patients [66]. Dose prescription was 60–65 Gy to the 80–90% isodose line. A brainstem volume equal to or less than 1 cm^3 was exposed at a dose of 10 Gy, with a maximum point dose (0.035 cm^3) of 30%

of the prescription dose. The pain relief rates were 87%, 82%, and 76% at 12, 24, and 36 months, respectively.

Lovo EE et al. [67] reported the outcome of 14 patients with TN treated with SRS to the Centromedian and Parafascicular Complex of the contralateral thalamus. They used a gamma angle regularly fixed at 90° and using a 4 mm collimator and the prescribed dose was 140 Gy to Dmax. Almost all patients (90%) reported some form of relief. Six patients (60%) reached the threshold of 50% pain relief. For four patients (40%), the procedure failed because the pain never improved.

Kundu B et al. [68] recently performed a retrospectively evaluation of patients with TN undergoing LINAC-SRS. The patients were evaluated with Barrow Neurological Institute (BNI) pain score and after a median follow-up of 5 months, 72% of the patients showed an improvement in this pain evaluation.

A retrospective study by Fraioli et al. of 45 patients compared SRS to FSRT using a linear accelerator. The authors compared 40 Gy in a single fraction and 72 Gy in six fractions. Patients treated with FSRT showed a higher pain recurrence rate than SRS (27.3% versus 8.3%) [79].

Finally, reirradiation with SRS can still be used in case of recurrence, with a pain relief rate of 50% as showed in two retrospective studies [80,81]. The reirradiation was usually delivered between 15.7 and 26.1 months after the first irradiation, with a dose in the range 70–80 Gy.

3.4.3. Brain Arteriovenous Malformations

Brain arteriovenous malformations (AVMs) are the persistence of a direct link between an artery and a vein; the nidus is located where small arteries and veins are connected [82]. Brain AVMs are rare (incidence estimated between 1.12 and 1.42 cases per 100,000 person-years) and mostly occur in young patients [83].

AVMs may be found as an incidental finding. They may be associated with intracranial hemorrhage, seizures, headaches, and/or neurological deficits. CT and MRI angiography are useful for the accurate diagnosis and nidus definition. Treatment approaches for AVMs are neurosurgery, embolization, and intracranial SRS. Stereotactic radiosurgery has a clinical obliteration rate of 60–80% [69–72]. The clinical benefit of SRS induced obliteration appear after 3 years or more. The success rate was increased with smaller volume (up to 30 cm^3), lower Spetzler-Martin grade, higher dose, and steeper dose gradient. Embolization performed before SRS provided significantly lower obliteration rates than SRS alone (at 3 years: 41% versus 59%, respectively; $p < 0.00001$) [84,85]. Some AVM locations (functional areas, as thalamus) are related to poorer outcomes and require multimodal management [86]. Fractionated intracranial SBRT is poorly used in patients with AVM, mainly due to the low obliteration and morbidity rates initially reported [87]. Most recent results had an obliteration rate of 50% using fractionated intracranial SBRT delivering equivalent 2 Gy fraction doses higher than 70 Gy [88]. Symptomatic and permanent radiation-induced side effects have been described in 8 to 11% and in 1 to 4% of patients treated with SRS, respectively [71]. Stereotactic radiosurgery can be repeated when no complete obliteration is reported after the initial treatment. This is likely to occur in patients with AVMs larger than 10 cm^3 and/or with high Spetzler-Martin grade [71]. Overall, the risks of hemorrhage and radionecrosis have to be considered and the treatment decision needs to be taken in a multidisciplinary setting in which the situation of each patient is assessed individually.

3.4.4. Graves Ophthalmopathy

Graves ophthalmopathy (GO) is the most frequent extrathyroidal manifestation of Graves' disease. Although GO is severe in only 3–5% of affected individuals, quality of life is severely impaired even in patients with mild GO. RT is a well-established method of treatment for GO. The main rationale is its anti-inflammatory effect and the high radiosensitivity of T lymphocytes and orbital fibroblasts. Although there are several reports about the benefits of RT [73–75], optimal initial treatment and its combination with steroids is still controversial. Various RT regimens with different doses and fractionations have been used: 16 or 20 Gy delivered in 8–10 fx (five days/week) is usually considered the standard [89]. A consensus statement from the European Group on Graves' Orbitopathy does not recommend doses higher than 20 Gy [90]. Recent studies evaluated altered fractionation RT for GO. In a randomized study, Kahaly et al. [76] compared the efficacy and tolerability of three RT regimens of 1 Gy given weekly for 20 weeks for a total dose of 20 Gy; 1 Gy given daily for 2 weeks for a total dose of 10 Gy; and 2 Gy given daily for 2 weeks for a total dose of 20 Gy for patients with moderately severe GO. The authors concluded that whereas all regimens provided similar response rates, the protracted regimen had a better effectiveness and tolerance. Cordoso et al. [77] demonstrated the efficacy of orbital RT with a total dose of 10 Gy, fractionated in 1 Gy once a week over 10 weeks in 18 patients with GO.

Overall, favorable responses have been reported in 60% of cases. The best responses were noted for inflammatory signs and recent onset of extraocular muscle involvement. RT is well tolerated and safe and a careful selection of patients is necessary.

4. Discussion

All the above-mentioned disease can benefit from RT with interesting and promising results. At the same time, several pitfalls need to be managed before RT can be considered a safe and efficacious technique to manage non-oncological disorders.

Specifically, more work is needed in the field of radiobiology to correctly quantify the RT damage to organs at risk, especially in the risk of secondary cancers [91–93]. As the development of RT induced secondary malignancies represents one of the most important late side-effects, this topic has significantly influenced treatment decision-making and limited the use of RT for non-oncological disorders. Considering the increased life expectancy and the higher number of older patients, as well as the advanced in treatment delivery, all efforts should be made to prevent the incidence of tumors induced by radiation and to correctly estimate the individual risk.

At the same time, the number of prospective trials investigating the field of non-oncological RT is extremely poor and needs to be improved in the next years (see Table 5).

Table 5. Ongoing prospective clinical trial on non-oncological radiotherapy.

NCT Number	Disease	Design	Location
NCT04722263	Keloids	Single arm, interventional pilot study (15 patients). RT: 15 Gy in 3 fractions.	Montefiore Medical Center, New York, US
NCT04853433	Keloids	Single arm, interventional pilot study (15 patients). The primary endpoint will be toxicity.	Montefiore Medical Center, New York, US
NCT04122313	Dupuytren's Disease	Prospective, Cohort study. Participants will be treated according to a standard treatment pathway, followed by post-operative radiation. RT: 15 Gy in 5 fx, followed by a 6–8 weeks break then a second identical course. Total dose: 30Gy.	University of Minnesota, US
NCT04424628	Gonarthrosis and Coxarthrosis	Non-inferiority study in which the investigators compare two low-dose radiotherapy schemes. Arm A will be treated at 3 Gy (0.5 Gy/fraction, 3 fractions/week), and patients in arm B will be treated at 6 Gy (1 Gy/fraction, 3 fractions/week).	GenesisCare, Malaga, Spain
NCT02708810	Trigeminal Neuralgia	To determine the feasibility of frameless Virtual Cone trigeminal neuralgia radiosurgery at a single institution prior to multi-institutional enrollment.	Hazelrig-Salter Radiation Oncology Center, Birmingham, Alabama, US
NCT03995823	Cerebral Arteriovenous Malformations	Prospective study including 50 patients with cerebral AVMs treated with GRKS to evaluate the sensitivity for nidus obliteration of MRI.	Department of Neurosurgery, Medical University of Vienna, Austria
NCT04843683	Cardiac Arrhythmias	Prospective, single-center, phase II trial that will be monitoring the safety and efficacy of using stereotactic ablative radiotherapy (SBRT) to treat arrhythmias.	University Health Network, Toronto, Canada

Table 5. *Cont.*

NCT Number	Disease	Design	Location
NCT04392193	Cardiac Arrhythmias	Proton Particle Therapy for Cardiac Arrhythmia Extracorporeal Energy Source Ablation of Cardiac Tissue: A First Stage Early Feasibility Study	Mayo Clinic, Rochester, Minnesota, US
NCT04984265	Cardiac Arrhythmias (Chagas)	SBRT in Chagas Disease Ventricular Tachycardia. A single 25 Gy dose will be delivered to the PTV.	University of Sao Paulo General Hospital, Sao Paulo, Brazil
NCT04642963	Cardiac Arrhythmias	Single arm, aimed at investigate safety requirements for clinical use. A single 25 Gy dose will be delivered to the PTV.	Medical University of Silesia, Katowice, Poland
NCT04833712	Cardiac Arrhythmias	The study aims to investigate the safety and preliminary efficacy of stereotactic radiotherapy for pulmonary vein isolation to treat refractory atrial fibrillation	Attikon University Hospital, Chaidari, Greece
NCT04486339	Cardiac Arrhythmias	Pulmonary Vein Isolation Using Stereotactic Radiotherapy System for the Treatment of Refractory Atrial Fibrillation	Xinhua Hospital, School of Medicine, Shanghai Jiao Tong University, Shangai, China
NCT03867747	Cardiac Arrhythmias	Radiosurgery for the Treatment of Refractory Ventricular Extrasystoles and Tachycardias (RAVENTA)	University Clinic Mannheim, Mannheim, Baden-Württemberg, Germany
NCT04162171	Cardiac Arrhythmias	Cohort Study—SBRT for VT Radioablation	Nova Scotia, Canada
NCT04066517	Cardiac Arrhythmias	STRA-MI-VT study is a spontaneous, open-label, not randomized, prospective clinical trial. The objective of the study is to evaluate the safety and efficacy of SBRT in strictly selected patients with refractory VT.	Istituto Europeo di Oncologia, IRCCS, Milan, Italy
NCT04612140	Cardiac Arrhythmias	Clinical trial: Patients with previously failed conventional RF catheter ablation will be randomized to radiosurgery (active treatment group) or repeated catheter ablation (control treatment group).	University Hospital Ostrava, Czechia

Abbreviations: SBRT: stereotactic body radiotherapy, RF: radiofrequency, VT: ventricular tachycardia, PTV: planning target volume, GRKS: gamma knife radiosurgery, AVM: Arteriovenous Malformations.

Most of the enrolling trials are investigating the new technique of Heart SBRT for arrhythmia disorders, whereas other diseases that can be managed with RT are not properly investigated. This aspect needs to be properly analyzed to test the role of modern RT in non-oncological fields and to sponsor the results among other specialists that do not know this approach. To this end, especially the scientific societies should undertake to study and spread the use of this type of RT, also involving specialists from other disciplines, following the path of DEGRO German Society.

In this context, it is noteworthy that the spread use of hypofractionation RT schedules is involving several cancer diseases and will decrease the number of fractions of patients at the LINACs [94–96] and so decreasing the reimbursements for RT centers. The knowledge and the use of non-oncological RT, thus, could counterbalance the expected loss of patients and fractions.

Scientific improvements of RT delivery, at the same time, needs to be tested in this particular field of disorders, as several other frontiers could be crossed in the next years that no-one can imagine now, such as the use of heart SBRT.

To the best of our knowledge, the present paper is one of the most recent overviews on RT for non-malignant diseases. Notwithstanding the narrative nature of the review, our results may strongly support the need for further investigation and represent a starting point for future clinical research, also in the context of RT Scientific Societies.

We finally recognize the limits of our review, in that it lacks explicit criteria for article selection and does not evaluate selected articles for validity.

By setting up the ambitious goal of addressing all these challenges, non-oncological RT can become a clinical reality in the next years and a higher number of prospective trials can be designed and conducted in the near future, with the endorsement of RT Scientific Societies.

Author Contributions: Conceptualization, V.N., E.D., R.G. (Roberta Grassi), A.S., F.G., G.D.M., V.S.M. and L.D.; methodology, V.N., E.D., R.G. (Roberta Grassi), A.S., F.G., G.D.M., V.S.M. and L.D.; formal analysis, F.C., L.B., V.S., C.G., I.D. (Isacco Desideriand), F.D.F., I.D. (Ida D'Onofrio), R.G. (Roberto Grassi), A.R. and S.C.; investigation V.N., E.D., R.G. (Roberta Grassi), A.S., F.G., G.D.M., V.S.M. and L.D.; resources, F.C., L.B., V.S., C.G., I.D. (Isacco Desideriand), F.D.F., I.D. (Ida D'Onofrio), R.G. (Roberto Grassi), A.R. and S.C.; data curation F.C., L.B., V.S., C.G., I.D. (Isacco Desideriand), F.D.F., I.D. (Ida D'Onofrio), R.G. (Roberto Grassi), A.R., and S.C.; writing—original draft preparation, V.N., E.D., R.G. (Roberta Grassi), A.S., F.G., G.D.M., V.S.M. and L.D.; writing—review and editing, F.C., L.B., V.S., C.G., I.D. (Isacco Desideriand), F.D.F., I.D. (Ida D'Onofrio), R.G. (Roberto Grassi), A.R. and S.C.; visualization, V.N., E.D., R.G. (Roberta Grassi), A.S., F.G., G.D.M., V.S.M., L.D.; supervision, F.C., L.B., V.S., C.G., I.D. (Isacco Desideriand), F.D.F., I.D. (Ida D'Onofrio), R.G. (Roberto Grassi), A.R. and S.C.; project administration, F.C., L.B., V.S., C.G., I.D. (Isacco Desideriand), F.D.F., I.D. (Ida D'Onofrio), R.G. (Roberto Grassi), A.R. and S.C. All authors have read and agreed to the published version of the manuscript.

Funding: This research received no external funding.

Institutional Review Board Statement: Not applicable.

Informed Consent Statement: Not applicable.

Data Availability Statement: Not applicable.

Conflicts of Interest: The authors declare no conflict of interest.

References

1. Sung, H.; Ferlay, J.; Siegel, R.L.; Laversanne, M.; Soerjomataram, I.; Jemal, A.; Bray, F. Global Cancer Statistics 2020: GLOBOCAN Estimates of Incidence and Mortality Worldwide for 36 Cancers in 185 Countries. *CA Cancer J. Clin.* **2021**, *71*, 209–249. [CrossRef] [PubMed]
2. Nakamura, K.; Sasaki, T.; Ohga, S.; Yoshitake, T.; Terashima, K.; Asai, K.; Matsumoto, K.; Shioyama, Y.; Honda, H. Recent advances in radiation oncology: Intensity-modulated radiotherapy, a clinical perspective. *Int. J. Clin. Oncol.* **2014**, *19*, 564–569. [CrossRef] [PubMed]
3. Park, J.M.; Wu, H.G.; Kim, H.J.; Choi, C.H.; Kim, J.I. Comparison of treatment plans between IMRT with MR-linac and VMAT for lung SABR. *Radiother. Oncol.* **2019**, *14*, 105. [CrossRef] [PubMed]
4. Garibaldi, C.; Essers, M.; Heijmen, B.; Bertholet, J.; Koutsouveli, E.; Schwarz, M.; Bert, C.; Bodale, M.; Casares-Magaz, O.; Gerskevitch, E.; et al. The 3(rd) ESTRO-EFOMP core curriculum for medical physics experts in radiotherapy. *Radiother. Oncol.* **2022**, *170*, 89–94. [CrossRef]
5. Hua, C.H.; Mascia, A.E.; Seravalli, E.; Lomax, A.J.; Seiersen, K.; Ulin, K. Advances in radiotherapy technology for pediatric cancer patients and roles of medical physicists: COG and SIOP Europe perspectives. *Pediatr. Blood Cancer* **2021**, *68* (Suppl. S2), e28344. [CrossRef]
6. Seegenschmiedt, M.-H.; Makoski, H.-B.; Trott, K.; Brady, L. *Radiotherapy for Non-Malignant Disorders*; Springer: Berlin/Heidelberg, Germany, 2008.
7. Seegenschmiedt, M.H.; Micke, O.; Muecke, R. Radiotherapy for non-malignant disorders: State of the art and update of the evidence-based practice guidelines. *Br. J. Radiol.* **2015**, *88*, 20150080. [CrossRef]
8. Seegenschmiedt, M.H.; Micke, O.; Niewald, M.; Mücke, R.; Eich, H.T.; Kriz, J.; Heyd, R. DEGRO guidelines for the radiotherapy of non-malignant disorders: Part III: Hyperproliferative disorders. *Strahlenther. Onkol.* **2015**, *191*, 541–548. [CrossRef]
9. Krijthe, B.P.; Kunst, A.; Benjamin, E.J.; Lip, G.Y.; Franco, O.H.; Hofman, A.; Witteman, J.C.; Stricker, B.H.; Heeringa, J. Projections on the number of individuals with atrial fibrillation in the European Union, from 2000 to 2060. *Eur. Heart J.* **2013**, *34*, 2746–2751. [CrossRef]
10. Zoni-Berisso, M.; Filippi, A.; Landolina, M.; Brignoli, O.; D'Ambrosio, G.; Maglia, G.; Grimaldi, M.; Ermini, G. Frequency, patient characteristics, treatment strategies, and resource usage of atrial fibrillation (from the Italian Survey of Atrial Fibrillation Management [ISAF] study). *Am. J. Cardiol.* **2013**, *111*, 705–711. [CrossRef]
11. Hart, R.G.; Eikelboom, J.W.; Brimble, K.S.; McMurtry, M.S.; Ingram, A.J. Stroke prevention in atrial fibrillation patients with chronic kidney disease. *Can. J. Cardiol.* **2013**, *29*, S71–S78. [CrossRef]
12. Cuculich, P.S.; Schill, M.R.; Kashani, R.; Mutic, S.; Lang, A.; Cooper, D.; Faddis, M.; Gleva, M.; Noheria, A.; Smith, T.W.; et al. Noninvasive Cardiac Radiation for Ablation of Ventricular Tachycardia. *N. Engl. J. Med.* **2017**, *377*, 2325–2336. [CrossRef]
13. Kurzelowski, R.; Latusek, T.; Miszczyk, M.; Jadczyk, T.; Bednarek, J.; Sajdok, M.; Gołba, K.S.; Wojakowski, W.; Wita, K.; Gardas, R.; et al. Radiosurgery in Treatment of Ventricular Tachycardia—Initial Experience Within the Polish SMART-VT Trial. *Front. Cardiovasc. Med.* **2022**, *9*, 874661. [CrossRef]

14. Wight, J.; Bigham, T.; Schwartz, A.; Zahid, A.T.; Bhatia, N.; Kiani, S.; Shah, A.; Westerman, S.; Higgins, K.; Lloyd, M.S. Long Term Follow-Up of Stereotactic Body Radiation Therapy for Refractory Ventricular Tachycardia in Advanced Heart Failure Patients. *Front. Cardiovasc. Med.* **2022**, *9*, 849113. [CrossRef]
15. Lee, J.; Bates, M.; Shepherd, E.; Riley, S.; Henshaw, M.; Metherall, P.; Daniel, J.; Blower, A.; Scoones, D.; Wilkinson, M.; et al. Cardiac stereotactic ablative radiotherapy for control of refractory ventricular tachycardia: Initial UK multicentre experience. *Open Heart* **2021**, *8*, e001770. [CrossRef]
16. Piccolo, C.; Vigorito, S.; Rondi, E.; Piperno, G.; Ferrari, A.; Pepa, M.; Riva, G.; Durante, S.; Conte, E.; Catto, V.; et al. Phantom study of stereotactic radioablation for ventricular tachycardia (STRA-MI-VT) using Cyberknife Synchrony Respiratory Tracking System with a single fiducial marker. *Phys. Med.* **2022**, *100*, 135–141. [CrossRef]
17. Bonaparte, I.; Gregucci, F.; Surgo, A.; Di Monaco, A.; Vitulano, N.; Ludovico, E.; Carbonara, R.; Ciliberti, M.P.; Quadrini, F.; Grimaldi, M.; et al. Linac-based STereotactic Arrhythmia Radioablation (STAR) for ventricular tachycardia: A treatment planning study. *Jpn J. Radiol.* **2021**, *39*, 1223–1228. [CrossRef]
18. Kovacs, B.; Mayinger, M.; Schindler, M.; Steffel, J.; Andratschke, N.; Saguner, A.M. Stereotactic radioablation of ventricular arrhythmias in patients with structural heart disease—A systematic review. *Radiother. Oncol.* **2021**, *162*, 132–139. [CrossRef]
19. Akdag, O.; Borman, P.T.S.; Woodhead, P.; Uijtewaal, P.; Mandija, S.; van Asselen, B.; Verhoeff, J.J.C.; Raaymakers, B.W.; Fast, M.F. First experimental exploration of real-time cardiorespiratory motion management for future stereotactic arrhythmia radioablation treatments on the MR-linac. *Phys. Med. Biol* **2022**, *67*, 065003. [CrossRef]
20. Kautzner, J.; Jedlickova, K.; Sramko, M.; Peichl, P.; Cvek, J.; Ing, L.K.; Neuwirth, R.; Jiravsky, O.; Voska, L.; Kucera, T. Radiation-Induced Changes in Ventricular Myocardium After Stereotactic Body Radiotherapy for Recurrent Ventricular Tachycardia. *JACC Clin. Electrophysiol.* **2021**, *7*, 1487–1492. [CrossRef]
21. Di Monaco, A.; Gregucci, F.; Bonaparte, I.; Troisi, F.; Surgo, A.; Di Molfetta, D.; Vitulano, N.; Quadrini, F.; Carbonara, R.; Martinelli, G.; et al. Paroxysmal Atrial Fibrillation in Elderly: Worldwide Preliminary Data of LINAC-Based Stereotactic Arrhythmia Radioablation Prospective Phase II Trial. *Front. Cardiovasc. Med.* **2022**, *9*, 832446. [CrossRef]
22. Franzetti, J.; Volpe, S.; Catto, V.; Conte, E.; Piccolo, C.; Pepa, M.; Piperno, G.; Camarda, A.M.; Cattani, F.; Andreini, D.; et al. Stereotactic Radiotherapy Ablation and Atrial Fibrillation: Technical Issues and Clinical Expectations Derived From a Systematic Review. *Front. Cardiovasc. Med.* **2022**, *9*, 849201. [CrossRef]
23. Baird, K.S.; Crossan, J.F.; Ralston, S.H. Abnormal growth factor and cytokine expression in Dupuytren's contracture. *J. Clin. Pathol.* **1993**, *46*, 425–428. [CrossRef]
24. Igarashi, A.; Nashiro, K.; Kikuchi, K.; Sato, S.; Ihn, H.; Fujimoto, M.; Grotendorst, G.R.; Takehara, K. Connective tissue growth factor gene expression in tissue sections from localized scleroderma, keloid, and other fibrotic skin disorders. *J. Investig. Dermatol.* **1996**, *106*, 729–733. [CrossRef]
25. Eng, T.Y.; Abugideiri, M.; Chen, T.W.; Madden, N.; Morgan, T.; Tanenbaum, D.; Wandrey, N.; Westergaard, S.; Xu, K.; Jane Sudmeier, L. Radiation Therapy for Benign Disease: Keloids, Macular Degeneration, Orbital Pseudotumor, Pterygium, Peyronie Disease, Trigeminal Neuralgia. *Hematol. Oncol. Clin. N. Am.* **2020**, *34*, 229–251. [CrossRef]
26. Lee, J.W.; Seol, K.H. Adjuvant Radiotherapy after Surgical Excision in Keloids. *Medicina* **2021**, *57*, 730. [CrossRef]
27. Goutos, I.; Ogawa, R. Brachytherapy in the adjuvant management of keloid scars: Literature review. *Scars Burn Heal.* **2017**, *3*, 2059513117735483. [CrossRef]
28. Guinot, J.L.; Rembielak, A.; Perez-Calatayud, J.; Rodríguez-Villalba, S.; Skowronek, J.; Tagliaferri, L.; Guix, B.; Gonzalez-Perez, V.; Valentini, V.; Kovacs, G. GEC-ESTRO ACROP recommendations in skin brachytherapy. *Radiother. Oncol.* **2018**, *126*, 377–385. [CrossRef]
29. Jiang, P.; Geenen, M.; Siebert, F.A.; Bertolini, J.; Poppe, B.; Luetzen, U.; Dunst, J.; Druecke, D. Efficacy and the toxicity of the interstitial high-dose-rate brachytherapy in the management of recurrent keloids: 5-year outcomes. *Brachytherapy* **2018**, *17*, 597–600. [CrossRef]
30. Kim, K.; Son, D.; Kim, J. Radiation Therapy Following Total Keloidectomy: A Retrospective Study over 11 Years. *Arch. Plast. Surg.* **2015**, *42*, 588–595. [CrossRef]
31. Shen, J.; Lian, X.; Sun, Y.; Wang, X.; Hu, K.; Hou, X.; Sun, S.; Yan, J.; Yu, L.; Sun, X.; et al. Hypofractionated electron-beam radiation therapy for keloids: Retrospective study of 568 cases with 834 lesions. *J. Radiat. Res.* **2015**, *56*, 811–817. [CrossRef]
32. Emad, M.; Omidvari, S.; Dastgheib, L.; Mortazavi, A.; Ghaem, H. Surgical excision and immediate postoperative radiotherapy versus cryotherapy and intralesional steroids in the management of keloids: A prospective clinical trial. *Med. Princ. Pract.* **2010**, *19*, 402–405. [CrossRef] [PubMed]
33. Malaker, K.; Vijayraghavan, K.; Hodson, I.; Al Yafi, T. Retrospective analysis of treatment of unresectable keloids with primary radiation over 25 years. *Clin. Oncol.* **2004**, *16*, 290–298. [CrossRef] [PubMed]
34. Lo, T.C.; Seckel, B.R.; Salzman, F.A.; Wright, K.A. Single-dose electron beam irradiation in treatment and prevention of keloids and hypertrophic scars. *Radiother. Oncol.* **1990**, *19*, 267–272. [CrossRef]
35. Borok, T.L.; Bray, M.; Sinclair, I.; Plafker, J.; LaBirth, L.; Rollins, C. Role of ionizing irradiation for 393 keloids. *Int. J. Radiat. Oncol. Biol Phys.* **1988**, *15*, 865–870. [CrossRef]
36. Van de Kar, A.L.; Kreulen, M.; van Zuijlen, P.P.M.; Oldenburger, F. The results of surgical excision and adjuvant irradiation for therapy-resistant keloids: A prospective clinical outcome study. *Plast. Reconstr. Surg.* **2007**, *119*, 2248–2254. [CrossRef]

37. Arneja, J.S.; Singh, G.B.; Dolynchuk, K.N.; Murray, K.A.; Rozzelle, A.A.; Jones, K.D. Treatment of recurrent earlobe keloids with surgery and high-dose-rate brachytherapy. *Plast. Reconstr. Surg.* **2008**, *121*, 95–99. [CrossRef]
38. Van Leeuwen, M.C.E.; Stokmans, S.C.; Bulstra, A.J.; Meijer, O.W.M.; van Leeuwen, P.A.M.; Niessen, F.B. High-dose-rate brachytherapy for the treatment of recalcitrant keloids: A unique, effective treatment protocol. *Plast. Reconstr. Surg.* **2014**, *134*, 527–534. [CrossRef]
39. Jiang, P.; Baumann, R.; Dunst, J.; Geenen, M.; Siebert, F.A.; Niehoff, P.; Bertolini, J.; Druecke, D. Perioperative Interstitial High-Dose-Rate Brachytherapy for the Treatment of Recurrent Keloids: Feasibility and Early Results. *Int. J. Radiat. Oncol. Biol. Phys.* **2016**, *94*, 532–536. [CrossRef]
40. Hafkamp, C.J.H.; Lapid, O.; Dávila Fajardo, R.; van de Kar, A.L.; Koedooder, C.; Stalpers, L.J.; Pieters, B.R. Postoperative single-dose interstitial high-dose-rate brachytherapy in therapy-resistant keloids. *Brachytherapy* **2017**, *16*, 415–420. [CrossRef]
41. Kadhum, M.; Smock, E.; Khan, A.; Fleming, A. Radiotherapy in Dupuytren's disease: A systematic review of the evidence. *J. Hand Surg. Eur. Vol.* **2017**, *42*, 689–692. [CrossRef]
42. Betz, N.; Ott, O.J.; Adamietz, B.; Sauer, R.; Fietkau, R.; Keilholz, L. Radiotherapy in early-stage Dupuytren's contracture. Long-term results after 13 years. *Strahlenther. Onkol.* **2010**, *186*, 82–90. [CrossRef]
43. Seinen, J.M.; Niebling, M.G.; Bastiaannet, E.; Pras, B.; Hoekstra, H.J. Four different treatment strategies in aggressive fibromatosis: A systematic review. *Clin. Transl. Radiat. Oncol.* **2018**, *12*, 1–7. [CrossRef]
44. Mozena, J.D.; Hansen, E.K.; Jones, P.C. Radiotherapy for Plantar Fibromas (Ledderhose Disease). *J. Am. Podiatr. Med. Assoc.* **2022**, *112*, 19-008. [CrossRef]
45. Hautmann, M.G.; Beyer, L.P.; Süß, C.; Neumaier, U.; Steger, F.; Putz, F.J.; Kölbl, O.; Pohl, F. Radiotherapy of epicondylitis humeri: Analysis of 138 elbows treated with a linear accelerator. *Strahlenther. Onkol.* **2019**, *195*, 343–351. [CrossRef]
46. Rogers, S.; Eberle, B.; Vogt, D.R.; Meier, E.; Moser, L.; Gomez Ordoñez, S.; Desborough, S.; Riesterer, O.; Takacs, I.; Hasler, P.; et al. Prospective Evaluation of Changes in Pain Levels, Quality of Life and Functionality After Low Dose Radiotherapy for Epicondylitis, Plantar Fasciitis, and Finger Osteoarthritis. *Front. Med.* **2020**, *7*, 195. [CrossRef]
47. Hautmann, M.G.; Beyer, L.P.; Hipp, M.; Neumaier, U.; Steger, F.; Dietl, B.; Evert, K.; Kölbl, O.; Süß, C. Re-irradiation for humeral epicondylitis: Retrospective analysis of 99 elbows. *Strahlenther. Onkol.* **2020**, *196*, 262–269. [CrossRef]
48. Micke, O.; Ugrak, E.; Bartmann, S.; Adamietz, I.A.; Schaefer, U.; Bueker, R.; Kisters, K.; Heinrich Seegenschmiedt, M.; Fakhrian, K.; Muecke, R. Radiotherapy for calcaneodynia, achillodynia, painful gonarthrosis, bursitis trochanterica, and painful shoulder syndrome—Early and late results of a prospective clinical quality assessment. *Radiat. Oncol.* **2018**, *13*, 71. [CrossRef]
49. Álvarez, B.; Montero, A.; Aranburu, F.; Calvo, E.; de la Casa, M.; Valero, J.; Hernando Requejo, O.; Lopez, M.; Ciérvide, R.; García-Aranda, M.; et al. Radiotherapy for ostheoarticular degenerative disorders: When nothing else works. *Osteoarthr. Cartil. Open* **2019**, *1*, 100016. [CrossRef]
50. Mahler, E.A.M.; Minten, M.J.; Leseman-Hoogenboom, M.M.; Poortmans, P.M.P.; Leer, J.W.H.; Boks, S.S.; van den Hoogen, F.H.J.; den Broeder, A.A.; van den Ende, C.H.M. Effectiveness of low-dose radiation therapy on symptoms in patients with knee osteoarthritis: A randomised, double-blinded, sham-controlled trial. *Ann. Rheum Dis* **2019**, *78*, 83–90. [CrossRef]
51. Ott, O.J.; Jeremias, C.; Gaipl, U.S.; Frey, B.; Schmidt, M.; Fietkau, R. Radiotherapy for benign achillodynia. Long-term results of the Erlangen Dose Optimization Trial. *Strahlenther. Onkol.* **2015**, *191*, 979–984. [CrossRef]
52. Rudat, V.; Tontcheva, N.; Kutz, G.; Orovwighose, T.O.; Gebhardt, E. Long-term effect and prognostic factors of a low-dose radiotherapy of painful plantar calcaneal spurs: A retrospective unicenter study. *Strahlenther. Onkol.* **2021**, *197*, 876–884. [CrossRef]
53. Hautmann, M.G.; Neumaier, U.; Kölbl, O. Re-irradiation for painful heel spur syndrome. Retrospective analysis of 101 heels. *Strahlenther. Onkol.* **2014**, *190*, 298–303. [CrossRef]
54. Niewald, M.; Müller, L.N.; Hautmann, M.G.; Dzierma, Y.; Melchior, P.; Gräber, S.; Rübe, C.; Fleckenstein, J. ArthroRad trial: Multicentric prospective and randomized single-blinded trial on the effect of low-dose radiotherapy for painful osteoarthritis depending on the dose-results after 3 months' follow-up. *Strahlenther. Onkol.* **2022**, *198*, 370–377. [CrossRef]
55. Navaser, M.; Ghaffari, H.; Mashoufi, M.; Refahi, S. Linac-based radiotherapy for epicondylitis humeri. *EXCLI J.* **2020**, *19*, 296–300. [CrossRef]
56. Arenas, M.; Sabater, S.; Hernández, V.; Rovirosa, A.; Lara, P.C.; Biete, A.; Panés, J. Anti-inflammatory effects of low-dose radiotherapy. Indications, dose, and radiobiological mechanisms involved. *Strahlenther. Onkol.* **2012**, *188*, 975–981. [CrossRef]
57. Alvarez, B.; Montero, A.; Hernando, O.; Ciervide, R.; Garcia, J.; Lopez, M.; Garcia-Aranda, M.; Chen, X.; Flores, I.; Sanchez, E.; et al. Radiotherapy CT-based contouring atlas for non-malignant skeletal and soft tissue disorders: A practical proposal from Spanish experience. *Br. J. Radiol.* **2021**, *94*, 20200809. [CrossRef]
58. Ploumis, A.; Belbasis, L.; Ntzani, E.; Tsekeris, P.; Xenakis, T. Radiotherapy for prevention of heterotopic ossification of the elbow: A systematic review of the literature. *J. Shoulder Elbow Surg.* **2013**, *22*, 1580–1588. [CrossRef]
59. Galietta, E.; Gaiani, L.; Giannini, C.; Sambri, A.; Buwenge, M.; Macchia, G.; Deodato, F.; Cilla, S.; Strigari, L.; Fiore, M.; et al. Prophylactic Radiotherapy of Hip Heterotopic Ossification: A Narrative Mini Review. *In Vivo* **2022**, *36*, 533–542. [CrossRef]
60. Rauch, C.; Semrau, S.; Fietkau, R.; Rampp, S.; Kasper, B.; Stefan, H. Long-term experience with fractionated stereotactic radiotherapy in pharmacoresistant epilepsy: Neurological and MRI changes. *Epilepsy Res.* **2012**, *99*, 14–20. [CrossRef]
61. Liang, S.; Liu, T.; Li, A.; Zhao, M.; Yu, X.; Qh, O. Long-term follow up of very low-dose LINAC based stereotactic radiotherapy in temporal lobe epilepsy. *Epilepsy Res.* **2010**, *90*, 60–67. [CrossRef]

62. Bartolomei, F.; Hayashi, M.; Tamura, M.; Rey, M.; Fischer, C.; Chauvel, P.; Régis, J. Long-term efficacy of gamma knife radiosurgery in mesial temporal lobe epilepsy. *Neurology* **2008**, *70*, 1658–1663. [CrossRef] [PubMed]
63. Barbaro, N.M.; Quigg, M.; Broshek, D.K.; Ward, M.M.; Lamborn, K.R.; Laxer, K.D.; Larson, D.A.; Dillon, W.; Verhey, L.; Garcia, P.; et al. A multicenter, prospective pilot study of gamma knife radiosurgery for mesial temporal lobe epilepsy: Seizure response, adverse events, and verbal memory. *Ann. Neurol.* **2009**, *65*, 167–175. [CrossRef] [PubMed]
64. Smith, Z.A.; Gorgulho, A.A.; Bezrukiy, N.; McArthur, D.; Agazaryan, N.; Selch, M.T.; de Salles, A.A. Dedicated linear accelerator radiosurgery for trigeminal neuralgia: A single-center experience in 179 patients with varied dose prescriptions and treatment plans. *Int J. Radiat. Oncol. Biol. Phys.* **2011**, *81*, 225–231. [CrossRef] [PubMed]
65. Rashid, A.; Pintea, B.; Kinfe, T.M.; Surber, G.; Hamm, K.; Boström, J.P. LINAC stereotactic radiosurgery for trigeminal neuralgia -retrospective two-institutional examination of treatment outcomes. *Radiat. Oncol.* **2018**, *13*, 153. [CrossRef]
66. Romanelli, P.; Conti, A.; Redaelli, I.; Martinotti, A.S.; Bergantin, A.; Bianchi, L.C.; Beltramo, G. Cyberknife Radiosurgery for Trigeminal Neuralgia. *Cureus* **2019**, *11*, e6014. [CrossRef]
67. Lovo, E.E.; Torres, B.; Campos, F.; Caceros, V.; Reyes, W.A.; Barahona, K.C.; Cruz, C.; Arias, J.; Alho, E.; Contreras, W.O. Stereotactic Gamma Ray Radiosurgery to the Centromedian and Parafascicular Complex of the Thalamus for Trigeminal Neuralgia and Other Complex Pain Syndromes. *Cureus* **2019**, *11*, e6421. [CrossRef]
68. Kundu, B.; Brock, A.A.; Garry, J.G.; Jensen, R.L.; Burt, L.M.; Cannon, D.M.; Shrieve, D.C.; Rolston, J.D. Outcomes using linear accelerator stereotactic radiosurgery for the treatment of trigeminal neuralgia: A single-center, retrospective study. *Surg. Neurol. Int.* **2022**, *13*, 246. [CrossRef]
69. Starke, R.M.; Kano, H.; Ding, D.; Lee, J.Y.; Mathieu, D.; Whitesell, J.; Pierce, J.T.; Huang, P.P.; Kondziolka, D.; Yen, C.P.; et al. Stereotactic radiosurgery for cerebral arteriovenous malformations: Evaluation of long-term outcomes in a multicenter cohort. *J. Neurosurg.* **2017**, *126*, 36–44. [CrossRef]
70. Ding, D.; Starke, R.M.; Kano, H.; Mathieu, D.; Huang, P.P.; Kondziolka, D.; Feliciano, C.; Rodriguez-Mercado, R.; Almodovar, L.; Grills, I.S.; et al. Stereotactic Radiosurgery for ARUBA (A Randomized Trial of Unruptured Brain Arteriovenous Malformations)-Eligible Spetzler-Martin Grade I and II Arteriovenous Malformations: A Multicenter Study. *World Neurosurg.* **2017**, *102*, 507–517. [CrossRef]
71. Patibandla, M.R.; Ding, D.; Kano, H.; Xu, Z.; Lee, J.Y.K.; Mathieu, D.; Whitesell, J.; Pierce, J.T.; Huang, P.P.; Kondziolka, D.; et al. Stereotactic radiosurgery for Spetzler-Martin Grade IV and V arteriovenous malformations: An international multicenter study. *J. Neurosurg.* **2018**, *129*, 498–507. [CrossRef]
72. Matsuo, T.; Kamada, K.; Izumo, T.; Hayashi, N.; Nagata, I. Linear accelerator-based radiosurgery alone for arteriovenous malformation: More than 12 years of observation. *Int. J. Radiat. Oncol. Biol. Phys.* **2014**, *89*, 576–583. [CrossRef]
73. Matthiesen, C.; Thompson, J.S.; Thompson, D.; Farris, B.; Wilkes, B.; Ahmad, S.; Herman, T.; Bogardus, C., Jr. The efficacy of radiation therapy in the treatment of Graves' orbitopathy. *Int. J. Radiat. Oncol. Biol. Phys.* **2012**, *82*, 117–123. [CrossRef]
74. Kouloulias, V.; Kouvaris, J.; Zygogianni, A.; Mosa, E.; Georgakopoulos, I.; Theodosiadis, P.; Antypas, C.; Platoni, K.; Tolia, M.; Beli, I.; et al. Efficacy and toxicity of radiotherapy for Graves' ophthalmopathy: The University of Athens experience. *Head Neck Oncol.* **2013**, *5*, 12.
75. Li Yim, J.F.; Sandinha, T.; Kerr, J.M.; Ritchie, D.; Kemp, E.G. Low dose orbital radiotherapy for thyroid eye disease. *Orbit* **2011**, *30*, 269–274. [CrossRef]
76. Kahaly, G.J.; Rösler, H.P.; Pitz, S.; Hommel, G. Low- versus high-dose radiotherapy for Graves' ophthalmopathy: A randomized, single blind trial. *J. Clin. Endocrinol. Metab.* **2000**, *85*, 102–108. [CrossRef]
77. Cardoso, C.C.; Giordani, A.J.; Wolosker, A.M.; Souhami, L.; Manso, P.G.; Dias, R.S.; Segreto, H.R.; Segreto, R.A. Protracted hypofractionated radiotherapy for Graves' ophthalmopathy: A pilot study of clinical and radiologic response. *Int. J. Radiat. Oncol. Biol. Phys.* **2012**, *82*, 1285–1291. [CrossRef]
78. Tuleasca, C.; Régis, J.; Sahgal, A.; de Salles, A.; Hayashi, M.; Ma, L.; Martínez-Álvarez, R.; Paddick, I.; Ryu, S.; Slotman, B.J.; et al. Stereotactic radiosurgery for trigeminal neuralgia: A systematic review. *J. Neurosurg.* **2018**, *130*, 733–757. [CrossRef]
79. Fraioli, M.F.; Strigari, L.; Fraioli, C.; Lecce, M.; Lisciani, D. Preliminary results of 45 patients with trigeminal neuralgia treated with radiosurgery compared to hypofractionated stereotactic radiotherapy, using a dedicated linear accelerator. *J. Clin. Neurosci.* **2012**, *19*, 1401–1403. [CrossRef]
80. Helis, C.A.; Lucas, J.T., Jr.; Bourland, J.D.; Chan, M.D.; Tatter, S.B.; Laxton, A.W. Repeat Radiosurgery for Trigeminal Neuralgia. *Neurosurgery* **2015**, *77*, 755–761; discussion 761. [CrossRef]
81. Park, K.J.; Kondziolka, D.; Berkowitz, O.; Kano, H.; Novotny, J., Jr.; Niranjan, A.; Flickinger, J.C.; Lunsford, L.D. Repeat gamma knife radiosurgery for trigeminal neuralgia. *Neurosurgery* **2012**, *70*, 295–305. [CrossRef]
82. Barreau, X.; Marnat, G.; Gariel, F.; Dousset, V. Intracranial arteriovenous malformations. *Diagn. Interv. Imaging* **2014**, *95*, 1175–1186. [CrossRef]
83. Abecassis, I.J.; Xu, D.S.; Batjer, H.H.; Bendok, B.R. Natural history of brain arteriovenous malformations: A systematic review. *Neurosurg. Focus* **2014**, *37*, E7. [CrossRef]
84. Xu, F.; Zhong, J.; Ray, A.; Manjila, S.; Bambakidis, N.C. Stereotactic radiosurgery with and without embolization for intracranial arteriovenous malformations: A systematic review and meta-analysis. *Neurosurg. Focus* **2014**, *37*, E16. [CrossRef]

85. Russell, D.; Peck, T.; Ding, D.; Chen, C.J.; Taylor, D.G.; Starke, R.M.; Lee, C.C.; Sheehan, J.P. Stereotactic radiosurgery alone or combined with embolization for brain arteriovenous malformations: A systematic review and meta-analysis. *J. Neurosurg.* **2018**, *128*, 1338–1348. [CrossRef]
86. Madhugiri, V.S.; Teo, M.K.C.; Westbroek, E.M.; Chang, S.D.; Marks, M.P.; Do, H.M.; Levy, R.P.; Steinberg, G.K. Multimodal management of arteriovenous malformations of the basal ganglia and thalamus: Factors affecting obliteration and outcome. *J. Neurosurg.* **2018**, *131*, 410–419. [CrossRef]
87. Karlsson, B.; Lindqvist, M.; Blomgren, H.; Wan-Yeo, G.; Söderman, M.; Lax, I.; Yamamoto, M.; Bailes, J. Long-term results after fractionated radiation therapy for large brain arteriovenous malformations. *Neurosurgery* **2005**, *57*, 42–49. [CrossRef]
88. Knippen, S.; Putz, F.; Semrau, S.; Lambrecht, U.; Knippen, A.; Buchfelder, M.; Schlaffer, S.; Struffert, T.; Fietkau, R. Predictors for occlusion of cerebral AVMs following radiation therapy: Radiation dose and prior embolization, but not Spetzler-Martin grade. *Strahlenther. Onkol.* **2017**, *193*, 185–191. [CrossRef]
89. Micke, O.; Seegenschmiedt, M.H. Consensus guidelines for radiation therapy of benign diseases: A multicenter approach in Germany. *Int. J. Radiat. Oncol. Biol. Phys.* **2002**, *52*, 496–513. [CrossRef]
90. Bartalena, L.; Baldeschi, L.; Dickinson, A.J.; Eckstein, A.; Kendall-Taylor, P.; Marcocci, C.; Mourits, M.P.; Perros, P.; Boboridis, K.; Boschi, A.; et al. Consensus statement of the European group on Graves' orbitopathy (EUGOGO) on management of Graves' orbitopathy. *Thyroid* **2008**, *18*, 333–346. [CrossRef]
91. Iorio, G.C.; Salvestrini, V.; Borghetti, P.; de Felice, F.; Greco, C.; Nardone, V.; Fiorentino, A.; Gregucci, F.; Desideri, I. The impact of modern radiotherapy on radiation-induced late sequelae: Focus on early-stage mediastinal classical Hodgkin Lymphoma. A critical review by the Young Group of the Italian Association of Radiotherapy and Clinical Oncology (AIRO). *Crit. Rev. Oncol. Hematol.* **2021**, *161*, 103326. [CrossRef]
92. Facoetti, A.; Barcellini, A.; Valvo, F.; Pullia, M. The Role of Particle Therapy in the Risk of Radio-induced Second Tumors: A Review of the Literature. *Anticancer Res.* **2019**, *39*, 4613–4617. [CrossRef] [PubMed]
93. Paganetti, H.; Depauw, N.; Johnson, A.; Forman, R.B.; Lau, J.; Jimenez, R. The risk for developing a secondary cancer after breast radiation therapy: Comparison of photon and proton techniques. *Radiother. Oncol.* **2020**, *149*, 212–218. [CrossRef] [PubMed]
94. Royce, T.J.; Qureshi, M.M.; Truong, M.T. Radiotherapy Utilization and Fractionation Patterns During the First Course of Cancer Treatment in the United States From 2004 to 2014. *J. Am. Coll. Radiol.* **2018**, *15*, 1558–1564. [CrossRef] [PubMed]
95. Spencer, K.; Defourny, N.; Tunstall, D.; Cosgrove, V.; Kirkby, K.; Henry, A.; Lievens, Y.; Hall, P. Variable and fixed costs in NHS radiotherapy; consequences for increasing hypo fractionation. *Radiother. Oncol.* **2022**, *166*, 180–188. [CrossRef]
96. Hunter, D.; Mauldon, E.; Anderson, N. Cost-containment in hypofractionated radiation therapy: A literature review. *J. Med. Radiat. Sci.* **2018**, *65*, 148–157. [CrossRef]

Journal of
Personalized
Medicine

MDPI

Review

Dendrimers, an Emerging Opportunity in Personalized Medicine?

Anne-Marie Caminade [1,2]

[1] Laboratoire de Chimie de Coordination (LCC), CNRS UPR8241, 205 Route de Narbonne, CEDEX 4, 31077 Toulouse, France; anne-marie.caminade@lcc-toulouse.fr

[2] LCC-CNRS, Université de Toulouse, CNRS, 31077 Toulouse, France

Abstract: Dendrimers are highly branched macromolecules tailorable at will to fulfil precise requirements. They have generated a great many expectations and a huge number of publications and patents in relation to medicine, including in relation to personalized medicine, but have resulted in very poor clinical translation up to now. As clinical trials are the first steps in view of developing new compounds for (a personalized) medicine, this review focusses on the clinical trials carried out with dendrimers. Many of these clinical trials have been recently posted (2020–2022); thus, only very few concern phase 3. The safety and efficiency of essentially two main types of dendrimers, based on polylysine and polyamidoamide scaffolds, have been assessed up to now. These dendrimers were tested with the aim of treating mainly bacterial vaginosis, cancers, and COVID-19.

Keywords: dendrimers; clinical trials; phases 1, 2, and 3; polylysine; polyamidoamine

Citation: Caminade, A.-M. Dendrimers, an Emerging Opportunity in Personalized Medicine? *J. Pers. Med.* **2022**, *12*, 1334. https://doi.org/10.3390/jpm12081334

Academic Editor: Jun Fang

Received: 2 August 2022
Accepted: 16 August 2022
Published: 19 August 2022

1. Introduction

Dendrimers [1] are a special class of synthetic macromolecules, constituted of branches built, step by step, around a central multifunctional core. Each layer of branching points creates a new "generation". Most of the properties of dendrimers depend on the type of their terminal functions (Figure 1). Dendrimers are often considered as 3-dimensional soft nanoparticles [2], in opposition to hard metal nanoparticles. Despite the fact that nature has favored branching structures at all levels, from galaxies to trees and to dendritic cells, examples of branching at the molecular level are extremely rare. One can cite glycogen, a branched polymer of glucose [3], and also some cases of branched lignin [4], but none of them have a precisely highly branched structure, as do dendrimers. Such unusual structure has generated a many expectations for dendrimers: a huge number of publications and patents exist in relation to medicine, including in relation to personalized medicine (some recent reviews about dendrimers in relation to personalized medicine: [5–12]) but have resulted in very poor clinical translation up to now [13].

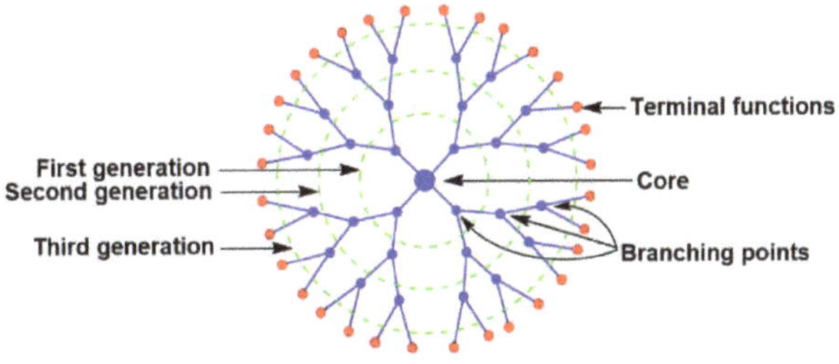

Figure 1. Schematized third-generation dendrimer, showing its main structural features.

It must be said that these synthetic macromolecules, which do not resemble anything natural or biological, arouse a certain number of fears. The very first dendrimers were synthesized in 1978 [14], and their synthesis was mainly developed in the 80th [15–17] and the 90th [18–21]. However, since that time, only very few clinical trials have concerned dendrimers, as illustrated by the fact that the site "ClinicalTrials.gov" (accessed on 2 August 2022) [22] contains 423,241 research studies (2 August 2022), of which only about 26 concern dendrimers, many of them having been posted in 2020–2022. As clinical trials are the first steps in view of developing new compounds for (a personalized) medicine, this review will focus on the clinical trials carried out with dendrimers. Information is mainly taken from ClinicalTrials.gov (accessed on 2 August 2022) (USA) and in part from clinicaltrialsregister.eu (European Union) [23]. It should be noted that the word "dendrimer" is not always used in these clinical trials. A code to define the structure is used in most cases, which renders the search relatively difficult.

2. Clinical Trials with Dendrimers Based on Poly-L-Lysine

The very first large dendrimers synthesized and patented were based on poly-L-lysine [15], and the first example of clinical trials with dendrimers concerned poly-L-lysine dendrimers. They were first functionalized with 24 macrocyclic complexes of gadolinium on the surface (Figure 2A), affording a high-relaxivity agent for MRI (magnetic resonance imaging) [24]. This compound, named Gadomer-17 or SH L 643 A (Schering AG, Berlin, Germany), was first injected to healthy volunteers for imaging their aortoiliac region and then to patients with coronary artery disease, affording an improved detection [25]. However, there is no recent result published with this compound.

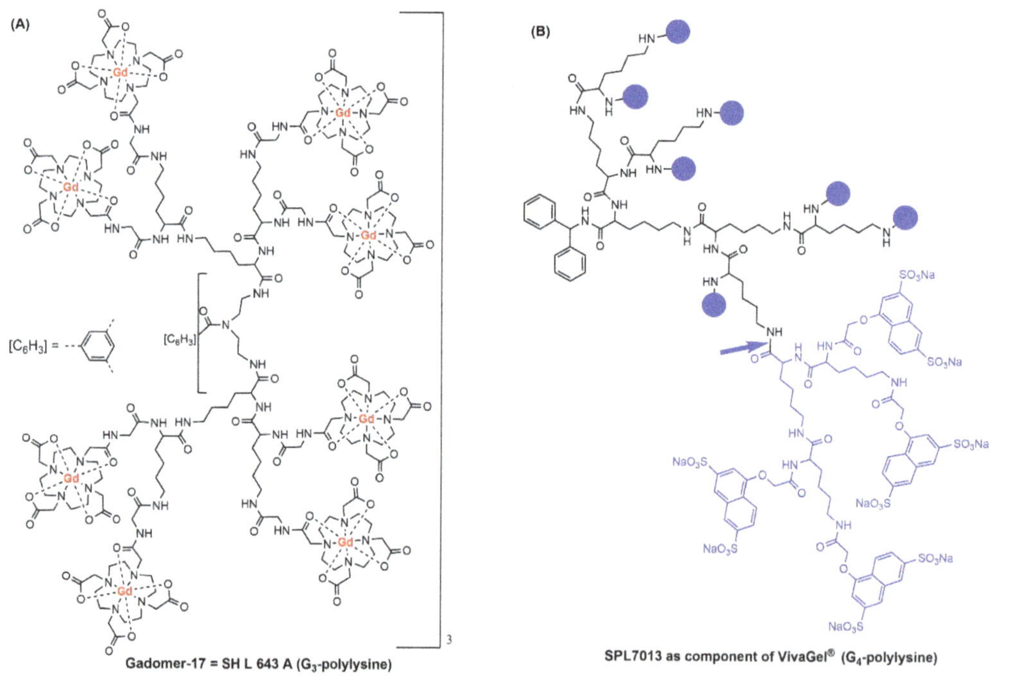

Figure 2. Chemical structure of two poly-L-Lysine dendrimers. (**A**) Gadomer-17 used as MRI agent. (**B**) SPL7013, also named Astodrimer sodium, the active ingredient in the mucoadhesive gel VivaGel®; the blue bowls correspond to the structure that is shown in blue after the blue arrow.

A second example of poly-L-lysine dendrimer was proposed by Starpharma (Australia) as an active ingredient in gels against sexually transmitted diseases, as it has antiviral properties and is able to block bacteria when formulated as a mucoadhesive gel [26]. It is based on a generation-four dendrimer bearing 32 Sodium 1-(Carboxymethoxy)naphthalene-

3,6-disulfonate on the surface, named SPL-7013 as well as Astodrimer sodium (Figure 2B). The formulated gel, named VivaGel®, has been tested in at least 13 clinical trials for assessing first the safety, tolerability, PK, and acceptability of vaginal administration. The first conclusion of a randomized, double-blind, placebo-controlled, dose-escalation trial conducted in 2004 at the Royal Adelaide Hospital (Australia) was that VivaGel®applied vaginally once daily for 7 days at concentrations of 0.5% to 3% was safe and well-tolerated in women, with no evidence of systemic toxicity or absorption [27]. Another phase 1 clinical trial (registered as NCT00331032, 29 May 2006) was carried out in the USA and Kenya as a randomized, placebo-controlled trial. The conclusion was that genitourinary adverse events and colposcopy findings were consistent with mild epithelial irritation and inflammation, in particular when using a 3% dosage [28]. It was also shown in this study that several mucosal immune parameters were increased compared to the placebo [29]. Another phase 1 clinical trial (registered as NCT00442910, 5 March 2007) was carried out almost at the same time in the USA (San Juan, Puerto Rico and Tampa, Florida). The conclusion was that VivaGel® was generally well-tolerated and comparable with the placebo although there was a higher incidence of low-grade related genital adverse effects, as in the previous study [30]. Another conclusion concerned the need for improvement of gel formulations to seek better acceptability [31]. The tolerance by men [32] and its safety for them (Phase 1 NCT00370357, 31 August 2006) were also assessed [33].

The antiviral effects were early assessed, in particular against genital herpes (Phase 1 NCT00331032, 29 May 2006) and against HIV (Phase 1 NCT00370357, 31 August 2006, and Phase 2 NCT00740584, 25 August 2008) [34]. However, the clinical trials were then oriented towards the study of the anti-bacterial properties of SPL-7013 [35], in particular against bacterial vaginosis (BV). A phase 2 clinical trial (NCT01201057, 14 September 2010) was conducted in a double-blind, multicenter, randomized, placebo-controlled, dose-ranging study for the treatment of bacterial vaginosis. It was shown that VivaGel (named Astodrimer in this study and in the followings) once daily for 7 days was superior to placebo and was well-tolerated at 1% dose, supporting a role for such gel as an effective treatment for bacterial vaginosis [36]. The dose range and the efficiency of the treatment against bacterial vaginosis was then assessed in several phase 3 clinical trials. Several publications have presented in details the outcomes of these clinical trials [37]. Two phase 3 clinical trials were carried out in the USA for one (NCT01577238, 13 April 2012) and in the USA, Germany, and Belgium for the second one (NCT01577537, 16 April 2012, and 2012-000752-33, 22 June 2012). Both studies confirmed that Astodrimer is an effective, safe, and well-tolerated treatment for women with bacterial vaginosis [38]. Another phase 3 study (NCT02237950, 12 September 2014, and 2014-000694-39, 21 October 2014) demonstrated that the gel administered every second day for 16 weeks was effective and superior to placebo for prevention of recurrent bacterial vaginosis in women with a history of recurrent BV [39]. A recent review emphasized that such formulation inhibits growth of bacteria associated with BV via a novel mechanism of action, compared with conventional antibiotics, by blocking the attachment of bacteria to cells and inhibiting the formation of and disrupting biofilms [40]. Table 1 displays the different types of clinical trials carried out with VivaGel®.

VivaGel® is now available in different countries under different names, as gel against BV, and also as lubricant for condoms, as it has been shown in laboratory studies to inactivate HIV, herpes simplex virus (HSV), and human papillomavirus (HPV) [26]. A recent and important sequel of the antivirus properties of dendrimer SPL-7013 concerns its efficiency against COVID-19. It has been reformulated in the form of a nasal spray to prevent COVID-19 infections under the name VIRALEZE™ [41]. Tests were carried out on healthy volunteers at an Australian clinical trials facility [42].

Another example of poly-L-lysine dendrimer in the port-folio of Starpharma is numbered AZD0466. It is a fifth-generation poly-L-lysine dendrimer bearing two types of terminal functions, 32 PEG_{2100} molecules and 32 molecules of a dual Bcl-2/Bcl-x_L inhibitor (anti-tumor agent) named AZD4320, attached through the linker X = N-Me (this compound is also named SPL-8977). Other different linkers have been used to modify the release rate of the drug (X = CH_2 for SPL-8931, X = S for SPL-8932, and X = O for SPL-8933) (Figure 3). The slowest release was observed with SPL-8931, which displayed no cardiac toxicity, contrarily to the compound having a rapid release. Combination of modeling and experiments finally converged to the choice of AZD0466 (SPL-8977, X = N-Me) as the best choice [43]. This dendrimer was tested on mouse xenografts of malignant pleural mesothelioma. AZD0466 was as effective as the standard-of-care chemotherapy, Cisplatin, and displayed significantly reduced degree of thrombocytopenia [44]. AZD0466 was also shown suitable to be delivered subcutaneously; it can act as a circulating drug depot accessing both the lymphatic and blood circulatory systems in mice to fight against lymphoma [45].

Table 1. Clinical trials with VivaGel®. Information taken from https://ClinicalTrials.gov (accessed on 2 August 2022) for NCT numbers and from https://www.clinicaltrialsregister.eu (accessed on 2 August 2022) for the other numbers.

Study Number	Date Posted	Phase	Aims
NCT00331032	29 May 2006	1	Safety and tolerability (female)
NCT00370357	31 August 2006	1	Safety (male)
NCT00442910	5 March 2007	1	Safety and acceptability (female)
NCT00490152	22 June 2007	1	Adherence, acceptability (female)
NCT00740584	25 August 2008	1/2	Retention and duration of activity (female)
NCT01201057	14 September 2010	2	Efficacy against BV [1] (female)
NCT01437722	21 September 2011	2	Prevention of recurrence of BV [1] (female)
NCT01577238	13 April 2012	3	Treatment of BV [1] (female)
NCT01577537	16 April 2012	3	Treatment of BV [1] (female)
2012-000752-33	22 June 2012	3	Treatment of BV [1] (female)
NCT02236156	10 September 2014	3	Prevention of recurrence of BV [1] (female)
NCT02237950	12 September 2014	3	Prevention of recurrence of BV [1] (female)
2014-000694-39	21 October 2014	3	Prevention of recurrence of BV [1] (female)

[1] Bacterial vaginosis.

In view of these interesting pre-clinical results in mice, at least three clinical trials are presently underway with AZD0466. NCT04214093 (30 December 2019) is a phase 1, first in human study in patients with advanced solid tumors, lymphoma, multiple myeloma, or hematologic malignancies. NCT04865419 (29 April 2021) is a phase 1 (AZD0466 alone) and phase 2 (AZD0466 in combination with the drug voriconazole) study in patients with hematological malignancies. NCT05205161 (25 January 2022) is a phase 1/phase 2 in patients with non-Hodgkin lymphoma.

The same concept of dually functionalized polylysine dendrimers was also applied to the docetaxel-dendrimer conjugate SPL8783, tested in the phase 1 clinical trial EudraCT Number 2016-000877-19 (13 July 2016) in patients with advanced solid tumors [46]. Another example in the same field concerns cabazitaxel–dendrimer conjugate CTX-SPL9111, named DEP®-SN38 [46]. The phase 1/2 clinical trial EudraCT Number 2017-003424-76 also concerned patients with advanced solid tumors. The phase 1/2 clinical trial EudraCT Number 2019-001318-40 was carried out with the same aims [23].

In relation with polylysine dendrimers, a less defined "Dendrigraft" fifth-generation polylysine [47] was mixed with nitro-imidazole-methyl-1,2,3-triazol-methyl-di-(2-pycolyl) amine. This mixture was named ImDendrim and was found suitable for the encapsulation of rhenium and radioactive ^{188}Re (Figure 4). This potential radiopharmaceutical agent displayed anti-tumoral activity in liver cancer in mice [48]. The clinical trial NCT03255343 (21 August 2017) with [^{188}Re]rhenium-ImDendrim concerns patients with non-operable liver tumor with resistance to other classical chemotherapy.

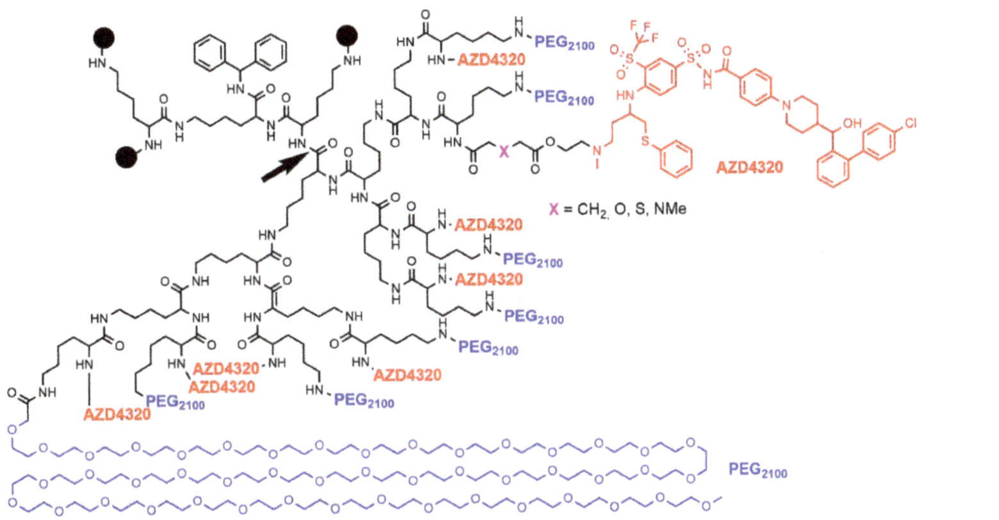

Figure 3. Structure of dendrimer AZD0466 (X = N-Me). The black bowls represent the structure that is shown after the black arrow.

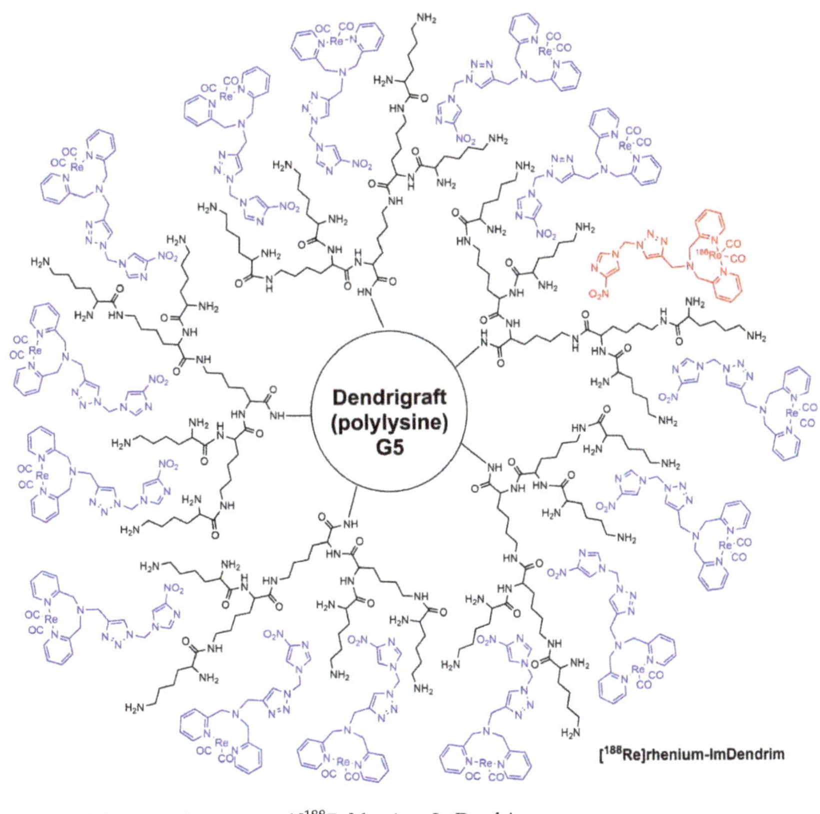

Figure 4. Schematized structure of [^{188}Re]rhenium-ImDendrim.

3. Clinical Trials with PAMAM Dendrimers

PAMAM (PolyAMidoAMine) dendrimers [16] are the most widely used type of dendrimers. In the native form, they have NH_2 terminal functions as poly-L-lysine dendrimers; thus, they can be easily modified at will. However, the risk of reverse Michael addition, particularly in acidic conditions, inducing the breakage of the structure, has precluded for a long time the use of PAMAM dendrimers in clinical trials [37]. Nevertheless, PAMAM dendrimers have been recently tested in clinical trial, essentially proposed by Ashvattha Therapeutics [49]. All were obtained by first modifying the terminal functions to hydroxyl, then by stochastically grafting other functions. The first dendrimer, labelled OP-101, is a fourth-generation PAMAM stochastically functionalized with 40 OH and 24 N-acetyl cysteine functions. Figure 5 displays the full structure of one of the possible isomers; Figure 6A displays a very simplified structure of the same compound.

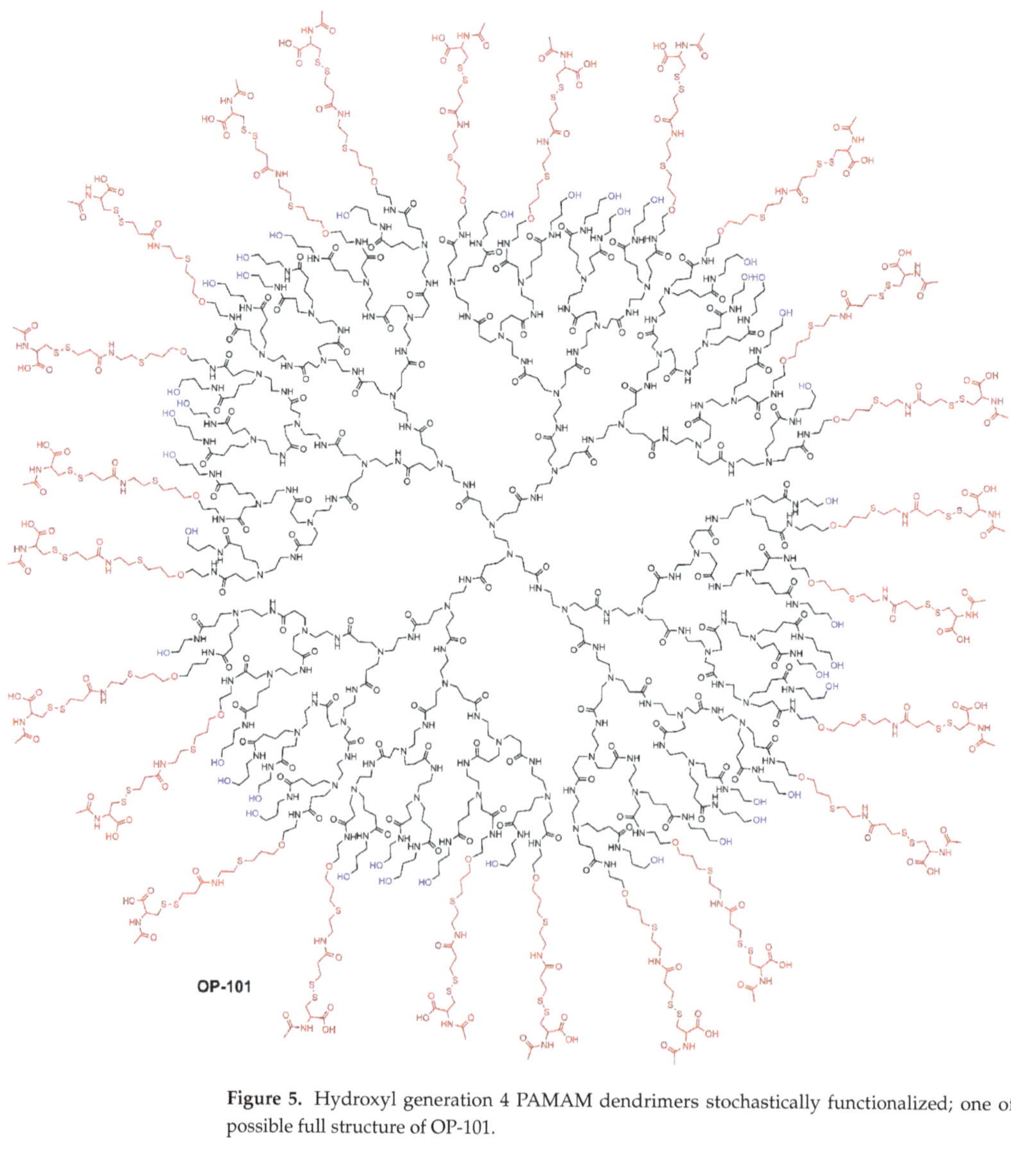

OP-101

Figure 5. Hydroxyl generation 4 PAMAM dendrimers stochastically functionalized; one of the possible full structure of OP-101.

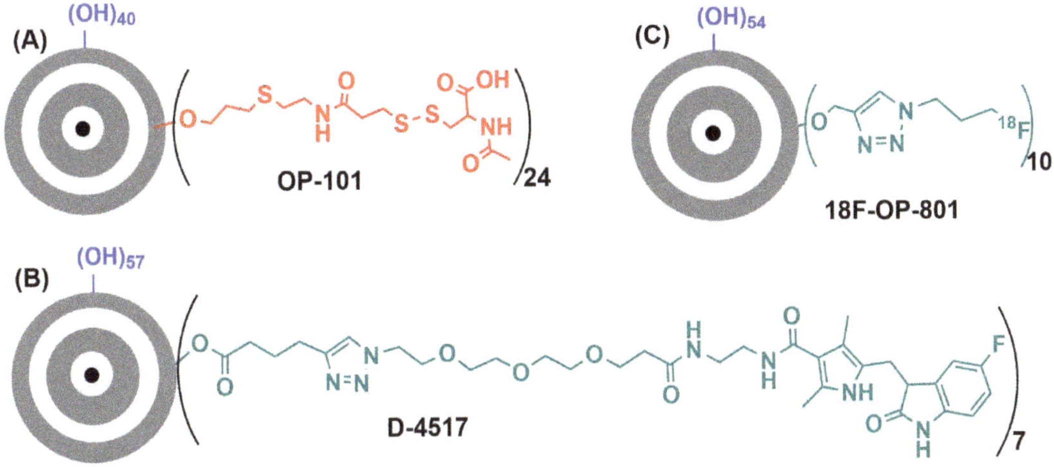

Figure 6. (**A**) Simplified structure of OP-101. (**B**) Simplified structure of D-4517. (**C**) Simplified structure of 18F-OP-801.

It was shown that this compound OP-101 has anti-inflammatory and anti-oxidant activity in microglia cells [50]. The first clinical trial, NCT03500627 (18 April 2018), was a phase 1 study to evaluate the safety, tolerability, and pharmacokinetics of OP-101 injected intravenously. The second trial, NCT04321980 (26 March 2020), evaluated the subcutaneous administration of OP-101. The objectives of the phase 2 study NCT04458298 (7 July 2020) are to evaluate the efficacy of OP-101 in patients with severe COVID-19 and its effect in reducing pro-inflammatory cytokines biomarkers. No drug-related adverse events were reported. OP-101 reduced morbidity and mortality in hospitalized patients with severe COVID-19. Indeed, at day 60, only three of seven patients given placebo were surviving, whereas they were 14 of 17 for OP-101–treated patients [51].

A second example of PAMAM hydroxyl dendrimers stochastically functionalized concerned the grafting of known drugs suitable for treating diverse eye diseases. A recent paper published this concept using triamcinolone-acetonide as drug conjugated to the dendrimer and tested in models of age-related macular degeneration [52]. However, the drug presently under clinical trials is a generation-four PAMAM hydroxyl dendrimer partly functionalized with Sunitinib, an inhibitor of vascular endothelial growth factor receptors (simplified structure in Figure 6B). The grafting through an ester bond enables the slow release of the drug. This dendrimer, named D-4517-2 [53], has been evaluated in the phase 1 clinical trial NCT05105607 (3 November 2021) for safety and tolerability after subcutaneous injection. The phase 2 NCT05387837 (24 May 2022) intends to evaluate safety, tolerability, and pharmacokinetics in patients with neovascular (wet) age-related macular degeneration or subjects with diabetic macular edema.

A third example of PAMAM-hydroxyl dendrimer concerns the stochastic grafting of a few radionuclides, in particular ^{18}F, for specific positron emission tomography (PET) imaging and radiotherapy, with the aim of detecting inflammatory sites or tumors [54]. This specific dendrimer is named 18F-OP-801 (simplified structure in Figure 6C) and is used in the phase 1 clinical trial NCT05395624 (27 May 2022) to evaluate the safety, pharmacokinetics, and biodistribution after intravenous administration to patients with amyotrophic lateral sclerosis.

Another clinical trial with PAMAM dendrimers is NCT04262076 (10 February 2020), which concerns the use of a PAMAM dendrimer with Pulpine for the remineralization of carious-affected dentin. However, the generation and the type of surface functions of the

PAMAM dendrimer used are unknown in this study carried out by the Al-Azhar University (Egypt). No result has been posted.

4. Clinical Trials with Other Types of Dendrimers

Two clinical trials against COVID-19 are currently on-going, both based on the same cationic peptide dendritic structure having lysine as branching elements, named KK-46, whose structure is shown in Figure 7 [55]. This compound is used as carrier of siRNA modified for silencing SARS-CoV-2 (siCoV) by inhibiting its replication [56]. The association of KK-46 with siCoV is named MIR 19®. Trial number NCT05208996 (26 January 2022) is a phase 1 clinical trial to assess the safety and tolerability of siCoV/KK46 in healthy volunteers. Trial number NCT05184127 (11 January 2022) is a phase 2 clinical trial to assess the efficacy and safety of MIR 19® via 14 days of treatment of hospitalized patients with symptomatic moderate COVID-19.

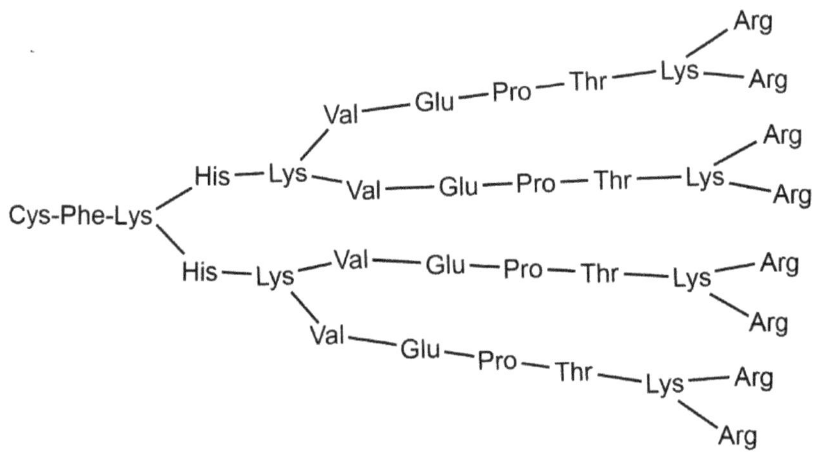

Figure 7. Structure of the cationic peptide dendrimer KK-46, usable as carrier of siCoV.

5. Discussion

This review is the first one fully dedicated to clinical trials carried out with dendrimers. It can be seen by the dates on which they have been posted that many of them are extremely recent (dated 2020–2022); thus, most of them concern phase 1 assays. The only dendrimer under phase 3 clinical trials is the polylysine used as an active ingredient of VivaGel® (Figure 2B) to treat bacterial vaginosis. It must be emphasized that the phases 1 with this dendrimer were carried out in 2004, 2006, and 2007, whereas the phases 3 began in 2012 and 2014, almost 10 years after the beginning of phases 1. Furthermore, VivaGel® is a gel, which does not penetrate the skin or the bloodstream, and thus cannot cause serious health problems. On the contrary, most of the dendrimers recently used in phase 1 clinical trials were injected either intravenously or subcutaneously. As practically no results have been posted with most of them, it is difficult to foresee if and when phase 3 clinical trials could be carried out with them. It is even more difficult to foresee if they could be used in personalized medicine.

However, dendrimers have already found application in personalized medicine to enhance the efficiency of diagnosis tools [57]. They are used as 3D spacers in chips for bioassays between the solid surface and the probe to increase the binding efficiency of the probe. Indeed, an increased distance from the solid surface enables the trapping of the target as efficiently as in solution, as illustrated in Figure 8. Such a concept was applied very early with PAMAM dendrimers for enhancing the radial partition in immunoassays. It is used in particular in the Stratus®CS Analyzer * for acute care diagnostics, which provides quantitative cardiac assays for fast, accurate evaluation of patients presenting suspected myocardial ischemia [58]. Another type of bioassays is carried out with phosphorhydrazone dendrimers [59]. Different types of chips are elaborated depending on the disease to be analyzed [60]. DendrisCHIP®RD concerns the simultaneous detection of 10 bacteria commonly involved in respiratory infections. DendrisCHIP®OA allows the detection of 10 bacteria responsible for deep and osteoarticular infections [61]. DendrisCHIP®DP is dedicated to the identification of fungi involved in dermatophytosis, whereas DendrisCHIP®SD is dedicated to the identification of 10 pathogens (bacteria, yeasts, and viruses) responsible for sexually transmitted infections [60].

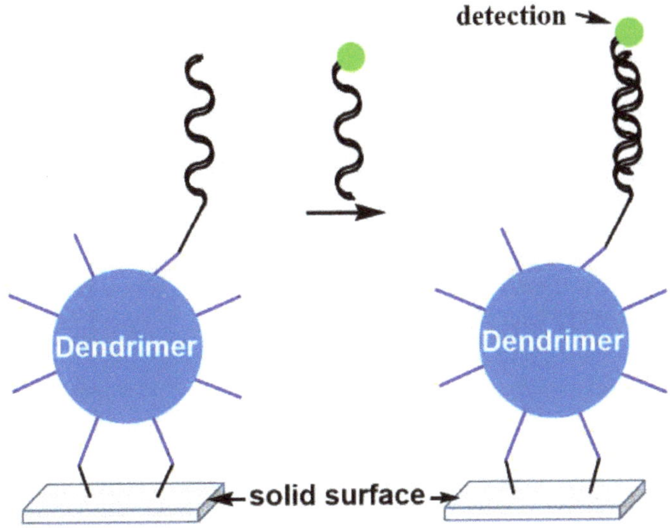

Figure 8. Principle of the use of dendrimers in bioassays.

6. Conclusions and Perspectives

The huge number of dendrimers already synthesized and the even larger number of dendrimers that could be synthesized as active *per se* or as carriers of drugs should have generated breakthrough in medicine, including in personalized medicine. However, this is not the case, with two exceptions that are bioassays for the specific detection of diseases and gels for the local treatment of bacterial vaginosis. The hope of expanding the number of dendrimers actually used in human health lies in the phase 1 clinical trials carried out recently and on others that are planned. One can cite in particular ongoing and future assays against influenza A and B virus; against COVID-19 [26]; against neurology, ophthalmology, inflammatory diseases, and neuro-oncology [49]; and to modulate the activity of the immune system to treat chronic inflammatory diseases [62,63]. The field of theranostic (diagnosis and therapy with the same "object") could be developed with multifunctional dendrimers (see for instance 18F-OP-801). Thus, there are still many pending works, but I do believe that dendrimers can be considered as an emerging opportunity in medicine and in part in personalized medicine.

Funding: This research received no external funding.

Institutional Review Board Statement: Not applicable.

Informed Consent Statement: Not applicable.

Data Availability Statement: Not applicable.

Acknowledgments: Thanks are due to the CNRS.

Conflicts of Interest: Anne-Marie Caminade is a shareholder of IMD-Pharma.

References

1. Caminade, A.M.; Turrin, C.O.; Laurent, R.; Ouali, A.; DelavauxNicot, B. *Dendrimers: Towards Catalytic, Material and Biomedical Uses*; Wiley: Chichester, UK, 2011; pp. 1–538.
2. Majoral, J.-P.; Zablocka, M.; Ciepluch, K.; Milowska, K.; Bryszewska, M.; Shcharbin, D.; Katir, N.; El Kadib, A.; Caminade, A.-M.; Mignani, S. Hybrid phosphorus-viologen dendrimers as new soft nanoparticles: Design and properties. *Org. Chem. Front.* **2021**, *8*, 4607–4622. [CrossRef]
3. Roach, P.J.; Depaoli-Roach, A.A.; Hurley, T.D.; Tagliabracci, V.S. Glycogen and its metabolism: Some new developments and old themes. *Biochem. J.* **2012**, *441*, 763–787. [CrossRef]
4. Balakshin, M.; Capanema, E.A.; Zhu, X.H.; Sulaeva, I.; Potthast, A.; Rosenau, T.; Rojas, O.J. Spruce milled wood lignin: Linear, branched or cross-linked? *Green Chem.* **2020**, *22*, 3985–4001. [CrossRef]
5. Mekuria, S.L.; Ouyang, Z.; Song, C.; Rodrigues, J.; Shen, M.; Shi, X. Dendrimer-Based Nanogels for Cancer Nanomedicine Applications. *Bioconjug. Chem.* **2022**, *33*, 87–96. [CrossRef]
6. Moorthy, H.; Govindaraju, T. Dendrimer Architectonics to Treat Cancer and Neurodegenerative Diseases with Implications in Theranostics and Personalized Medicine. *ACS Appl. Bio Mater.* **2021**, *4*, 1115–1139. [CrossRef]
7. Mishra, V.; Yadav, N.; Saraogi, G.; Tambuwala, M.M.; Giri, N. Dendrimer based nanoarchitectures in diabetes management: An overview. *Curr. Pharm. Des.* **2019**, *25*, 2569–2583. [CrossRef]
8. Patel, P.; Meghani, N.; Kansara, K.; Kumar, A. Nanotherapeutics for the Treatment of Cancer and Arthritis. *Curr. Drug Metab.* **2019**, *20*, 430–445. [CrossRef]
9. Setrajcic-Tomic, A.J.; Popovic, J.K.; Vojnovic, M.; Dzambas, L.D.; Setrajcic, J.P. Review of core-multishell nanostructured models for nano-biomedical and nano-biopharmaceutical application. *Bio-Med. Mater. Eng.* **2018**, *29*, 451–471. [CrossRef]
10. Xia, Y.; Matham, M.V.; Su, H.B.; Padmanabhan, P.; Gulyas, B. Nanoparticulate Contrast Agents for Multimodality Molecular Imaging. *J. Biomed. Nanotechnol.* **2016**, *12*, 1553–1584. [CrossRef]
11. Cole, J.T.; Holland, N.B. Multifunctional nanoparticles for use in theranostic applications. *Drug Deliv. Transl. Res.* **2015**, *5*, 295–309. [CrossRef]
12. Muthu, M.S.; Leong, D.T.; Mei, L.; Feng, S.S. Nanotheranostics—Application and Further Development of Nanomedicine Strategies for Advanced Theranostics. *Theranostics* **2014**, *4*, 660–677. [CrossRef] [PubMed]
13. Svenson, S. The dendrimer paradox—High medical expectations but poor clinical translation. *Chem. Soc. Rev.* **2015**, *44*, 4131–4144. [CrossRef] [PubMed]
14. Buhleier, E.; Wehner, F.; Vögtle, F. "Cascade-" and "Nonskid-chain-like" syntheses of molecular cavity topologies. *Synthesis* **1978**, *78*, 155–158. [CrossRef]
15. Denkewalter, R.G.; Kolc, J.; Lukasavage, W.J. Macromolecular Highly Branched Homogeneous Compound Based on Lysine Units. US Patent 4,289,872, 15 September 1981.
16. Tomalia, D.A.; Baker, H.; Dewald, J.; Hall, M.; Kallos, G.; Martin, S.; Roeck, J.; Ryder, J.; Smith, P. A new class of polymers—Starburst-dendritic macromolecules. *Polym. J.* **1985**, *17*, 117–132. [CrossRef]
17. Newkome, G.R.; Yao, Z.Q.; Baker, G.R.; Gupta, V.K. Micelles. 1. Cascade molecules—A new approach to micelles—A [27]-arborol. *J. Org. Chem.* **1985**, *50*, 2003–2004. [CrossRef]
18. Wooley, K.L.; Hawker, C.J.; Frechet JM, J. Hyperbranched macromolecules via a novel double-stage convergent growth approach. *J. Am. Chem. Soc.* **1991**, *113*, 4252–4261. [CrossRef]
19. Zhou, L.L.; Roovers, J. Synthesis of novel carbosilane dendritic macromolecules. *Macromolecules* **1993**, *26*, 963–968. [CrossRef]
20. de Brabander van den Berg, E.M.M.; Meijer, E.W. Poly(Propylene Imine) Dendrimers—Large-Scale Synthesis by Hetereogeneously Catalyzed Hydrogenations. *Angew. Chem.-Int. Edit. Engl.* **1993**, *32*, 1308–1311. [CrossRef]
21. Launay, N.; Caminade, A.M.; Lahana, R.; Majoral, J.P. A general synthetic strategy for neutral phosphorus-containing dendrimers. *Angew. Chem.-Int. Edit. Engl.* **1994**, *33*, 1589–1592. [CrossRef]
22. Clinical Trials USA. Available online: https://ClinicalTrials.gov (accessed on 2 August 2022).
23. Clinical Trials EU. Available online: https://www.clinicaltrialsregister.eu (accessed on 2 August 2022).
24. Runge, V.M.; Heverhagen, J.T. Advocating the Development of Next-Generation High-Relaxivity Gadolinium Chelates for Clinical Magnetic Resonance. *Investig. Radiol.* **2018**, *53*, 381–389. [CrossRef]
25. Herborn, C.U.; Schmidt, M.; Bruder, O.; Nagel, E.; Shamsi, K.; Barkhausen, J. MR Coronary Angiography with SH L 643 A: Initial Experience in Patients with Coronary Artery Disease. *Radiology* **2004**, *233*, 567–573. [CrossRef] [PubMed]
26. Starpharma. Available online: https://starpharma.com (accessed on 2 August 2022).
27. Holmes, W.R.; Maher, L.; Rosenthal, S.L. Attitudes of men in an Australian male tolerance study towards microbicide use. *Sexual Health* **2008**, *5*, 273–278. [CrossRef] [PubMed]
28. Chen, M.Y.; Millwood, I.Y.; Wand, H.; Poynten, M.; Law, M.; Kaldor, J.M.; Wesselingh, S.; Price, C.F.; Clark, L.J.; Paull, J.R.A.; et al. A Randomized Controlled Trial of the Safety of Candidate Microbicide SPL7013 Gel When Applied to the Penis. *Jaids-J. Acquir. Immune Defic. Syndr.* **2009**, *50*, 375–380. [CrossRef]
29. O'Loughlin, J.; Millwood, I.Y.; McDonald, H.M.; Price, C.F.; Kaldor, J.M.; Paull, J.R.A. Safety, Tolerability, and Pharmacokinetics of SPL7013 Gel (VivaGel®): A Dose Ranging, Phase I Study. *Sex. Transm. Dis.* **2010**, *37*, 100–104. [CrossRef]

30. Cohen, C.R.; Brown, J.; Moscicki, A.B.; Bukusi, E.A.; Paull, J.R.A.; Price, C.F.; Shiboski, S. A Phase I Randomized Placebo Controlled Trial of the Safety of 3% SPL7013 Gel (VivaGel®) in Healthy Young Women Administered Twice Daily for 14 Days. *PLoS ONE* **2011**, *6*, e16258. [CrossRef]
31. Moscicki, A.B.; Kaul, R.; Ma, Y.; Scott, M.E.; Daud, I.I.; Bukusi, E.A.; Shiboski, S.; Rebbapragada, A.; Huibner, S.; Cohen, C.R. Measurement of Mucosal Biomarkers in a Phase 1 Trial of Intravaginal 3% StarPharma LTD 7013 Gel (VivaGel) to Assess Expanded Safety. *J. Acquir. Immune Defic. Syndr.* **2012**, *59*, 134–140. [CrossRef]
32. McGowan, I.; Gomez, K.; Bruder, K.; Febo, I.; Chen, B.A.; Richardson, B.A.; Husnik, M.; Livant, E.; Price, C.; Jacobson, C.; et al. Phase 1 randomized trial of the vaginal safety and acceptability of SPL7013 gel (VivaGel) in sexually active young women (MTN-004). *Aids* **2011**, *25*, 1057–1064. [CrossRef]
33. Carballo-Dieguez, A.; Giguere, R.; Dolezal, C.; Chen, B.A.; Kahn, J.; Zimet, G.; Mabragana, M.; Leu, C.S.; McGowan, I. "Tell Juliana": Acceptability of the Candidate Microbicide VivaGel® and Two Placebo Gels Among Ethnically Diverse, Sexually Active Young Women Participating in a Phase 1 Microbicide Study. *AIDS Behav.* **2012**, *16*, 1761–1774. [CrossRef]
34. Price, C.F.; Tyssen, D.; Sonza, S.; Davie, A.; Evans, S.; Lewis, G.R.; Xia, S.; Spelman, T.; Hodsman, P.; Moench, T.R.; et al. SPL7013 Gel (VivaGel (R)) Retains Potent HIV-1 and HSV-2 Inhibitory Activity following Vaginal Administration in Humans. *PLoS ONE* **2011**, *6*, e24095. [CrossRef]
35. Madan, R.P.; Dezzutti, C.S.; Rabe, L.; Hillier, S.L.; Marrazzo, J.; McGowan, I.; Richardson, B.A.; Herold, B.C.; Microbicide Trials Network, B.; Team, M.T.N.P. Soluble Immune Mediators and Vaginal Bacteria Impact Innate Genital Mucosal Antimicrobial Activity in Young Women. *Am. J. Reprod. Immunol.* **2015**, *74*, 323–332. [CrossRef]
36. Waldbaum, A.S.; Schwebke, J.R.; Paull, J.R.A.; Price, C.F.; Edmondson, S.R.; Castellarnau, A.; McCloud, P.; Kinghorn, G.R. A phase 2, double-blind, multicenter, randomized, placebo-controlled, dose-ranging study of the efficacy and safety of Astodrimer Gel for the treatment of bacterial vaginosis. *PLoS ONE* **2020**, *15*, e0232394. [CrossRef] [PubMed]
37. McCarthy, T.D.; Karellas, P.; Henderson, S.A.; Giannis, M.; O'Keefe, D.F.; Heery, G.; Paull, J.R.A.; Matthews, B.R.; Holan, G. Dendrimers as drugs: Discovery and preclinical and clinical development of dendrimer-based microbicides for HIV and STI prevention. *Mol. Pharm.* **2005**, *2*, 312–318. [CrossRef]
38. Chavoustie, S.E.; Carter, B.A.; Waldbaum, A.S.; Donders, G.G.G.; Peters, K.H.; Schwebke, J.R.; Paull, J.R.A.; Price, C.F.; Castellarnau, A.; McCloud, P.; et al. Two phase 3, double-blind, placebo-controlled studies of the efficacy and safety of Astodrimer 1% Gel for the treatment of bacterial vaginosis. *Eur. J. Obstet. Gynecol. Reprod. Biol.* **2020**, *245*, 13–18. [CrossRef] [PubMed]
39. Schwebke, J.R.; Carter, B.A.; Waldbaum, A.S.; Agnew, K.J.; Paull, J.R.A.; Price, C.F.; Castellarnau, A.; McCloud, P.; Kinghorn, G.R. A phase 3, randomized, controlled trial of Astodrimer 1% Gel for preventing recurrent bacterial vaginosis. *Eur. J. Obstetrics Gynecol. Reprod. Biol. X* **2021**, *10*, 100121. [CrossRef] [PubMed]
40. Mendling, W.; Holzgreve, W. Astodrimer sodium and bacterial vaginosis: A mini review. *Arch. Gynecol. Obstet.* **2022**, *306*, 101–108. [CrossRef]
41. VIRALEZE. Available online: https://www.starpharma.com/viraleze/spl7013 (accessed on 2 August 2022).
42. Castellarnau, A.; Heery, G.P.; Seta, A.; Luscombe, C.A.; Kinghorn, G.R.; Button, P.; McCloud, P.; Paull, J.R.A. Astodrimer sodium antiviral nasal spray for reducing respiratory infections is safe and well tolerated in a randomized controlled trial. *Sci. Rep.* **2022**, *12*, 10210. [CrossRef]
43. Patterson, C.M.; Balachander, S.B.; Grant, I.; Pop-Damkov, P.; Kelly, B.; McCoull, W.; Parker, J.; Giannis, M.; Hill, K.J.; Gibbons, F.D.; et al. Design and optimisation of dendrimer-conjugated Bcl-2/xL inhibitor, AZD0466, with improved therapeutic index for cancer therapy. *Commun. Biol.* **2021**, *4*, 112. [CrossRef]
44. Arulananda, S.; O'Brien, M.; Evangelista, M.; Jenkins, L.J.; Poh, A.R.; Walkiewicz, M.; Leong, T.; Mariadason, J.M.; Cebon, J.; Balachander, S.B.; et al. A novel BH3-mimetic, AZD0466, targeting BCL-XL and BCL-2 is effective in pre-clinical models of malignant pleural mesothelioma. *Cell Death Discov.* **2021**, *7*, 122. [CrossRef]
45. Feeney, O.M.; Ardipradja, K.; Noi, K.F.; Mehta, D.; De Rose, R.; Yuen, D.; Johnston, A.P.R.; Kingston, L.; Ericsson, C.; Elmore, C.S.; et al. Subcutaneous delivery of a dendrimer-BH3 mimetic improves lymphatic uptake and survival in lymphoma. *J. Control. Release* **2022**, *348*, 420–430. [CrossRef]
46. DEP®Docetaxel, DEP®Cabazitaxel, and DEP®Irinotecan. Available online: https://starpharma.com/drug_delivery (accessed on 1 August 2022).
47. ColCom. Available online: https://www.colcom.eu/ (accessed on 2 August 2022).
48. Yang, G.; Sadeg, N.; Belhadj-Tahar, H. New potential in situ anticancer agent derived from [^{188}Re]rhenium nitro-imidazole ligand loaded 5th generation poly-L-lysine dendrimer for treatment of transplanted human liver carcinoma in nude mice. *Drug Des.* **2017**, *6*, 1000144. [CrossRef]
49. Ashvattha Therapeutics. Available online: https://avttx.com/ (accessed on 2 August 2022).
50. Sharma, R.; Sharma, A.; Kambhampati, S.P.; Reddy, R.R.; Zhang, Z.; Cleland, J.L.; Kannan, S.; Kannan, R.M. Scalable synthesis and validation of PAMAM dendrimer-N-acetyl cysteine conjugate for potential translation. *Bioeng. Transl. Med.* **2018**, *3*, 87–101. [CrossRef] [PubMed]
51. Gusdon, A.M.; Faraday, N.; Aita, J.S.; Kumar, S.; Mehta, I.; Choi, H.A.; Cleland, J.L.; Robinson, K.; McCullough, L.D.; Ng, D.K.; et al. Dendrimer nanotherapy for severe COVID-19 attenuates inflammation and neurological injury markers and improves outcomes in a phase2a clinical trial. *Sci. Transl. Med.* **2022**, *14*, eabo2652. [CrossRef] [PubMed]

52. Kambhampati, S.P.; Bhutto, I.A.; Wu, T.; Ho, K.; McLeod, D.S.; Lutty, G.A.; Kannan, R.M. Systemic dendrimer nanotherapies for targeted suppression of choroidal inflammation and neovascularization in age-related macular degeneration. *J. Control. Release* **2021**, *335*, 527–540. [CrossRef]
53. Cleland, J.; Sharma, R.; Appiani, S. Dendrimer Compositions and Methods for Drug Delivery to the Eye. WO 2021/113662, 10 June 2021.
54. Cleland, J.L.; Sharma, R.; Sun, M.; Appiani La Rosa, S. Radiolabeled ether dendrimer conjugates for PET imaging and radiotherapy. WO 2022/094327, 5 May 2022.
55. Khaitov, M.R.; Shilovskii, I.P.; Kozhikhova, K.V.; Kofiadi, I.A.; Smirnov, V.V.; Koloskova, O.O.; Sergeev, I.V.; Trofimov, D.Y.; Trukhin, V.P.; Skvortsova, V.I. Combination Antiviral Formulation against SARS-CoV-2 Comprising SARS-CoV-2 Genome-Targeting siRNAs and Transfection-Enhancing Cationic Peptide Dendrimer. RU2746362 C1 2021-04-12, 11 March 2021.
56. Khaitov, M.; Nikonova, A.; Shilovskiy, I.; Kozhikhova, K.; Kofiadi, I.; Vishnyakova, L.; Nikolskii, A.; Gattinger, P.; Kovchina, V.; Barvinskaia, E.; et al. Silencing of SARS-CoV-2 with modified siRNA-peptide dendrimer formulation. *Allergy* **2021**, *76*, 2840–2854. [CrossRef]
57. Majoral, J.P.; Francois, J.M.; Fabre, R.; Senescau, A.; Mignani, S.; Caminade, A.M. Multiplexing technology for in vitro diagnosis of pathogens: The key contribution of phosphorus dendrimers. *Sci. China-Mater.* **2018**, *61*, 1454–1461. [CrossRef]
58. Siemens-Healthineers. Available online: https://www.siemens-healthineers.com/cardiac/cardiac-systems/stratus-cs-acute-care (accessed on 2 August 2022).
59. Trevisiol, E.; Le Berre-Anton, V.; Leclaire, J.; Pratviel, G.; Caminade, A.M.; Majoral, J.P.; Francois, J.M.; Meunier, B. Dendrislides, dendrichips: A simple chemical functionalization of glass slides with phosphorus dendrimers as an effective means for the preparation of biochips. *N. J. Chem.* **2003**, *27*, 1713–1719. [CrossRef]
60. Dendris. Available online: https://dendris.fr/medical-diagnostic/ (accessed on 2 August 2022).
61. Bernard, E.; Peyret, T.; Plinet, M.; Contie, Y.; Cazaudarré, T.; Rouquet, Y.; Bernier, M.; Pesant, S.; Fabre, R.; Anton, A.; et al. The DendrisCHIP® Technology as a New, Rapid and Reliable Molecular Method for the Diagnosis of Osteoarticular Infections. *Diagnostics* **2022**, *12*, 1353. [CrossRef]
62. Caminade, A.-M.; Turrin, C.-O.; Poupot, R. Curing inflammatory diseases using phosphorous dendrimers. *Wiley Interdiscip. Rev.-Nanomed. Nanobiotechnol.* **2022**, e1783. [CrossRef]
63. IMD-Pharma. Available online: http://www.imd-pharma.com/ (accessed on 2 August 2022).

MDPI
St. Alban-Anlage 66
4052 Basel
Switzerland
Tel. +41 61 683 77 34
Fax +41 61 302 89 18
www.mdpi.com

Journal of Personalized Medicine Editorial Office
E-mail: jpm@mdpi.com
www.mdpi.com/journal/jpm

www.ingramcontent.com/pod-product-compliance
Lightning Source LLC
Chambersburg PA
CBHW061822210926
51850CB00006B/782